The Delphi Method
Techniques and Applications

Contributors

Marvin Adelson	(pp. 433–462)
Samuel Aroni	(pp. 433–462)
Klaus Brockhoff	(pp. 291–321)
J. Douglas Carroll	(pp. 402–431)
Norman C. Dalkey	(pp. 236–261, 327–337, 387–401)
Lawrence H. Day	(pp. 168–194)
Alex Ducanis	(pp. 288–290)
Selwyn Enzer	(pp. 195–209)
Nancy H. Goldstein	(pp. 210–226)
Olaf Helmer	(pp.xix-xx)
Irene Anne Jillson	(pp. 124–159)
Robert Johansen	(pp. 517–534, 550–562)
Chester G. Jones	(pp. 160–167)
Julius Kane	(pp. 369–382)
Harold A. Linstone	(pp. 3–12, 15–16, 75–83, 229–235, 325–326, 385–386, 489–496, 573–586, 589–614)
John Ludlow	(pp. 102–123)
Richard H. Miller	(pp. 517–534)
Ian I. Mitroff	(pp. 17–36)
Norman W. Mulgrave	(pp. 288–290)
Charlton R. Price	(pp. 497–516)
D. Sam Scheele	(pp. 37–71)
M. Scheibe	(pp. 262–287)
J. Schofer	(pp. 262–287)
James A. Schuyler	(pp. 550–562)
Thomas B. Sheridan	(pp. 535–549)
M. Skutsch	(pp. 262–287)
John W. Sutherland	(pp. 463–486)
Murray Turoff	(pp. 3–12, 15–16, 17–36, 75–83, 84–101, 229–235, 325–326, 338–368, 385–386, 489–496, 563–569, 589–614)
Jacques F. Vallee	(pp. 517–534)
Myron Wish	(pp. 402–431)

The Delphi Method
Techniques and Applications

Edited by

Harold A. Linstone
Portland State University

Murray Turoff
New Jersey Institute of Technology

With a Foreword by

Olaf Helmer
University of Southern California

1975

Addison-Wesley Publishing Company
Advanced Book Program
Reading, Massachusetts

London · Amsterdam · Don Mills, Ontario · Sydney · Tokyo

Delphi — The Tholos, Courtesy of the Greek National Tourist Office
(End papers for hardbound edition only.)

The illustration on the cover is taken from a silver stater (about 350 B.C.) formerly in the collection of John Pierpont Morgan. It depicts the Greek god Apollo Pythios seated on the sacred *omphalos* at Delphi. Apollo, son of Zeus and Leto, became master of Delphi upon slaying the dragon Pythos there. He was renowned not only for his youth and perfect beauty, but even more for his ability to foresee the future. The omphalos, located in the Temple of Apollo, was a conically shaped stone representing the navel of the earth and the center of the universe. Thus Delphi assumed the status of the most revered oracular site in Greece.

Library of Congress Cataloging in Publication Data
Main entry under title:

The Delphi method.

"Delphi bibliography": p.
Includes index.
1. Delphi method. I. Linstone, Harold A.
II. Turoff, Murray.
T174.D44 — 001.4′33 — 75-25650
ISBN 0-201-04294-0
ISBN 0-201-04293-2 pbk.

ABCDEFGHIJ-MA-798765

Manufactured in the United States of America

Contents

Biographical Data

THE EDITORS

Harold A. Linstone received his B. S. degree in 1944 from the City College of New York and his Ph. D. in Mathematics from the University of Southern California in 1954. He has been a member of the RAND Corporation, a Senior Scientist at Hughes Aircraft Company, and Associate Director of Corporate Development Planning at Lockheed Aircraft Corporation. From 1965 to 1973, he was also Adjunct Professor of Industrial and Systems Engineering at the University of Southern California. There he introduced courses in "Technological Forecasting" and "Planning Alternative Futures." The latter won USC's Dart Award for Innovation in Teaching in 1970. He has presented numerous seminars on Technological Forecasting and Long-Range Planning in the United States and Europe and started the Center for Technological and Interdisciplinary Forecasting at Tel-Aviv University. Since 1970, he has been Professor of Systems Science at Portland State University and Director of its new interdisciplinary Ph. D. Program in this field. He is Senior Editor of the international journal, *Technological Forecasting and Social Change*.

Murray Turoff received his B. S. degree in 1958 from the University of California at Berkeley in Physics and Mathematics, and his Ph. D. in Physics from Brandeis University in 1965. He has been a Systems Engineer for IBM, a member of the Science and Technology Division of the Institute for Defense Analyses and a Senior Operations Research Analyst for the Office of Emergency Preparedness in the Executive Offices of the President. In 1972-73 he introduced and taught a course in "Technological Forecasting" at American University. Since 1973 he has been an Associate Professor of Computer and Information Science at

The Editors (continued) the New Jersey Institute of Technology. He is also Associate Director of the Center for Technology Assessment. Professor Turoff has conducted or designed a number of major Delphi studies and is noted for introducing the Policy Delphi concept and the first computer-based Delphi, as well as a number of Management Information Systems based upon Delphi-like communication concepts. Currently he is engaged in a major research program at NJIT concerned with the use of computers to augment human communication capabilities.

THE CONTRIBUTORS

Marvin Adelson is Professor in the School of Architecture and Urban Planning at the University of California at Los Angeles. He also directs the Creative Problem-Solving Program for undergraduates at UCLA.

Samuel Aroni is Professor of Architecture/Urban Design and Acting Dean of the School of Architecture and Urban Planning at the University of California at Los Angeles. Dr. Aroni has published widely in the fields of structures, concrete materials, statistical methods, future research, building systems, and housing.

Klaus Brockhoff is Professor of Business Administration at the University of Kiel (Germany) and an Associate to the European Institute for Advanced Studies in Management in Brussels (Belgium). He has had extensive interest in R&D management and has published numerous papers as well as a book on related topics.

J. Douglas Carroll is a Member of the Technical Staff in the Acoustical and Behavioral Research Center at Bell Telephone Laboratories. Dr. Carroll specializes in methodological research on multidimensional scaling, clustering, and related multivariate techniques for the behavioral sciences.

Norman C. Dalkey	is associated with the Systems Engineering Department at the University of California at Los Angeles. He was a pioneer in Delphi research during his long tenure at the RAND Corporation.
Lawrence H. Day	is Staff Supervisor–Business Planning in the Headquarters Planning of Bell Canada. He holds an M. B. A. degree from McMaster University in Hamilton, Ontario.
Alex Ducanis	is Chairperson, Division of Specialized Professional Development, School of Education, University of Pittsburgh.
Selwyn Enzer	is Associate Director of the Center for Futures Research at the University of Southern California. Previously he was Senior Research Fellow and Treasurer of the Institute for the Future. He has extensive experience in the development and application of the basic tools of long-range forecasting, particularly the Delphi and cross-impact techniques.
Nancy H. Goldstein	has an M. A. Degree in Government from the University of Massachusetts, and experience as an information systems specialist with the Office of Emergency Preparedness in Washington, D. C.
Olaf Helmer	is Harold Quinton Professor of Futures Research at the University of Southern California. His field is the development of methods of long-range forecasting and planning. He is one of the originators of the Delphi and cross-impact techniques.
Irene Anne Jillson	is a policy analyst who has implemented a policy Delphi in the drug abuse field, and has assisted in developing the content for a Canadian Delphi conferencing system. She is a research consultant with the Johns Hopkins University School of Hygiene and Public Health. She is also President, Policy Research, Inc., Baltimore, Maryland.
Robert Johansen	is a Research Fellow at the Institute for the Future, Menlo Park, California. He has a doctorate from Northwestern University and has recently been involved in a research effort to assess the social effects

of communicating via a computer conferencing system.

Chester G. Jones is an Operations Analyst at the Aeronautical Systems Division of the United States Air Force at Wright-Patterson Air Force Base, Ohio. He received his doctorate in 1971 from Ohio State University.

Julius Kane is Professor of Mathematics and Mathematical Ecology at the University of British Columbia. He received his doctorate in Applied Mathematics from New York University and has extensive experience in mathematical modeling and simulation.

John Ludlow is Assistant Professor of Management at Northern Michigan University. He was previously Program Director of the B-1 Strategic Bomber Program for the United States Air Force.

Richard H. Miller is Research Associate at Applied Communication Research. He has an M. A. Degree in Communications Research from Stanford University and is currently working on projects involving scientific and technical information dissemination for the National Science Foundation.

Ian I. Mitroff is a professor in The Graduate School of Business, Interdisciplinary Department of Information Science, the Department of Sociology, and the Philosophy of Science Center, University of Pittsburgh.

Norman W. Mulgrave is Associate Professor of Educational Development and School Psychology at the University of Pittsburgh School of Education.

Charlton R. Price is a sociologist and management consultant to government and industry. He is currently associated with the Program of Policy Studies in Science and Technology at The George Washington University and with Social Engineering Technology, Inc., Los Angeles, California.

D. Sam Scheele is President of Social Engineering Technology, a Los Angeles firm which designs high-performance human services and produces plans for the future-sensitive development of institutions and individuals.

M. Scheibe is associated with R. H. Pratt Associates in Kensington, Maryland. He was formerly in the Department of Civil Engineering at Northwestern University.

J. Schofer is associated with the Department of Civil Engineering at Northwestern University.

James A. Schuyler is Director of Computers and Teaching, a research project into Computer-Aided-Instruction, at Northwestern University.

Thomas B. Sheridan is Professor of Mechanical Engineering at the Massachusetts Institute of Technology and current President of the IEEE Systems, Man, and Cybernetics Society. He supervises research in group decision and man-machine systems and is co-author with W. R. Ferrell of "Man-Machine Systems" (MIT Press, 1974).

M. Skutsch is with the Department of Geography, University of Dar-es-Salaam, Tanzania, East Africa. The work in this book was undertaken while he was in the Department of Industrial Engineering and Management Science at Northwestern University.

John W. Sutherland is Professor of Mathematics and Psychology at the University College, Rutgers University. At the time his paper was written, he was Visiting Professor of Systems Science at Portland State University.

Jacques F. Vallee is Senior Research Fellow at the Institute for the Future. He received his Ph. D. Degree in Computer Science from Northwestern University. From 1969 to 1972 he was with Stanford University as the Manager of Information Systems, and subsequently a Research Engineer at Stanford Research Institute. At the Institute for the Future he is leading an effort on the design and implementation of an on-line computer interrogation and conferencing system which is intended to facilitate interaction among geographically remote participants.

Myron Wish is a Member of the Technical Staff in the Acoustical and Behavioral Research Center of Bell Laboratories. He received his Ph. D. in psychology from The University of Michigan in 1966. Before coming to Bell Labs in 1968 he taught psychological measurement and statistics at Teachers College, Columbia University. His current research interests are in the development of new methods for measuring interpersonal communication and for comparing audio and audiovisual telecommunication with face-to-face communication.

Foreword by

OLAF HELMER

A week seldom goes by in which I do not receive a request for information on the Delphi method, and I am delighted that at last a compendium has been put together that fills the clear need indicated by such inquiries.

Delphi has come a long way in its brief history, and it has a long way to go. Since its invention about twenty years ago for the purpose of estimating the probable effects of a massive atomic bombing attack on the United States, and its subsequent application in the mid-sixties to technological forecasting, its use has proliferated in the United States and abroad. While its principal area of application has remained that of technological forecasting, it has been used in many other contexts in which judgmental information is indispensable. These include normative forecasts; the ascertainment of values and preferences; estimates concerning the quality of life; simulated and real decisionmaking; and what may be called "inventive planning," by which is meant the identification (including invention) of potential measures that might be taken to deal with a given problem situation and the assessment of such proposed measures with regard to their feasibility, desirability, and effectiveness.

Despite these many applications, Delphi still lacks a completely sound theoretical basis. This is due, largely, to the fact that Delphi, by definition, is concerned with the utilization of experts' opinions and that experts are rarely available as experimental laboratory subjects. Delphi experience, therefore, derives almost wholly either from studies carried out without proper experimental controls or from controlled experiments in which students are used as surrogate experts. It is still an open question which of the many results obtained through this latter kind of experimentation carry over to the case of real experts, and I hope that further investigations in this area will bc undertaken.

Further solidification of the Delphi technique, based on careful experimentation, clearly would be desirable, especially in view of the far-reaching applications to which the method in principle lends itself. Among these, particular attention is deserved by the following two, one of which has to do with inputs into research, the other with the utilization of research output: The first such application consists in the employment of Delphi surveys to provide judgmental input data for use in studies in the

The Delphi Method: Techniques and Applications, Harold A. Linstone and Murray Turoff (eds.) ISBN 0-201-04294-0; 0-201-04293-2 social-science area, in cases where hard data are unavailable or too costly to obtain. For instance, in a study of socioeconomic strategies for a developing country, reliance on judgmental information supplied by area specialists may be an adequate substitute for hard, but nonexistent, data. The other major application of Delphi, of which numerous examples already exist, is to the process of gathering of expert opinions among the nationwide "advice community" on which governmental decisionmakers frequently rely. In this mode of application, Delphi can be of considerable utility, both by systematizing the process and by lending greater objectivity to its "adversary" aspects. (I am glad to note, in this connection, that the concept of a "D-net," which I have advocated over the years, has reached the point of serious experimentation and may see full-scale implementation within the next few years. A D-net is a remote-conferencing network via computer terminals, which can, in particular, be used in a Delphi mode to tap the expertise of the participating respondents.)

These uses of Delphi, to supply "soft" data in the social sciences and to provide decisionmakers with ready access to specialized expertise, are of great potential importance. They place considerable demands on the integrity of the method and of its practitioners. This volume, I hope, will provide professional designers and users of Delphi surveys with much-needed background information.

I. Introduction

I. Introduction

HAROLD A. LINSTONE and MURRAY TUROFF

General Remarks

It is common, in a book of this kind, to begin with a detailed and explicit definition of the subject—the Delphi technique. However, if we were to attempt this, the reader would no doubt encounter at least one contribution to this collection which would violate our definition. There is in addition a philosophical perspective that when something has attained a point at which it is explicitly definable, then progress has stopped; such is the view we hold with respect to Delphi.

In 1969 the number of Delphi studies that had been done could be counted in three digits; today, in 1974, the figure may have already reached four digits. The technique and its application are in a period of evolution, both with respect to how it is applied and to what it is applied. It is the objective of this book to expose the richness of what may be viewed as an evolving field of human endeavor. The reader will encounter in these pages many different perspectives on the Delphi method and an exceedingly diverse range of applications.

For a technique that can be considered to be in its infancy, it would be presumptuous of us to present Delphi in the cloak of a neatly wrapped package, sitting on the shelf and ready to use. Rather, we have adopted the approach, through our selection of contributions, of exhibiting a number of different objects having the Delphi label and inviting you to sculpt from these examples your own view and assessment of the technique. For, if anything is "true" about Delphi today, it is that in its design and use Delphi is more of an art than a science.

However, as editors, we would be remiss if there were not some common thread underlying the articles brought together in this volume. As long as we restrict ourselves to a very general view, it is not difficult to present an acceptable definition of the Delphi technique which can be taken as underlying the contributions to this book:

> Delphi may be characterized as a method for structuring a group communication process so that the process is effective in allowing a group of individuals, as a whole, to deal with a complex problem.

To accomplish this "structured communication" there is provided: some feedback of individual contributions of information and knowledge; some assessment of the group judgment or view; some opportunity for individuals to revise views; and some degree of anonymity for the individual responses. As the reader will discover, there are many different views on what are the "proper," "appropriate," "best," and/or "useful" procedures for accomplishing the various specific aspects of Delphi. We hope that the reader will find this book a

The Delphi Method: Techniques and Applications, Harold A. Linstone and Murray Turoff (eds.)
ISBN 0-201-04294-0; 0-201-04293-2

rich menu of procedures from which he may select his own repast if he should seek to employ the Delphi technique.

When viewed as communication processes, there are few areas of human endeavor which are not candidates for application of Delphi. While many people label Delphi a forecasting procedure because of its significant use in that area, there is a surprising variety of other application areas. Among those already developed we find:

- Gathering current and historical data not accurately known or available
- Examining the significance of historical events
- Evaluating possible budget allocations
- Exploring urban and regional planning options
- Planning university campus and curriculum development
- Putting together the structure of a model
- Delineating the pros and cons associated with potential policy options
- Developing causal relationships in complex economic or social phenomena
- Distinguishing and clarifying real and perceived human motivations
- Exposing priorities of personal values, social goals

It is not, however, the explicit nature of the application which determines the appropriateness of utilizing Delphi; rather, it is the particular circumstances surrounding the necessarily associated group communication process: "Who is it that should communicate about the problem, what alternative mechanisms are available for that communication, and what can we expect to obtain with these alternatives?" When these questions are addressed, one can then decide if Delphi is the desirable choice. Usually, one or more of the following properties of the application leads to the need for employing Delphi:

- The problem does not lend itself to precise analytical techniques but can benefit from subjective judgments on a collective basis
- The individuals needed to contribute to the examination of a broad or complex problem have no history of adequate communication and may represent diverse backgrounds with respect to experience or expertise
- More individuals are needed than can effectively interact in a face-to-face exchange
- Time and cost make frequent group meetings infeasible
- The efficiency of face-to-face meetings can be increased by a supplemental group communication process
- Disagreements among individuals are so severe or politically unpalatable that the communication process must be refereed and/or anonymity assured
- The heterogeneity of the participants must be preserved to assure validity of the results, i.e., avoidance of domination by quantity or by strength of personality ("bandwagon effect")

Hence, for the application papers in this book the emphasis is not on the results of a particular application but, rather, on discussion of why Delphi was used and how it was implemented. From this the reader may be able to transpose the considerations to his own area of endeavor and to evaluate the applicability of Delphi to his own problems.

Those who seek to utilize Delphi usually recognize a need to structure a group communication process in order to obtain a useful result for their objective. Underlying this is a deeper question: "Is it possible, via structured communications, to create any sort of collective human intelligence[1] capability?" This is an issue associated with the utility of Delphi that has not as yet received the attention it deserves and the reader will only find it addressed here indirectly. It will, therefore, be a subjective evaluation on his part to determine if the material in this book represents a small, but initial, first step in the long-term development of collective human intelligence processes.

Characteristics of the Delphi

The Delphi process today exists in two distinct forms. The most common is the paper-and-pencil version which is commonly referred to as a "Delphi Exercise." In this situation a small monitor team designs a questionnaire which is sent to a larger respondent group. After the questionnaire is returned the monitor team summarizes the results and, based upon the results, develops a new questionnaire for the respondent group. The respondent group is usually given at least one opportunity to reevaluate its original answers based upon examination of the group response. To a degree, this form of Delphi is a combination of a polling procedure and a conference procedure which attempts to shift a significant portion of the effort needed for individuals to communicate from the larger respondent group to the smaller monitor team. We shall denote this form *conventional Delphi.*

A newer form, sometimes called a "Delphi Conference," replaces the monitor team to a large degree by a computer which has been programmed to carry out the compilation of the group results. This latter approach has the advantage of eliminating the delay caused in summarizing each round of Delphi, thereby turning the process into a real-time communications system. However, it does require that the characteristics of the communication be well defined before Delphi is undertaken, whereas in a paper-and-pencil Delphi exercise the monitor team can adjust these characteristics as a function of the group responses. This latter form shall be labeled *real-time Delphi.*

Usually Delphi, whether it be conventional or real-time, undergoes four distinct phases. The first phase is characterized by exploration of the subject under discussion, wherein each individual contributes additional information he feels is pertinent to the issue. The second phase involves the process of

[1]We refer to "intelligence" in this context as including attitudes and feelings which are part of the process of human motivation and action.

reaching an understanding of how the group views the issue (i.e., where the members agree or disagree and what they mean by relative terms such as importance, desirability, or feasibility). If there is significant disagreement, then that disagreement is explored in the third phase to bring out the underlying reasons for the differences and possibly to evaluate them. The last phase, a final evaluation, occurs when all previously gathered information has been initially analyzed and the evaluations have been fed back for consideration.

On the surface, Delphi seems like a very simple concept that can easily be employed. Because of this, many individuals have jumped at trying the procedure without carefully considering the problems involved in carrying out such an exercise. There are perhaps as many individuals who have had disappointing experiences with a Delphi as there are users who have had successes. Some of the common reasons for the failure of a Delphi are:

- Imposing monitor views and preconceptions of a problem upon the respondent group by overspecifying the structure of the Delphi and not allowing for the contribution of other perspectives related to the problem
- Assuming that Delphi can be a surrogate for all other human communications in a given situation
- Poor techniques of summarizing and presenting the group response and ensuring common interpretations of the evaluation scales utilized in the exercise
- Ignoring and not exploring disagreements, so that discouraged dissenters drop out and an artificial consensus is generated
- Underestimating the demanding nature of a Delphi and the fact that the respondents should be recognized as consultants and properly compensated for their time if the Delphi is not an integral part of their job function

In addition to the latter problems associated with the Delphi technique another class of criticisms directed at Delphi is often raised in the literature. These are the "virtual" problems that do not in themselves affect the utility of the technique.[2] Typical of these is the question of how to choose a "good" respondent group. This is, in fact, a problem for the formation of any group activity—committees, panels, study groups, etc. One has this problem no matter what communication mode is used; therefore, while it is a real and significant problem, it is not a problem unique to Delphi. However, the nature of certain applications does, in fact, dictate special consideration of this problem and it is discussed in a number of articles. Another virtual problem frequently arises when a particular Delphi design for a particular application is taken as representative of all Delphis, whereupon it is then observed that this design does not work for some other application. The problem here is that of making too explicit and restrictive a definition for Delphi. A third virtual

[2]See, for example, Gordon Welty, "A Critique of the Delphi Technique," *Proceedings of the American Statistical Association*, 1971 Social Statistics Section.

problem is the honesty of the monitor team, and it is of the same concern as the honesty of any study or analysis group. In fact, there is probably more likelihood in most instances of exposure of misrepresentation in a Delphi summary than in a typical group study report. Finally, misunderstandings may arise from differences in language and logic if participants come from diverse cultural backgrounds. Since we consider these virtual issues to be somewhat irrelevant to Delphi per se, we have made no attempt to give them special attention within this book. Other problems will be discussed in Chapter VIII.

It is quite clear that in any one application it is impossible to eliminate all problems associated with Delphi. There is, for example, a natural conflict in the goal of allowing a wide latitude in the contribution of information and the goal of keeping the communication process efficient. It is the task of the Delphi designer to minimize these problems as much as possible and to balance the various communication "goals" within the context of the objective of the particular Delphi and the nature of the participants. Arriving at a balanced design for the communication structure is still very much an art, even though there is considerable experience on how to ask and summarize various types of questions.

It can be expected that the use of Delphi will continue to grow. As a result of this, one can observe that a body of knowledge is developing on how to structure the human communication process for particular classes of problems. The abuse, as well as the use, of the technique is contributing to the development of this design methodology.

Table 1 compares the properties of normal group communication modes and the Delphi conventional and real-time modes. The major differences lie in such areas as the ability of participants to interact with the group at their own convenience (i.e., random as opposed to coincident), the capacity to handle large groups, and the capability to structure the communication. With respect to time considerations, there is a certain degree of similarity between a committee and a conventional Delphi process, since delays between committee meetings and Delphi rounds are unavoidable. Also, the real-time Delphi is conceptually somewhat analogous to a randomly occurring conference call with a written record automatically produced. It is interesting to observe that within the context of the normal operation of these communication modes in the typical organization—government, academic, or industrial—the Delphi process appears to provide the individual with the greatest degree of individuality or freedom from restrictions on his expressions. The items highlighted in the table will be discussed in more detail in many of the articles in this book.

While the written word allows for emotional content, the Delphi process does tend to minimize the feelings and information normally communicated in such manners as the tone of a voice, the gesture of a hand, or the look of an eye. In many instances these are a vital and highly informative part of a communication process. Our categorization of group communication processes is not meant to imply that the choice for a particular objective is limited, necessarily, to one

TABLE 1
Group Communication Techniques

	Conference Telephone Call	Committee Meeting	Formal Conference or Seminar	Conventional Delphi	Real-Time Delphi
Effective Group Size	Small	Small to medium	Small to large	Small to large	Small to large
Occurrence of Interaction by Individual	Coincident with group	Coincident with group	Coincident with group	Random	Random
Length of Interaction	Short	Medium to long	Long	Short to medium	Short
Number of Interactions	Multiple, as required by group	Multiple, necessary time delays between	Single	Multiple, necessary time delays between	Multiple, as required by individual
Normal Mode Range	Equality to chairman control (flexible)	Equality to chairman control (flexible)	Presentation (directed)	Equality to monitor control (structured)	Equality to monitor control or group control and no monitor (structured)

TABLE 1 (continued)

Group Communication Techniques

	Conference Telephone Call	Committee Meeting	Formal Conference or Seminar	Conventional Delphi	Real-Time Delphi
Principal Costs	Communications	—Travel —Individuals' time	—Travel —Individuals' time —Fees	—Monitor time —Clerical —Secretarial	—Communications —Computer usage
	Time-urgent considerations	Forced delays		Forced delays	Time-urgent considerations
Other Characteristics	—Equal flow of information to and from all —Can maximize psychological effects		—Efficient flow of information from few to many	—Equal flow of information to and from all —Can minimize psychological effects —Can minimize time demanded of respondents or conferees	

communication mode. As the readers will see from some of the contributions to this book, there are instances where it is desirable to use a mix of these approaches.

The Evolution of Delphi

The Delphi concept may be viewed as one of the spinoffs of defense research. "Project Delphi" was the name given to an Air Force-sponsored Rand Corporation study, starting in the early 1950's, concerning the use of expert opinion.[3] The objective of the original study was to "obtain the most reliable consensus of opinion of a group of experts ... by a series of intensive questionnaires interspersed with controlled opinion feedback."

It may be a surprise to some that the subject of this first study was the application of "expert opinion to the selection, from the point of view of a Soviet strategic planner, of an optimal U. S. industrial target system and to the estimation of the number of A-bombs required to reduce the munitions output by a prescribed amount." It is interesting to note that the alternative method of handling this problem at that time would have involved a very extensive and costly data-collection process and the programming and execution of computer models of a size almost prohibitive on the computers available in the early fifties. Even if this alternative approach had been taken, a great many subjective estimates on Soviet intelligence and policies would still have dominated the results of the model. Therefore, the original justifications for this first Delphi study are still valid for many Delphi applications today, when accurate information is unavailable or expensive to obtain, or evaluation models require subjective inputs to the point where they become the dominating parameters. A good example of this is in the "health care" evaluation area, which currently has a number of Delphi practitioners.

However, because of the topic of this first notable Delphi study, it took a later effort to bring Delphi to the attention of individuals outside the defense community. This was the "Report on a Long-Range Forecasting Study," by T. J. Gordon and Olaf Helmer, published as a Rand paper in 1964.[4] Its aim was to assess "the direction of long-range trends, with special emphasis on science and technology, and their probable effects on our society and our world." "Long-range" was defined as the span of ten to fifty years. The study was done to explore both the methodological aspects of the technique and to obtain substantive results. The authors found themselves in "a near-vacuum as far as tested techniques of long-range forecasting are concerned." The study covered

[3]N. Dalkey and O. Helmer, "An Experimental Application of the Delphi Method to the Use of Experts," *Management Science* **9**, No. 3 (April 1963), p. 458.

[4]Rand Paper P-2982. Most of the study was later incorporated into Helmer's *Social Technology*, Basic Books, New York, 1966.

six topics: scientific breakthroughs; population control; automation; space progress; war prevention; weapon systems. Individual respondents were asked to suggest future possible developments, and then the group was to estimate the year by which there would be a 50 percent chance of the development occurring. Many of the techniques utilized in that Delphi are still common to the pure forecasting Delphis being done today. That study, together with an excellent related philosophical paper providing a Lockean-type justification for the Delphi technique,[5] formed the foundation in the early and mid-sixties for a number of individuals to begin experimentation with Delphi in nondefense areas.

At the same time that Delphi was beginning to appear in the open literature, further interest was generated in the defense area: aerospace corporations and the armed services. The rapid pace of aerospace and electronics technologies and the large expenditures devoted to research and development leading to new systems in these areas placed a great burden on industry and defense planners. Forecasts were vital to the preparation of plans as well as the allocation of R&D (research and development) resources, and trend extrapolations were clearly inadequate. As a result, the Delphi technique has become a fundamental tool for those in the area of technological forecasting and is used today in many technologically oriented corporations. Even in the area of "classical" management science and operations research there is a growing recognition of the need to incorporate subjective information (e.g., risk analysis) directly into evaluation models dealing with the more complex problems facing society: environment, health, transportation, etc. Because of this, Delphi is now finding application in these fields as well.

From America, Delphi has spread in the past nine years to Western Europe, Eastern Europe, and the Far East. With characteristic vigor the largest Delphi undertaken to date is a Japanese study. Starting in a nonprofit organization, Delphi has found its way into government, industry, and finally academe. This explosive rate of growth in utilization in recent years seems, on the surface, incompatible with the limited amount of controlled experimentation or academic investigation that has taken place. It is, however, responding to a demand for improved communications among larger and/or geographically dispersed groups which cannot be satisfied by other available techniques. As illustrated by the articles in this book, aside from some of the Rand studies by Dalkey, most "evaluations" of the technique have been secondary efforts associated with some application which was the primary interest. It is hoped that in coming years experimental psychologists and others in related academic areas will take a more active interest in exploring the Delphi technique.

While many of the early articles on Delphi are quite significant and liberally mentioned in references throughout this book, we have chosen to concentrate

[5]O. Helmer and N. Rescher, "On the Epistemology of the Inexact Sciences," Project Rand Report R-353, February 1960.

on work that has taken place in the past five years and which represents a cross section of diverse applications.

Although the majority of the Delphi efforts are still in the pure forecasting area, that application provides only a small part of the contents of this volume.

Chapters II and III of this book consist of articles which provide an overview of the Delphi technique, its utility, the underlying philosophy, and broad classes of application.

Chapter IV takes up recent studies in the evaluation of the technique. Precision and accuracy are considered in this context. Between the reviews, articles, and associated references, the reader should obtain a good perspective on the state of the art with respect to experimentation.

Chapters V and VI describe some of the specialized techniques that have evolved for asking questions and evaluating responses. Foremost among them is cross-impact analysis (Chapter V). This concept reflects recognition of the complexity of the systems dealt with in most Delphi activities, systems where "everything interacts with everything." In essence, these sections explore the quantitative techniques available for deeper analysis of the subjective judgments gathered through the Delphi mechanism.

The effect computers can have on Delphi and speculations on the future of the technique itself are discussed in Chapter VII. The book concludes with a summary of pitfalls which can serve the practitioner as a continuing checklist (Chapter VIII).

We have striven to avoid making this volume a palimpsest of previously published papers: all but four of the articles have been especially prepared for this work. The four reprinted articles were selected from the journal *Technological Forecasting and Social Change*, a rich lode of material on Delphi. The extensive bibliography in the Appendix provides a guide to those who wish to probe the subject further. It is thus our hope that this volume will serve the reader as a useful reference work on Delphi for a number of years.

II. Philosophy

II.A. Introduction

HAROLD A. LINSTONE and MURRAY TUROFF

Any human endeavor which seeks recognition as a professional or scientific activity must clearly define the axioms upon which it rests. The foundation of a discipline, as the foundations of a house, serves as a guide and basis for the placement of the building blocks of knowledge gathered through research and development activities. It is the definition, exposure, and investigation of the philosophical foundation that distinguishes a scientific profession from other endeavors.

In a well-established scientific endeavor, the foundation is made explicit so that one is able to recognize when the resulting structure can no longer be properly supported and a reexamination of the fundamentals is in order. A classic example of this was the impact of quantum mechanics on the foundations of physics. With respect to new disciplines, such as the investigation of Delphi methodology, the situation is one where not enough of the structure has been blueprinted to discriminate which of many possible foundations supply the "best" underpinnings.

The early attempt by Helmer and Rescher in their classic paper "On the Epistemology of the Inexact Sciences" proposed one foundation, largely of a Lockean nature, which was very adequate for the typical technological forecasting applications for which Delphi has been popular. However, in recent years extensions to Delphi methodology have demonstrated a need for a broader basis. Certainly the theme of this book, which largely views Delphi as the process of structuring human communications, further enhances this position.

The first article by Mitroff and Turoff, examines what the various classic or "pure mode" epistemologies of Western philosophy have to offer for insight into the Delphi process. The philosophies covered are those represented by Locke, Leibniz, Kant, Hegel, and Singer. It largely follows the morphological structure of philosophical inquiry first proposed by C. West Churchman in his "Design of Inquiring Systems." As with any young discipline, it should not come as a surprise that such a rich diversity of foundation axioms may be used to give form and shape to Delphi. In a sense this is an expression of the yet untapped potential for future development of the technique.

The second article, by Scheele, illustrates how a user of Delphi may compose for his own view and application of Delphi a very specific philosophical foundation. The author, being primarily concerned in many of his applications with the perceptions of individuals as they may relate to marketing problems, adapts elements of the Lockean, Kantian, and Singerian philosophies and merges them with the existentialist concept of subjective or negotiated reality. The result is a foundation for a design precisely matched to the user's unique needs.

Throughout the book one will find in the various articles explicit or implicit support for a mode or manner of applying Delphi which rests on the philosophies brought out in these two papers. It is interesting to note that a recent sociological perspective views Delphi as a ritual.[1] Primitive man always approached the future ritualistically, with ceremonies involving utensils, liturgies, managers, and participants. The Buckminster Fuller World Game, Barbara Hubbard's SYNCON, as well as Delphi, can be considered as modern participatory rituals. The committee-free environment and anonymity of Delphi stimulate reflection and imagination, facilitating a personal futures orientation. Thus, the modern Delphi is indeed related to its famous Greek namesake.

[1]A. Wilson and D. Wilson, "The Four Faces of the Future," New York, Grove Press, 1974.

II. B. Philosophical and Methodological Foundations of Delphi

IAN I. MITROFF and MURRAY TUROFF

It takes two of us to create a truth, one to
utter it and one to understand it.
Kahlil Gibran

Introduction

The purpose of this article is to show that underlying *any* scientific technique, theory, or hypothesis there is always *some* philosophical basis or theory about the nature of the world upon which that technique, theory, or hypothesis fundamentally rests or depends. We also wish to show that there is more than one fundamental basis which can underlie any technique, or in other words, that there is no one "best" or even "unique" philosophical basis which underlies any scientific procedure or theory. Depending upon the basis which is presumed, there results a radically different developmental and application history of a technique. Thus in this sense, the particular basis upon which a scienfific procedure depends is of fundamental practical importance and not just of philosophical interest.

We human beings seem to have a basic talent for disguising through phraseology the fundamental similarities that exist between common methodologies of a different name. As a result, we often bicker and quarrel about such superficial matters as whether this or that name is appropriate for a certain technique when the real issue is whether the philosophical basis or system of inquiry that underlies a proposed technique or methodology is sound and appropriate. We are indeed the prisoners of our basic images of reality. Not only are we generally unaware of the different philosophical images that underlie our various technical models, but each of us has a fundamental image of reality that runs so deep that often we are the last to know that we hold it. As a result we disagree with our fellow man and we experience inner conflict without really knowing why. What's worse, we ensure this ignorance and conflict by hiding behind catchwords and fancy names for techniques. The field of endeavor subsumed under the name of Delphi is no less remiss in this respect than many other disciplines. Its characteristic vocabulary more often obscures the issues than illuminates them.

The Delphi Method: Techniques and Applications, Harold A. Linstone and Murray Turoff (eds.)
ISBN 0-201-04294-0; 0-201-04293-2

One of the basic purposes of our discussion is to bring these fundamental differences and conflicts of methodology up to the surface for conscious examination so that, one hopes, we can be in a better position to choose explicitly the approach we wish to adopt. In order to accomplish this we consider a number of fundamental historical stances that men have taken toward the problem of establishing the "truth content" of a system of communication signals or acts. More precisely, the purpose of this article is to examine the variety of ways and mechanisms in which men have chosen to locate the criteria which would supposedly "guarantee" our "true and accurate understanding" of the "content" of a communication act or acts. We will also show that every one of these fundamental ways differs sharply from the others and that each of them has major strengths as well as major weaknesses. The moral of this discussion will be that there is *no one "single best way"* for ensuring our understanding of the content of a set of communication acts or for ascribing validity to a communication. The reason is that there is no one mode of ensuring understanding or for prescribing the validity of a communication that possesses all of the desired characteristics that one would like any preferred mode to possess. As we wish to illustrate, this awareness itself constitutes a kind of strength. To show that there is no one mode that can satisfy our every requirement, i.e., that there is no one mode that is *best* in all senses and for all circumstances, is not to say that each of these modes does not appear to be "better suited" for some special set of circumstances.

Since these various modes or characteristic models for ensuring validity basically derive from the history of Western philosophy, another objective of this article is also to show what philosophy and, especially, what the philosophy of science specifically and concretely has to offer the field of Delphi design. For example, one of the things we wish to show is which among these various philosophical modes have been utilized to date (and how) and which have been neglected. When there has been little or no utilization of a particular philosophical basis then we may infer existing gaps in the development of the Delphi to date.

Before we describe each of thesse philosophical modes or systems more fully, we can rather easily and simply convey the general spirit of each of them by means of the following exercise. Suppose we are given a set of statements or propositions by some individual or group which pretend to describe some alleged "truth." Then each of our philosophical systems (hereafter referred to as an Inquiring System, or IS) can be simply differentiated from one another in terms of the kind of characteristic question(s) that each would address either to the statement itself or to the individual (group) making the statement or assertion. Each question in effect embodies the major philosophical criterion that would have to be met before that Inquiring System would accept the propositions as valid or as true.

The Leibnizian analyst or IS would ask something like:

> How can one independently of any empirical or personal considerations give a *purely rational* justification of the proposed proposition or assertion? Can one build or demonstrate a rational model which underlies the proposition or assertion? How was the result deduced; is it precise, certain?

The Lockean analyst or IS would ask something like:

> Since for me data are always prior to the development of formal theory, how can one independently of any formal model justify the assertion by means of some objective data or the consensus of some group of expert judges that bears on the subject matter of the assertions? What are the supporting "statistics"? What is the "probability" that one is right? Are the assertions a good "estimate" of the true empirical state of affairs?

The Kantian analyst or IS would ask something like:

> Since data and theory (models) *always* exist side by side, does there exist *some combination* of *data* or expert judgment *plus* underlying *theoretical justification* for the data that would justify the propositions? What *alternative* sets of propositions exist and which best satisfy my objectives and offer the strongest combination of data plus model?

The Hegelian (Dialectical) analyst or IS would ask something like:

> Since every set of propositions is a reflection of a more general theory or *plan* about the nature of the world as a *whole system,* i.e., *a world-view,* does there exist some alternate sharply differing world-view that would permit the serious consideration of a completely opposite set of propositions? Why is this opposing view not true or more desirable? Further, does this conflict between the plan and the counterplan allow a third plan or world-view to emerge that is a *creative synthesis* of the original plan and counterplan?

Finally, the Singerian analyst or IS would ask:

> Have we taken a broad enough perspective of the basic problem? Have we from the very beginning asked the right question? Have we focused on the right objectives? To what extent are the questions and models of each inquirer a reflection of the unique *personality* of each inquirer as much as they are felt to be a "natural" characteristic or property of the "real" world?"

Even at this point in the discussion, it should be apparent that as a body these are very different kinds of questions and that each of them is indicative of a fundamentally different way of ascribing content to a communication. It should also be apparent, and it should really go without saying, that these do not exhaust the universe of potential questions. There are many more philosophical positions and approaches to "validity" than we could possibly hope to deal with in this article. These positions do represent, however, some of the most significant basic approaches and, in a sense, pure-modes from which others can be constructed.

The plan of the rest of this article is briefly as follows: first, we shall describe each inquirer in turn and in general terms, but we hope in enough detail to give the reader more of a feel for each system; second, along with the description of each inquirer, we shall attempt to point out the influence or lack of influence each philosophy of inquiry has had on the Delphi technique; and third, we shall attempt to point out some general conclusions regarding the nature and future of the Delphi technique as a result of this analysis.

It should be borne in mind as we proceed that the question of concern is not how we can determine or agree on the meaning of "truth" with "perfect or complete certainty," for put in this form, the answer is clearly that we cannot know anything with "perfect certainty." We cannot even know with "perfect certainty" that "we cannot know anything with 'perfect certainty.'" The real question is *what* can we know and, even more to the point, how we can *justify* what we think we can know. It is on this very issue that the difference between the various Inquiring Systems arises and the utility or value of the Delphi technique depends.

Inquiring Systems (IS)

The process of inquiry, whether it be for a single individual or a group of individuals, may be "represented" by a very general system. We start with some *assumed* "external event" or "raw data set" which for the moment we *consider* to be a characteristic property of the "real world," i.e., we *assume* the data set "exists" in the "external world." (As we shall see in a moment, this amounts to assuming a Lockean IS beginning. The point is that we can't even begin to describe the "world" and our "knowledge" of "it" without having to invoke some "conceptualization," i.e., some Inquiring System characterization, of "it.") Next we apply some transformation and/or filter to the "raw data" in order to get it into the "right form" for input to some model. The model, which may be *any sort* of structured process, is represented by a set of rules which may be either in the form of an algorithm or a set of heuristic principles. The model acts on the input to transform it from the state of "input data" to the state of "output information." This output information may in turn be passed through another filter or transform to put it in the right form so that a decisionmaker can recognize and utilize it as "information" or as a "policy recommendation." In terms of this general configuration, the various IS can be differentiated from one another with respect to (1) the priority assigned to the various systems components, i.e., which components are regarded as more important or fundamental by one IS than by another, and (2) the degree of interdependence assigned to the various systems components by each IS.

Our objective in the following discussion will be to draw a sufficient distinction between the philosophical Inquirying System (IS) concepts so that we can place alternative Delphi design methodologies into this perspective.

Lockean IS

As first pioneered by Dalkey, Helmer, and Rescher at Rand, the Delphi technique represents a prime example of Lockean inquiry. Indeed, one would be hard pressed to find a better contemporary example of a Lockean inquirer than the Delphi.

The philosophical mood underlying the major part of *empirical science* is that of Locke. The sense of Lockean IS can be rather quickly and generally grasped in terms of the following characteristics:

(1) Truth is *experiential,* i.e., the truth content of a system (or communication) is associated *entirely* with its empirical content. A model of a system is an *empirical model* and the truth of the model is measured in terms of *our* ability (a) to reduce every complex proposition down to its simple empirical referents (i.e., simple observations) and (b) to ensure the validity of each of the simple referents by means of the widespread, freely obtained *agreement* between different human observers.

(2) A corollary to (1) is that the truth of the model does not rest upon *any* theoretical considerations, i.e., upon the prior assumption of any theory (this is the equivalent of Locke's *tabula rasa*). Lockean inquirers are opposed to the prior presumption of theory, since in their view this exactly reverses the justifiable order of things. Data are that which are prior to and justify theory, not the other way around. The only general propositions which are accepted are those which can be justified through "direct observation" or have already been so justified previously. In sum, *the data input sector is not only prior to the formal model or theory sector but it is separate from it as well. The whole of the Lockean IS is built up from the data input sector.*

In brief, Lockean IS are the epitome of *experimental, consensual* systems. On any problem, they will build an empirical, inductive representation of it. They start from a set of elementary empirical judgments ("raw data," observations, sensations) and from these build up a network of ever expanding, increasingly more general networks of factual propositions. Where in the Leibnizian IS to be discussed shortly the networks are theoretically, deductively derived, in a Lockean IS they are empirically, inductively derived. The guarantor of such systems has traditionally been the function of human agreement, i.e., an empirical generalization (or communication) is judged "objective," "true," or "factual" if there is "sufficient widespread agreement" on it by a group of "experts." The final information content of a Lockean IS is identified almost exclusively with its empirical content.

A prime methodological example of Lockean thinking can be found in the field of statistics. Although statistics is heavily Leibnizian in the sense that it devotes a considerable proportion of its energies to the formal treatment of data and to the the development of formal statistical models, there is a strong if not almost pure Lockean component as well. The pure Lockean component manifests itself in the attitude that although statistical methods may "transform" the "basic raw data" and "represent" "it" differently, statistical methods themselves are presumed not to create the "basic raw data." In this sense, the "raw data" are presumed to be *prior to* and *independent of* the formal (theoretical) statistical treatment of the data. The "raw data" are granted a prior existential status. Another way to put this is to say that there is little or no match between the theory that the observer of the raw data has actually used (and has had to use) in order to collect his "raw data" in the first place and the theory (statistics) he has used to analyze it in the second place A typical Lockean point of view is the assertion that one doesn't need any theory in order to collect data first, only to analyze it subsequently.

As mentioned at the beginning of this section, the Delphi, at least as it was originally developed, is a classic example of a Lockean inquirer. Furthermore, *the Lockean basis of Delphi still remains the prime philosophical basis of the technique to date.*

As defined earlier Delphi is a procedure for structuring a communication process among a large group of individuals. In assessing the potential development of, say, a technical area, a large group (typically in the tens or hundreds) are asked to "vote" on when they think certain events will occur. One of the major premises underlying the whole approach is the assumption that a large number of "expert" judgments is required in order to "treat adequately" any issue. As a result, a face-to-face exchange among the group members would be inefficient or impossible because of the cost and time that would be involved in bringing all the parties together. The procedure is about as pure and perfect a Lockean procedure as one could ever hope to find because, first, the "raw data inputs" are the opinions or judgments of the experts; second, *the validity of the resulting judgment of the entire group is typically measured in terms of the explicit "degree of consensus" among the experts.* What distinguishes the Delphi from an ordinary polling procedure is the feedback of the information gathered from the group and the opportunity of the individuals to modify or refine their judgments based upon their reaction to the collective views of the group. Secondary characteristics are various degrees of anonymity imposed on the individual and collective responses to avoid undesirable psychological effects.

The problems associated with Delphi illustrate the problems associated with Lockean inquiry in general. The judgments that typically survive a Delphi procedure may not be the "best" judgments but, rather, the compromise position. As a result, the surviving judgments may lack the significance that extreme or conflicting positions may possess.

The strength of Lockean IS lies in their ability to sweep in rich sources of experiential data. In general, the sources are so rich that they literally overwhelm the current analytical capabilities of most Leibnizian (analytical) systems. The weaknesses, on the other hand, are those that beset all empirical systems. While experience is undoubtedly rich, it can also be extremely fallible and misleading. Further, the "raw data," "facts," or "simple observables" of the empiricist have *always* on deeper scientific and philosophical analysis proved to be exceedingly complex and hence further divisible into other entities themselves thought to be indivisible or simple, ad infinitum. More troublesome still is the almost extreme and unreflective reliance on agreement as the sole or major principle for producing information and even truth out of raw data. The trouble with agreement is that its costs can become too prohibitive and agreement itself can become too imposing. It is not that agreement has nothing to recommend it. It is just that agreement is merely one of the many philosophical ways for producing "truth" out of experiential data. The danger with agreement is that it may stifle conflict and debate when they are needed most. As a result, Lockean IS are best suited for working on well-structured problem situations for which there exists a *strong consensual* position on "the nature of the problem situation." If these conditions or assumptions cannot be met or justified by the decisionmaker—for example, if it seems too risky to base projections of what, say, the future will be like on the judgments of experts—then no matter how strong the agreement between them is, some alternate system of inquiry may be called for.

While the consensus-oriented Delphi may be appropriate to technological forecasting it may be somewhat inappropriate for such things as technology assessment, objective or policy formulation, strategic planning, and resource allocation analyses. These latter applications of Delphi often or should involve the necessity to explore or generate *alternatives,* which is very different from the generation of consensus.

Leibnizian IS

The philosophical mood underlying the major part of theoretical science is that of Leibniz. The sense of Leibnizian inquiry can be rather quickly and generally captured in terms of the following characteristics:

(1) Truth is *analytic*; i.e., the truth content of a system is associated *entirely* with its formal content. A model of a system is a formal model and the truth of the model is measured in terms of *its* ability to offer a theoretical explanation of a wide range of general phenomena and in terms of *our* ability as model-builders to state clearly the formal conditions under which the model holds.

(2) A corollary to (1) is that the truth of the model does not rest upon *any external considerations,* i.e., upon the raw data of the external world. Leibnizian inquirers regard empirical data as an inherently risky base upon which to

found universal conclusions of any kind, since from a finite data set one is never justified in inferring any general proposition. The only general propositions which are accepted are those that can be justified through purely rational models and/or arguments. Through a series of similar arguments, *Leibnizian IS not only regard the formal model component as separate from the data input component but prior to it as well.* Another way to put this is to say that *the whole of the Leibnizian IS is contained in the formal sector and thus it has priority over all the other components.*

In short, Leibnizian IS are the epitome of formal, symbolic systems. For *any* problem, they will characteristically strive to reduce it to a formal mathematical or symbolic representation. They start from a set of elementary, primitive "formal truths" and from these build up a network of ever expanding, increasingly more general, formal propositional truths. The *guarantor* of such systems has traditionally been the precise specification of what shall count as a proof for a derived theorem or proposition; other guarantor notions are those of internal consistency, completeness, comprehensiveness, etc. The final information content of Leibnizian IS is identified almost exclusively with its symbolic content.

A prime example of Leibnizian inquiry is the field of Operations Research (OR) in the sense that the major energies of the profession have been almost exclusively directed toward the construction and exploration of highly sophisticated formal models. OR is a prime example of Leibnizian inquiry not because there is no utilization of external data whatsoever in OR models but because in the training of Operations Researchers significantly more attention is paid to teaching students how to build sophisticated models than in teaching them equally sophisticated methods of data collection and analyses. There is the implication that the two activities are separable, i.e., that data can be collected independently of formal methods of analysis.

Delphi by itself is not a Leibnizian inquirer and is better viewed from the perspective of some of the alternative Inquiring Systems. However, many of the views and assertions made with respect to the Delphi technique involve Leibnizian arguments. Delphi has, for example, been accused of being very "unscientific." When assertions of this type are examined one usually finds the underlying proposition rests on equating what is "scientific" to what is "Leibnizian." This is a common misconception that has also affected other endeavors in the social, or so-called soft, sciences where it is felt that the development of a discipline into a science must follow some preordained path leading to the situation where all the results of the discipline can be expressed in Leibnizian "laws." We have today in such areas as economics, sociology, etc., schools of research dedicated to the construction of formal models as ends in themselves.

In Delphi we find a similar phenomenon taking place where models are constructed for the purpose of describing the Delphi process and for determining *the* "truth" content of a given Delphi. (See, for example, the articles in Chapter IV.) One model hypothesizes that the truth content of a Delphi result (often measured as the error) increases as the size of the Delphi group increases.

This concept is often used to guide the choice of the size of the participant group in a Delphi. Other formal models have been proposed to measure an individual's "expertise" as a function of the quantity of information supplied and the length of associated questions. All such models, which are *independent of the content* of what is being communicated but look for structured relationships in the process of the communication, are attempts to ascribe Leibnizian properties to the Delphi process. The existence of such models in certain circumstances do not in themselves make the Delphi technique any more or less "scientific." They are certainly useful in furthering our understanding of the technique and should be encouraged. However, they are based upon assumptions, such as the superiority of theory over data and the general applicability of formal methods of reasoning, which are quite suspect with respect to the scope of application of the Delphi technique and the relative experimental bases upon which most of these models currently rest. The utility of Delphi, at least in the near future, does not appear to rest upon making Delphi appear or be more Leibnizian but, rather, in the recognition of what all the IS models can contribute to the development of the Delphi methodology. Our current understanding of human thought and decision processes is probably still too rudimentary to expect generally valid formal models of the Delphi process at this time.

For which kinds of problem situations are Leibnizian analyses most appropriate? First of all, the situations must be sufficiently "well understood" and "simple enough" so that they can be modeled. Thus, Leibnizian IS are best suited for working on clearly definable (i.e., well-structured) problems for which there exists an analytic formulation as well as solution. Second, the modeler must have strong reasons for believing in the assumptions which underlie Leibnizian inquiry, e.g., that the model is universally and continually applicable. In a basic sense, the fundamental guarantor of Leibnizian inquiry is the "understanding" of the model-builder; i.e., he must have enough faith in his understanding of the situation to believe he has "accurately" and "faithfully" represented it.

Note that there is no sure way to prove or justify the assumptions underlying Leibnizian inquiry. The same is true of all the other IS. But then this is not the point. The point is to show the kinds of assumptions we are required to make if we wish to employ Leibnizian inquiry so that if the decisionmaker or modeler is unwilling to live with these assumptions he will know that another IS may possibly be called for.

Kantian IS

The preceding two sections illustrate the difficulties that arise from emphasizing one of the components of a tightly coupled system of inquiry to the detriment of other components. Leibnizian inquiry emphasizes theory to the detriment of data. Lockean inquiry emphasizes data to the detriment of theory. When these attitudes are translated into professional practice, what often results is the development of highly sophisticated models with little or no concern for the difficult problems associated with the collection of data or the seemingly endless proliferation of data with little regard for the dictates of currently existing models.

The recent controversy surrounding the attempts of Forrester and Meadows[1] to build a "World Model" is a good illustration of the strong differences between these two points of view. In our opinion, the work of Forrester and Meadows represents an almost pure Leibnizian approach to the modeling of large, complicated systems. The Forrester and Meadows model is, in effect, data independent. One can criticize the model on pure Leibnizian grounds, e.g., whether the internal theory and structure of the model are sound with respect to current economic and social theory, and some of the critics have chosen to do this. However, it would seem to us that more often than not the critics have chosen to offer a Lockean critique, i.e., that some other way, say using accurate statistical data, is a better way to build a sound forecast model of the world. While this is a legitimate method of criticism, to a large extent it only further exacerbates the differences between the two approaches and hence misses the real point. To us the real point is not whether the Forrester-Meadows approach is *the* correct Leibnizian approach, or whether there is *a* correct Lockean approach, but rather, whether *any* Leibnizian or Lockean approach acting independently of the other could ever possibly be "correct." Forrester and Meadows seek to justify (guarantee?) their approach through the robustness and richness of their *model,* and their Lockean critics attempt to establish the validity of their approach through the priority and "regularity" of the statistical *data* to which they appeal. Perhaps if the debate proves anything, it raises the serious question as to whether an advanced modern society can continue to rely on purely Leibnizian or Lockean efforts for its planning. In order to evaluate the relative merits of separate Leibnizian or Lockean inquirers, it is necessary to go to a competing philosophy which incorporates both, such as the Kantian inquirer.

The sense of Kantian inquiry can be rather quickly grasped through the following set of general characteristics:

(1) Truth is *synthetic;* i.e., the truth content of a system is not located in either its theoretical or its empirical components, but in *both.* A model of a system is a synthetic model in the sense that the truth of the model is measured in terms of the model's ability (a) to associate every theoretical term of the model with some empirical referent and (b) to show that (how) underlying the collection of every empirical observation related to the phenomenon under investigation there is an associated theoretical referent.

(2) A corollary to (1) is that neither the data input sector nor the theory sector have priority over one another. Theories or general propositions are built

[1]Meadows, Dennis "Limits to Growth" 1972 Universe Books.

up from data, and in this sense theories are dependent on data, but data cannot be collected without the prior presumption of some theory of data collection (i.e., a theory of "how to make observations," "what to observe," etc.), and in this sense data are dependent on theories. *Theory and data are inseparable.* In other words, Kantian IS require some coordinated image or plan of the system as a whole before any sector of the system can be worked on or function properly.

These hardly begin to exhaust all the features we identify with Kantian inquiry. A more complete description would read as follows: Kantian IS are the epitome of multimodel, synthetic systems. On any problem, they will build at least two alternate representations or models of it. (If the alternate representations are complementary, we have a Kantian IS; if they are antithetical, we have a Hegelian IS, as described in the next section.) The representations are partly Leibnizian and partly Lockean; i.e., Kantian IS make explicit the strong interaction between scientific theory and data. They show that in order to collect some scientific data on a problem a posteriori one *always* has had to presuppose the existence of some scientific theory a priori, no matter how implicit and informal that theory may be. Kantian IS presuppose at least two alternate scientific theories (this is their Leibnizian component) on any problem or phenomenon. From these alternate Leibnizian bases, they then build up at least two alternate Lockean fact nets. The hope is that out of these alternate fact nets, or representations of a decisionmaker's or client's problem, there will be one that is "best" for representing his problem. The defect of Leibnizian and Lockean IS is that they tend to give only one explicit view of a problem situation. Kantian IS attempt to give *many* explicit views. The guarantor of such systems is the degree of fit or match between the underlying theory (theoretical predictions) and the data collected under the presumption of that theory plus the "deeper insight" and "greater confidence" a decisionmaker feels he has as a result of witnessing many different views of his problem.

The reason Kantian IS place such a heavy emphasis on alternate models is that in dealing with problems like planning for the future, the real concern is how to get as many perspectives on the nature of the problem as possible. Problems which involve the future cannot be formulated and solved in the same way that one solves problems in arithmetic, i.e., via a single, well-structured approach. There seems to be something fundamentally different about the class to which planning problems belong. In dealing with the future, we are not dealing with the concrete realities of human existence, but, if only in part, with the hopes, the dreams, the plans, and the aspirations of men. Since different men rarely share the same aspirations, it seems that the best way to "analyze" aspirations is to compare as many of them against one another as we can. If the future is 99 percent aspiration or plan, it would seem that the best way to get a handle on the future is *to draw forth explicitly as many different aspirations or plans for the future as possible.* In short, we want to examine as many different alternate futures as we can.

In the field of planning, Normative Forecasting, Planning Programming Budgeting Systems (PPBS), and Cost-Effectiveness or Cost-Benefit Analyses are all examples of Kantian inquiry, although these are such low-level Kantian inquirers as to be almost more Leibnizian in nature than Kantian. The Kantian element that these various approaches share is the fact that they are all concerned with *alternate paths,* or *methods,* of getting from a present state to a future state characterized by certain objectives, needs, or goals. When these various planning vehicles have failed, it is not just because we are dealing with an inherently fuzzy problem—indeed that is the basic nature of the problem—but because we have failed to produce alternatives that are true alternatives and to show that the data, models, and objectives cannot be separated for purposes of planning.

In recent years, there have been a number of Delphi studies which in contrast to the original Lockean-based consensus Delphis begin "to take on" more actively the characteristics of Kantian inquiry. The initial Delphis were characterized by a strong emphasis on the use of consensus by a group of "experts" as *the* means to converge on a single model or position on some issue. In contrast, the explicit purpose of a Kantian Delphi is to elicit alternatives so that a comprehensive overview of the issue can take place. In terms of communication processes, while a "consensus," or Lockean, Delphi is better suited to setting up a communication structure among an already informed group that possesses the same general core of knowledge, a Kantian, or "contributory," Delphi attempts to design a structure which allows many "informed" individuals in different disciplines or specialties to contribute information or judgments to a problem area which is much broader in scope than the knowledge that any one of the individuals possesses. This type of Delphi has been applied to the conceptualization of such problems as: (1) the definition of a structural model for material flows in the steel industry (see Chapter III, C, 3); (2) the examination of the current and the potential role of the mentally retarded in society (see Chapter VI, D); (3) the forecasting of the future characteristics of recreation and leisure (see Chapter VI, D); and (4) the examination of the past history of the internal combustion engine[1] for a clue to significant events possibly affecting its future. While all of these Delphis had specific forecasting objectives, none of them could be achieved if all the parties to the Delphi were drawn from the same specialized interest group. The problems were broader than that which could be encompassed by any single discipline or mode of thinking. For example, the examination of the role of the mentally retarded in our society is neither the exclusive problem nor the sole province of any special group. Educators, psychiatrists, parents, and teachers all have different and valid perspectives to contribute to the definition of the "problem." Consensus on a single definition is not the goal, at least not in the

[1]proprietary Delphi in 1969 by Kenneth Craver of Monsanto Company.

initial stages, but rather, the eliciting of many diverse points of view and potential aspects of the problem. In essence, the objective is establishing how to fit the pieces of a jigsaw puzzle together, and even determining if it is one or many puzzles.

The problem of conceptualizing goals and objectives is not an explicit part of the three inquiry processes we have discussed so far. That is, the Leibnizian and Lockean IS are not explicitly goal directed. For example, Leibnizian IS assume that the same rational model is applicable no matter what the problem and the objectives of the decisionmaker or who it is that has the problem. In contrast, the Kantian IS is explicitly goal oriented, i.e., it hopes by presenting a decisionmaker with several alternative models of his problem to better clarify both the problem and the nature of the objectives, which after all are part of the "problem."

Kantian inquiry is best suited to problems which are inherently ill-structured, i.e., the kinds of problems which are inherently difficult to formulate in pure Leibnizian or Lockean terms because the nature of the problem does not admit of a clear consensus or a simple analytic attack. On the other hand, the Kantian inquiry is not especially suited for the kinds of problems which admit of a single clear-cut formulation because here the proliferation of alternate models may not only be costly but time consuming as well. Kantian inquiry may also overwhelm those who are used to "the single best model" approach to any problem. Of course this in itself is not necessarily bad if it helps to teach those who hold this belief that there are some kinds of problems for which there is no one best approach. Social problems inherently seem to be of this kind and thus to call for Kantian approach. The concept of "technology assessment" as a vehicle for determining the relationships between technology and social consequences would also seem to imply the necessity of at least a Kantian approach. Many efforts which fall under the heading of "assessments" have proved to be inadequate because they were conducted on pure Leibnizian or Lockean bases.

Hegelian, or Dialectical, IS

The idea of the Hegelian, or Dialectical, IS can be conveyed as follows:

(1) Truth is *conflictual*; i.e., the truth content of a system is the result of a highly complicated process which depends on the existence of a *plan* and a *diametrically opposed counterplan*. The plan and the counterplan represent strongly divergent and opposing conceptions of the whole system. The function of the plan and the counterplan is to engage each other in an unremitting debate over the "true" nature of the whole system, in order to draw forth a new plan that will, one hopes, reconcile (synthesize, encompass) the plan and the counterplan.

(2) A corollary to (1) is that by itself the data input sector is totally

meaningless and only becomes meaningful, i.e., "information," by being coupled to the plan and the counterplan. Further, it is postulated that there is a particular input data set which can be shown to be consistent with both the plan and the counterplan; i.e., by itself this data set supports neither naturally, but there is an interpretation of the data such that it is consistent with both the plan and the counterplan. It is also postulated that without both the plan and the counterplan the meaning of the data is incomplete, i.e., partial. Thus, under this system of inquiry, the plan and the counterplan which constitute the theory sector are prior to the input sector and indeed constitute opposing conceptions of the whole system. Finally, it is also assumed that *on* every *issue of importance, there can be found or constructed a plan and a counterplan; i.e., a dialectical debate can be formulated with respect to* any *issue.* On any issue of importance there will be an intense division of opinion or feeling.

Hegelian, or Dialectical, IS are the epitome of *conflictual*, synthetic systems. On any problem, they build *at least two, completely antithetical,* representations of it. Hegelian IS start with either the prior existence (identification) of or the creation of two strongly opposing (*contrary*) Leibnizian models of a problem. These opposing representations constitute the contrary underlying assumptions regarding the theoretical nature of the problem. Both of these Leibnizian representations are then applied to the *same* Lockean data set in order to demonstrate the crucial nature of the underlying theoretical assumptions, i.e., that the same data set can be used to support either theoretical model. The point is that data are not information; *information is that which results from the interpretation of data*. It is intended that out of a dialectical confrontation between opposing interpretations (e.g., the opposing "expert" views of a situation), the underlying assumptions of both Leibnizian models (or opposing policy experts) will be brought up to the surface for conscious examination by the decisionmaker who is dependent upon his experts for advice. It is also hoped that as a result of witnessing the dialectical confrontation between experts or models, the decisionmaker will be in a better position to form his own view (i.e., build his own model or become his own expert) on the problem that is a "creative synthesis" of the two opposing views.

In considering the resource allocation and decision processes which govern our society and institutions, the role of the "expert" has become somewhat confused and clouded. In a historical perspective the emergence of systems analysts, efficiency or productivity experts, and operation researchers can be viewed as the establishment of a new group of advocates. They advocate decisions, policies, and actions which may optimize certain unique measures such as benefits, costs, efficiency, etc. However, their training does not enable them to reflect on all the factors which the decisionmaker must account for in the process of reaching a decision. Perhaps part of the problem we have had in the past is a misconception that the "expert" has the only view pertinent to the decision and our error in our not attempting to balance and place in perspective the views arising from political, sociological, psychological and ethical considerations which may advocate alternative options. Perhaps "experts" can be better used by the decision processes if they are viewed from the perspective of the Hegelian inquirer as just *one component* of the decision analysis process. This view of the use of expertise underlies concepts such as the Policy Delphi.

Whereas, in the Lockean IS the guarantor of the validity of a proposition is agreement, in the Hegelian it is intense conflict, i.e., the presumption that conflict will expose the assumptions underlying an expert's point of view that are often obscured *precisely because of* the agreement between experts. Hegelian IS are best suited for studying "wickedly" ill-structured problems. These are the problems that, precisely because of their ill-structured nature, will produce intense debate over the "true" nature of the problem. Conversely, Hegelian IS are extremely unsuited to well-structured, clear-cut problems because here conflict may be a time-consuming nuisance.

Except for the Policy Delphi concept of Turoff (see Chapter III, B, 1,3), the use of conflict as a methodology is conspicuously absent in the field of technological forecasting or in Delphi studies in general. In the Policy Delphi the communication process is designed to produce the best underlying pro or con arguments associated with various policy or resource allocation alternatives. In a non-Delphi mode of communication (e.g., face to face), one of the most interesting applications can be found in the activity of corporate or strategic planning. In an important case study, Mason[2] literally pioneered the development of what may be termed the Dialectical Policy inquirer. The situation encountered by Mason was one in which the nature of the problem prevented traditional well-structured technical approaches to planning (i.e., Leibnizian and Lockean methods) from being used.

In the company situation studied by Mason, there were two strongly opposing groups of top executives who had almost completely contrary views about the fundamental nature and management of their organization. They were faced with a crucial decision concerning the future of their company. It was literally a life-and-death situation, since the decision would have strong repercussions throughout all of their company's activities. The two groups each offered fundamentally differing plans as to how to cope with the situation. Neither of the plans could be proved or "checked out" by performing any technical study, since each plan rested on a host of assumptions, many of them unstated, that could probably never be verified in their entirety even if time to do this were available, which it wasn't. Indeed, if the executives wanted to be around in the future to check on how well their assumptions turned out, they

---

[2]Richard Mason "A Dialectical Approach to Strategic Planning," *Management Science* 15, No. 8 (April 1969).

had to make a decision in the present. It was at this point that the company agreed to let Mason try the Dialectical Policy inquirer to see if it could help resolve the impasse and suggest a way out.

After careful study and extensive interviews with both sides, Mason assembled both groups of executives and made the following presentation to them: First, he laid out side by side on opposite halves of a display board what he took to be the underlying assumptions on which the two groups were divided. Thus, for every assumption on the one side there was an opposing assumption for the other side. It is important to appreciate that this had never been done before. Prior to Mason's contact, both groups had never fully developed their underlying positions. They were divided, to be sure, but they didn't know precisely how and why. In this sense Mason informed both groups about what they "believed" individually. Next, Mason took a typical set of characteristic operating data on the present state of the company (profit, rate of return on investment, etc.) and showed that every piece of data could be used to support either the plan or the counterplan; i.e., there was an interpretation of the data that was consistent with both plans. Hence, the real debate was never really over the surface data, as the executives had previously thought, but over the underlying assumptions. Finally, as a result of witnessing this, both groups of executives were asked if they, not Mason, could now formulate a new plan that encompassed their old plans. Fortunately in this case they could and because of the intense and heated debate that took place, both groups of executives felt that they had achieved a better examination of their proposed course of action than normally occurred in such situations.

Of course, it should be noted that such a procedure does not guarantee an optimal solution. But then, the DIS (Dialectical Inquiring System) is most applicable for those situations in which the problem cannot be formulated in pure Leibnizian terms for which a unique optimal solution can be derived. DIS are most appropriate for precisely those situations in which there is no better tool to rely on than the opinions of opposing experts. If the future is 99 percent opinion and assumption, or at least in those cases where it is, then the DIS may be the most appropriate methodology for the "prediction" and "assessment" of the future.

It is important to appreciate that the DIS and Policy Delphis differ fundamentally from other techniques and procedures that make use of conflict. In particular, they differ greatly from an ordinary courtroom debate or adversary procedure. In an ordinary courtroom debate, both sides are free to introduce whatever supporting data and opposing arguments they wish. Thus, the two are often confounded. In a DIS, Hegelian inquirer or Policy Delphi, the opposing arguments are kept strictly apart from the data so that the crucial function of the opposing arguments can be explicitly demonstrated. This introduces an element of artificiality that real debates do not have, but it also introduces a strong element of structure and clarity that makes this use of conflict much more controlled and systematic. In essence, the Hegelian Inquiry process dictates a conceptual communication structure which relates the conflict to the data and the objectives. Under this conception of inquiry, conflict is no longer antithetical to Western science's preoccupation with objectivity; indeed, conflict actually serves objectivity in this case. This will perhaps be puzzling to those who have been brought up on the idea that objectivity is that upon which men can agree and not on what they disagree. While the Hegelian inquirer does not always lead to a new agreement (i.e., a new plan), the resulting synthesis or new agreement, when it occurs, is likely to be stronger than that obtained by the other inquirers.

Singerian-Churchmanian IS

Singerian IS are the most complicated of all the inquirers encountered thus far and hence the most difficult to describe fully. Nevertheless, we can still give a brief indication of their main features as follows:

(1) Truth is *pragmatic*; i.e., the truth content of a system is relative to the overall goals and objectives of the inquiry. A model of a system is *teleological*, or explicitly goal-oriented, in the sense that the "truth" of the model is measured with respect to its ability to define (articulate) certain systems objectives, to propose (create) several alternate means for securing these objectives, and finally, at the "end" of the inquiry, to specify new goals (discovered only as a result of the inquiry) that remain to be accomplished by some future inquiry. Singerian inquiry is thus in a very fundamental sense nonterminating though it is response oriented at any particular point in time; i.e., Singerian inquirers never give final answers to any question although at any point they seek to give a highly refined and specific response.

(2) As a corollary to (1), Singerian IS are the most strongly coupled of all the inquirers. No single aspect of the system has any fundamental priority over any of the other aspects. The system forms an inseparable whole. Indeed, Singerian IS take *holistic thinking* so seriously that they constantly attempt to sweep in new variables and additional components to broaden their base of concern. For example, it is an explicit postulate of Singerian inquiry that the *systems designer is a fundamental part of the system*, and as a result, he must be explicitly considered in the systems representation, i.e., as one of the system components. *The designer's psychology and sociology are inseparable from the system's physical representation.*

Singerian inquirers are the epitome of synthetic multimodel, *interdisciplinary* systems. In effect, Singerian IS are meta-IS, i.e., they constitute a theory about all the other IS (Leibnizian, Lockean, Kantian, Hegelian). Singerian IS include all the previous IS as submodels in their design. Hence, Singerian inquiry is a theory about how to manage the application of all the other IS. In effect, Singerian inquiry has been illustrated throughout this chapter in our descriptions of the inquirers, for example, in our previous representations of the inquirers and in our discussions of which kinds of problems the inquirers are best-suited to study. A different theory of inquiry would have described each of the preceding inquirers differently.

Singerian IS contain some rather distinctive features which none of the other IS possess. One of their most distinctive features is that they speak almost exclusively in the language of commands, for example, "*Take* this model of the system as the "true" model (or the true model within some error limits $\pm\epsilon$)." The point is that all of the models, laws, and facts of science are only approximations. All of the "hard facts" and "firm laws" of science, no matter how "well-confirmed" they are, are only hypotheses, i.e., they are only "facts" and "laws" providing we are willing to accept or make certain strong assumptions about the nature of the reality underlying the measurement of the facts and the operation of the laws. The thing that serves to legitimize these assumptions is the command, in whichever form it is expressed, to *take* them seriously, e.g., "*Take* this is as the true model underlying the phenomenon in question so that with this model as a background we can do such-and-such experiments." Thus, for example, the Bohr model of the atom is not a "factually real description of the atom," but if we regard it as such, i.e.,. if we *take* it as "true," then we can perform certain experiments and make certain theoretical predictions that we would be unable to do without the model. What Singerian inquirers do is to draw these hidden commands out of every system so that the analyst is, he hopes, in a better position to choose carefully the commands he wishes to postulate. Although it is beyond the scope of this chapter, it can be shown how this notion leads to an interesting reconciliation between the scientist's world of facts (the language of "is") and the ethicist's world of values (the language of "ought"). In effect, *Singerian inquiry shows how it is possible to sweep ethics into the design of every system*. If a command underlies every system, it can be shown that *behind every technical-scientific system is a set of ethical presuppositions*.

Another distinctive feature is that Singerian IS greatly expand on the potential set of systems designers and users. In the extreme, the set is broadened to include all of mankind, since in an age of larger and larger systems nearly everyone is affected by, or affects, every other system. While the space is not available here to discuss the full implications of this proposition, it can be shown that *every Singerian IS is dependent upon the future for its complete elucidation.* If the set of potential users for which a system exists is broadened to include all of mankind, then this implies that every system must be designed to satisfy not only the objectives of the present but also the objectives of the future. Thus, a Singerian theory of inquiry is explicitly concerned with the future and is by definition involved with the forecasting of the future. Singerian IS attempt to base their forecast of the future on the projections of as many diverse disciplines, professions, and types of personalities as possible.

Singerian inquiry has been conspicuously absent from the field of Delphi design; hence, unfortunately, we cannot talk about any current applications of Singerian IS to Delphi. There are hints of Singerian overtones in those few Delphis that ask people for the contrast in their real views and the views they would state publicly. However, none of these has ever explored the underlying values and psychology to the extent of warranting a Singerian label. Nevertheless, we can say something about what a Singerian Delphi would look like.

Of all the many features that Singerian inquiry could potentially add to Delphi design, one of the primary ones would be a general broadening of the class of designers. That is, at some point the participants should not merely participate in a Delphi but be swept into its design as well. In a Singerian Delphi, one of the prime features of the exercise would not only be to add to our "substantive knowledge" of the subject matter under investigation, but just as much, to add to the participants' knowledge of themselves. How do the participants change as the result of participating in a Delphi? Are their conceptions of policy formation, and of who and what constitutes an "expert," the same afterwards as before? How is it possible to sweep the participants more actively and more consciously into the design of the Delphi? What are the values and/or psychology that led me and my fellow respondents to answer with this view? These are only a very few of the many issues with which a Singerian-designed Delphi would be concerned, and as a result, would thus act to build into the design of the Delphi the potential for pursuing these questions systematically. In short, a Singerian-based Delphi is concerned with raising and building explicitly into the design of the technique *the self-reflective question: How do I learn about myself in the act of studying others and the world*? Why is it that some minds think they can best learn about the world and the contents of other minds (i.e., their communications) by formal models only? Why do others believe they can best learn through empirical consensual means, and others still, through multiple synthetic or conflictual means? And finally, why do Singerians want to spend so much time studying the others? What kind of mind is it that studies others? Perhaps, perverse; most certainly, reflective—the very spirit that moved the first pioneers of the Delphi technique to want to study how and under which circumstances a *group* of reflective minds was better than one.

Concluding Remarks

In many ways a brief commentary on the strengths and weaknesses of Singerian inquiry provides the most fitting summary to this chapter.

The strength of Singerian inquiry is that it gives the broadest possible modeling of any inquirer on any problem. The weakness is the potentially prohibitive cost involved in comprehensive modeling efforts. However, given the increased fear and concern with our environment, we may no longer have the choice but to pay the price. We may no longer be able to afford the continued "luxury" of building large-scale Leibnizian and Lockean technological models that are devoid of serious and explicit ethical considerations and which fail to raise the self-reflective question. We certainly no longer seem able to afford the faulty assumption that there is only one philosophical base upon which a technique can rest if it is to be "scientific." Indeed if our conception of inquiry is "fruitful" (notice, not "true" or "false" but "productive") then to be "scientific" would demand that we study something (model it, collect data on it, argue about it, etc.) from as many diverse points of view as possible. In this sense strict Leibnizian and Lockean modes of inquiry are "unscientific" because they inhibit this effort, a conclusion which we are sure most of our "scientific" colleagues would be surprised to find and even more reluctant to accept. But then, believing in conflict as we do, we might have a good debate on the matter. If one were to design a Delphi to investigate the matter, which Delphi inquirer design do *you* think we (you) *ought* to use?

References

The references listed below are intended to provide the reader with general reviews, further background, and some specific examples of topics covered in the article. On the subject of Inquiring Systems the best place to seek further explanation would be:

C. West Churchman, *The Design of Inquiring Systems*, Basic Books, New York, 1971.

Those interested in attempts to construct formal mathematical representations of Inquiring Systems are directed to the following three articles:

Ian I. Mitroff, "A Communication Model of Dialectical Inquiring Systems—A Strategy for Strategic Planning," *Management Science*, **17**, No. 10 (June 1971), pp. B634-B648.

Ian I. Mitroff and Frederick Betz, "Dialectical Decision Theory: A Meta-Theory of Decision Making," *Management Science* **19**, No. 1 (September 1972), pp. 11-24.

Ian I. Mitroff, "Epistemology as a Basis for Building a Generalized Model of General Policy-Sciences Models," *Management Science*, special issue on "The Philosophy of Science of Management Science," to appear.

This chapter is, in large part, a specialization of an earlier more general article:

Ian I. Mitroff, and Murray Turoff, "Technological Forecasting and Assessment: Science and/or Mythology?" *Technological Forecasting and Social Change,* **5**, No. 1 (1973).[3]

[3]A condensed version of the above directed to an engineering audience appeared in the March 1973 issue of *Spectrum*, which is the magazine of the Institute of Electronic and Electrical Engineers.

II. C. Reality Construction as a Product of Delphi Interaction

D. SAM SCHEELE

The mind is but a barren soil;
a soil which is soon exhausted, and
will produce no crop, or only one,
unless it be continually
fertilized and enriched
with foreign matter.

Sir Joshua Reynolds

Reality is a name we give our collections of tacit assumptions about what is. We bring along these realities to give meaning to our interactions. Each of us maintains several of these realities—at least one for every significant set of others in our lives. We have domestic realities, parental-family realities, professional realities, sexual realities, organizational realities, stylistic realities.... Since this article is about Delphi design, the important thing is not how many different realities we each have, but that one important product of each Delphi panel is the reality that is defined through its interaction.

Realities can be described as presumed agreements which give meaning to our thoughts and make reasonable our actions in each setting. Most of these agreements about reality are implicit, and are merely confirmed and elaborated by our acts and conversations. Sometimes our interactions subtly modify these realities. Occasionally, a group's reality is actively renegotiated or even constructed *de novo* for a new situation. Delphi inquiries might produce any of these results. The purpose of this essay is to suggest ways of managing Delphi interactions in order to create intentionally a reality that will prompt the appropriate kinds of active interventions.

I believe with others that there is nothing more practical than good theory. Much of what is related in this article is theory, but theory in search of application, by the reader, out of his understanding, to produce results in his specific contexts. I have used a number of examples to illustrate how realities are asserted, modified, reconceptualized during a Delphi interaction. Each example will require detailed consideration. This is painful if the reader is interested merely in an overview. Therefore, I have tried to write the discursive text so that it makes sense even if the reader skip the examples. Each example is a freeze-dried caricature of a set of rich interactions. To reconstitute, the

The Delphi Method: Techniques and Applications, Harold A. Linstone and Murray Turoff (eds.)
ISBN 0-201-04294-0; 0-201-04293-2

reader must supply the cerebral juices and attention, such connective interpretation being necessary to make static diagrams into what might pass for interaction.

Another problem with examples is that they represent only a small fraction of the myriad of potential applications for this approach to Delphi inquiry. Clearly it would be incestuous to build a design rationale solely on generalizations drawn from available applications. Taking an expository approach based on cases would foster an already widespread predilection—method and technique in search of applications. This would be the antithesis of my primary recommendation: that the particular qualities of the circumstances that prompt and define the inquiry be used as a basis for the Delphi design. Further, let me suggest that the results of a Delphi be seen as the product of a carefully designed and managed interaction and not answers to a set of abstract questions that are obtained by following prescribed methods. Hence, a slogan for this essay: Concepts from doing.

This paper might have been called: What to think about when considering, designing, and managing (even interpreting) a Delphi. The reader will find many propositions asserted that require a reflection and reinterpretation for application to his particular undertaking. The extensive illustrations are intended to enable the reader to develop a feel for the importance of details of style and tone in presenting materials to panelists. These illustrations should not be thought of as "cases" to emulate, but as necessary to describe the more general pedagogic points about the importance of self-conscious presentation of information in suggestive, but open-ended, frameworks to facilitate the negotiation of realities. Most of the illustrations are based on Delphis we have conducted. In several cases the substantive content of the illustration has been changed because of the proprietary nature of the inquiry or the possibility that our intent would be misinterpreted if the material were to be seen out of context. Also included is some of our thinking which occurred when we did not do a Delphi.

The italic text provides a setting for some of the illustrative examples. Each illustration depicts a synthesis of the interaction between the panelists with summarization, juxtaposition, interpretation, reconjecture by the Delphi monitor. The role of the diagrammatic presentation of the examples is described in the illustration below (Fig. 1). The intention is not to create order or to impose a unique conceptualization. Neither are the diagrams supposed to be balanced, "well-designed," synthesizing abstractions, or even documentations. In some of the Delphis, the major part of the panelists' comments were sent back on tape cassettes. The emphasis is a personal verisimilitude with the process of undertaking conceptual forays. Most of the panelists' thinking processes cannot be directly shared, so we have attempted to depict for the group some typical points of view out of which to define a reality of relevance.

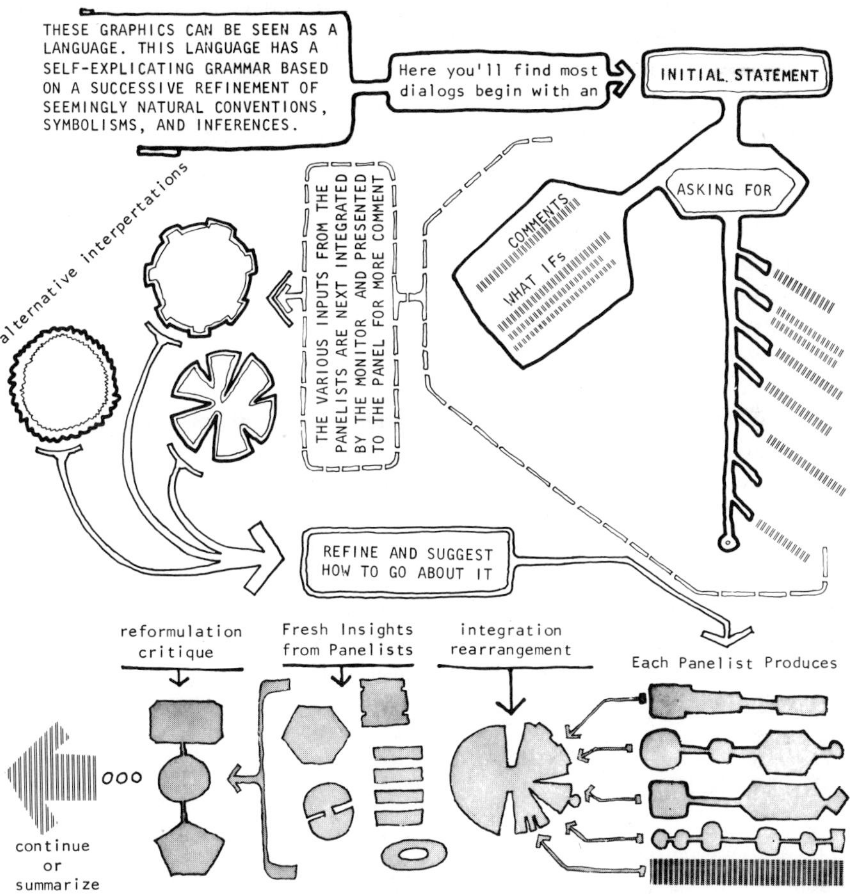

Fig. 1. Role of the diagrammatic presentation of examples.

The particular graphic style adopted in this paper is a personal one. It developed out of use. It is intended to support the process of thinking and not simply to represent completed conceptualizations. Graphic aids can be useful in stimulating your own thinking and organization of ideas. Start by trying your hand at whatever seems cogent to you and make adjustments as you go. There are a rationale and some techniques that have evolved in the use of graphics in thinking, but their explanation would make another book. The quality of the drawings is intended to convey the liveliness of the concepts and their important properties effectively and quickly. This is difficult to do in writing or with more formal diagrams. Also this drawing style encourages participation, and the organization of the diagrams usually readily admits of modification or extension. In addition to aiding individual thinking, common graphic constructions, or explicit group memories, are useful in moderating discussions. Here the shorthand of positional relationships and the insightfulness of successive interpretations and alterations prove very productive. Since Delphi inquiries are group processes, these kinds of graphic representations have proved equally valuable in Delphi applications. The figures may not mean the same thing to you as to me, but your explanation is accessible, and therefore should be more useful to you than mine would be.

Some of the points may seem trivial—like, "use bright colors" or "state ideas in emotive language"—but their impact is significant. Others—like "provide a concrete situational context" or "depict an explicit theoretical framework"—have resulted from trying to de-abstract Delphi inquiries. It is important to deal with the different assumptions of panelists, monitor,[1] and sponsor which, when left alone, limit the potential fruitfulness of the Delphi interaction. Most of the methodological insights suggested here have resulted from efforts at designing other kinds of subjective information processes, such as diffusion of innovations, learning by doing, technology transfer, policy development, management of creativity, and design of service systems. Look before you leap, but eventually leap.

Concepts of Reality

The impact of one conceptualization of a situation upon others and the influence of the various constructions of reality assumed by the Delphi panelists generate what I believe are the most significant results from any Delphi inquiry. Panelists can be made aware of these seemingly subtle differences in the nature of the realities they presume in the course of the interactions. The panel can then produce a common reality for the situation at hand as a result of their participation.[2] How this comes about is determined by the monitor of the Delphi interaction. If panelists are reluctant to make specific contributions or if a very wide, almost unrelatable, array of conjectures is produced, one suspects that there are great differences in the meaning each panelist is attributing to the way the inquiries are stated in the materials provided to stimulate response and the way panelists expect the results to be used. This

---

[1]Monitor is a term I use for the person or group conducting the Delphi inquiry, i.e., preparing the materials, interpreting the responses, integrating the insights, presenting the results, etc.

[2]For a longer exposition of, and details about, this concept see S. Lyman and M. B. Scott, *A Sociology of the Absurd*, Appleton-Century-Crofts, New York, 1970, and H. N. Lee, *Percepts, Concepts, and Theoretical Knowledge*, Memphis State University Press, Memphis 1973.

ambiguity often might be what you want—productive of interesting premises. An array of differently bracketed realities that include a particular object-event-concept is often useful. For example:

Four young adults who are retarded enter a restaurant with older couple. Panelists were asked to select likely responses for restaurant manager, waiters, and other patrons from a list provided. Many panelists added their own. The panel included parents of, and professionals who work with the retarded, as well as individuals to simulate response of general community. The responses of the panelists could be mapped:

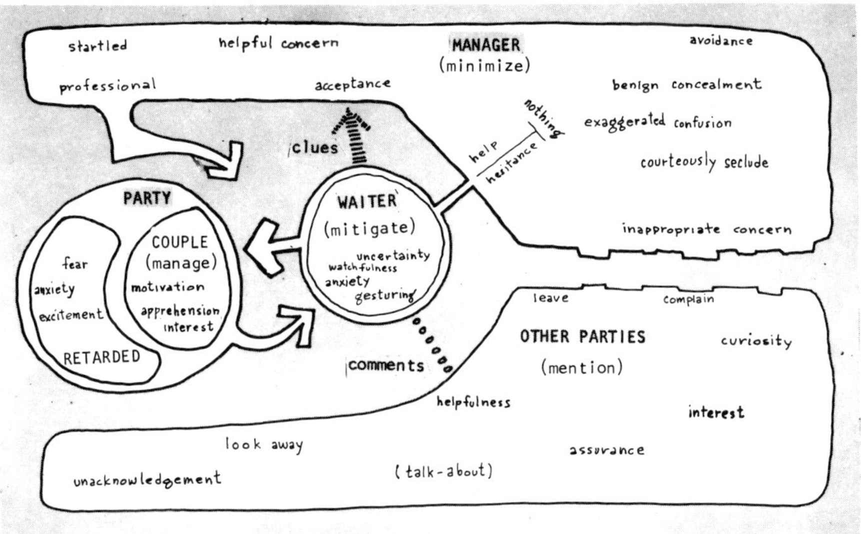

Fig. 2-A

Later, panelists were asked how repeated contact with the retarded would affect the key actors:

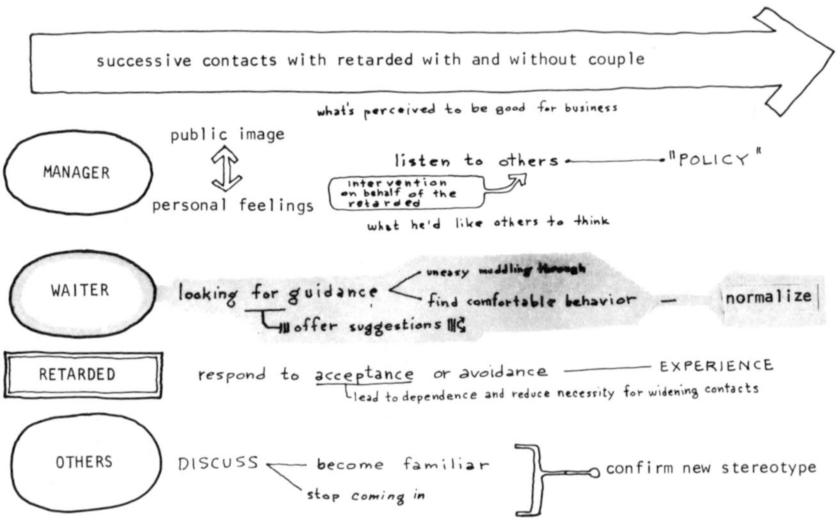

Fig. 2. What is going on in the restaurant.

On the other hand, you may want to create greater focus and consensus. If this is so you can begin with examples for interpretation instead of general questions. This enables you to direct attention in subsequent rounds to contrasts between the assumptions imbedded in the initial situations and panelists' contributions. Differing reality constructs can produce divergence from even seemingly unambiguous statements. Focusing attention on differences in the reality constructs will usually yield either a more refined and widely agreed-upon definition of the appropriate context or clearer and more precise distinctions between competing contexts—possibly leading to an estimation of the relative probabilities of each, or a search for present options that could influence the circumstances.

In the preceding discussion the notion of socially constructed or intentionally negotiated realities was employed. This concept grew out of the work of

Husserl,[3] Merleau-Ponty,[4] Heidegger,[5] and others, which led to the formulation of a phenomenological epistemology that is now being applied by neo-symbolic interactionalists and ethnomethodologists. The concept of a negotiated reality can be related to Mitroff and Turoff's discussion of the philosophical bases for inquiry systems in the preceding article. This discussion describes a range of inquiry systems (IS) using as differentiating labels the name of the principal philosopher whose concepts undergird each approach. The array includes the Leibnitzian, Lockean, Kantian, Hegelian, and Singerian IS's. Since these categories are well defined there, the philosophical premise for an IS based on a view of reality as a context-specific product of interaction will be described in relation to this framework. First, to select a label consistent with the others will be slightly misleading, because any one name tends to obscure the contributions of others and imply that the ideas are largely set. Dubbing this territory of philosophic exploration after Merleau-Ponty seemed the least misleading. To make a contrast with the Singerian analyst, the Merleau-Pontyean is concerned with the particular reality created by the "bracketing" of an event or idea out of the great din of experience, rather than explicating a pragmatic reality that can be used to define possible actions. Truth to the Merleau-Pontyean is agreement that enables action by confirming or altering "what is normal" or to be expected. By contrast, the Singerian views truth as an external articulation of systems to define goals and options for action. Reality is viewed by the Merleau-Pontyean as the product created out of intentions and actions instead of an external basis for intelligent actions. To reiterate Mitroff and Turoff, the importance is not which philosophy is "correct," but which is appropriate to the kinds of situations one is attempting to impact. The Merleau-Pontyean inquiry system seems applicable to situations either where a redefinition of contextual reality can facilitate the generation of new options, or where the acceptance of a new reality must be negotiated to create the impetus for technical or social change—as, for example, in defining as "progress" a reduced or more limited consumption that would permit reallocations instead of "progress" as lower unit production cost to support increased demand. This philosophical point of view leads to viewing the future as a situation where both the dominant reality and the technology are invented as well as inherited, and where culture is transformed as well as transmitted.

Merleau-Ponty and others suggest reality be viewed in a new way: as a currently prevailing shared assumption about a specific situation. This implies that reality is the product of our experience and not external to it. In a

[3]Edmund Husserl, *Ideas: General Introduction into Phenomenology* (trans. W. R. Boyce), Allen & Unwin, London, 1931.

[4]Remy C. Kant, *Merleau-Ponty and Phenomenology*, in *Phenomenology* (Kockelmans, ed.), Doubleday, Garden City, N. Y., 1967.

[5]Calvin O. Schrag, *Phenomenology, Ontology and History in the Philosophy of Martin Heidegger*, Revue Internationale de Philosophie, vol 2 (1858).

commonsense view, reality is a collection of observable things and occurrences which is animated by a society of individuals. Although we are not usually aware of these distinctions, our everyday realities can be seen as created by us out of the meanings we give things and events.[6] Since we do not exist alone, we are continuously asserting and having validated or challenged our definition of "what's going on" or "what it's all about." Our collection of situational definitions constitutes our reality. We select realities from our repertoire that seem appropriate in order to know how to act, attribute meaning, and interpret behavior. This means that instead of continuously discovering more of an external verity—"the reality out there"—we are, wittingly or not, continuously adding, verifying, or revising our "shelf-stock" or versions of what is normal or to be expected in particular circumstances. We each have a shelf-stock of realities that have been produced by our interactions.

Earlier the basic philosophic question had been "What is the structure of social reality?" Now phenomonological insights have transformed this question into "What realities have been or are being socially constructed?"

What does this mean for conducting Delphis? *First*, since the results of a Delphi are produced by interaction, albeit highly structured, the results can be said to constitute a reality construct for the group.[7] Because processes of successive refinement, like the Delphi, strongly tend to induce convergence and agreement, the monitor of a Delphi should purposely introduce ambiguities, even disruptions. These might take the form of "angle" items[8] to challenge and redefine reality as well as "quirk" items[9] to act as catalysts to explore the limits of the reality. For example:

Mass transit could compete with private vehicles by offering more than lower cost—particularly in enhancing the use of commuting time by offering:

Round 1: *What types of services*?

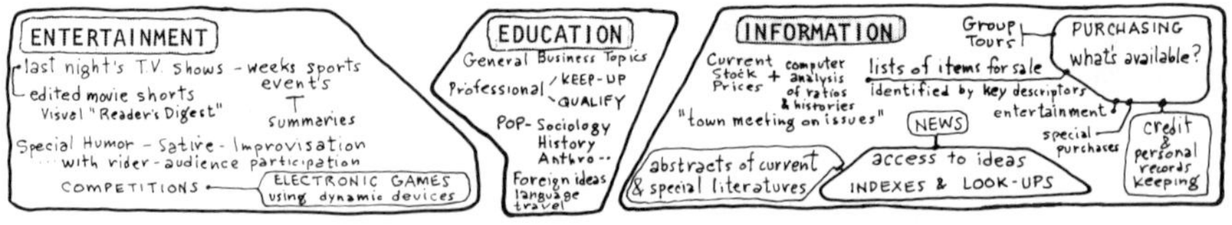

Fig. 3-A

[6]Jack Douglas, *Understanding Everyday Life*, Aldine, New York, 1970.

[7]Hans Peter Drietzel, *Recent Sociology No. 2—Patterns of Communicative Behavior*, Macmillan, New York, 1970.

[8]Example: "Suppose you had invented the better mousetrap and people had beaten a path to your door—would they buy?"

[9]Example: the famous rejoinder, "And how does that relate to the Jewish question?" or Stephen Potter's functional equivalent, "...but what of the South?"

Round 2: *Can't money change hands—transit as market place*? (*angle*)—
Relate services to attracting and serving youth? (*quirk*)

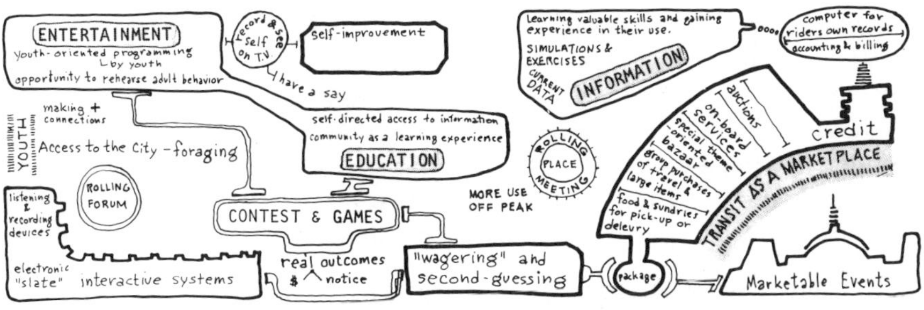

Round 3: *How might the service be organized and supplied*?

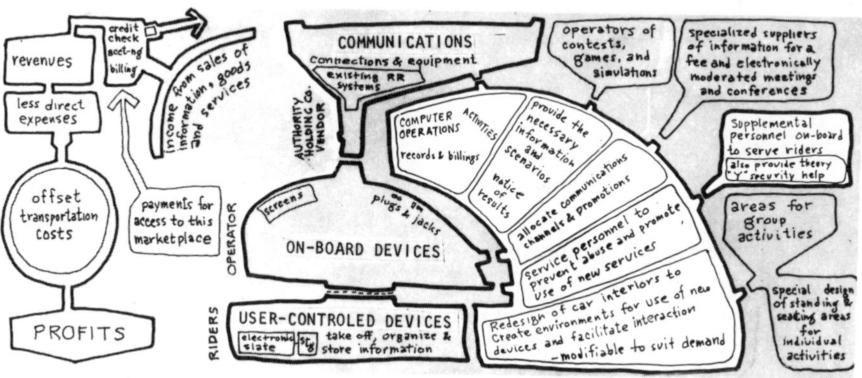

Fig. 3. Activities to enrich mass transit experience.

Second, since the knowable reality is in competition with the other conceptions, including the idea of reality as a negotiable construct, the unknown or unexplained cannot simply be attributed to greater degrees of complexity (i.e., the "more data and better instruments" gambit). This means that further efforts to obtain information, such as a Delphi, must go beyond attempts to unravel what has often been assumed as merely additional complexities. Instead, information should be sought that can shape reality, such as identifying new considerations or introducing new options. This means that the systems being described are viewed as indeterminate, arbitrary, delimited, multiplistic, even convenient fictions if this facilitates discourse, but not as complex Cartesian clockworks. In conducting a Delphi then, "what if" and "why not" items might be introduced or highlighted if suggested by a panelist to prompt consideration of new alternatives.

Third, the reality we construct can be expected to be different by at least as much in the future as our technology will be advanced or our society restructured. This is almost always overlooked by forecasters and other futurists. Predictions may well occur as forecast, but their occurrence will not necessarily mean the same thing then as it now seems that it would. You can note people's naïve understanding of this in their response to the prediction of new occurrences with the statement, "...but I guess they [the people of the future] will be ready for it by then."

Fourth, expect reality to continue to be negotiated. This means that the kinds of realities within which occurrences will be given meaning and be understood will vary from those prevailing at present. To a large extent changes in reality shape the kind of attention, consideration, and effort that will be spent on developing a new idea. They also determine whether the new concept seems plausible, desirable, and feasible. Further, both interest in, and advocacy of, a new concept, along with precipitous events that come to be associated with it, can shape reality to the extent that a new concept becomes one of the ways that reality is defined. Our present notion of the "urban crisis" is an example of a concept that has become imbedded in our realities.

A Delphi inquiry, then, might explore two sides of the negotiation of reality with regard to a specific occurrence: (1) how alternative realities might affect the meaning of the occurrence and how likely each is and (2) how public perceptions of the occurrence, the interests and activities of its proponents, and the concepts and ideology that come to be associated with it will shape the reality that is negotiated.

In many cases it also may be useful to consider the possibility of precipitating events. These events (seen from hindsight) have often dramatically altered the collective awareness of consciousness of the society. For example, context-specific realities were shaped by the assassinations of the Kennedys and King for gun-control measures, Nader and "Unsafe at Any Speed" for auto safety campaigns, and the Soviet Sputnik for the U.S. space program and educational redirection.

Let us look at how these considerations can be handled as part of a Delphi inquiry. To grasp how different prevailing realities might affect a particular topic, one can posit several alternative "reality gestalts." Encourage participants to supply their own realities. But I believe examples are necessary to clarify this notion initially—even if these proto-realities are overly "caricatured." The example below illustrates this kind of inquiry.

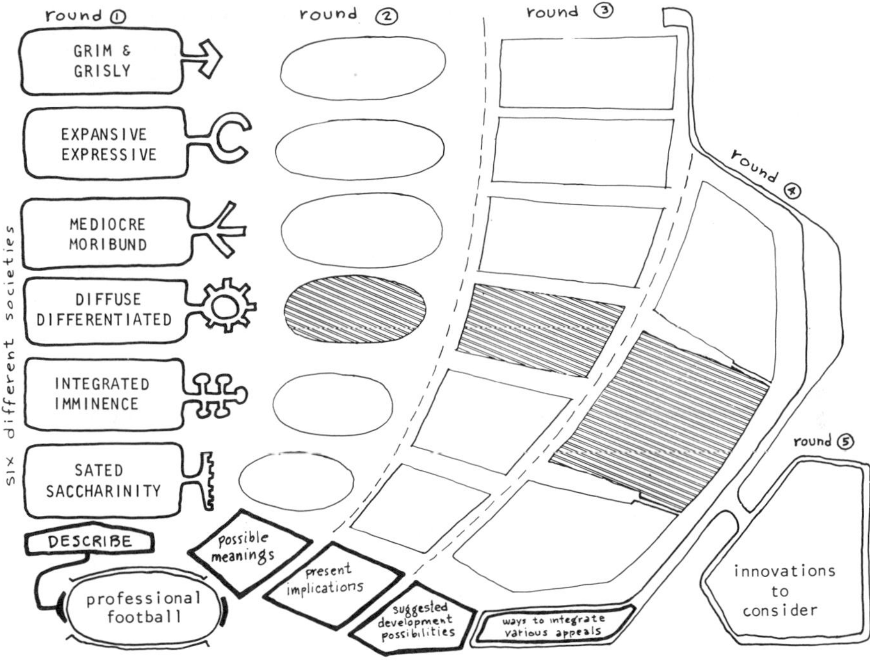

Fig. 4. Professional spectator sports—prevailing views.

Also, you might want to probe the participants' insights regarding items that might be significant in the construction of realities in the future. Here it is useful to suggest items that you believe are important as possible examples. Even include candidate-precipitating events—although their nature and timing are by definition unanticipatable. Knowing that the meaning of events is, so to say, up for grabs can sensitize managers to the use of public-attitude-shaping tools. Introducing these perturbations will begin the discussion and evoke additions and comments from the panelists. An important corollary point is: the meaning of the future in the reality of those taking actions (of necessity in the present) may change. It has dramatically in the past ten years or this book wouldn't have been considered for publication, possibly not even written. One's vision or concept of the future has not always shaped reality in the same way. It is as important to know in what ways images of the future are given meaning as to have a full complement of alternative futures (see Chapter VI, Section D).

The medieval glory-beyond-the-travail-down-here view of the future would prompt a different kind of action from the expectation of a better tomorrow through hard work and technology which has characterized the industrial revolution.

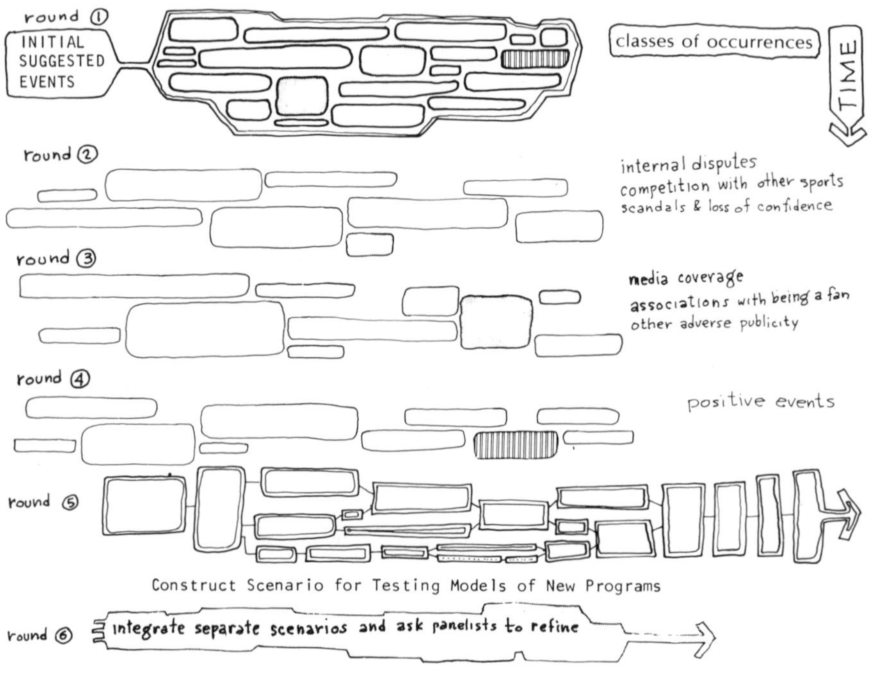

Fig. 5. Professional spectator sports—reality shapers.

Mental Climates and Styles of Thinking

The dominant reality of the "modern" world has been called the positivist-functionalist view. In this view reality is "out there" (we are in it, but it is not in us), discoverable by improved access and greater attention and presumed to be ultimately knowable—given sufficient time and diligence. Lukacs[10] and Panofsky[11] call these global realities the *habitus mentalis* of an era. A *habitus*

[10]Georg Lukacs, *History and Class Conciousness* (trans. Rodney Livingstone), M.I.T. Press Cambridge, Mass., 1971.

[11]Erwin Panofsky, *Meaning in the Visual Arts*, Doubleday, Garden City, N.Y., 1955.

mentalis has at least these three characteristics:

- an epistemology—shared assumptions about what is valid knowledge or "truth," how it can be discovered, and what constitutes acceptance or "proof."
- a social theory—concepts that a society uses to explain its own workings to itself, including the identifiers for constituting aggregates.
- a guidance process—methods and ground rules worked out and accepted by the members of a society—what Harold Lasswell describes as "who gets what, when, where, and how." At different times this is also called "politics" or "management."

At a macro scale the dominant *habitus mentalis* of society is being negotiated. Ours is now. I would like to summarize some of the changes we expect in the postindustrial society. To begin, I'd like to share a name we have coined for the postindustrial society that is descriptive of it, instead of merely indicating *after* industrial. This new label is "the idiomergent era." This is a neologism. *Idio* is a Greek root indicating uniqueness, separateness, distinctness, as in the word idiosyncrasy, and in its most peculiar sense, idiot. It has been combined with the Latin root *mergere*, to cause to be swallowed up, to immerse, to combine or coalesce, to lose identity. The result describes a more separate and a unique kind of interdependent aggregation. It seems that the idiomergent era will be characterized by both extreme variety and increased uniqueness in individual behavior and organizational purposes and at the same time be dependent on the greater integration of the individual and organization into the structures of a variety of specialized communities of interest, producing networks of cohesiveness for the society as a whole. Thus the term "idio-mergent" denotes greater individuality and autonomy combined with a deeper involvement of the individual in fluid social arrangements, innovative organizational structures, and experimental personal relationships. This describes change in the vanguard of society: a new *habitus mentalis* that will spawn new realities.

The *habitus mentalis* for the idiomergent era will no doubt affect the design of Delphi inquiries. Exactly how depends on the response of practitioners and monitors to their perceptions of what is relevant and their selection and advocacy of ideas that come to be in good currency. I will try to share my construct for this new mental climate. You realize that I have no way of knowing the future, nor am I merely asserting an explanation to serve as a basis for judgment. These and many other possible descriptions of what this image of the future is can be offered. My view is that I am engaged with you in negotiating what to expect from Delphis. Use it as you will.

Before getting to differences in the epistemology, social theory, and guidance process of the thinking style of the idiomergent era and what these augur for the design of Delphi inquiries, I would like to suggest another feature of this era—an increasing concern with the future. Greater interest in the future has resulted from a series of changes in our dominant social reality. These changes seem relevant to consider in designing systems for assembling useful information. First, change became accepted as normal. Reality became dynamic, but this alone could not make the future more important now; things would just work out differently. Second, technological development and social and physical mobility have increased the number of options available to decisionmakers. This made their reality wider, but not necessarily extended deeper into the future. Third, the number of self-evident paths, inevitable selections, and unquestioned preferences was reduced by an illusion of choice fostered by great increases in the availability of education—both schooling and media-supported awareness. This ballooning of information has expanded reality to include the possible, the imagined, the dreamed, the fantasied, and all other "residents of the future in good standing."

The itch for a fix on the future afflicts both people and organizations. Individuals want to know more about the future so they can choose places to live, things to experience, roles to adopt that will be highly desirable when the future occurs. Businesses analyze future opportunity areas and establish organizations to prepare to exploit them. Institutions assess their roles in the future and begin reconceptualizations. Government investigates the future and identifies many unacceptable contingencies that require more controls. In most cases the future reality is portrayed as an unfolding panorama of large-scale changes in general circumstances and situations whose meanings are anchored in the present. The inference to be made for the sponsor of an inquiry into the future was, "If I know more about what's going to happen, I can get ready for it." It was like spying on or getting the drop on the future. These futures were not the world of will, action, and events; but broad backdrops of scattered possibilities from which a series of succeeding "presents" would coalesce on the moving pointer of time. The message was that to "realistically" take effective actions the future must be considered. This future dimension of reality, now so important, has been chiefly portrayed as a passive set of probable possibilities.

The growing recognition that the "meaning" of things and events is not determined, but in fact refers to social agreements that have changed in the past, and which can be renegotiated, even for each circumstance, marks the beginning of the idiomergent era. The reality of the future as a broad techno-cultural script is too ambiguous to support particular undertakings. Attempts have and will be made to bracket specialized future realities that contain only highly relevant possibilities—but in most instances the number of considerations in even a "bracketed" future will not support finite analysis to select a single course of action. But the process can be reversed. Decisionmakers can attempt to negotiate a specific reality for the future which, if they are successful, can come to be widely enough shared that it is realized through successive choices. The future will be invented. Thus in the idiomergent era you can expect attention to focus on the selection of methods to create and portray new realities that have the potential to be actualized through tacit agreement between the intentions of those affected.

In order not to lend credence to the prevailing presumption that Delphi inquiries are concerned only with the future, let us return to look at ways that

Delphis can be used to contribute to the store of knowledge, enlarge the society's understanding of itself, and improve the style of governance in an idiomergent culture. For each of these three characteristics of an era, the currently prevailing concepts will be contrasted with their idiomergent counterparts. This will be done rather telegraphically to provide a background for an example of an item in a Delphi inquiry. Elucidating these points more clearly would require another book.

Differences in Epistemology concern not only what is sought and accepted as knowledge, but how information is categorized and organized to support actions. As a man thinks, so is he.

Industrial	Idiomergent
Reality is external and knowable.	Reality is constructed and negotiable.
Hypotheses are general and are offered for validation.	Hypotheses are context-specific and serve to redirect the ongoing discovery process.
Categories of description are based on observable differences in measurable dimensions.	Categories for description include intentions and selections and are based on an esthetics of knowledge.
Anticipate future conditions by extrapolation of past performances and behaviors (exploratory images).	Invent future images out of creative integrations of expectations (normative gestalts).

To illustrate these epistemological differences, the best example seems to be the development of categorizations of consumers for use in market planning for new products. In the industrial era, consumers were defined demographically, geographically, and by aggregates of features that indicated a consumer's relative propensity to buy either specific types of items or to buy in a particular way. One idiomergent approach would be to develop first a model for the process of product introduction, next those categories of consumers likely to be involved, and then where and how to locate them and their probable numbers.

Differences in Social Theory are alternative views of how the society is organized, its tenants perpetuated, and its actions explained. Society is a name for us to call "what is."

Industrial	Idiomergent
Behavior follows laws that can be derived from situational observation.	Behavior occurs in activity sets that presume and assert meaning.
Actions are explained in terms of expression of internal motivations and conflicts.	Actions are taken in response to assessment of interactional consequences.
Organizations and roles are structures which define possible actions.	Actions and the need for their explication define roles and organizations.
Society is categorized by structural and functional properties to permit measurement and management.	Categorizations of society mark limits for special realities in order to facilitate communication and collaborative actions.

For the Delphi inquirer these differences in social theory suggest greater emphasis on finding out about the appropriateness of societal norms, roles, and institutions, limning how they came into being and, probably most importantly, suggest ways they might be reshaped. The example describes an application to one particular field.

Round 1: Please elaborate on the diagram below—additional relationships between the retarded and blind person living together to provide mutual support.

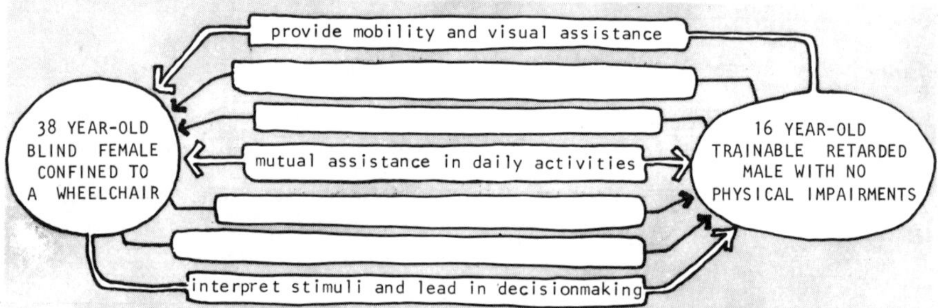

Fig. 6-A

Round 2: The relationships described by the panel have been aggregated below; suggest where indicated which significant others might contribute to make this relationship more workable, and what they might do. As a neighbor living in the same apartment building, what could you do to contribute to this relationship?

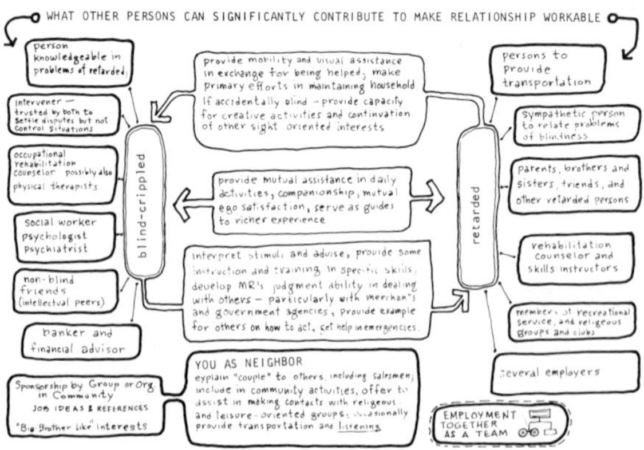

Fig. 6-B

Round 3: *Identify potential difficulties and problems that might be expected to be encountered.*

Round 4: *What might you propose to avoid, ameloriate, or resolve these difficulties?*

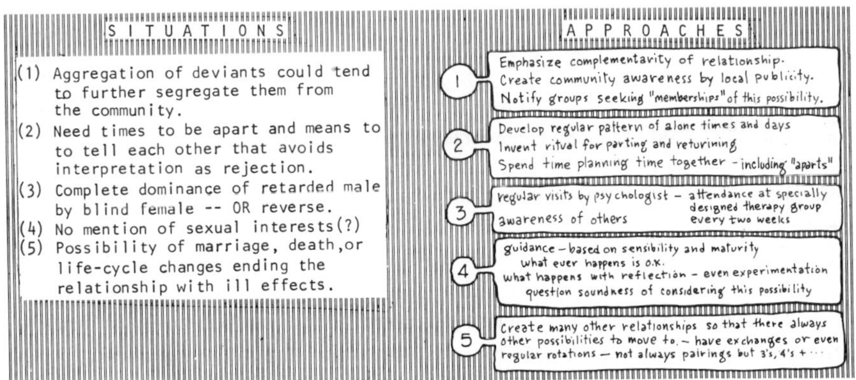

Fig. 6. Symbiotic life-support relationships for the retarded.

Differences in Guidance Processes used in a society reflect implicit, usually unwitting, agreements about the mutual concerns of groups (from pairs to organizations to segments of the society as a whole). These concerns and their mutuality change. Sometimes they change in response to awareness of threats and opportunities. At other times, improvements in techniques such as communications, media, information systems, or analytic schemes change the way fundamental relationships are perceived. Economically and politically the negotiation of power continues. The society's guidance process is not simply decreed—as by a constitution or framework of laws. It includes these, but it is built up rather than laid on. A society's guidance process embodies in conventions all of the expectations of individuals concerning how things ought to be "fairly" and "knowledgeably" decided in order to produce the greater good and what to do if it's not happening as it ought. Societal guidance is all we ourselves are not contending about and in which we assume someone else is watching out for our interests.

Industrial	Idiomergent
Selection of conceptual models based on analytic evaluations including ideological considerations.	Selection of approaches based on evaluation of prototypes and related implications viewed pragmatically.
Competitive determination of potential market for goods and services exchanged in a predominately private marketplace.	Entrepreneurial innovation to create markets for goods and services exchanged in a publicly moderated marketplace.
Representational selection and approval of proposed actions within general jurisdictions based on broad policies.	Collaborative development of proposals for action for particular situations with participative review of specific implications.
Management predicated on adherence to comprehensive plans developed to reach selected goals and objectives defined in limited dimensions to be periodically revised.	Management incorporating participation of those directly affected in the selection of means to respond to short-range opportunities and vulnerabilities identified by multidimensional plans produced by continuing review of an unfolding context.

The changes in style of societial guidance that I expect to characterize the emergence of the idiomergent era will contribute directly to the need for more Delphi and Delphi-like processes for collective consideration of topics. However, evaluation of the usefulness of Delphi inquiries will be made more on their suitability as a process for discussion and choice than on the concerns about the accuracy, insightfulness, or agreeableness of the results. The next example was chosen to illustrate how a Delphi can be used to focus consideration of a topic that has no established institutional advocates.

Round 1: *What actions are possible to reduce peak impact demand for transportation to reduce "overhead" costs of quality urban lifestyle?*

Fig. 7-A

Round 2: *Select most promising approach*(*es*)

Round 3: What interests and organizations would be involved in implementation and what interests do they have which could be appealed to?

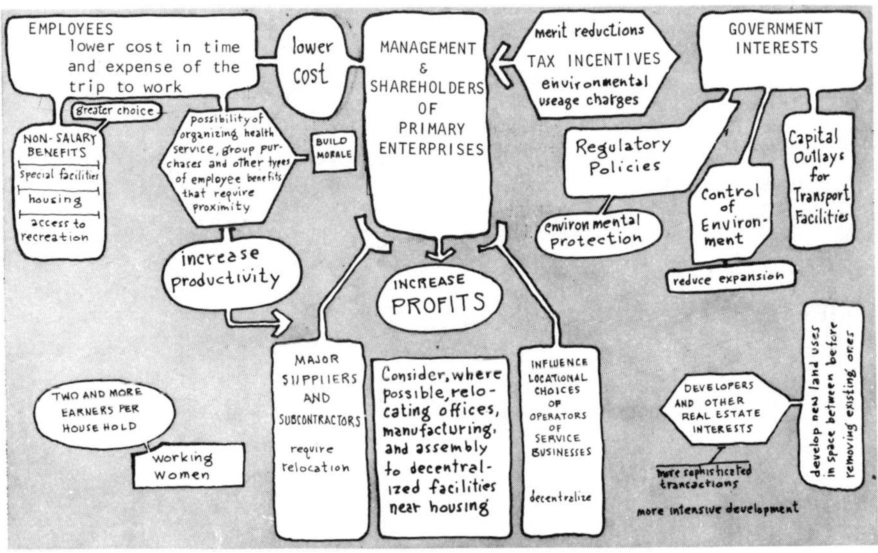

Fig. 7-B

Round 4: Suggest administrative mechanism to manage program to achieve objectives.

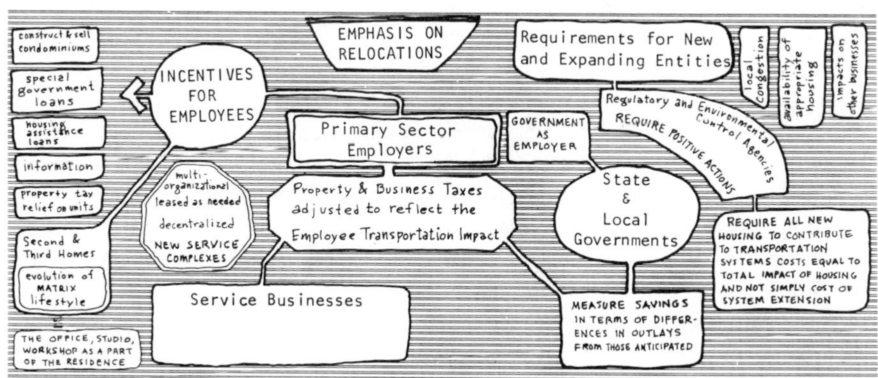

Fig. 7-C

Round 5: Consider implementation: What kind of demonstration program or other approach might be used to rally and mobilize support?

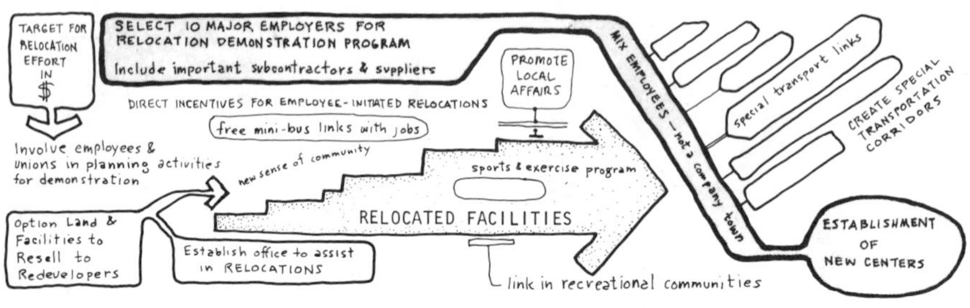

Fig. 7. Relocation incentives in urban transportation policymaking.

Group Influence on Reality Construction

We all have highly idiosyncratic experiences, ideas, and fantasies. While they actually are our own, their meaning is created in the crucible of their interaction with what is going on in our various contexts. For example, a dream shared with people you have recently met has a different impact on the reality being built than the same dream would have if described to an informal caucus during a psychiatry conference. In the first circumstance, your reality is being created from such elements as your image as a person, the meaning of a relationship with you, the place of personal interpretation in your respective motivational structures, and the bearing of dreams on events to come. In the second circumstance, the dream becomes material for negotiation of professional prowess, theoretical differences, clinical inferences, concepts of consciousness—even a trigger for discussions of what this discussion group or the profession is all about. You can imagine that sharing facts, feelings, and proposed actions would also impact differently on the reality being constructed in these contrasting contexts. Try imagining the contrasting reality that could be created in each of the two contexts just described (telling potential friends about yourself *vs.* a discussion about you as a topic in a professional group) for each of the following: the recent death of a relative; your interest in traveling to India; an idea for creating a "kitty" for vacationing by selling the work of local artists to your fellow employees.

Delphi inquiries are conducted with groups where the individual participants might be expected to view the group differently, particularly since they are often anonymous. For most, "being on a Delphi panel" has no particular meaning independent of the topic. Frequently the presentation of the inquiry items and other information is indefinite or ambiguous about the nature of the panel as a group. This directs the panel's attention and energy, in part, to the task of defining the reality of their relationship instead of creating richer realities for the domain of the proported topic. Further, the quality of the reality within which the individual inquiry items are elaborated varies with the ways the panelists view the meaning of the group and its findings, insights, recommendations.

This means that it is not enough that the panelists share a firm idea of the group's identity, but the perception of that identity tends to shape the nature of the individual messages, the quality of the interaction, and the character of what is produced. Table 1 (below) presents a crude taxonomy of the features of groups and some notion of how a group's conception of itself affects the products of Delphi interactions.

The monitor (or modulator) of a Delphi inquiry strongly shapes which conceptualization of the group is assumed by each participant. This usually implicit concept of the group dictates the mode of interaction that each finds appropriate and determines the reality each uses as a frame of reference in attempting to contribute. Several different assumptions can often be made depending on how individuals interpret the implications for reality of the messages in the communications and materials provided them to begin the interaction. Until each panelist is comfortable with his notion of the group's nature and believes it to be confirmed by responses, it is difficult to produce meaningful contributions. In the paragraphs below, I suggest some ways in which a Delphi monitor can shape the group's conceptualization of itself and adopt a mode of interaction that can produce results which will be congruent with their anticipated use.

Transactions will likely become the dominant mode of interaction if there is:

- abstract categorization of the panelists by expertise with inclusion of obviously relevant specialties.
- formal statement of items for consideration, possibilities for estimation, or hypotheses for assessment.
- reiteration of responses categorized by original items with few additions or deletions.
- a product expected of the inquiry as a pooled judgment that will have a "validity" believed to be greater than that of any individual.

In adopting this mode panelists assume that the participants have no other commonalities and expect none. For most this means they will adopt the most familiar model of interaction—probably "answering a questionnaire as

Table 1

Group	Examples	Mode of Interaction	Nature of Realities Produced
Collections of Individuals	Occupants, attendees, shoppers, passengers, gathering, respondents	Transactions	Perfunctory, patterned, provisional, ambiguous—in exceptional cases completely new and perturbing: possibly creating a new pattern.
Casual Groups	Friends, party, clique, players, diners, meetings, trips, class	Experiences	Unique but stylized, comfortable, shaped by random inputs with a strong interest in symmetry and completeness; occasionally evoke powerful new gestalt.
Purposive Group	Colleagues, families, members, associates, collaborators, teams	Episodes	Tightly structured, matter-of-fact, repetitive, presumed heavy with meaning, resist redefinition when alterations (usually catastrophic) occur
Affiliative Groups	Unions, professions, swingers, activists, supporters, delegates	Event	Ritualistic, symbiotic, one-dimensional, subtle shift effected to build power or influence, attempt to be prescriptive, sometimes unsuccessful and produces schism
Defined Group	Students, aged, work force, minorities, dieters, communities	Affairs	Amorphosis, working against stereotype, exploring for unity, proclamatory, infrequently build basis for becoming affiliation
Agents for Society	Citizens, representatives, concerned people, reformers, preservationalists, ecologists	Occurrences	Global, dogmatic, historical, mythological, polemical, sometimes deteriorating into diffuse "glossolalic" diatribe

'accurately' as possible." Exchanges will tend to be formal. They will center on "proof" or "refutation" of ideas identified by the monitor. There will usually be a statistical integration of the group's assessments. Anonymity will likely be used to reinforce the panelists' self-concept as impartial observers in an "analytic" reality. Panelists who are uncomfortable with this reality will devote great attention to selected issues, often digressions from the principal considerations.

Variations in the transactional mode can be effected by (1) beginning the Delphi with open-ended, agendaless items in search of meaning (to explore the agendas and gestalts of the future held by the panelists); (2) selecting the extraneous and derivative responses of the panelists for feedback and later exploring the interpretation of the discourse with the panelists (to identify latent options, issues, and considerations); (3) extracting the explanatory premises from the panelists' responses after two rounds, then focus on estimation, followed by attempts to elaborate a system of relationships (to build theorics and models from premises derived from content rather than technique); and (4) starting with a preliminary design of a thing or program, proceed to iterate critique and redesign to produce a better design by successive interaction of multiple viewpoints, or to derive by interpretation of the "designer-panelists'" responses which considerations seemed important or even critical to the design.

Obviously, there are many other manipulations that could be developed, but these are sufficient to illustrate the type of variations that can be created for particular applications. A Delphi should not be undertaken to validate concepts which you already have developed and refined; panelists want to make significant contributions and these will seldom build meaningfully on highly elaborated initial concepts.[12] Delphi inquiries are obviously applicable to much more than obtaining pooled judgements about particular options.

Experiences become the prevailing mode of interaction when:

- the panelists are familiar with each other and identify with the subject or sponsor of the Delphi inquiry.
- the involvement is for a fixed duration, with known consequences that follow a familiar pattern.
- original items serve primarily as jumping-off places for further inquiry.
- the generic form of expected product of the inquiry is clearly indicated.

Inherent in the assumption of this mode of interaction is a commitment to the anticipated process and results. The principal interest of the panelists is the particular qualities or insights each seeks to contribute. Even when anonymity of panelists is maintained, their messages tend to be informal and approval-seeking (other-directed conjectures) instead of raising formal distinctions and

[12]Presenting a less-than-complete concept takes guts. To each panelist there will be "obvious" omissions. You will be severely criticized. But, I believe a stance of calculated naïveté produces the best results, if you live through it.

striving for accuracy and defensibility. Panelists in later rounds will attempt to make contributions that will affirm their brilliance or uniqueness. The result can be omission of important but obvious points and dangling insights (the monitor can point out these). When the panel is largely drawn from a single discipline or field of application, the interaction quickly moves out to the ethereal zone instead of enriching the context for action. Modifications by the experiential mode include: (1) try to organize the panel into "teams" to represent particular interest groups or points of view (to preview negotiations or develop highly detailed but still preliminary agreements for consideration); (2) present initially a clear and detailed description of the expected product of the inquiry; or a unique and special situation that calls out for great imagination and close collaboration to be described first so that the prevailing rules for judging the suitableness of contributions are partially suspended (to spark creativity and develop a relatively complete conceptualization of a viable new approach).

Episodes are the characteristic mode of interaction for groups that:

- are made of individuals who have a significant and continuing relationship, possibly more than one.
- deal with familiar topics or operate in well-known ways.
- are more influenced by consensual implications than external factors in making choices.
- are more concerned with the perceived quality of the interaction than the product—in fact the idea of a product, as such, may not even be considered.

This mode of interaction has the highest emotional content and the most potential for prompting action based on the insights produced during the Delphi inquiry, which is by definition supplemental to the group's normal interactions. In these cases you should consider face-to-face group processes. Other techniques (such as program-planning method—PPM, nominal grouping, multiattribute utility assessment) may be as productive, require less time, and have more spillover benefits than a Delphi. It is also imaginable that a group of strangers could sufficiently internalize instructions and materials that would be dramatic enough to induce an episodic style of interaction for an inquiry. Such an inquiry would have to be virtually a simulation. The substantive insights produced by a small purposive group can be expected to be highly specific to their context. The insights produced by such groups regarding the personal and emotional dimensions of the topics are probably not generally indicative. In fact, it may be necessary to create a simulated purposive group in order to probe the interpersonal and emotional dimensions of circumstances that are themselves the subject of conjectures or competing proposals.

Variations on this mode of Delphi inquiry are limited because alteration in the premises usually induces the group to adopt a new reality resulting in a less highly bonded mode of interaction. The most interesting results are produced in instances when the group's strongly shared reality is collectively modified in response to new insights and even to incorporation of new members. However, this collective modification of reality is not always completed successfully and it can be expected several realities will prevail. This can result in dissolution of the group, a breaking down of the group process, or the production of highly emotional and idiosyncratic contributions.

Events, often with more form than content, distinguish the interaction of groups which:

- are made up of individuals who see themselves as having a commonality worth recognizing, but not becoming related or involved on other matters.
- center their interests on the specific issues of their commonality and its continuance (survival) and enhancement.
- are guided in their interactions by "the way things are done"—hierarchical considerations and formal divisions of responsibilities within the group.
- make choices to allow greatest individual diversity while maintaining the essential solidarities that define the group.

The inherent one-dimensionality of the interactions of these groups exhibits itself most clearly in the importance of agenda-setting. Whether highly informal—"wait, should we be discussing this"—or employing a very formally maintained docket, these groups find it essential to screen topics for relevance to the organizing commonality. Issue focus has become more necessary as society becomes more diverse, in order to keep an affiliative group from fractionating into several puposive-type groups, each of which would maintain multidimensional commonalities. To survive and grow, however, such affiliative groups usually have to deal with threats and opportunities that are on the perimeter (if not outside) the group's reason for being. This means that most of the messages exchanged can be assigned to one of two distinct classes: *proforma* statements necessary to group maintenance and expository statements intended to modify the group's prevailing notion of their common interest. There will be very few exchanges about things that affect them as individuals or alter the group's relationship to the society.

Delphi panels who think of themselves as affiliative groups tend to proceed in two-step responses to the items in the initial inquiry. First, they make narrow and direct responses to each item. Second, they have additional considerations. The whole panel is usually asked whether these should be dealt with. If the panel or some substantial subgroup agrees that the additional items should be added, then they are taken up. Consequently, much of the reality that is created has to do with the meaning of considering certain topics and not the meanings derived from actual consideration of those topics. A Delphi should never be attempted with a "single-purpose" group until they have agreed that there is something that needs their attention which would not take care of itself.

Affairs are the interactions of groups characterized by:

- being asked to consider items that apparently are of interest.
- having an abstract categorical definition of their membership.
- continuing attempts to give direction and create meaning for the interaction impelled by recognizing the insufficiency of the group's label to serve as an adequate definition for their involvement.
- a tendency for speakers to become "spokesmen," to indulge in formalism, and to imply collective interests for the group.

The beginnings of these groups are abstract categorizations which have meaning for the inquirer or the explainer, but seldom for the group itself. So when asked almost anything, these groups scramble in several directions in search of cohesive interests. The direction of a question or inquiry presumes a meaningful response. A variation of this occurs when the inquiry is directed to explore differences within the group. In this instance the inquiry presumes distinct factions. The group will often develop these "factions" in formulating responses, but the factions are transitory. A corollary dynamic in the interactions of these groups is the prefacing of comments with remarks which either attempt to identify the speaker with the whole group or a major faction, or will uniquely distinguish him and his contentions. These interactions tend to be characterized by a series of declarations—"Run it up the flagpole and see who salutes." Occasionally in short interactions, and eventually in almost all, a series of declarations will coalesce a faction (occasionally including the whole group), which will concentrate thereafter on the construction of a position and purpose. Once constituted, such factions interact as large affiliative groups, usually in a way that produces schisms and dissolution.

Because researchers, administrators, and most of the rest of us have been almost conditioned to think of society as divided into distinct kinds of "labeled" groups, the response of abstractly defined groups (e.g., doctors, women, radicals, educated) is frequently sought in inquiries which end in producing results that do not satisfy the inquirer. One way to remedy this is not to expect a coherent position to emerge, but to look for a pattern of diversity in the group's assertions and the ways in which these points are qualified. These groups cannot examine issues that apparently concern them, because, to paraphrase Gertrude Stein's description of Oakland, "there isn't any 'they' there." Instead, such groups can be used to design product or service "lines" or to formulate arrays of options that seem complete—covering most of the considerations. Also, these groups can be used to define and map the factions implied by a number of related specific proposals for action. Panels constituted in this way are used all too frequently. As an alternative, consider making up panels that have two or three distinctive subgroups and manage feedback selectively.

Occurrences are what goes on with groups where:

- the membership is made up of individuals who believe they are acting for the interests of the society in general as they see it.
- topics serve as occasions for participants to expound their general ideologies.
- most messages refer to particular cosmologies about the way things are supposed to be or are appended to statements that describe world-views.
- the envisioned outcome of a set of interactions is a collection of more widely agreed-on precepts that will guide the group, its constituents, and, if possible, all others.

All panels on occasion become forums to prescribe for the ills of society. These groups virtually insist on this mode. In most instances, this is counterproductive. This is because in most circumstances it is necessary to produce a fine-grained, narrowly bracketed reality to support specific actions. The reason for arranging this kind of a panel is to obtain a broad consensus, but not on everything. Since nothing is ever *done* in general, but only in a particular instance, introduction of situation-specific facts and considerations can help to focus the panel on agreeing about society's interest in the instance at hand, instead of on society's interest generally in things of this kind.

Another reason to arrange a broadly based general panel is to develop a contextual mapping that would describe the overlapping large-scale realities which underlie different parts of a society's response to any complex issue. Here it is useful to inquire about "for instances" that are likely to be seen by different elements of society as analogous to the situation being inquired into. "Well, it's just like thus-and-so..." and similar phrases are powerful connectives to ready-made meanings for a proposed idea or approach. These loose, even unrational, associations tend to bring along existing emotional loadings which panelists in this mode tend to deny. In a similar way, apparent metaphors (e.g., like David and Goliath) and apt personifications (e.g., rape of the ecology) should also be explored. This can be done by asking about which ones suggest themselves or provide several examples to be selected from, modified, or added to. Identifying analogies and metaphors is a more fruitful way to define subtle differences in global constructs than asking individuals to examine their core values and beliefs and describe them directly. While there are problems with interpretation of these complex symbols, the interpretations can be iterated for refinement, and also for validation. Since the "picture" is changing, and "observation" using a Delphi inquiry is not a snapshot, there is a gross limit on the fineness of the picture of a whole society that can be produced. But, even faint shadows on pale gray are helpful in many situations.

Do not do a Delphi with representatives or spokesmen for society when you want information for particular practitioners; use experts "about society" who are respected by the practitioners. In fact, you may want a split panel containing both experts and practitioners to create a joint reality shared by both that can be potentially extended to society.

What I Meant to Say

Most of the foregoing is tentative and much of it highly abstract. It is even embryonic. I share it for whatever use you can make of it and apologize for its density. There are obviously many hypotheses here for research on the process of Delphi interaction that go well beyond the issue of "accuracy" or "reliability." My only suggestion is that you try to supply or create examples from your own experience that seem to explicate these generalities. I attempt to summarize the main points below. In the next—and final—part of this article, I will offer a few pragmatic suggestions for the designer—monitor of Delphi inquiries. Obviously, not all of these strategies and techniques should be attempted simultaneously.

Summary

(1) Each Delphi interaction produces a shared reality which is initially formulated by the panelists from their expectations and the style of presentation used in the initial materials; this particular reality is elaborated and modified by the succeeding interactions.

(2) During the process of interaction the panelists' responses can be expected to deal with personal esteem, group self-concept, and relevant world-views as well as to convey substantive ideas, forecasts, and estimates.

(3) Patterns of styles of interaction can be fostered, retarded, even transformed in order to produce results that will have greater insight, more usefulness, higher impact.

(4) The size and shape of the reality within which things come to be viewed is more important than the specific substantive descriptions produced by the panelists.

(5) The believability and significance to the user of the results of a Delphi inquiry depend as much on the user's perception of the clarity, compellingness, and fit of the reality implied and possibly defined by the results as on the perceived quality of the information.

Some Design Considerations

I shall now briefly focus on some design considerations to help you develop a better Delphi and orchestrate the interaction. These suggestions do not constitute a checklist. I hope they will trigger some ideas and preparations that otherwise would not have occurred to you. Each could separately be the subject of a much longer discussion. Supplying these discussions is left to you, the reader, when you pursue particular applications.

Making a Delphi Context Specific

The conceptual productiveness of the interplay of intentions and circumstances is lost in many Delphi processes because of the lack of concreteness in the context. The mental life of most individuals involves attribution of an intimately known context, even when this context is the world of abstract ideas. However, the subjects of most inquiries, including Delphis, are stated in terms of general propositions; those things that apply only to an instance are defined as being of no significance for action (those things yet to be determined, decided, or done). So the difficulty lies in the extraction of generalizable propositions from particular instances.

I can offer two approaches; there are undoubtedly more. The first approach is to develop a concrete example with detailed features that are typical of a more general case. Another source of specific contexts is to create hypothetical constructs which make a distinct break with most present circumstances and are almost fanciful, e.g., on another planet. Because the hypothetical construct will differ in important respects from the panel's conventional circumstances, it can often provoke new thinking, retard discussions of routine experiences that are rooted in presently accepted and unquestioned conventions, and create an unclaimed and common experience base that can facilitate collaborative (*vs.* competitive or compliant) interaction. Emergencies and other kinds of special circumstances can also be used as hypothetical constructs within which possibilities and insights can be developed for later transfer to the more general everyday circumstances.

Once a case is selected, proceed with the Delphi, focusing on the detailed case initially. Then shift the inquiry by asking the panel to create generalizations—possibly proposing some yourself and iterating the ones they proposed. You can next elaborate these general propositions by asking,"What other circumstances do you know or can you imagine might prevail, and within each would these propositions be valid?" The second approach is to develop several special cases for initial inquiry using subpanels. Later, the results of each subpanel can be shared with the whole. Then, ask for general propositions. What is forthcoming can then be refined and elaborated with examples.

The Domain of Time as a Context

Another shortcoming of "looking at things in general" is that time is assumed as an undifferentiated flow—that one minute or hour is as good as any other. If asked, we all would acknowledge perceptual differences—time dragging or racing. But, unless our attention is specifically directed to it, we tend to overlook structural distinctions in time. The structural features that limit the allocation of time are of increasing importance as time becomes a scarce resource. When inquiring "What will people do?" "How can resources be used?" "Which group is being designed for?" the time domains become critical concepts. Time domains can be categorized or partitioned in many different ways, as can any continuum. Conventions in the dominant culture have created workdays (and, increasingly, work evenings and nights), regular times off, lunches, weekends, holiday weekends (recently added to by Congressional action), annual vacation times, "the holidays," special events, and so forth. There are also times in the so-called life cycle—youth, teenage, young married. Time domains can be treated as the subject of policy and design, for example, staggered work schedules and new work patterns (4 ten-hour days, 6 twelve-hour days with a week off, etc.) You can also invent new "times" for things, like the enculturation of daily exercise periods. Figure 8, below, deals with several of these considerations and relates them to the selection of market opportunities.

Demand for recreational services is limited more by the time consumers have available for their pursuit than by the expenses involved. Time is experienced in periods that have each come to have their own meaning and expectations. To create new markets for recreational services, the meanings of particular periods or domains can be altered and new periods can be created and given meaning.

Round 1: *Suggest new time domains you think could be developed and indicate how existing domains could be revised or differentiated to create new markets for recreational services.*

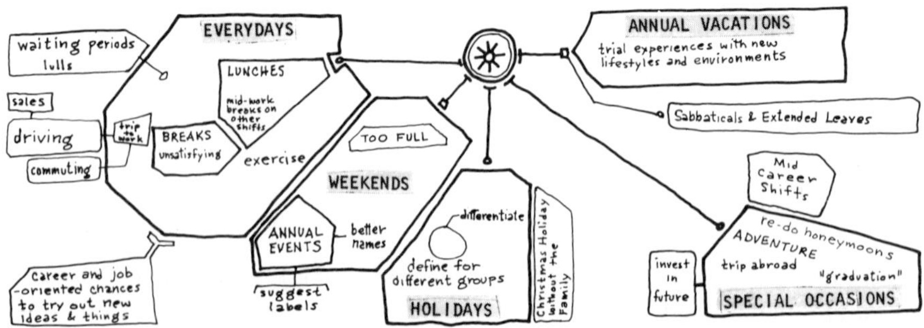

Fig. 8-A

Round 2: Identify time domains where additional involvements for working adults without children would be most welcome and suggest recreational pursuits.

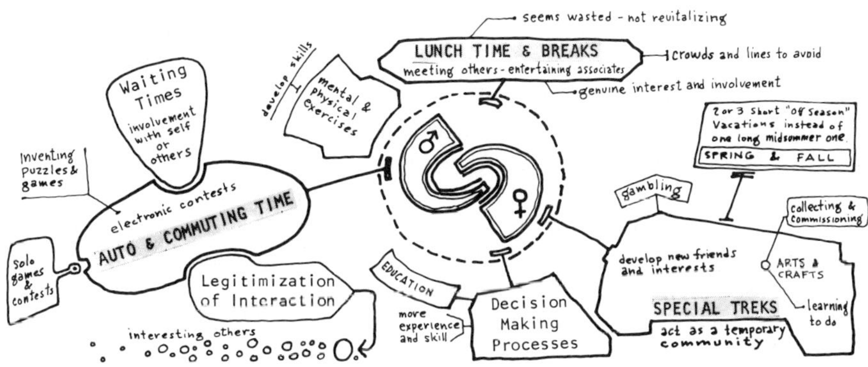

Round 3: Select market opportunities you believe are the most attractive and suggest specific types of products and services and actions to create expectations and aggregate markets.

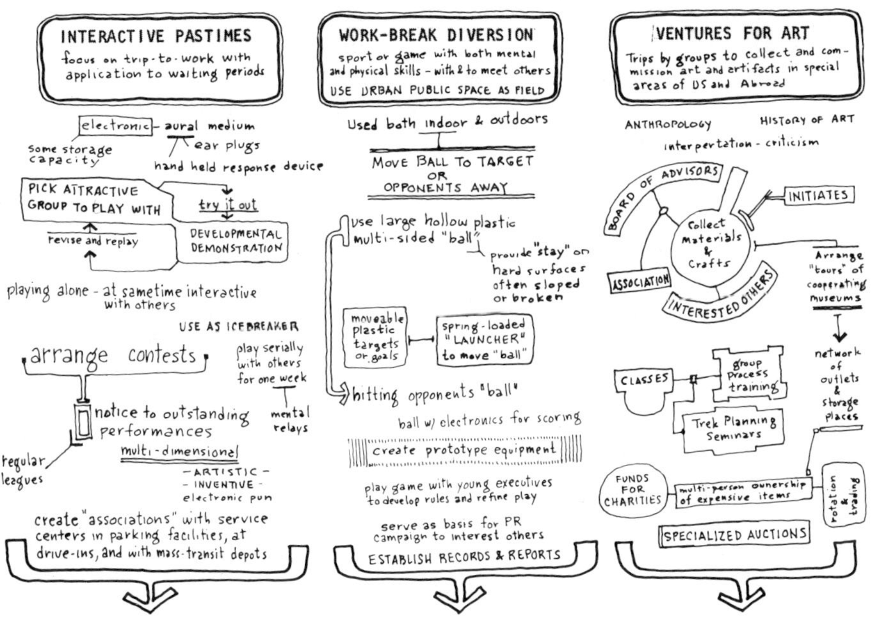

Fig. 8 Time domains and market opportunities.

Getting a Product Out of Results

Every Delphi inquiry can be expected to produce results of some kind. Usually, if they are thought about at all, results are seen as things to be "captured" from their existence "out there somewhere" by the Delphi process. Instead, try to visualize, or think about, the results of a Delphi as being created by the interaction. Just as the important properties of a building are not simply those of the sum of its "stones," the results of a Delphi are not just the individual items produced in the interaction but the reality comprised by the whole. I believe that a Delphi will be more productive if the monitor sees his role as producing results and not as "surveying" things that are already there.

Here are some more design tips:

Creating panels is usually the first task, unless they are a given—as with an organization that wants to study its own future in a participating manner. Even then you may want to augment the group. Three kinds of panelists are ingredients for creating a successful mix: *stakeholders*, those who are or will be directly affected; *experts*, those who have an applicable specialty or relevant experience; and *facilitators*, those who have skills in clarifying, organizing, synthesizing, stimulating... plus, when it seems appropriate, individuals who can supply alternative global views of the culture and society. The proportion of a panel from each category should be tailored for each application. *Note*: There are almost never enough women on panels.

There are no general rules of thumb for creating panels. For example, where options and interests are clear but acceptance of direction and action is fractionated, stakeholders might predominate. If it is clear who has to act, but not clear how, a heavy salting of experts may be best. Where issues, relationships, and values are unclear, a preponderance of facilitators may prove most useful. Pay attention to the minority as well as the majority views expressed by each type of panelist. Also, watch that panelists contribute as you expect—experts, in particular, drift into acting as stakeholders in aggressive moments and facilitators in passive ones.

After abstractly thinking about the panel, it comes down to names. Most of the time you will not know everyone you want. Sometimes you can start with a small group of potential panelists and begin by discussing possible names and searching as a group for interesting and appropriate candidates. This is a better strategy than searching lists of relatively unknown names by categories. Frequently, staff members of professional societies and other "people brokers" can be utilized to suggest panelists. Often the most fruitful part of a Delphi process is assembling a panel, not simply for the particular effort, but because it enhances your contacts. At least, try to develop two alternative tactics for identifying panelists and consider their trade-offs before embarking on the Delphi.

Stimulating response of any kind from quality panelists is not easy. Getting quality inputs is even harder. Few people like "questionnaires." And, from their point of view, engaging in abstract speculation with people you do not know or whom you want to impress about a subject that is not central to your interest (and where your performance is unrelated to your survival) is hardly compelling. Motivation requires more than a good cause, a pep talk, and thanks. If prospective panelists tend to be uninterested, find a "worthy" or prestigious sponsor, or make participation of significant publicity value. Then a token payment or "honorarium" will stimulate interested responses. Payments alone, even of fairly sizable sums, will not assure quality participation. It is difficult to pay panelists based on performance since it is difficult to agree on what is expected; panelists generally believe that "what I do is what's enough." Attractive and potentially stimulating peers are probably the most powerful incentive for participation. Consider gifts and "in kind" rewards for participants, because often a sponsor can provide these at modest cost—particularly travel. Still another way to organize participants is to employ a two-step approach: suggest that key participants use their staff, students, or others they can induce to participate to formulate the specific responses. The key participant then collects, reviews, and submits them for each round. Here publicity and recognition of importance by other individuals are powerful incentives.

Once initial participation is secured, the next important consideration is the quality of the materials, their tone, style, and presentation. Use lots of color. Give the materials style. What you send out reflects the significance of the inquiry and the value that is placed on it. Use emotive language and vernacular expressions to engage panelists and convey importance of results—not another abstract study. Detailed situational descriptions, mini-scenarios, bits of conversation—all help. Other media besides print can be used effectively to make response easier and more scintillating—even speedier. These include tape cassettes and even local "interviewers" in cases where the Delphi sponsor or manager has a large potential manpower pool that could be employed at low cost. Don't forget to pay attention to creating an audience for the results and communicating this to panelists. Examples of good and bad responses are also useful, particularly for discouraging stereotypical remarks and obvious but unhelpful "insights."

Orchestrating interaction requires attention to details in the panelists' responses and a feedback of an overall movement, countervailing forces, or whatever macro-observations are appropriate to describe what seems to be going on between and with the individual items. Highlight divergence and consensus, even when both are true for different sets of responses to a quirk item. Be sure to cheer at least one item from each panelist—everyone needs to be appreciated. Introduce ambiguous factors if discourse focuses on individual items when you would rather explore relationships between items. You can also "stipulate" certain constraints when particular items have evoked massive, but trivial discussions of immaterial distinctions. Where appropriate, supply tentative theoretical constructs and cosmological frameworks. Indicate the way responses are being categorized for review and whether your interest centers on enriching the decision environment or identifying and refining options for consideration. Also, you can employ cross-impact analysis (see Chapter V), comparisons with analogous situations, and other techniques to provide a basis for evaluating responses in a different light. Again, at the risk of overemphasis in this essay, as interaction progresses, share with the panelists the reality construct(s) you believe they are referring to and shaping by their responses.

Interpretation and summation of responses is never complete, because the panelists do not send in their "heads" but only their responses. A complete record of the interactions is unwieldy and diffuse and difficult to use. It is important to begin "interpreting" responses during the interactions, even at the start. This makes interpretation subject to review by the panelists and can include their refinements, which I have found most insightful. Summation can be made from several points of view—using the one best for each major consideration in the inquiry. It is often useful to point out nil findings, omissions, and ignored items. Try to capture and describe the reality that was negotiated by the panelists and the monitor, because this provides a perspective for understanding and indicates the application foreseen for the results.

Communication of the results may not be felt by some to be a legitimate concern for researchers. Findings, it is supposed, have their own importance. However, the process of acquiring "understanding" carries with it an impetus for action that is not conveyed by presenting the conclusions alone. Most Delphi users, I believe, intend to influence the formulation of policy or the making of decisions. Doing this requires attention to the communication process. Well-organized, lucid, attractively presented reports help—but they are not enough. To augment the report you should attempt to: (1) create involvement—one way being to include the intended users or members of their staffs as panelists in the Delphi, or else to engage users in parallel deliberations that focus on the same issues, infusing from time to time interim considerations developed by the other Delphi panel; (2) generate interest—this usually means fanning respectable controversy and building a climate of expectation for, and awareness of, the Delphi inquiry among those who are important to the users, particularly colleagues and constituents; (3) present the results interactively—this can be done by organizing a simulation exercise where the panelists, the staff of the Delphi monitor, or some of both take roles that are significant in the user's operating environment and then enact a series of situations that can demonstrate which Delphi results are applicable, and when and how they might be used. It may be useful to define a communication process as part of the initial conceptualization of the Delphi design so that the implications for its communication are anticipated in the way the interaction of the panel is orchestrated.

I cannot imagine that all the recommendations in this Chapter will prove unassailable or always useful. I hope they will spark some interest and provoke a few insights. This will require a lot more of you than of me. I have benefited in putting this down, but you will have to work to make it apply to your situation in order to profit. To quote Henry James: "Many things have to be said obscurely before they can be said clearly."

III. General Applications

III. A. Introduction

HAROLD A. LINSTONE and MURRAY TUROFF

In this chapter we sample the rich menu of applications. The purposes of the Delphis are as varied as the users. Seven authors focus on specific planning tasks in the areas of government and business. Additional studies are also sketched in this introduction ("Comments on Other Studies").

Government Planning

The four articles covering this field address national, regional, and organizational planning problems. Turoff deals with the basic concept of the Policy Delphi and reviews several efforts of this type. Ludlow's concern is resource management in the Great Lakes region and a major aim is to establish improved communications between technical experts and interested citizens. Jillson focuses on the use of Delphi as an integral instrument in national drug abuse policy formulation, operating from three distinct perspectives: "top-down", "bottom-up", and "issue oriented". Jones explores priorities in System Concept Options for the U.S. Air Force Laboratories, with emphasis on comparing views of four in-house organizations competing for funds.

All four move beyond the use of Delphi as a forecasting tool and stress its value as a communications system for policy questions. A policy question is defined here as one involving vital aspects, such as goal formation, for which there are no overall experts, only advocates and referees. Its resolution must take into consideration the conflicting goals and values espoused by various interest groups as well as the facts and staff analyses. It should be clearly understood that Delphi does not substitute for the staff studies, the committee deliberations, or the decision-making. Rather, it organizes and clarifies views in an anonymous way, thereby facilitating and complementing the committee's work.

Whereas Turoff's panelists constitute a homogeneous group, Ludlow seeks to establish a communication process between the potential users of new knowledge and a team of interdisciplinary researchers. He raises a point which is of concern for Delphi studies generally. The probability used is of the personal or subjective type; it can be interpreted as a "degree of confidence". Scientists and engineers are brought up on a different kind of probability—frequency of

The Delphi Method: Techniques and Applications, Harold A. Linstone and Murray Turoff (eds.)
ISBN 0-201-04294-0; 0-201-04293-2

occurrence, i.e., the limit of the ratio of the number of successes to the total number of trials as the latter approaches infinity. Thus the frequency type of probability assumes repeatability of the experiment (e.g., tossing a coin). But the subjective probability has meaning even if an event can occur only once. A boxing match is a one-time event; the odds usually associated with such a match indicate the "degree of confidence" in the outcome on the part of informed bettors. Both definitions are mathematically valid and have been used to develop distinct probability theories. Businessmen intuitively use the "degree of confidence" concept and therefore have no built-in resistance when faced with it in Delphi questionnaires.

Ludlow also presents an evaluation of Delphi by the three participating groups—technicians, behaviorists and decision-makers. Not surprisingly the latter prove to be the strongest proponents of the technique. They are, after all, the one group which must regularly seek a consensus and usually has to make decisions on complex issues without adequate information.

Miss Jillson's article is a progress report on a study designed to develop drug abuse policy options, explore the applicability of the Policy Delphi to questions of social policy generally, and determine the practicability of using it on both an "as-needed" and "on-going" basis (i.e., indefinite duration). The participants include researchers, administrators, and policymakers—both in the field and in impacted areas (e.g., police chiefs). A unique feature is the use of the three perspectives. In the "top-down" approach, the objectives for the next five years are emphasized; the "bottom-up" approach deals with factors which control transition between various states or levels of drug use and employs a matrix format; the "issue-oriented" approach crystallizes statements of policy issues in "should/should not" form.

Jones' Delphi reflects the typical consensus or Lockean oriented approach in its design to gain consensus among representatives or organizations subject to different pressures in their competition for limited financial resources. He uses 61 senior managerial and technical personnel (both military and civilian) representing most departments in the four organizations. Different organizational viewpoints are apparent although significant self-interest biases are not detectable. This effort contrasts very nicely with the Kantian nature of Ludlow, the Hegelian approach of Turoff, and the mixed Kantian-Hegelian work of Jillson.

Business and Industry

In a corporate environment Delphi fills several roles. Bell-Canada's Lawrence Day lists three: educational device for senior management, environmental trend background material for technological planners in research laboratories, and trading material for use with planner-counterparts in other organizations.

If the corporation is large and diversified it may have the analytical staff to run the study and the expertise to form the panels "in house". TRW used 140 of the "the most imaginative and creative members of TRW's technical staff of more than 7,000 graduate scientists and engineers". It should be noted that in the hierarchical environment of a business, the rate of response or participation tends to be higher than average. A university professor may feel no compunction about ignoring questionnaires or giving perfunctory or dilatory answers; an employee has stronger incentives to cooperate in a company exercise. Goldstein's experience in this regard is echoed in numerous other cases.

The ability to conduct a Delphi without bringing the respondents together physically is another advantage in the large corporation with units spread over a wide geographical area (e.g., multinational corporations). Overseas personnel can be drawn into a Delphi with relative ease and at minimal cost.

Day covers four Delphis performed by Bell-Canada's Business Planning Group: Educational Technology, Medical Technology, Business Information Processing Technology, and Home Communications Services.

In cases where the corporation does not have the expertise in either the subject of the forecast or the Delphi procedure, it may turn to a professional consulting organization. Enzer's article describes a Delphi on the subject of plastics undertaken for a client by the Institute for the Future. The field of materials for the future is a particularly difficult one for the forecaster. First, the number of possibilities is overwhelming (e.g., tailormade plastics). Second, in a hierarchy (e.g., relevance tree) which has metasystems at the top, followed by systems, subsystems and components, materials are close to the bottom. This means that a given material can branch upward in many ways, i.e., it can find use in a multitude of old and new systems by 1985. Third, each substitution of a new material for an old one may prove sensitive to slight variations in the relative prices, with major deviations resulting in the 1985 forecasts. These factors strongly suggest the desirability of using at least two entirely different methods of forecasting to provide a reasonable degree of confidence.

Nancy Goldstein gives us a hint of this in her article on a Delphi covering steel and ferroalloy materials. She compares the Delphi results with those of a conventional panel study conducted simultaneously. There is considerable agreement in the forecasts. Ideally one would seek a comparison between methods which are more radically different than a conventional panel and a Delphi. However, this particular comparison makes one startling methodological point: the conventional panel study brooks no uncertainty and no areas of disagreement in its forecasts; it presents a definitive view and a set of conclusions. The Delphi, on the other hand, reflects the uncertainties and highlights the differences among its participants. It is more concerned with exploring minds than setting down precise recommendations (cf. Scheele, Chapter II, C).

The Delphis by Day, Enzer, and Goldstein are largely Lockean in nature. However, the use of panels of differing backgrounds by Day has Kantian aspects, and Goldstein suggests Hegelian overtones.

Another use of Delphi has recently evolved in business in connection with risk analysis.[1] It concerns the uncertainties associated with new projects or investments. Normally decisions must be made in the absence of adequate information. The potential market for a new product is uncertain and the development costs may exceed the engineers' estimates. Marketing personnel in an operating unit of the corporation frequently exhibit a glowing optimism which neglects to credit the competition with high intelligence or quick reaction capability. Engineers tend to assume that the cost of a complex product is a linear function (i.e., sum) of the cost of the components which comprise it. They neglect the interactions which result in nonlinear behavior: the total cost is much greater than the sum of the parts. Thus the cost is grossly underestimated. One recent study of a large number of defense-oriented development projects indicated approximately a 50 percent chance of 50 percent cost overrun.[2]

Delphi may be used to advantage to provide input to the risk analysis. The most critical part of such analysis is the subjective probability distribution assumed for the uncertainties. Delphi can serve to probe the views of personnel connected with the project as well as outsiders (e.g., corporate offices or other units), senior executives as well as junior engineers and scientists. The anonymity is particularly valuable in a highly structured environment where individuals may feel constrained in expressing their own views.

Comments on Other Studies

Many applications of Delphi are carried on as integral parts of planning projects or as staff work of a proprietary nature. Therefore, a considerable amount of very good Delphi work has not been published in the open literature or in a form adequately explaining the process used. Before proceeding to the separate application papers it is worthwhile to note a number of Delphi studies of a unique nature which are not yet otherwise documented explicitly for those interested in the technique itself.

The Delphi method has been applied extensively in the medical area. The initial work by Bender *et al*[3] was largely of a straightforward forecasting variety. However, a number of Dr. Bender's Delphis did deal with estimating

---

[1] See D. B. Hertz, "Risk Analysis in Capital Investment," *Harvard Business Review*, Jan-Feb. 1964, p. 95, and D. H. Woods, "Improving Estimates That Involve Uncertainty," *Harvard Business Review*, Jul-Aug. 1966, p. 91.

[2] S. H. Browne, "Cost Growth in Military Development and Production Programs," unpublished, Dec. 1971.

[3] Bender, Strack, Ebright & Von Haunalter, *Delphic Study Examines Developments in Medicine, Futures*, June 1969. George Teeling-Smith, *Medicine in the 1990's*, Office of Health Economics, England, October 1969.

the necessity and desirability of potential medical research accomplishments. And it did not take very long before the application was broadened to include unique objectives other than future projections.

Dr. John W. Williamson, of John Hopkins University's School of Hygiene and Public Health, has utilized Delphi extensively for estimating historical data. Typical questions deal with determining the incidence of a given disease and the estimated rate of success in utilizing various treatment methods. This is, of course, an area where current reporting practices do not give reliable data owing to differing standards across the country and the effect of multiple complications, e.g., death due to pneumonia while ill with cancer. Usually Dr. Williamson would ask respondents for their best estimate of a number, then a low and high value which would *shock* them. Also requested would be an estimate of their confidence in the estimate and a statement whether their estimate was based upon a particular source, such as an article they had read. In a number of these exercises questions were asked which dealt with the results of unpublished new clinical studies. In this manner one could observe how well the Delphi panelists actually did on part of the exercise and utilize this insight to gain an impression of their capability for providing answers to the rest of the exercise.

An excellent and, perhaps, classic example of this is a study Williamson did at the Philips Electric Corporation Plant in Eindhoven, Holland, in 1970. Approximately 50 doctors who are involved with the company's medical program for the 36,000 employees participated in the Delphi. The first part of the Delphi asked the physicians to estimate the percent of male employees absent from work due to sickness during differing intervals of time. The population was further divided by young, old, blue collar, white collar. This required sixteen estimates from each doctor. When the real data was collected from the computer files three months later it was found that 12 of the 16 estimates were within 10 percent error and the other 4 within 30 percent error. Of course, this represented only a small portion of the exercise as the real objective was to determine what effect various potential changes to the health care program would have on the absenteeism rate. However, one could see that the physicians involved had a good feel for the situation as it existed. Also it was possible to examine how well various subgroups did, e.g., general practitioners *vs.* specialists. Dr. Williamson has conducted four major studies of the above type (involving validity checks) with a total of approximately 200 respondents over the past four years.

Professor Alan Sheldon of the Harvard Medical School, together with Professor Curtis McLaughlin of the University of North Carolina Business School, did a Delphi in 1970 on the Future of Medical Care. A unique feature of this Delphi was the process of combining the events evaluated by the respondents into scenarios in the form of typical newspaper articles. The respondents were then asked to propose additions or modifications to the scenarios and give their reaction to the scenario as a whole. This concept of utilizing the vote on individual items to group events into scenarios classed by such things as likelihood and/or desirability has become a standard technique.

Also with respect to scenarios it has become fairly common to provide the respondents on a forecasting Delphi with a scenario or alternative scenarios providing a reference point on considerations outside the scope of the Delphi but having impact on the subject of the inquiry. For example, in forecasting the future of a given industry the respondents might be given a "pessimistic," "optimistic," and "most likely" scenario on general economic conditions and asked that their estimates for any question be based on each alternative in turn.

While there have been a number of Delphis on the general future of medical care, a recent Delphi by Dr. Peter Goldschmidt of the Department of Hygiene and Public Health, Johns Hopkins University, dealt with the future of health care for a specific geographical entity, Ocean City, Maryland. The problem in health planning in this case is the tremendous influx of vacation people in the summer months. In order to examine the future growth of the Ocean City area and its resulting medical needs, it was felt the Delphi panel should include individuals who resided in the area and simultaneously worked in endeavors related to the mainstream of the local economy—recreation. Therefore, the Delphi involved long-time residents, hoteliers, bar owners, real estate dealers, and civic officials as well as the usual "experts" such as the regional planning people from local government and industry. This widening, or broadening, of the concept of "experts" to that of "informed" is becoming quite customary in the application of Delphi. In this particular Delphi, Dr. Goldschmidt was able to check the "intuition" of his respondents by comparing their estimates on vacation populations in Ocean City currently with estimates he could analytically infer from the processing load history of the sewage plant that serves the area. As in Williamson's case the results were quite good.

A superb example of the Delphi technique was carried out by Richard Longhurst as a master's thesis at Cornell University.[4] The Delphi attempted to assess the impact of improved nutrition, family income, and prenatal care on pregnancy outcome in terms of birth weight and the resulting I.Q. and intellectual development of young children. The resulting output were of a form useful for incorporation into cost-benefit analyses of government programs to improve the nutrition of pregnant mothers and young children. This is, of course, an excellent example where data exist to indicate that malnutrition in the mother or young child has some degree of impact on the long-term intellectual capabilities of a child. However, this evidence is not of direct quantitative utility to the type of analyses an economist would like to perform in evaluating a government program.

The respondent group was divided into two subpanels. One in the area of

[4] Richard Longhurst, "An Economic Evaluation of Human Resources Program with Respect to Pregnancy Outcome and Intellectual Development," M.S. thesis, Cornell University, Ithaca, N.Y. December 1971.

intellectual development comprised experts in child psychology and development; the other, in the area of pregnancy outcomes, was composed of experts in pregnancy, nutrition, and medical care. The group was given a specific group of low-income mothers in a depressed urban area as the population they were concerned with. This was a real group on which a good deal of data on socioeconomic status were available. In the first round the respondents were asked to sort out the relative importance of environmental components that might be manipulated by the introduction of a government program. The second round presented a set of feasible intervention programs which related to the factors brought out in the first round. They were then asked to estimate for each program the resulting incidence of low birth weight and the average I.Q. score of 5-year-old children resulting from the pregnancies under the program. They provide the baseline data on these parameters for the target group as it currently existed. Certain programs were estimated to reduce the incidence of low birth weight from 15% to 10% and to raise the five-year I.Q. scores from 85 to 100 points. The shift in I.Q. can then be used to shift the average education and earning power of the children when they are adults. This then is translatable into dollar benefits that can be used to compare the merits of alternative programs in this area. The Delphi itself involved respondents for three rounds and questionnaires were tightly designed to take about fifteen minutes to fill out.

The area of trying to translate scientific knowledge into an informed judgment on evaluating and analyzing decision options is clearly a potential one for the Delphi method.

Another effort in health care planning is the work of Professor David Gustafson at the University of Wisconsin. This work has been tied into the Governor's Health Planning and Policy Task Force effort. One of the Delphis Professor Gustafson conducted dealt with delineation of the current barriers to the performance of research and development in the health services area—the rather interesting topic of trying to clarify what the real "problem" is. The respondents were asked to delineate barriers of three types: (1) solution development barriers; (2) problem selection barriers; (3) evaluation barriers. For each barrier the group developed comments, implications, and possible reactions or corrective measures. A vote was taken on the significance of each barrier. This was an excellent example of utilizing Delphi to try and isolate the significant part of the problem. Very often, in planning areas, preconceptions by one individual lead to tremendous efforts on the wrong problem. The specification of a particular problem usually predetermines its method of investigation and at times its conclusions.

The use of Delphi for regional planning has probably become popular because of the feeling that there is a necessity to establish better communications among many individuals with diverse backgrounds.

In this area a significant number of Delphis have been conducted by various Canadian government agencies such as Health and Welfare, Department of

Public Works, Department of the Environment, and the Postal Service, to name a few. Most of these are being done by internal staff and very often they tend to be short, focused on a very specific issue, and require a diverse backgroud of respondents. A good example is one done in 1974 by Madhu Agowal on "The Future of Citizen Participation in Planning Federal Health Policy." The Delphi sought to explore and delineate specific options for citizen participation and to determine the consequences of such programs.

There has been very active use of the Delphi in the educational establishment and a survey of that work may be found in an article by Judd.[5] Curiously almost all educational Delphis have been confined to administrative matters and hardly considered as a teaching tool. It is not surprising that educationalists are enthusiastic about the method. There is a high degree of participative planning in higher education. Authoritarianism is eschewed to such an extent that anarchy sometimes results. There is also an entrenched bureaucracy which feeds on well-structured procedures and questionnaires of all kinds.

Delphi is used for several aspects of administrative planning: general goals, curricula, campus design, and development of teacher ratings and cost-benefit criteria. Judd describes many of the problems encountered in the use of Delphi in this environment.

However, to find a clue to what may prove to be the most serious difficulty, we must turn to the conclusion of a (non-Delphi) survey of school administrators conducted recently by R. Elboim-Dror[6] in Israel on the subject of education in the year 2000:

> The lack of creative imagination as revealed in this study, the limited number of new alternatives and innovating ideas expressed by the subjects, and especially the students (of school administration), are a serious sign.

In such an atmosphere Delphi can be as barren as most of the paperwork which traditionally suffocates educational bureaucracies. When the educational field begins to see Delphi in the deeper context discussed in Chapter II, when it starts to consider Delphi as an educational tool as well as a planning tool, then it may be able to escape this trap.

If there is a single message in the philosophy of Singer (see p. 33), it is that the past and the present are often as hard to interpret, or conjecture about, as the future. It is therefore not surprising to see the Delphi method applied to these areas as well as to the future. A recent Delphi devoted to examining the past is an exercise by Professor Russell Fenske of the University of Wisconsin. This involved about twenty-five leading researchers in the field of operations research who utilized the Delphi to review the "state of the art of industrial

---

[5] R. C. Judd, "The Use of Delphi in Higher Education," *Technological Forecasting and Social Change*, **4**, 173 (1973).

[6] R. Elboim-Dror, "Educators' Image of the Future," paper presented at the Third World Future Research Conference, Bucharest, September 1972.

operations research": Each participant could propose significant contributions to the literature in the areas of theory, applications, and economic impact. The group then voted on each item for importance and impact. Also gathered were brief comments on the significance of an item and suggestions for important areas of future research.

Michael Marien has used the Delphi process to elicit from a panel of 14 futurists the most significant books (a "hot list") on the future. Considering the volume of material being produced in most professional areas, these particular applications of Delphi are likely to become quite popular in the future.

A similar concept has been applied in some corporate or organizational uses of Delphi, where the study examines historical performance or factors that have affected the market place for a particular item. The objective is usually to focus on 50 to 100 key items out of hundreds of candidates so that a concise summary of the historical perspective can be prepared for management. This is largely the process of getting a group to filter out the signal of real information from the multitude of communications or noise that may exist on a particular complex topic. This concept is very similar to what Professor James Bright refers to as the monitoring function in technological forecasting, and an excellent example of a Delphi on "Events Leading to the Limitation or Elimination of the Internal Combustion Engine" forms the basis for an exercise in his recent book.[7] The example, based upon a Delphi conducted in 1969 by a chemical company, was, to the best of our knowledge, the first which dealt exclusively with evaluating the past.

A much deeper systemic study of the past is envisioned in the "retrospective futurology" approach which applies dynamic programming to historic societies such as the city-state of Athens. The "hyper-sophisticated polling of experts" mentioned by Wilkinson in conjunction with this concept[8] strongly suggests the Delphi method.

---

[7] James Bright, *A Brief Introduction to Technology Forecasting: Concepts and Exercises*, Pemaquid Press, Austin, Texas, 1972.

[8] J. Wilkinson, R. Bellman, and R. Garaudy, "The Dynamic Programming of Human Systems," Occasional Paper, Center for the Study of Democratic Institutions, MSS Information Corp., New York, 1973, p. 29.

III. B. 1. The Policy Delphi

MURRAY TUROFF[1]

A Seer upon perceiving a flood should be the first to climb a tree.
Kahlil Gibran

Introduction

The Policy Delphi was first introduced in 1969 and reported on in 1970.[1] It represented a significant departure from the understanding and application of the Delphi technique as practiced to that point in time. Delphi as it originally was introduced and practiced tended to deal with technical topics and seek a consensus among homogeneous groups of experts. The Policy Delphi, on the other hand, seeks to generate the strongest possible opposing views on the potential resolutions of a major policy issue. In the author's view, a policy issue is one for which there are no experts, only informed advocates and referees. An expert or analyst may contribute a quantifiable or analytical estimation of some effect resulting from a particular resolution of a policy issue, but it is unlikely that a clear-cut (to all concerned) resolution of a policy issue will result from such an analysis; in that case, the issue would cease to be one of policy. In the face of the policy issue, systems analysis, operations research, and other related disciplines can do no more than supply a factual basis for advocacy. The expert becomes an advocate for effectiveness or efficiency and must compete with the advocates for concerned interest groups within the society or organization involved with the issue.

The Policy Delphi also rests on the premise that the decisionmaker is not interested in having a group generate his decision; but rather, have an informed group present all the options and supporting evidence for his consideration. The Policy Delphi is therefore a tool for the analysis of policy issues and not a mechanism for making a decision. Generating a consensus is not the prime objective, and the structure of the communication process as well as the choice of the respondent group may be such as to make consensus on a particular resolution very unlikely. In fact, in some cases the sponsor may even request a design which inhibits consensus formulation.

---

[1]Murray Turoff, "The Design of a Policy Delphi," *Technological Forecasting and Social Change* **2**, No. 2 (1970).

The Delphi Method: Techniques and Applications, Harold A. Linstone and Murray Turoff (eds.)
ISBN 0-201-04294-0; 0-201-04293-2

The Committee and the Delphi Process

Traditionally the approach in most organizations to the examination and exploration of policy issues has been the committee process. Certainly it is well documented by a number of writers on the functioning of government organizations, that the committee system is a structure that evolved initially to promote the advocacy process associated with policy analyses.[2,3] The committee approach brings people together across organizational lines in order that all views at similar organizational levels in the whole organization may be represented, and a meaningful view arrived at after the differing interests have been adequately expressed and advocated. It is the contention here, however, that from a pragmatic viewpoint, the committee approach in government and most other organizations no longer functions as effectively in the realm of policy formulation as it once may have.

Many organizations today have become bigger, serve more functions, and span a much wider range of complex interacting functions. Committees that truly represent all interests on an issue are often quite large and unwieldy. By the time one has reached the point of twenty or more people constrained to reach a view in a limited amount of time, a complete and free exchange of views among all concerned is often too time consuming or impossible within the scope of the allocated effort for the job.

With increasing size of organizations, the ratio of the number of people at the top echelons to those in the lower echelons has decreased over the years, particularly in government. This implies that those at the top must spend more time devoted to day-to-day management functions and less time for committee participation on the longer-range issues associated with policy. As a result, the responsibility for committee participation falls more and more on lower-level people. Individuals at the lower levels are less likely to be advocates of anything until they have had ample time to clear it with their supervisors. This often forces the committee into added weeks of delay whenever any new point is made and usually results in the early or premature termination of new considerations that might result from the advocacy process.

If an organization is top heavy a similar problem also develops. Power becomes too diffuse and no one feels he has the authority or jurisdiction to act as an advocate on the broader issues that usually arise at the policy level. There are so many narrowly defined functional responsibilities that everyone is taking care not to tread on their neighbors' territory.

The complexity of issues today usually calls for a great deal of additional staff to supplement the committee process. More often than not, this time or support is not allocated to or available for committee participants. In an atmosphere of budget cuts, belt tightening, and competition for limited funds, it may appear

[2]Charles F. Schultze, "The Politics and Economics of Public Spending," Brookings Institution, Washington, DC., 1968.

[3]Numerous references to Lindblom's writings on committee processes appear in the work cited in reference 2.

advantageous not to advocate, not to be noticed, and especially not to be held accountable for views, promises, or positions which require effort to document or substantiate. In addition, in most organizations today, we have individuals who are not familiar with many of the new decision aids coming out of operations research and systems analyses but who have an intuitive feel for the complexities of the particular business or function the organization is involved in. We also have a good many individuals who have been trained in many of the modern management techniques and who are sometimes a little too confident that these approaches can be applied to every problem. The lack of effective communication between these two groups has brought about the ineffectiveness of many committee exercises.

It is the above factors, or any combinations of these factors, which have motivated attempts to seek substitutes for the committee process. Contrary to the above, the earlier writings on Delphi have usually presented a separate but canonical set of problems associated with committees that tend to reflect psychological characteristics of committee processes:

- The domineering personality, or outspoken individual that takes over the committee process
- The unwillingness of individuals to take a position on an issue before all the facts are in or before it is known which way the majority is headed
- The difficulty of publicly contradicting individuals in higher positions
- The unwillingness to abandon a position once it is publicly taken
- The fear of bringing up an uncertain idea that might turn out to be idiotic and result in a loss of face

Given a small committee of around ten individuals with sufficient time to consider and explore the issues, and some assurance that the privacy of their respective remarks will be respected outside of the committee room, it is doubtful that any of the above issues would greatly inhibit the process. However, as the size of the committee increases, the time available decreases, and the organizational considerations listed above present themselves, the psychological problems will also come into play.

Delphi, however, is not a replacement for the committee process. The proposition presented here is that the Policy Delphi can be utilized to revise the effectiveness of the committee approach.

A Policy Delphi can be given to anywhere from ten to fifty people as a precursor to a committee activity. Its goal in this function is once again not so much to obtain a consensus as to expose all the differing positions advocated and the principal pro and con arguments for those positions. In many policy areas, a larger number of respondents, in the area of twenty or more, is commensurate with the number of differing interests that must often be considered in the increasingly complex issues facing organizations.

Once the Delphi has been accomplished, a small workable committee can utilize the results to formulate the required policy. This then is the author's view of the role of the Policy Delphi—a mechanism for reviving the advocacy process in organizations through improving the effectiveness of lateral policy formulating committees. In this way, Policy Delphis operate as precursors to the committee activity.[4]

The Policy Delphi, therefore, is not in any way a substitute for studies, analyses, staff work, or the committee. It is merely an organized method for correlating views and information pertaining to a specific policy area and for allowing the respondents representing such views and information the opportunity to react to and assess differing viewpoints. Because the respondents are anonymous, fears of potential repercussions and embarrassment are removed and no single individual need commit himself publicly to a particular view until after the alternatives have been put on the table. Even in those cases where the Policy Delphi uses only the committee or sponsoring body as the respondent group, it has the advantage of eliminating the principal bottleneck in the committee procedure by providing a clear delineation of specific differing views, thereby providing an opportunity for the committee members to prepare their respective cases adequately.

A Policy Delphi should be able to serve any one or any combination of the following objectives:

- To ensure that all possible options have been put on the table for consideration
- To estimate the impact and consequences of any particular option
- To examine and estimate the acceptability of any particular option

The ability of the Delphi technique to improve current practices for handling the first objective seems quite clear. Whether or not it can meet or fulfill any portion of the other objectives probably depends on whether the design team can distinguish the motivation of the respondents in making a particular judgment on an option. More specifically, when a difference in judgment does occur on an option, is it based upon uncertainty and/or lack of information with respect to consequences, or is it based upon differences among the self-interests as represented by the respondent group? If the Delphi can be designed to make this distinction it should be able to serve these latter objectives of examining and distinguishing consequences and acceptabilities. Because in some cases people are not fully aware of the motivating factors behind their views, the exposing of these factors could require fairly sophisticated approaches, such as multidimensional scaling.

The Mechanics of a Policy Delphi

A Policy Delphi is a very demanding exercise, both for the design team and for the respondents. There are six phases that can be identified in the communica-

[4] Jerry B. Schneider, "The Policy Delphi: A Regional Planning Application," *Technological Forecasting and Social Change* **3**, No. 4 (1972).

tion process that is taking place. These are:

(1) Formulation of the issues. What is the issue that really should be under consideration? How should it be stated?

(2) Exposing the options. Given the issue, what are the policy options available?

(3) Determining initial positions on the issues. Which are the ones everyone already agrees upon and which are the unimportant ones to be discarded? Which are the ones exhibiting disagreement among the respondents?

(4) Exploring and obtaining the reasons for disagreements. What underlying assumptions, views, or facts are being used by the individuals to support their respective positions?

(5) Evaluating the underlying reasons. How does the group view the separate arguments used to defend various positions and how do they compare to one another on a relative basis?

(6) Reevaluating the options. Reevaluation is based upon the views of the underlying "evidence" and the assessment of its relevance to each position taken.

In principle the above process would require five rounds in a paper-and-pencil Delphi procedure. However, in practice most Delphis on policy try to maintain a three- or four-round limit by utilizing the following procedures: (1) the monitor team devoting a considerable amount of time to carefully preformulating the obvious issues; (2) seeding the list with an initial range of options but allowing for the respondents to add to the lists; (3) asking for positions on an item and underlying assumptions in the first round.

With the above simplifications it is possible to limit the process to three rounds. However, new material raised by the respondents will not get the same complete treatment as the initial topics put forth by the monitor team. Still, very successful Delphis have been carried out within a three-round format. Ultimately, however, the best vehicle for a Policy Delphi is a computerized version of the process in which the round structure disappears and each of these phases for a given issue is carried through in a continuous process.[5]

It is also necessary on a Policy Delphi that informed people representative of the many sides of the issues under examination are chosen as participants. These individuals will not be willing to spend time educating the design team, by way of the Delphi, on the subject matter of concern. The respondents must gain the feeling that the monitors of the exercise understand the subject well enough to recognize the implications of their abbreviated comments. Therefore, the initial design must ensure that all the "obvious" questions and subissues have been included and that the respondent is being asked to supply the more subtle aspects of the problem.

In some instances, the respondent group may overconcentrate its efforts on

---

[5]Murray Turoff, "Delphi Conferencing," *Technological Forecasting and Social Change*, **3**, No. 2 (1972).

some issues to the detriment of the consideration of others. This may occur because the respondent group finally obtained was not as diversified as the total scope of the exercise required it should be. With proper knowledge of the subject material, the design team can stimulate consideration of the neglected issues by interjecting comments in the summaries for consideration by the group. It is a matter of the integrity of the design team to use this privilege sparingly to stimulate consideration of all sides of an issue and not to sway the respondent group toward one particular resolution of an issue. If, however, the respondent team is as diversified as required by the material, there should be no need to engage in this practice.

A Policy Delphi deals largely with statements, arguments, comments, and discussion. To establish some means of evaluating the ideas expressed by the respondent group, rating scales must be established for such items as the relative importance, desirability, confidence, and feasibility of various policies and issues. Furthermore, these scales must be carefully defined so that there is some reasonable degree of assurance that the individual respondents make compatible distinctions between concepts such as "very important" and "important." This is further complicated by the fact that many of the respondents may not have to think through their answers in order to remain consistent in answering different parts of the questionnaire.

The Delphi technique is not just another polling scheme, and the practices that are standard in polling should not be transferred to Delphi practice without close scrutiny of their applicability. Consider, for example, a poll of different groups in an organization asking for their budget projections over the next five years. This is a comparatively straightforward request which does not ask any one group to place itself in context or to worry about consistency with other groups in the organization. A Delphi on the same subject would ask each group to make projections for every group's budget and, in addition, to project separately a feasible total budget for the organization as a whole.

The normal budget process in an organization is essentially a poll. A few research laboratories have in recent years attempted a budget review process via the Delphi mode, but unfortunately these are never reported in the literature because of the proprietary nature of the subject material. In principle, it would appear that the Delphi offers more opportunity for people to support budget items outside of their current management function and often to obtain a better appreciation of the budget trade-offs that have to be made.

There are many different voting scales that have been utilized on policy type Delphis; however, there are four scales, or voting dimensions, that seem to represent the minimum information that must be obtained if an adequate evaluation is to take place. On the resolutions to a policy issue it is usually necessary to assess both desirability and feasibility. One will usually find a significant number of items which are rated desirable and unfeasible or undesirable and feasible. These types of items will usually induce a good deal of discussion among the respondents and may lead to the generation of new options. The underlying assumptions or supporting arguments are usually evaluated with respect to importance and validity or confidence. In this case a person may think an invalid item is important (because others believe it to be true) or that a true item is rather unimportant. It is usually unwise to attempt to ask for a vote on more than two dimensions of any item. However, if one has established a significant subset of items utilizing these scales then further questions can be introduced focusing on the significant subset. For example, there is the possibility of taking desirable options and asking the probability for each, given certain actions are taken.

Typical examples of these scales follow. Note that no neutral answer is allowed other than No Judgment (which is always allowed on any question). A neutral position offers very little information in policy debates and it is usually desirable to force the respondent to think the issue out to a point where he can take a nonneutral stance. In other words, the lack of a neutral point promotes a debate which is in line with developing pros and cons as one primary objective. This design choice has sometimes upset those who feel consensus is the only valid Delphi objective.

Desirability (Effectiveness or Benefits)

Very Desirable	Will have a positive effect and little or no negative effect extremely beneficial justifiable on its own merit
Desirable	will have a positive effect and little or no negative effect beneficial justifiable as a by-product or in conjunction with other items
Undesirable	will have a negative effect harmful may be justified only as a by-product of a very desirable item, not justified as a by-product of a desirable item
Very Undesirable	will have a major negative effect extremely harmful not justifiable

Feasibility (Practicality)

Definitely Feasible	no hindrance to implementation no R&D required no political roadblocks acceptable to the public

Possibly Feasible	some indication this is implementable some R&D still required further consideration or preparation to be given to political or public reaction
Possible Unfeasible	some indication this is unworkable significant unanswered questions
Definitely Unfeasible	all indications are negative unworkable cannot be implemented

Importance (Priority or Relevance)

Very Important	a most relevant point first-order priority has direct bearing on major issues must be resolved, dealt with, or treated
Important	is relevant to the issue second-order priority significant impact but not until other items are treated does not have to be fully resolved
Slightly Important	insignificantly relevant third-order priority has little importance not a determining factor to major issue
Unimportant	no priority no relevance no measurable effect should be dropped as an item to consider

Confidence (In Validity of Argument or Premise)

Certain	low risk of being wrong decision based upon this will not be wrong because of this "fact" most inferences drawn from this will be true
Reliable	some risk of being wrong willing to make a decision based on this but recognizing some chance of error some incorrect inferences can be drawn

Risky	substantial risk of being wrong not willing to make a decision based on this alone many incorrect inferences can be drawn
Unreliable	great risk of being wrong of no use as a decision basis

The first and foremost problem in conducting a Policy Delphi occurs with the initial steps in the process. If the respondents feel strongly about the issues, and this should be the case, they will generate a large amount of written material. If they are provided a certain number of items to deal with on the first round then each of them will make approximately the same number of written comments or additions in response. These must be abstracted carefully and duplications among the respondents eliminated. On the average, the written material in the questionnaire for the second round will be five to ten times that of the first round.

After the votes are taken on the second round, the material should be rearranged by the average vote on the third round. In other words, referring to the preceding scales, the options should be reordered by Desirability and the supporting arguments reordered for each option by Importance. When the votes are in, the resulting summary for the third round should clearly point out which items exhibited polarized distributions, which ones exhibited a flat distribution across the whole range, skewed distributions, or on which items only a very small sample of the respondents were able to make a judgment. For these items, additional comments should be solicited. If possible, the revote should be put off until a fourth round when everyone can see the additional remarks. In a three-round exercise a revote is taken on the third round.

In many cases it may be desirable to keep track of certain subgroups making up the respondent group as a whole. This provides a mechanism to check whether polarized views reflect the affiliations or the backgrounds of the respondents. Depending on the application, this information can be fed back to the group. Schneider[4] in his article on Policy Delphis proposed a very concise "Measure of Polarization" among the subgroups. Take all two-by-two combinations of subgroups and add the absolute value difference of the average vote on a given item. This sum of first differences is now an index which provides an appropriate ranking of the degree to which differences exist for each item relative to the group of items as a whole. The same measure may be applied to each individual who voted on the item when a subgroup breakdown is not appropriate. Note that in opposition to average and standard deviation this measure is a strong function of the number who voted when applied on an individual basis.

Some additional guidelines on carrying out the Policy Delphi process are as follows:

- The number of professionals acting as the design-monitor team must be at least two so that one can check the other. Ideally, one should be knowledgeable in the problem at hand (but not precommitted) and the other should have editorial talents.
- A month or more is needed to develop the first-round questionnaire. In addition to the questionnaire, a factual summary of background material is usually supplied, and in some cases single or multiple sets of scenarios specifying certain items the respondents are to assume as given are provided for the purpose of evaluating the issues. Typically these scenarios deal with future economic conditions such as the rate of inflation. Sometimes it is more appropriate to introduce a set of alternative assumptions making up scenarios and let the respondents form a group scenario by voting on the validity of each.
- Each questionnaire should be pretested on coworkers who have not been involved in the design. There is a very high probability that this will identify items that are stated in a confusing manner.
- Take care to avoid compound statements to be voted upon. The question "if *A* and *B* are true" should be broken into two separate items. The exception is statements of the form "if *A* then *B*," which are quite useful in some situations.
- The respondents, if new to Delphi, will respond with compound and sometimes lengthy comments. Therefore it is a good idea to show them some examples of the form you would like comments to take, in terms of being short, specific, and singular in nature.
- If there is a trade-off between the ease of summarizing the results and the ease of the respondents in providing the answers and understanding the results, the choice should always favor the respondent.
- The respondents should be allowed to suggest changes in the wording of items which should then be introduced as new items. Experience has shown that the vote on a policy item is very sensitive to wording. Because of this property, the material on Policy Delphi can mushroom in size and represents considerably more effort than the traditional forecasting Delphi oriented to largely quantitative response after the first round.
- When asking for revotes on an item, the individual respondent should be shown his original vote. The respondent should also be provided two copies of the questionnaire so that he may retain one for later reference or to do rough work. He can type his answers on the other copy if he is concerned with security. On Policy Delphis security can be a problem with respect to convincing the respondents that it will be maintained. The design team should set up a procedure where they themselves cannot identify the returns with the individuals involved.
- The respondents must be convinced that they are participating in an exercise which involves a peer group. Therefore it is usually necessary on the letter of

invitation to indicate the types of backgrounds reflected in the participant group. In some cases, a list of the respondents involved can be provided if there is no other effective way to convince the group of the significance of the exercise.

As can be seen, there are many things to be considered in running a Policy Delphi, or any other Delphi for that matter. The Delphi concept seems so simple that many people have thought it an easy thing to do. Consequently there have probably been more poorly done Delphis than ones that have been well done.

One additional aspect of the Policy Delphi which usually argues for four or more rounds arises in the situation where the respondents feel very strongly about their respective views. In such a case they sometimes have an attitude where they cannot imagine that there are rational and intelligent people who hold a contrary view. Even with a vote on the first round on a given issue, the reaction of this type of respondent to the vote presented on the second round is that the individuals holding the opposite view to his just don't understand the problem completely. A few simple comments will clear up their ignorance. It is only until the third round comes back that this type of respondent feels the shock resulting from a realization that the other side also feels it has some valid points to be made. Therefore, it is only at the third round that this type of respondent begins to put a great deal of careful effort into the points he is making and to consider more carefully what the other side is saying. The material generated out of this type of process could have a significant impact on the group views if carried back in a fourth round.

The selection of respondents is one of the most difficult tasks. However, this problem applies to any committee or study effort. The sponsor is likely to have a certain candidates in mind. The design team should try to structure the problem in order to get a comprehensive coverage of the topic. It is also a good idea to mix in a couple of lateral thinkers and devil's-advocate types, just on a matter of general principle—i.e. those individuals who always manage to come up with the unexpected.

It is possible on a Policy Delphi to observe two very different phenomena taking place. One is when the exercise starts with disagreement on a topic and ends with agreement. This can be very useful to those sponsoring the study if it does occur, but, as has been said, is not a necessary result. Another process is to start with agreement on a topic and end with disagreement. In a sense this can be viewed as an educational process taking place among the respondents who suddenly realize, as a result of the process, that the issue is not as clear-cut or simple as they may have thought. Unfortunately, to this point in time there has not been sufficient exploration of the use of the Delphi technique as an educational process. Schneider[4] also discusses this point. As he pointed out, Delphi could be used by a planning agency to interface more effectively with representatives of the community and serve an educational function for both groups.

Another unexplored use of the Policy Delphi is the investigation of the performance of past policy actions. Too many organizations do not have an appropriate mechanism for taking stock of what they have accomplished. Understanding of what has occurred is often lacking and can lead to future mistakes in policy formulation.

Examples of Policy Delphis

One of the first Delphis that bordered on being policy oriented was an exercise undertaken in 1968 by the National Industrial Conference Board. It was titled "An Experimental Public Affairs Forecast." It involved 70 people representing the following areas of expertise:

Economy, Business, and Labor	17
Science, Technology, and Change	9
Government, Law, and Politics	6
Resources	9
Education and Training	5
Communications	8
Culture, Family, and Behavior	12
International Security	4

The vast majority had titles of chief executive or director. All were considered by the Conference Board to be distinguished in their field.

The overall objective of the study was to obtain a rank ordered list of National Priorities or Areas of Major Concern to the Nation, areas which could create major public problems in the seventies and eighties and should receive attention by U. S. leadership. The top ten in that list in order of priority were: (1) division in U. S. society; (2) international affairs; (3) education; (4) urban areas; (5) law and order; (6) science, technology, management of change; (7) economy; (8) resources; (9) values; (10) population.

The Delphi was completed before the presidential campaign and one may note a degree of correspondence between the priorities set by this exercise and the Republican campaign themes. While the Delphi dealt with policy considerations, it was largely oriented to putting the pieces of the problem together by collecting information and views from a diverse set of respondents. Therefore it largely reflected a Kantian-type exercise. The bulk of the material produced was a collection of commentaries on the problem areas with some estimate of when particular problems would arise. Each item was handled in terms of the following categories of information:

- description of the item
- description of public reaction to the item
- beginning date of maximum impact on U. S.
- intensity of impact on U. S.
- opportunity for leadership to change the expected

The Delphi appeared to be quite adequate in meeting the needs of its sponsors; however, the exercise has never been described in the literature so one can only infer this from the final report, which unfortunately did not receive public distribution beyond those immediately involved and some individuals working in the Delphi area at that time. One major fault of the study was the decision made by the staff people not to abstract the comments of the panelists but to retain the full text. In part this decision was probably influenced by the distinguished nature of the respondent group. The result was a very large volume of material which is a little painful to wade through to gather the particular nuggets of wisdom that were produced. One goal of a Delphi design shouid, therefore, always be to obtain a filtering of the essential from the superfluous.

The next Delphi in the policy area was one conducted by Emory Curtis as a consultant to San Mateo County in California. This one involved around 80 community people representative of the many different constituent groups making up the public body. A great deal of effort went into obtaining a broad-based distribution of respondents. They were provided a large number of policy options dealing with the structure and functions of the county government, and asked to vote on these for relative agreement on a seven-point agreement scale. Additional items were added as a result of the first round. However, the one shortcoming of this exercise was the lack of exploration of the factors underlying disagreement when it did occur. The exercise produced some new options and exhibited consensus where it occurred but provided no mechanism for effectively resolving disagreement. However, it represented one of the first attempts to use the Delphi in policy areas related to community government.

In 1970 a Delphi was conducted by the Office of Emergency Preparadness and the Rand Corporation on the subject of Civil Defense Policy. This Delphi introduced a number of unique features. It exhibited the structure of a Hegelian inquiring system as opposed to the earlier Lockean- and Kantian-type Delphis. First, it recognized in the design that strong disagreement already existed on a number of the issued involved. For a number of items the respondents were asked to choose sides by circling "could/could not," "should/should not," these being choices in the wording of the items. They were also asked to develop the strongest arguments on the various sides of a given issue. The sponsor was not interested in having the group make a decision for him, but in having the group develop, compare, and evaluate the best possible arguments on each side of an issue.

The details of this exercise are well documented in the literature.[1] As typical of these types of Delphis, the respondents generated about eight times the amount of material they were initially given on the first round, which contained some seventy items for evaluation. Basically policy options were evaluated on scales of desirability and feasibility, while supporting points were evaluated on importance or validity. This Delphi was really the first to incorporate a structured debating-type format, which appears to be the useful approach for the exploration of policy issues.

In 1970 Professor J. B. Schneider at the University of Washington adopted the same approach to the exploration of transportation planning as it applied to highway development in the Seattle area. His report of the exercise[4] is an excellent example of applying these techniques to urban planning problems. He also contributed some very useful observations on the methodology for handling disagreement in that particular context.

Following the same line of development, Joel Goodman of the College of Marine Studies, University of Delaware, conducted a policy-type Delphi on the Coastal Zone Land Use Planning Issue. This involved a large number of people representing government business, public groups, and specialists. This exercise was done in 1971 and 1972. It converged the following types of items into different sections of the questionnaires: respondent characteristics; respondent attitudes; arguments pro and con; general policy and budget items; specific policy issues; specific programs; strategic issues.

Some sample questions from that exercise follow:

- As an individual, list in order of priority your 5 principal concerns with respect to the way in which the coastal zone is developing.

  (a) Health hazard
  (b) Unsightly buildings
  (c) Dirty water (visual appearance)
  (d) Too much land going to waste
  (e) Too crowded
  (f) Not enough housing
  (g) Not enough boating facilities
  (h) Not enough camp ground
  (i) Beaches too narrow
  (j) Too many fisherman
  (k) Other

- Why are you as an individual concerned with pollution in the coastal zone and its effects upon the marine environment? Check up to three responses and signify relative importance by numbering principal reason as "1."

  (a) biological danger
  (b) potential loss of recreational opportunity, i.e., swimming, boating, etc.
  (c) potential loss of aesthetic values, i.e., vistas, landscape, etc.
  (d) potential loss of income or revenues
  (e) community involvement
  (f) other (specify)

Indicate by check mark who should assume responsibility for establishing:

	Quality limits for coastal zone environment (1)	Use patterns for coastal zone shore lands (2)	Use patterns for coastal zone submerged lands (3)	Use patterns for coastal waters (4)
a. State				
b. County				
c. Local Community				
d. Other (Specify)				

- If standards for the quality of the marine environment are to be maintained, then the authority and responsibility for regulation should be vested in:

SELECT ONE

a. A state agency within the executive branch

b. A county agency

c. Individual municipalities

d. Criteria established by state; regulation by municipalities and counties

e. A new organization responsible to——— with elected/appointed officials (Fill in blank and select one or the other means of acquiring the officials.)

A large number of the current Delphis have started to incorporate policy issues even when that was not the primary concern. Such issues have the psychological advantage of making the exercise of more interest to the respondents. The policy orientation has been introduced in some different ways. Instead of asking individuals to extrapolate data into the future in terms of their best estimate of what they think will occur, a policy approach would be to ask what would be a desirable and possible extrapolation as well as an undesirable and possible extrapolation. Based upon those estimates one can ask what are the factors that could make the curve go one way or the other.

A Delphi study conducted by the Federal Department of Public Works in Canada illustrates the incorporation of policy options into an essentially non-Policy Delphi. The department's major role is providing accommodation for federal civil servants, and the Delphi was undertaken as part of a model for forecasting government employment with the purpose of determining future accommodation needs. But the department's mandate extends beyond simply providing buildings to house federal employees. It is concerned with the total work environment of the civil service.

Consequently, the Public Works Delphi also explored the existing procedure for space allocation, which at present is based on the average salary of all employees using that space, and asked respondents to comment on that process. In the first round, after reviewing the present process, the respondents were asked to list what they felt were the strengths and weaknesses in the process, and asked to suggest possible options for change. In the second round these options were voted upon according to *Desirability* and *Feasibility*, keeping in mind that a particular option could be desirable and unfeasible at the same time, or vice versa. Some examples of suggestions for change according to *Desirability* were:

- formula approaches, if used, must reflect the quality of space as well as the quantity
- relate space to function not salary
- more emphasis on multipurpose facilities
- DPW should lead the way in educating agencies in new building concepts

The Delphi also looked at possible parameters for measuring building performance that would go beyond the usual cost/benefit measures, such as the ratio of rentable square feet to total square feet. Specific suggestions or concepts for consideration fell into the following categories:

- psychological and motivational impact on employees
- transportation to building
- aesthetic value of building
- community and public service
- energy and environment

The respondents were asked to vote on the *Desirability* and *Feasibility* of specific suggestions and to suggest ways in which some of these concepts could be measured..

Public Works then used the Delphi exercise not only to fulfill its immediate objective of forecasting federal government employment but also to explore policy options relating to its mandate of fulfilling broader social, economic, and environmental objectives.

Another excellent example of a Delphi mixing policy issues with future forecasts was one done by the Canadian Department of Health and Welfare on the Future of Genetic Counseling Services in Canada. The exercise involved some sixty respondents ranging from research geneticists to public health workers. The design was well balanced between "technical issues" of what was possible at what point in time and "policy issues" of who could, would, or should do what. In this latter area, the issue of a genetic registry and its potential abuses as well as uses were explored. The Delphi used the same sort of scales that were mentioned earlier. However, it tended to redefine a scale such as "importance" for each question it was used on. This had the merit of minimizing what the respondent had to remember, since each question was largely self-contained. It also minimized the chance of confusion by placing the scale within the context of the particular question. Furthermore, it allowed more variety in the sequencing of questions. Most other designs, by grouping questions of a given type under one explanation, can produce a feeling of monotony as the respondent goes through the exercise.

The Problems of a Policy Delphi

We have already mentioned the danger that a Policy Delphi can be misinterpreted as a decisionmaking tool as opposed to a decision-analysis tool. Everyone at heart is a decisionmaker, or wishes to be, and it is all too easy on the part of the designer to appeal to this unrequited desire. It should be a matter of intellectual honesty for designers to make clear just what the objective of the exercise is. If we have a problem in organizations today, especially governmental ones, it is that the responsibility for a given decision is not clearly focused on one individual. A decision should be made by one individual, and the role of the Policy Delphi and other tools is to provide the best possible information and ensure that all the options are on the table. To do this the Delphi must explore dissension. Both Dalkey and Helmer in the early writings on Delphi expressed the need to establish clearly the existent basis for observed dissension. However, this implies a good deal more work for the design team and has often been neglected in the majority of the early exercises. When a strong minority view exists and is not explored, the dissenters will often drop out, leading to an "artificial" consensus on the final product.

Once a Policy Delphi has been started, there is no way to guarantee a specific outcome if it is to be an honest exercise. This is something the sponsor must be well aware of. Occasionally a sponsor, particularly in a policy exercise, will desire that the group not reach a consensus on any particular option. While it is consistent with the objective of a Policy Delphi to choose a respondent group such that a consensus is unlikely to occur, it can never be guaranteed that it will not be a result. However, there is a fine line between Delphi as an analysis tool and Delphi as an educational or persuasion device. It is possible to consider using a Delphi to educate at least a part of a respondent group on options they may not be aware of. Unfortunately, very little work has been done on the use of Delphi in an educational mode even though most designers would agree that educational processes take place in most exercises.

A Policy Delphi is a forum for ideas. In opening up the options for review, items may arise which can be disconcerting to members of the group. If a sensitive area is under review and an attempt has been made to have diverse representation in the group, then premature leakage of the results can occur. In such a case, individuals outside the exercise may misinterpret what is taking place. This problem of lifting items out of context occurs all the time in the committee process. A workable approach to this problem in the Delphi process is to incorporate members of the press into the respondent group when dealing with major public policy items.

As with any policy process, there are many ways to abuse the use of the Policy Delphi: the manner in which comments are edited, the neglect of items, the organization of the results. However, such a process is a rather dangerous game and not likely to go unnoticed by some segment of the respondents. There are very few greater wraths than that of a respondent who discovers himself to be engaged in a biased exercise. Furthermore, Delphi has reached the point where there is no longer any excuse on a professional basis for making many of the mistakes found in earlier exercises. The person seeking to undertake a Delphi today should be reasonably familiar with what has taken place in the field.

III. B. 2. Delphi Inquiries and Knowledge Utilization

JOHN LUDLOW

I. Introduction

The development of methods to obtain, refine, and communicate the informed judgments of knowledgeable people is one of the most crucial problems in planning and decisionmaking. The task is particularly challenging in the Michigan Sea Grant Program, which emphasizes a systems approach by a multidisciplinary group of researchers. Some of these researchers are experts in extremely specialized areas, representing a wide range of technical, economic, social, legal, and political disciplines.

From its inception the general goal of the Michigan Sea Grant Program[1] has been to provide the common management effort necessary to develop and bring to bear university expertise on short- and long-term resources management problems in the Great Lakes. The major approach of the program has been the development of basic information and predictive models for resolution of resource problems, followed by applications and/or demonstrations of such information and models to appropriate agencies and groups. Over 120 research and faculty personnel from practically every major school or college in the university are presently active in the program. Research and planning groups representing federal, state, and local government agencies, industry, and concerned citizen groups are also part of the problem-solving team.

The Grand Traverse Bay watershed region was selected as the focus of pilot efforts to develop research and planning methodologies that will be applicable in dealing with problems and opportunities of all the Great Lakes, and in particular Lake Michigan. In the area of finding mechanisms to improve the coordination of the Sea Grant effort at the University of Michigan it was decided to investigate the potential for utilizing the Delphi technique.

The Michigan Sea Grant Delphi inquiries[2] were designed to obtain and refine an interdisciplinary group of researchers' judgments about issues and developments that should be considered when planning for intelligent management of the water resources of the Great Lakes.

An important objective of the exercises was to convey the judgments of the

---

[1]The term "Sea Grant Program" was derived from the National Sea Grant College and Program Act, whose intent was to involve the nation's academic community in the practical problems and opportunities of the marine environment, including the Great Lakes.

[2]The term "Delphi inquiry" was propounded by Turoff and refers to the complete Delphi process. He observed that any particular Delphi design can be characterized in terms of the "inquiring systems" specified in Churchman's writings. See reference 1 at the end of this article.

The Delphi Method: Techniques and Applications, Harold A. Linstone and Murray Turoff (eds.)
ISBN 0-201-04294-0; 0-201-04293-2

researchers to the communities which are to benefit from the research. One approach toward this objective was to include on the panels—on the same basis as the researchers—people who were believed to be influential in the political processes through which regional planning is accomplished. Their knowledge of the issues and the region was beneficial to the deliberations, but more importantly, their participation was judged to be an effective way of communicating information to regional planners and decisionmakers.

Two of the three panels were made up of researchers who were designated as technicians and behaviorists. The third group was made up of concerned citizens who were designated as decisionmakers. In addition to forecasting, the method was used in several other roles involving the quantification of subjective judgments. The exercises were designed to be progressive and cumulative, with an emphasis on an orderly development of informed judgments.

The Delphi inquiries were one of several Michigan Sea Grant projects related to the general task of transmitting new knowledge to people and organizations in a way that results in effective use. Respondents in these exercises—a group with exceptional qualifications—served as the primary resource in evaluating the methodology.

The technical panel was composed of thirty-three individuals whose expertise was primarily in the physical sciences and who were divided about equally between Sea Grant researchers and faculty, graduate students, and others in the School of Engineering. A second panel included Sea Grant researchers who were not selected for the technical panel. Generally their academic backgrounds and interests were oriented more to the behavioral sciences, and for this reason they were labeled behaviorists. They represented a wide range of ages, academic disciplines, and university schools and laboratories. Participants for the third panel were randomly selected from groups of Grand Traverse Bay area residents believed to be influential in the following fields: civics, business, planning, politics, natural resources, government, education.

The names associated with the panels, although somewhat arbitrary, are reasonably consistent with the roles each group would be expected to play in planning the management of regional water resources. The technical panel operated independently of the other two panels and its output was fed into the deliberations of two broader-based panels, which operated independently in the earlier rounds and as a combined panel in the final round. The nature of their participation is summarized in Table 1.

In order to provide continuity, a person's judgments on the previous round were used whenever he or she could not respond on a particular round.

Several significant modifications and refinements in the basic Delphi methodology were tested in the Michigan Delphi inquiries. These changes were motivated by the perceived threat of a manipulated consensus, the desire for constrained or conditional judgments, and recognition of desirable aspects of interpersonal methods not obtainable using the Delphi technique exclusively. The concept of informed judgments as contrasted with expert opinion provided the rationale for the inclusion of politicians and concerned citizens on the

Table 1
Participation in Michigan's Sea Grant Delphi Probes

Activity	Technical Panelists	Behaviorists	Decision-makers
Contact established with Delphi administrator	33	16	22
Unavailable after the start of the Delphi probe	6	0	3
Written comments and evaluations made on at least one round	28	11	21
Written comments and evaluations made on 3 or more rounds	14	6	9
Written comments and evaluations on final round	20	9	11
Written evaluation of methodology or evaluation interview	29	12	16

panels; it also provided an opportunity to exploit an inherent characteristic of the method—*to inform* during the process of soliciting judgments.

II. Outline of the Procedures: Social, Political, and Economic Trends

The portion of the Delphi inquiry concerned with social, political, and economic trends was designed to provide respondents on the broader-based panels with some basic reference points in making subsequent judgments regarding future social and technical developments.

The information package for round one presented the trends for eight measures which have commonly been used to indicate the social and economic development of a region. Curves were plotted from 1950 to 1970, taking advantage of the 1970 census and the standardized enumeration procedures of the Bureau of the Census. Panel members were asked to extend the curves through 1990 and to indicate the numerical values for 1980 and 1990 [2].

In the second round, curves representing the medians and interquartile ranges were provided for the panelists, as well as pertinent comments submitted by respondents on the previous round. Panelists were asked to reconsider their estimates, and if any of the new estimates were outside the designated consensus range for the previous round they were asked to support their position briefly.

On this round the graphs of three additional statistical measures were introduced for consideration. A cumulative summary of the group response was provided in the information package for round three to serve as background information for other panel deliberations.

Important Developments and Requisite Technology

The Delphi method has had its greatest application and acceptance as a means of compiling a list of future technical events or developments and collecting subjective judgments regarding them. In the Michigan inquiries social, political, and economic developments were also solicited and evaluated so that panelists would be encouraged to consider all environments in making judgments regarding water quality, waste-water treatment systems, and research priorities.

The initial evaluation matrix for the technical panel did not present a list of potential developments, something which is usually done in order to facilitate participation and generate additional items. It was believed that this unstructured approach would result in a wider range of suggestions; however, the information feedback of the second round did include—in addition to items suggested by respondents—thirteen events that were taken from Delphi exercises conducted at Rand and the Institute for the Future. These events covered areas considered by the researcher to be of interest to the panel and were also good examples of how developments should be specified to avoid ambiguity, particularly with respect to occurrence or nonoccurrence.

The evaluation matrix for the third round provided the respondent with his estimates for the second round and a summary of the group's response. Comments submitted by respondents were also provided, as were the median estimates for technical and economic feasibility if they differed significantly. The evaluation matrix for the third round was designed so that a panel member could easily determine if his reassessed estimates for a specific development were outside the group's consensus range—arbitrarily identified as the group's median 25 percent and 75 percent estimates. If a respondent's latest estimate was outside the consensus range for the previous round he was asked to support this "extreme" position briefly.

The evaluation matrix for the fourth round presented a more comprehensive summary of the previous round than had been provided up to this point in the exercises. Statistical summaries were presented not only for all the respondents but also for those who rated their competence relatively high and for those in the latter group who indicated a familiarity with the Grand Traverse Bay area. In addition, the persons arguing for an earlier or later probability date than that indicated as the consensus were identified by a number which correlated to a list of biographical sketches.

On the final round of the technical-panel exercises, respondents were also asked to make specific conditional probability estimates for pairs of events that panel members had suggested were closely related. First they were to consider the effects of the occurrence of the conditioning event and then the effects of the nonoccurrence of the conditioning event (see Fig. 1). One of the objectives of this procedure was to encourage panelists to reexamine their estimates for individual events in the light of the influence and probabilities of related events. Analysis of all individual responses reveals that a relatively high percentage of respondents altered their final estimates for those developments included in the set of events which was subjected to conditional probability assessments. Since this was the third iteration of feedback and reassessment for many of these developments, it is not unreasonable to assume that the change in estimates primarily resulted from the evaluation of relationships among events—*relationships which previously had not been fully considered*. This assumption is further supported by the fact that these respondents made almost no changes in their estimates of other developments, which were not subjected to the specific routine of estimating conditional probabilities (but were given the benefit of the feedback of all of the other types of information used in these exercises). In view of the fact that the relationships among events were stressed throughout these exercises, any movement in the final estimates as a result of the consideration of specific conditioning effects is believed to be significant.

Developments and Events that Respondents Have Suggested Are Interrelated	Probability 1971-80	50% Probability Date
D-32 Requirement by the state, calling for tertiary treatment of municipal sewage for Traverse City		
Your Previous Estimates	80	1977
Panel Estimates, Round 3	75 (50-85)*	1977 (1975-80)
Those Who Rated Competence ≥ 3	83 (62-95)	1978 (1975-80)
Your Next Estimates for D-32		
D-31 Construction of a spray irrigation system for waste water disposal in the Grand Traverse Bay region		
Your Previous Estimates	50	1980
Panel Estimates, Round 3	50 (50-50)	1980 (1980-90)
Those Who Rated Competence ≥ 3	50 (15-50)	1980 (1978-80)
Your Next Estimates for D-31		
If you were certain that D-32 would occur before 1980, your estimates for D-31 would be		
If you were certain that D-32 would not occur in 1971-80, your estimates for D-31 would be		

* Interquartile Range

Fig. 1. Example of interrelated developments.

An analysis of the estimates of the technical panel showed that some respondents appeared to have considerable difficulty making probability estimates both for a fixed period (1971–80) and for fixed levels of probability (25, 50, and 75 percent). In some cases inconsistent estimates were made (for example, the probability of occurrence during 1971–80 was estimated to be greater than 50 percent, but the year associated with a 50 percent probability was later than 1980).

Fixed probabilities of 25, 50, and 75 percent were selected for personal probability assessments by the broader-based panelists for several reasons:

(1). There was strong agreement among the three groups involved in the exercises—technical, behavioral, and decisionmakers—on the words and phrases that they associated with the numerical probabilities of 25, 50, and 75 percent [3].

(2) Individual distributions provided the decisionmakers with more information than single probability estimates and were believed to be helpful to the estimator in making assessments that were consistent with his judgment [4].

(3) The 25, 50, and 75 percent levels of probability were ideal for using a betting rationale, that is, systematically dividing the future into equally attractive segments.

(4) It was believed that group medians associated with these fixed probabilities would provide an easily identifiable consensus range.

Since it was likely that many of the decisionmakers would have had little experience with the notion of personal probabilities, a guide for making personal estimates of probability was sent to all members of the broad panels—researchers as well as decisionmakers. The guide presented a systematic method for arriving at the timing estimates for each technical and social development. The assessor was asked to visualize a movable pointer below a sequence of numbers representing years, as in the diagram below. He was asked to move the pointer mentally so as to divide the future into two periods in which the development was equally likely to occur.

```
1                        1              1
9                        9              9
7                        8              9
1 2 3 4 5 6 7 8 9 0 1 2  3 4 5 6 7 8 9  0 Later
                         A
                         ↑
                  50% Probability
```

If the result appeared as it does in the diagram above, 1983 should be entered as the 50 percent probability date. It could also be described as the "1-to-1" odds or "even chance" date. If the pointer came to rest beyond 1990, "Later" would be recorded, and the assessor would go on to consider the next development. The assessor was then instructed how to divide up the results to estimate the "3-to-1" and "1-to-3" odds.

Because of the interest in technology transfer and knowledge utilization in Michigan's Delphi inquiries, there was a special interest in the judgment patterns of the technicians and decisionmakers, which were displayed as in Fig. 2. For each round the panel medians (connected by a solid line) and the interquartile ranges (connected by dashed lines) were shown. The rounds were numbered from left to right for the researchers and from right to left for the decisionmakers, to facilitate the comparisons. The average judgments of respondents in each group who rated their competence in the area being considered relatively high were indicated by asterisks. For most items generally, each group's median estimate for the final round was very close to the median estimates of those who considered themselves relatively competent in the subject. Also, the consensus—as measured by the interquartile range—narrowed and the average estimates of the two groups tended to come closer together. Some of the other patterns, while not ideal from the standpoint of movement toward a narrower consensus, provided a decisionmaker with information as to a course of further inquiry.

Sources of Pollution

A crucial consideration in planning for intelligent management of water resources is the identification of the most important sources of pollution. In making their judgments, panelists were asked to assume a future social and political environment consistent with present trends. However, it was expected that concurrent Delphi inquiries regarding important developments and requisite technology would influence their estimates.

On the first round the technical panel was provided with a list of sources of pollution and specific pollutants thought to be important. Panelists were requested to add other items that they felt would affect a body of water comparable to Grand Traverse Bay in the next twenty years. The collated responses identified seventeen additional sources of pollution and eighteen additional pollutants for the panel to consider. Since there were too many alternatives to present in a matrix designed to encourage the careful consideration of several evaluation factors, the primary objective of round two was to narrow the number of alternatives. The evaluation matrix of round three presented the ten most important sources of pollution as determined by a statistical summary of the estimates made in round two. Panelists were asked to distribute 100 points among the sources of pollution, according to each one's relative importance, for two future periods. The information feedback for the following round provided statistical summaries for Group A, all respondents; Group B, those who rated their competence on sources of pollution relatively

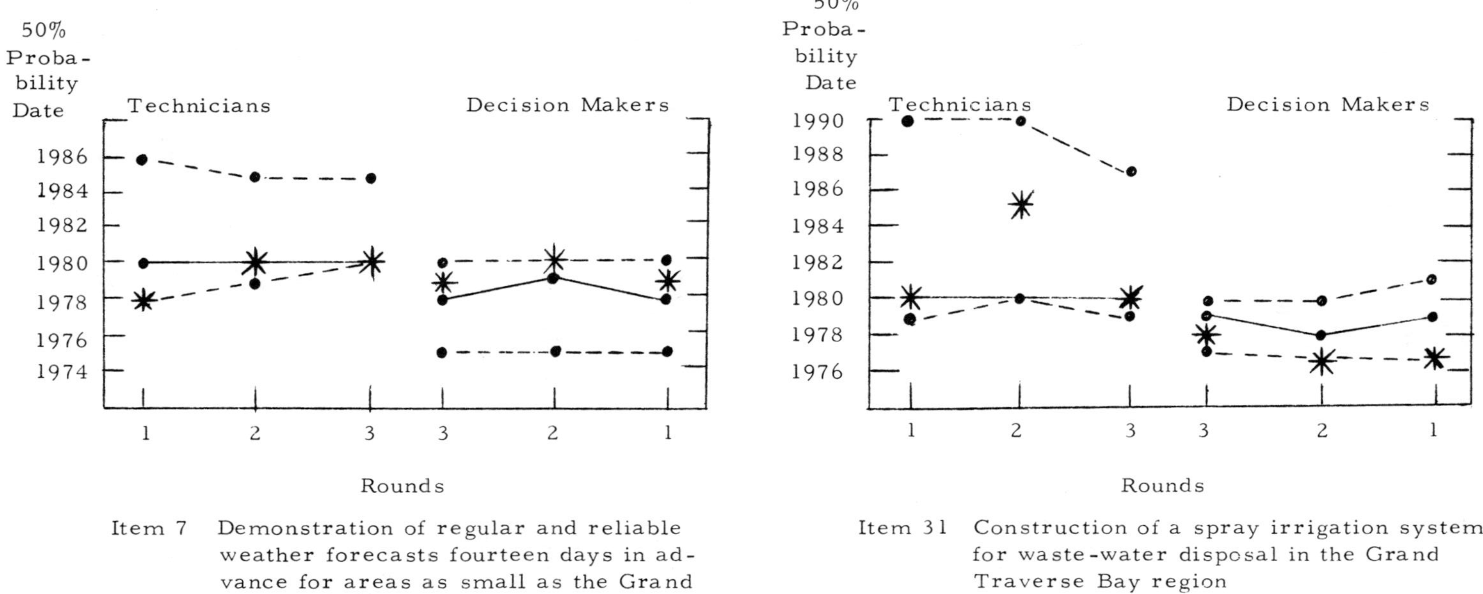

Item 7 Demonstration of regular and reliable weather forecasts fourteen days in advance for areas as small as the Grand Traverse Bay region

Item 31 Construction of a spray irrigation system for waste-water disposal in the Grand Traverse Bay region

✳ Median estimates of those who rated their competence relatively high.

●—● Median estimates of entire panel.

●- -● Boundaries for middle 50% of estimates.

Fig. 2. Statistical summaries—technicians and decisionmakers.

high; and Group C, respondents in Group B who were also relatively familiar with the Grand Traverse Bay watershed area. Although Group B differed considerably in size from Group C, the average estimates of the two were remarkably close. This finding might suggest that technical competence is a more important requisite for panel membership than familiarity with a specific region, an idea that could have important implications for interdisciplinary programs such as Sea Grant, in which research methodologies developed for a subregion are to be applied to a larger socioeconomic system.

In the broad panel exercises the evaluation matrix for round two was similar to the final matrix used in the technical panel. The evaluation matrix for the following round provided statistical summaries of the estimates of both the technical panel and the broader-based panels (Fig. 3).

A significant difference in the final estimates regarding the relative importance of the effluent from the Traverse City sewage system suggests that a series of estimates conditional on specific social, political, or technical developments could be used to determine the assumptions on which the evaluators based their estimates. The reason for the differences in estimates could also be sought through interviews and other means of communication.

Recommended Waste-Water Treatment and Disposal Systems

Many communities in the Great Lakes basin are confronted with decisions on waste-water treatment and disposal systems that will have important consequences for the future socioeconomic development of their region. This is a highly technical and complex issue, and decisionmakers must intuitively assess the judgments of experts in many specialized areas.

A systematic consideration of the available alternatives and the identification of areas of agreement and disagreement within and between the three general groups involved in these exercises may aid planners from this region as well as those from many other communities in the Great Lakes region facing similar problems and decisions.

Included in the technical panel's round-three information package was an evaluation matrix that listed six alternative waste-water treatment and disposal systems. Panel members were asked to suggest other alternatives and to evaluate each of them in terms of two different starting dates for the construction of the necessary facilities. Variances in the estimates were to be attributed to assumptions about the technology that would be available at the two starting dates. Panel members were instructed to give 100 points to their first choice for each time period and a portion of 100 points to the remaining alternatives according to their value relative to the first choice.

The round-four information package provided panel members with a summary of the estimates made in the third round. The evaluation matrix for

Ten Most Important Sources of Pollution	ABATEMENT FEASIBILITY 1971-80				RELATIVE IMPORTANCE					
	ECONOMIC Scale 1-5 1=Extreme subsidy required 2=Substantial subsidy 3=Moderate subsidy 4=Slight subsidy 5=Routine		TECHNICAL Scale 1-5 1=Extremely difficult 2=Very difficult 3=Moderately difficult 4=Slightly difficult 5=Routine		Panel members were asked to distribute 100 points among the most important sources of pollution during the two periods indicated.					
					1971-80			1981-90		
	Technical Panel	Broad Panel	Technical Panel	Broad Panel	Those on technical panel who rated familiarity 3 or more (13)†	Broad Panel (13)†	Final Estimates	Those on technical panel who rated familiarity 3 or more (12)†	Broad Panel (12)†	Final Estimates
Effluent, Traverse City sewage system	3(2-3)*	2(1-2)*	4(3-5)*	3(2-4)*	31	22		27	15	
Septic tanks in the region	2(2-3)*	5(2-5)*	3(3-4)*	4(3-4)*	15	13		10	14	
Storm water run-off, urban	2(2-4)*	3(1-3)*	3(2-5)*	3(2-4)*	12	9		9	9	
Food-processing wastes	5(3-5)*	4(4-5)*	3(3-4)*	3(3-4)*	10	13		7	8	
Industry wastes (less food processing and agriculture)	4(3-5)*	4(4-5)*	3(3-4)*	3(2-3)*	7	8		9	9	
Power production utilities	4(4-5)*	5(4-5)*	3(3-5)*	4(2-5)*	5	4		10	8	
Agriculture	4(2-5)*	4(3-4)*	4(3-4)*	3(2-4)*	9	13		9	12	
Oil spillage, bulk marine vessels	5(2-5)*	5(5-5)*	3(2-3.5)*	4(4-5)*	4	4		4	6	
Direct contact inputs, human use	4(1-5)*	5(4-5)*	3(2-3.5)*	3(2-5)*	4	6		8	9	
Ground water run-off carrying natural pollutants	2.5(1-4)*	3(1-5)*	3(1-4)*	2(1-5)*	3	8		8	10	

* Median--interquartile range in parentheses. Interquartile range contains the middle 50% of the estimates

† Number of respondents who contributed numerical estimates.

Fig. 3. Broad panel's third-round evaluation matrix—sources of pollution.

that round (Fig. 4) requested two evaluations for the six alternative waste-water treatment and disposal systems for two different starting dates. In the first evaluation the respondents were asked to consider all factors, in particular the technology available at the start of construction; in the second evaluation they were to consider only ten-year operating costs.

The broad panels used the same evaluation matrix as the technical panel in their final round of estimates, and they were also given a summary of the results of the technical panel's evaluation of all factors except for cost estimates. The broad panelists were advised that the technical panel probably emphasized technical factors in making their estimates. They were also told that the recommendations applied to a region similar to the Grand Traverse Bay area and could differ significantly if the technical panel had considered a specific situation.

A comparison of the average estimates of those on the technical panel who rated their competence relatively high with the average estimates of the respondents on the broad panels showed a very close agreement for both planning periods. This agreement was evident when panelists considered all factors and also when they considered ten-year operating costs, although the values assigned to each alternative relative to operating costs varied considerably from the values assigned when all factors were considered.

The judgments of the technical experts are believed to embody risk considerations applied to a general situation, whereas the judgments of the broad panels are thought to be more oriented to the benefits of alternative approaches for their specific region. Cost estimates include operating costs only; the consideration of investment costs and financing methods could be equally important to the decision maker.

The waste-water treatment and disposal system issue was undertaken primarily to educate the participants and to explore the problem of gathering a representative group of people and interesting them in the problem. The results could provide important material for gaming techniques and background information for deliberations using a variety of methods of information exchange and analysis.

Regional Opportunities, Problems, and Planning Strategies

A Delphi methodology was used to generate and evaluate suggestions regarding regional opportunities, problems, and planning strategies. The group summaries represent initial individual judgments in terms of a Delphi methodology in that these items were suggested on one round and evaluated on a subsequent round, but not subjected to iterative cycles of reassessments based on statistical feedback. However, many of the assessments were influenced by prior consideration of the following in other phases of the Delphi exercises:

Listed below are several waste-water treatment methods and several disposal techniques which respondents have suggested can be combined to make up six alternative waste-water treatment and disposal systems. You are requested to score each alternative system for its suitability in a region similar to the Grand Traverse Bay area. Two different dates are indicated for the start of construction. For each starting date make two evaluations. First, consider all factors, but particularly the technology available at the start of construction, and second, consider only 10-year operating costs after the system becomes operational. For each evaluation give 100 points to your first choice and a portion of 100 points to each alternative according to its value relative to your first choice.

WASTE-WATER TREATMENT		DISPOSAL	ALTERNATIVE	RECOMMENDED APPROACHES							
				Start July 1972				Start July 1976			
				Summaries		Next Estimates		Summaries		Next Estimates	
				Total Panel	Your Last Estimate	All Factors (1)	Cost Only (2)	Total Panel	Your Last Estimate	All Factors (1)	Cost Only (2)
1 Physico-chemical	A	Discharge to bay	1A	84				87			
2 Activated sludge or trickling filter biological treatment (sludge disposed of by incineration or land disposal)	A	Discharge to scenic river	2A	38				35			
	B	Discharge to bay	2B	82				84			
	C	Spray irrigation	2C	65				60			
3 Biological treatment followed by "tertiary" treatment	A	Discharge to scenic river	3A	55				49			
	B	Discharge to bay	3B	100				100			

Please mark the one phrase that comes closest to expressing your familiarity with waste-water treatment and disposal techniques and with the Grand Traverse Bay region.

Waste-Water Treatment/Disposal	G. T. B. Region	Descriptive Words for Numerical Rating Scale
(1) ☐	(1) ☐	(1) = Totally unfamiliar
(2) ☐	(2) ☐	(2) = Casually acquainted
(3) ☐	(3) ☐	(3) = Well acquainted
(4) ☐	(4) ☐	(4) = Generally familiar
(5) ☐	(5) ☐	(5) = Actively studying

Fig. 4. Evaluation matrix—recommended waste-water treatment and disposal systems.

(1) the trends of statistical measures which have traditionally been used to describe social and economic development; (2) the probabilities and importance associated with potential technical, social, economic, and political developments; (3) the relative importance of future sources of pollution; and (4) alternative waste-water treatment and disposal systems.

On the final round a list of suggestions regarding opportunities, problems, and planning strategies was presented to the broader-based panels. Panel members were asked to indicate whether an individual item should be singled out for special consideration by regional planners using a six-point scale and associated descriptive words as shown in Fig. 5. In interpreting the group means a value of 3.5 was viewed as the neutral point. The boundaries for the descriptive phrases are as shown.

strongly disagree	disagree	somewhat disagree	somewhat agree	agree	strongly agree
1	2	3	4	5	6

1.5 2.5 3.5 4.5 5.5

Fig. 5. Singling out individual item for special consideration.

Although the primary interest in these exercises was to identify areas of disagreement and the underlying reasons for them, a Delphi inquiry provides an accounting of the complete set of items that was considered by the respondents—an important concept when an interdisciplinary team of researchers is involved.

III. Evaluation of the Methodology

There is far from universal agreement on the merits of the Delphi techniques. Rand believes that Delphi marks the beginning of a whole new field of research, which it labels "opinion technology"[5]. However, a paper presented at the joint statistical meetings of the American Statistical Association in August 1971 described the Delphi techniques as the antithesis of scientific forecasting and of questionable practical credibility [6].

According to a recent *Wall Street Journal* article, the Delphi technique is gaining rather widespread use in technological forecasting and corporate planning, although the same article cautions:

> It's easy enough to see the shortcomings of the Delphi procedure; it's much harder to rectify them, as many are struggling to do. Remedial work must be done if the method is to be used in good conscience [7].

The Sea Grant Delphi exercises offered an exceptional opportunity for a critical evaluation of the Delphi techniques in an operational environment. The panelists—the main resource in evaluating the methodology—were interested in the improvement of techniques to integrate the judgments of a multidisciplinary research team and to convey its informed insights to society. Their evaluations were not biased by a strong emotional involvement in the success of the Delphi exercises, as has been true with many of the individual assessments of the method that have been published. From both a program budgeting standpoint and demands on researchers' time, the Delphi exercises competed with a wide variety of other methods for securing and disseminating information.

The primary instrument in evaluating the effectiveness of the method and its potential in other applications was a formal questionnaire. It was developed almost entirely by the respondents themselves using the Delphi technique of feeding back collated individual suggestions to generate additional suggestions. This procedure somewhat reduces the vulnerability of the questionnaire to the biases and shortcomings of the investigator. The six-point scale and associated descriptive words shown in Fig. 5 were used to quantify degrees of agreement and disagreement. To supplement the formal questionnaire, over thirty-five interviews with panelists were conducted.

Summaries were made for the three general groups participating in the Sea Grant Delphi exercises: technicians (Group I), behaviorists (Group II), and decisionmakers (Group III). For some issues the summaries for technical panelists under forty years of age and panelists with previous experience with the Delphi method were shown. Using the sample results, tests of significance were made to test the hypothesis that the distributions of the judgments of the Delphi method are homogeneous across the groups (the test procedure was based on the chi-square test statistic) and to test the null hypothesis that the means of the judgments of the population represented by the groups are identical (based on analysis of variance and the F-test). The results of these tests were used to support the discovery of basic differences in judgments made by different groups which had been formed on the basis of similar backgrounds and experiences. Their evaluations provided evidence that the method is effective not only in its designed role but in two other roles that are important and challenging from a management standpoint: encouraging greater involvement and facilitating communication between researchers and decisionmakers. The evaluations also showed that among the carefully selected samples of people the techniques were more highly regarded among groups which were formed on the basis of broad ranges in training and experience than among technicians—the group most administrators of the techniques have focused on.

The reliability of the method was demonstrated by the fact that the performance of the respondents, as measured by group statistical summaries, was similar for the three groups. Respondents from all three groups were generally willing to suggest future developments, sources of pollution, and research priorities; to utilize scaled descriptors to quantify subjective judgments; to accept a statistical aggregation of weights supplied by a group; and to reassess their judgments on the basis of feedback of information supplied by the group.

Some insight into the nature of the difference between judgments based on panelists' experience in the Sea Grant Delphi inquiries and the panelists' conception of an ideal application of the Delphi techniques can be gained by examining a cumulative summary of the evaluation of the effectiveness of the method in three specific roles shown in Table 2.

Table 2
Comparison of Evaluations Based on Experience in Sea Grant Exercises with Evaluations Based on the Potential of the Delphi Method

Group/Agreement Code=			Distribution of Judgments (proportion of total)								
			1	2	3	4	5	6	n	$\bar{x}$	σ
I	Technicians	Experience	.11	.14	.08	.44	.19	.03	36	3.6	1.340
		Potential	.05	.05	.16	.39	.30	.05	57	4.0	1.172
II	Behaviorists	Experience	.00	.00	.10	.65	.15	.10	20	4.3	.696
		Potential	.00	.00	.04	.15	.42	.32	24	5.1	.859
III	Decisionmakers	Experience	.00	.00	.00	.33	.37	.30	30	5.0	.809
		Potential	.00	.00	.00	.13	.50	.37	30	5.2	.679
I, II, & III	All Respondents	Experience	.05	.06	.06	.45	.24	.14	83	4.2	1.218
		Potential	.03	.03	.09	.27	.40	.19	111	4.6	1.101
	Respondents under 40	Experience	.07	.02	.07	.54	.22	.07	41	4.0	1.172
		Potential	.05	.00	.09	.21	.45	.20	56	4.6	1.218
	Previous Delphi Experience	Experience	.00	.00	.08	.49	.31	.13	39	4.5	.941
		Potential	.00	.00	.11	.17	.45	.28	47	4.9	.938

1: strongly disagree.
6: strongly agree.

The greatest difference in judgments based on experience and potential was found in the behaviorists' group, and the least among the decisionmakers. On the basis of the average judgments for all respondents the Sea Grant Delphi exercises corresponded rather closely to the panelists' conception of an ideal treatment, and there is a similar spread in the judgments of the subgroup which had previous experience with the method.

Table 3
Effectiveness of Delphi in Obtaining, Combining, and Displaying the Opinion of Informed People

Group/Agreement Code=			Distribution of Judgments (proportion of total)								
			1	2	3	4	5	6	n	$\bar{x}$	σ
I	Technicians	Experience	.11	.11	.06	.50	.22	.00	18	3.6	1.290
		Potential	.05	.05	.11	.42	.32	.05	19	4.1	1.177
II	Behaviorists	Experience	.00	.00	.00	.72	.29	.00	7	4.3	.488
		Potential	.00	.00	.00	.43	.43	.14	7	4.7	.756
III	Decisionmakers	Experience	.00	.00	.00	.22	.22	.56	9	5.3	.675
		Potential	.00	.00	.00	.10	.50	.40	10	5.3	.675
I, II, & III	All Respondents	Experience	.06	.06	.03	.47	.23	.15	34	4.2	1.274
		Potential	.03	.03	.06	.33	.39	.17	36	4.5	1.108
	Respondents under 40	Experience	.06	.00	.00	.69	.19	.06	16	4.1	1.025
		Potential	.05	.00	.00	.21	.58	.16	19	4.7	1.098
	Previous Delphi Experience	Experience	.00	.07	.00	.47	.27	.20	15	4.5	1.060
		Potential	.00	.00	.06	.19	.50	.25	16	4.9	.085

$P(F>4.6233)=.0165$
$P(\chi^2>8.3588)=.2130$
Combining Groups II and III: $P(\chi^2>3.959)=.0464$

1: strongly disagree.
6: strongly agree.

A summary of the evaluation of the effectiveness of the Delphi method in obtaining, combining, and displaying the opinions of informed people is shown in Table 3. It indicates that the technicians, on the average, agreed somewhat that the method was effective compared to alternative methods. However, there was considerable dispersion in their estimates; some respondents strongly disagree that the method was effective in this role. The behaviorists agreed that it was effective in the role, and the decisionmakers displayed strong agreement. Dispersion in the estimates of the two latter groups was much less than it was in the technicians' estimates. An analysis of variance on these data gives a value of 4.6233 for the F statistic. The probability of obtaining an F value larger than 4.6233 when the groups are identical is .0165. A chi-square analysis gives a value for chi-square of 8.3588. The probability of obtaining a larger value, when in fact the distributions come from a homogeneous population, is .2130—the descriptive level of significance. A combination of the judgments of the decisionmakers and the behaviorists resulted in $P(\chi^2 > 9.165) = .0025$.

Evaluation of the method's effectiveness in encouraging greater involvement in Sea Grant activities provided similar results. The average judgments of effectiveness were $\bar{X}_T = 4.0$, $\bar{X}_B = 4.7$, and $\bar{X}_{DM} = 5.1$, and the dispersion in the estimates of the technicians was much greater than for the other two groups. The descriptive levels of significance are .0026 based on the F distribution, and .0099 based on the chi-square distribution.

Average judgments regarding the method's effectiveness in communicating information to regional planners and decisionmakers were $\bar{X}_T = 4.1$, $\bar{X}_B = 5.4$, and $\bar{X}_{DM} = 5.3$. The judgments of the behaviorists were exceptionally high and may reflect this group's special concern for the psychological and sociological barriers associated with alternate methods. The descriptive level of significance is .0003 based on the F distribution, and .0108 based on the chi-square distribution.

For all three roles there appears to be little difference in the average estimates aggregated according to the respondent's age and the average estimates of all respondents. However, respondents with previous Delphi experience showed substantially higher average estimates of the method's effectiveness in obtaining, combining, and displaying subjective judgments and in encouraging involvement.

Average judgments regarding specific applications that are appropriate for a Delphi methodology are shown in Table 4. The support of specific applications is generally strongest among the decisionmakers and weakest among the technicians. Only the behaviorists and decisionmakers evaluated the method as an aid in decisionmaking, and both groups supported its use in this role.

In the evaluation of positive and negative aspects of the method that had been suggested by respondents, the panels agreed that there should be more emphasis on the idea that an expert should "not feel obligated to express an opinion on every issue." However, the Sea Grant Delphi inquiries stressed the concepts of a systems approach and multidisciplinary teams. Therefore it was desired that each respondent consider all items and attempt estimates on those with which he had some familiarity. A self-evaluation index was provided so that a panelist could assess his competence regarding each item. The competence index was a control factor in developing the statistical summaries that were part of the information feedback. This procedure allows an informed person to evaluate such things as relative importance and desirability—evaluations which he can make without being an expert in the area—and gives the administrator additional assurance that panelists considered items outside their specialized areas. In addition, there was some interest in comparing the estimates of experts and nonexperts on specific issues.

Table 4
Suitable Applications for a Delphi Methodology

Application	Technicians		Behaviorists		Decisionmakers	
	Mean	σ	Mean	σ	Mean	σ
Develop Criteria to Recruit Industries	4.3	.996	4.5	1.130	5.3	.675
Establish Program Priorities within a Committee	3.9	1.393	4.6	1.453	5.0	1.080
Long-range Planning by a University	3.8	1.393	4.1	1.453	4.8	1.080
Education of Practicing Politicians	4.1	.793	4.3	.707	4.6	1.424
Budgeting of an Interdisciplinary Program	3.9	1.353	3.7	.866	4.8	.789
Situations Dealing with the Future, Uncertainty, Conflicting Views	4.5	.943	4.3	1.165	4.5	.707
Decisionmaking Aid	–	–	5.1	.835	5.0	1.054

The suggestion that the method "can result in a manipulated and arbitrary consensus" received a neutral judgment from all three groups, perhaps indicating that the respondents felt this danger to be no greater than it would be in alternative techniques for securing group judgments. However, it is this administrator's opinion that the Delphi techniques could be a powerful tool for manipulating opinion and policy [8].

IV. Summary of Findings and Recommendations

The design and administration of a Delphi exercise in which the concept of a multidisciplinary team and a systems approach is desired can best be handled as a project within a professional research organization. The scope of the exercise is generally determined by the respondents, and, as interesting and unexpected issues are suggested, flexibility is needed in designing evaluation matrices and in determining the composition of the panels. Experts knowledgeable in specialized areas should be available on an adhoc basis to formulate questions and collate responses in order to minimize redundancy and ambiguity. The demand for their services in the course of a Delphi exercise is very uneven, as is the need for designing, editing, typing, and distribution services. There are significant start-up and learning costs associated with the Delphi techniques that can be justified only if the technique will become a routine management tool to be used on a continuing basis. This is particularly true if the benefits of computer processing are to be realized.

The following are some general observations that are consistent with the items suggested and evaluated by respondents in the Sea Grant exercises and with the information gained in personal interviews with panelists.

- Respondents will be more receptive if the techniques are tailored to specific groups on the basis of their training and experience.
- The administrator should consistently emphasize the distinction between the characteristics of a Delphi interrogation and those of conventional questionnaires and polls.
- Panelists—particularly those with technical backgrounds—must be *convinced* that judgments often have to be made about issues *before* all facets of the problems have been researched and analyzed to the extent they would like. (For these situations they must be persuaded that their subjective judgments may be a decisionmaker's most valuable source of information.)

There are several procedural recommendations that may be helpful to designers and administrators of future Delphi exercises.

- Interpersonal techniques, such as interviews and seminars, should be interspersed with the rounds of questionnaires and information feedback.
- The source of a suggested item should be identified (for example, panel member number and basic biographical information), taking care not to compromise the anonymity of specific inputs.
- Standardized scaled measures should be available to a respondent so that he can qualify his response to specific questions. Such measures are relative competence in a technical area, familiarity with a geographical region, or confidence in an estimate.
- If a multidisciplinary approach is desired, respondents should be encouraged to consider all items but to make estimates only on those scaled descriptive phrases with which he feels comfortable. For example, in these exercises it was helpful when respondents indicated their familiarity with a specialized area or the importance of an item even though they did not make probability estimates.
- The panelists should decide through their suggestions and evaluations what items should be considered. The criteria for retaining an item for further evaluation should be made clear at the outset of the exercise.

• Personal comments and arguments submitted by respondents should be part of the information feedback.

The Delphi inquiries have complemented the Michigan Sea Grant gaming-simulation activities by providing the following types of inputs:

• Data which can be helpful in describing social, economic, and political forces affecting the region's development during the next twenty years.
• Regional planning strategies, listed in order of preference for both university researchers and regional planners.
• Problems and issues which provide the link between the simulated regional area and a set of decision roles which are gamed [9].

Integration of a Delphi methodology with the Michigan Sea Grant gaming-simulation exercises will give them a more dynamic aspect and provide greater motivation for the participants. Some particularly interesting applications would be in cost-benefit analyses similar to those used in the Delphi inquiries to evaluate waste-water treatment and disposal systems, selection of research projects through an evaluation of effectiveness in terms of basic objectives and risk factors associated with various levels of funding, and the development of alternate scenarios for a region such as the Grand Traverse Bay watershed area.

According to Michael [10], Delphi inquiries and gaming-simulation exercises are techniques for introducing customers in a nonthreatening way to a more complex way of thinking and a better way of perceiving their needs in terms of the kind of knowledge we have. Knowledge utilization depends upon discovering the nature of the awareness of the problem among potential customers both as a set of variables and as a system of interrelationships; getting new knowledge absorbed by individuals and then by the organizations these individuals are in; and developing a capability and inclination to plan rather then employing an ad lib approach.

Concerned citizens were included as panelists in the Delphi inquiries not only for the purpose of informing them but also to accord the other panel members the benefits of the citizens' knowledge of the area, to take into account political and institutional considerations, and to communicate findings in such a way that the acceptance and implementation of policies and actions on which there appeared to be a reasonable consensus would be encouraged. The behavioral sciences provide support for this type of approach in effective communication [11,12].

The Michigan Delphi inquiries have provided some carefully formulated judgments of a multidisciplinary team of researchers and potential users of research data regarding: the importance and effects of technical, social, economic, and political developments; sources of pollution and recommended waste-water treatment and disposal systems; and regional opportunities, problems, and planning strategies. More important, a critical evaluation of the method has shown the potential of a Delphi inquiry for improving the dialogue between researchers and regional problem solvers. It has also provided empirical evidence to support further investigation of several innovations which may bring the methodology closer to Jantsch's idealized concept of a forecasting technique, wherein exploratory and normative components are joined in an iterative cycle in which informed citizens can work with researchers in planning for the future [13].

References

1. Murray Turoff, "Delphi and Its Potential Impact on Information Systems," Paper 81, Proceedings of the Fall Joint Computer Conference, Vol. 39. AFIPS Press, Washington, D. C., November 1971.
2. The techniques and procedures used in this series of interrogations and information feedback are similar to those described in "Some Potential Societal Developments—1970–2000" by Raoul De Brigard and Olaf Helmer, IFF Report R-7. Institute for the Future, Middletown, Conn., April 1970.
3. In one phase of the Delphi inquiries panelists were asked to assign numerical probabilities to commonly used words or phrases to indicate the likelihood of an event. The Delphi technique of reassessment based on the feedback of a group response was extremely effective in narrowing the dispersion of the estimates. Verbal phrases associated with numerical probabilities were believed to encourage respondents to think about a probability scale in similar terms and might be more appropriate than numerical probabilities in expressing the likelihood of socioeconomic developments.
4. The Michigan Delphi inquiries provided empirical evidence that the feedback and reassessment techniques which are inherent in the basic Delphi method reduced the number of inconsistencies in personal estimates of probability as the rounds progressed. It also indicated a tendency for a learning "curve" for respondents with respect to the technique itself.
5. "Forecasters Turn to Group Guesswork," *Business Week*, March 14, 1970.
6. Gordon A. Welty, "A Critique of the Delphi Technique" (summary of paper presented at the Joint Statistical Meetings of the American Statistical Association, Colorado State University, Fort Collins, Colorado, Aug. 23–26, 1971).
7. "Futuriasis: Epidemic of the '70s," *Wall Street Journal*, May 11, 1971.
8. For a discussion of the dangers associated with a Delphi devoted to policy issues see Turoff's article, above (Chapter III. B. 1.). For a discussion of deliberate distortion see "A Critique of Some Long-Range Forecasting Developments," by Gordon Welty (paper presented at 38th session of the International Statistical Institute, Washington, D. C., August 1971).
9. The gaming-simulation concept for the Sea Grant Program is presented in "Developing Alternative Management Policies," unpublished report, University of Michigan Sea Grant Office, 1971.
10. Donald Michael is presently program director, Center for Research on Utilization of Scientific Knowledge, Institute for Social Research, University of Michigan. These are a summary of his unpublished remarks to a site visit team of government officials and academicians in Ann Arbor, Michigan, March 4, 1972.
11. Douglas McGregor, "The Professional Manager" (New York: Harper & Row, 1967), p. 153: "...My conception of a two-way communication is that it is a process of mutual influence. If the communicator begins with the conviction that his position is right and must prevail the process is not transactional but coercive."
12. Peter F. Drucker, "Technology, Management and Society" (New York: McGraw-Hill Book Co., 1970), pp. 22–23: "... They must understand it because they have been through it, rather than accept it because it is being explained to them."
13. Erich Jantsch, "Technological Forecasting in Perspective," Organization for Economic Cooperation and Development, Paris, 1967.

Other References

Examples of the key instruments used in the Michigan Sea Grant Delphi inquiries can be found in "Substantive Results in the University of Michigan's Sea Grant Delphi Inquiry," by John D. Ludlow, Sea Grant Technical Report No. 23, University of Michigan, Ann Arbor, 1972, and in "Evaluation of Methodology in the University of Michigan's Sea Grant Delphi Inquiry," by John D. Ludlow, Sea Grant Technical Report No. 22, University of Michigan, Ann Arbor, 1972. Complete sets of the information packages can be found in "A Systems Approach to the Utilization of Experts in Technological and Environmental Forecasting," by John D. Ludlow, Ph.D. dissertation for the University of Michigan, 1971. Available through University Microfilms, Inc., Ann Arbor.

III. B. 3. The National Drug-Abuse Policy Delphi: Progress Report and Findings to Date

IRENE ANNE JILLSON

Introduction

Rationale

Although the abuse of drugs has been recognized in this country for nearly one hundred years, its popularity as a national problem has resurfaced relatively recently. In 1968, former President Nixon declared drug abuse "public enemy number 1"; from 1969 until 1974, some $2.4 billion in federal funds were obligated to combat the problem, and an industry was created. This expenditure represents funding of programs that were almost exclusively aimed at heroin abuse, rather than the broad spectrum of drug abuse. Since 1968, drug abuse has been the subject of intense public and private debate. Controversy over the government's appropriate response has ranged from debate regarding drug laws (to what extent different drugs should be controlled, and in what manner), to the basic question of whether or not alcohol programs are to be included with "other" drug programs on the federal level.

During the past ten years, the increased concern and expansion of drug-abuse prevention programs has resulted in a swelling of the ranks of professionals who have developed expertise in this field; however, the use of these experts in policy advice and formulation has been sporadic and unsystematic. At the same time, numerous research and evaluation studies of drug-abuse prevention programs themselves have been carried out. The degree to which resultant data from these studies can be, have been, or should be, used in decision-making at the nation level has never been resolved.

In the fall of 1973, it was clear that the problem of drug abuse had diminished in priority, and that substantial reductions in federal funding were imminent. This may be attributed primarily to the apparent abatement of the heroin epidemic, which had served as the stimulus for increased concern and program funding in the late 1960s. Although the crisis associated with the heroin epidemic may have passed, it is by no means clear that the broader problem of drug abuse has been resolved: polydrug use and alcoholism appear to be increasing; and the abuse of prescription drugs, once the "hidden drug problem," is surfacing in many communities. Such a time of decreased public

The Delphi Method: Techniques and Applications, Harold A. Linstone and Murray Turoff (eds.)
ISBN 0-201-04294-0; 0-201-04293-2

interest and funding for a yet-existing problem calls for careful consideration of the basic issues, and deliberation of the strategies to be followed, in order to maximize effective use of resources available.

Viewed from this perspective, the procedures utilized by Policy Delphi studies seemed most appropriate to explore national drug-abuse policy options for the next five years. The volatility of many of the issues, most of which involve fundamental value and sometimes moral choices; the diverse backgrounds of those who make or influence policy; the apparent differences in the positions held by various experts and groups; and the apparent inability of past policy studies to aid decision-makers led to the conceptualization and implementation of a national drug-abuse Policy Delphi study. The unusually high response rate, the degree of participation achieved to date, and the interest on the part of federal and state decision-makers has borne out this initial hypothesis.

History

The study described in this chapter was originally conceived in 1973, and designed during the fall of that year. Implementation began in December 1973; the first questionnaire was disseminated in March 1974. The first two questionnaires were developed under a contract funded by the National Institute of Drug Abuse.[1] Analysis of the data generated by the second questionnaire, and the further development of the use of the Delphi procedures in the exploration of national drug-abuse policy, as originally conceived, was sponsored by the National Coordinating Council on Drug Education.

The study analyzing the first two rounds was completed and published in December 1974.

Objectives of This Study

There are three primary objectives for this study:

- To develop a range of possible national drug-abuse policy options
- to explore applications of the Policy Delphi methodology to this and other areas of social policy
- to explore the possibilities of applying the technique on an as-needed basis and on an ongoing basis

Since the level of drug abuse in the United States is presumed to be both endemic and epidemic, and since strategies to respond to changes in use patterns need to be both immediate and long range, this study is concerned with ascertaining the feasibility of utilizing the Delphi technique to meet these needs of policy formulation and planning.

- *On an as-needed basis.* This would involve the use of a panel of experts who would respond to queries sent as the need arises. For example, if a decision-maker were to be

[1]NIDA Contract Number B2C-5352/HOIMA-2-5352.

informed that there was a dramatic decline in the number of patients entering treatment programs, a Delphi would be developed to determine the opinion of selected experts with regard to this particular trend.

- *On an ongoing basis.* If one agrees that there is an endemic level of drug abuse, then it would seem appropriate to develop an ongoing Delphi study of indefinite duration. This Delphi would be implemented such that questionnaires would be distributed at regular intervals. Since the panel would be of indefinite duration, membership might be fluid. Current trends in the field would be incorporated into the questionnaires, so that there would be a continuous flow of information for the use of the policymaker, and those operating programs of various disciplines in the field.

Study Design

A number of advisers—experts in the field of drug abuse, policy planning and analysis, and the Delphi technique in particular—assisted in developing the study design.To date, they have continued to provide assistance in all aspects of the study. Included are Dr. Norman Dalkey, Engineering Systems Department, University of California at Los Angeles, who originally was instrumental in developing the technique in the late 1950's and who continues to explore its use as part of his decision-theory research; and Dr. Murray Turoff, who has developed the Policy Delphi and is co-editor of this book. Dr. Peter Goldschmidt has assisted in planning and management of this study, and M. Alexander Stiffman has assisted in developing the analytic approach and designing the computer analysis. Dr. Raymond Knowles, Charlton Price, and Anthony Siciargo have assisted in pretesting the questionnaires.

The final study design was based on the premise that policy may be formulated from a number of different perspectives. In designing this Policy Delphi study we are exploring the formulation of national drug-abuse policy from three perspectives:

- the "top-down" approach—establishing national drug-abuse policy objectives to be achieved over the next five years, based on the respondents' value systems
- a "bottom-up" approach—identifying factors which control the transition between general population and various degrees of drug use; deciding which of these are important and can be affected by national drug-abuse policy; and determining appropriate policies to affect them
- an issue-oriented approach—deriving policies from issues which are the subject of current controversy or debate

The underlying thesis in this design is that different decision-makers may formulate policy predominately using one or another perspective, and that this results in distinct types of policy options and considerations which may appear attractive from one perspective, but turn out to be counterproductive from another. For example, setting an objective using the "top-down" approach may result in its achievement becoming a political issue, while an objective that is formulated as the result of a political issue may never be achieved because it is technically impossible even though its achievement is valued.

This top-down, bottom-up approach speaks to the general concept of the Policy Delphi, in that alternative policy options are drawn from a number of different vantage points (it is not simply a statement of objectives, for example) Additionally, we are not interested in consensus per se, but rather, in exploring alternatives, and *pro* and *contra* arguments for the alternatives.

For these reasons the outputs from the three approaches used in this study will be brought together so that the panel can identify more appropriate alternatives. We anticipate that this will prove more fruitful in terms of exploring national policy options than either approach alone. A description of the approach, round by round, is given in Table 1.

Respondent Population

A list of more than one hundred highly selected potential respondents was developed from among the most notable "experts" in the field and from those who directly impacted on the field (e.g., police chiefs). Invitations to participate in the study were sent to forty-five persons; the remaining names were held in reserve, as a second series of invitations was anticipated in order to secure twenty-five participants. In fact, thirty-eight individuals (84%) responded positively. Since that time, three respondents have withdrawn from the study owing to a change in career orientation from drug abuse. Needless to say, there was no necessity for a second series of invitations.

As the study progressed past the first two rounds, several additional respondents were added. These additional respondents were selected to represent areas of interest which had developed in the study, but for which respondents had not been initially selected. Experts in alcoholism were added to the panel, for example, because a significant proportion of existing panelists expressed the view that a national drug-abuse policy could not be considered separately from alcoholism. The addition of such experts will allow their views to be added to those of the present panelists, and so provide an appropriate additional perspective. The present panel consists of thirty-nine persons.

Our respondent group represents some of the most respected authorities in the field. They include the Deputy Director of the Alcohol, Drug Abuse and Mental Health Administration, a former director of the Bureau of Narcotics and Dangerous Drugs, officials from the Office of the Secretary, and Office of the Assistant Secretary for Health of H.E.W., notable researchers, treatment administrators, law-enforcement officials, and policymakers in the field of drug abuse. It should be emphasized that participation is voluntary, and that no honorarium is paid to respondents.

The Questionnaires

First Questionnaire

A number of study approaches were explored during the developmental phases preparatory to Round One. The final draft of the resultant questionnaire was pretested, and the entire forecasting section deleted when it was determined

Table 1

THE STUDY DESIGN

Perspective	First Questionnaire	Second Questionnaire	Third Questionnaire	Fourth Questionnaire	Fifth Questionnaire	Final Summary
Objectives ("top down")	List objectives List key indicators	Rate feasibility, and desirability of objectives Expand key indicators, perform initial rating	Re-rate feasibility and desirability of selected objectives	Identify any contradiction in objectives and policies formulated by the three approaches	Synthesize a consistent and realistic set of national drug abuse policy options	Write a brief summary of the national drug abuse policy options identified by the study, including the normative forecasts for the key indicators Identify future policy research needs
Transition Factors ("bottom up")	List factors, indicate direction		Final rating of factors Develop policy areas to affect important transition factors which can be influenced by national policy			
Policy Issue Statements ("political)	Rate selected policy issues; develop for and against arguments Add other important policy issues	Final rating of selected policy issues; rate for and against arguments Rate and give arguments for and against additional policy issue statements				

that the time to complete the questionnaire was decreased considerably by deleting this section.

The first-round questionnaire, disseminated in mid-March, consisted of three primary sections:

1. *Development of Objectives*. The respondents were asked to develop up to five national policy objectives in the field of drug abuse, given a five-year time frame; and to list up to three key indicators for each objective.
2. *Transition Matrix*. Respondents were given a simplified transition model which depicted the flow from one state of drug involvement to another and were asked to list the factors which promoted or inhibited the movement of people from one state to another. (A detailed description of this matrix is provided in a subsequent section.)
3. *Policy Issue Statements*. Twelve issues, in the form of should/should not statements, were posed. These had been culled from a potential list of twenty-five issues felt to be current and controversial.
4. *Additional Items*. These included a self-rated expertise question; two questions relating to expectations and objectives for participating in the study; a request for responses to a list of definitions; and a request for feedback on questionnaire design and/or content.

Twenty-four of the study's thirty-five respondents (69%) actually completed the Round One questionnaire. The breakdown of returns by category of respondent is shown in Table 2. It should be pointed out that mailing posed serious difficulties both in disseminating and in returning the questionnaires. In several cases it took two full weeks for the questionnaires to arrive by air mail to their destinations; several never arrived, and duplicate packages had to be sent out. For this reason, the deadline for completion of the first-round questionnaires was extended by two weeks.

The absolute range of time for completion of the first questionnaire was fifteen minutes to ten hours. The interquartile range was one to three hours; the median time for completion was $2\frac{1}{4}$ hours, which was approximately the median we had anticipated. The respondents included in our predesignated *policymakers* subpanel category spent a median of $3\frac{1}{2}$ hours; the median for all the remaining participants was $2\frac{1}{2}$ hours.

In addition to the substantive-issue questions, respondents were asked to self-rate their expertise in ten drug-related areas (see Table 3). These data were then used to cross-check the categories (subpanels) into which respondents had been placed, and to further analyze responses to particular items. There is one particular point of interest to be noted. As critical as evaluation is held to be in the formulation of national policy, not one of the policymakers rated himself herself as expert in this area.

Respondents were also asked to indicate their expectations for the study, and to list personal objectives with regard to participation. Eighty-two percent of those responding reported that they expected the study to be of direct benefit.

Table 2

PARTICIPATION AND RETURNS BY CATEGORIES OF RESPONDENTS (SUBPANEL)*

Subpanel	Invited	Agreed to Participate	Formally Withdrew	Delphi Panel	Completed Round One	Completed Round Two
Policy Makers	9	9	0	9	5	5
Researchers	12	9	0	9	9	9
Treatment Program Administrators	11	11	2	9	4	6
Criminal Justice Administrators	9	6	1	5	3	3
Other	4	3	0	3	3	2
Total	45	38	3	35	24	25

* As of 1 August, 1974. Does not include panelists added after that date.

Table 3

A SUMMARY OF SELF-RATED EXPERTISE BY AREA AND SUBPANEL*

Area of Expertise	Subpanel: Percent Who Rated Themselves "Expert"(1) Policy Makers	Researchers	C.J.S.(2) Adminis-trators	T.P.(2) Adminis-trators	Other	All Respondents
National Drug Abuse Policy Planning	<u>100</u>	33	33	0	100	50
Prevention	20	22	33	50	0	26
Intervention	0	33	33	75	0	30
Treatment	40	78	0	<u>100</u>	0	57
Law Enforcement	20	11	<u>100</u>	0	33	25
Research	20	<u>100</u>	0	50	0	50
Evaluation	0	78	0	50	33	44
Training	0	33	33	50	0	26
Education	20	56	33	50	0	39
Pharmacology	0	11	0	0	0	4
Number of Respondents	5	9	3	4	3	24

* As of 1 August, 1974. Does not include panelists added after that date.
(1) Percentage of those who responded.
(2) C.J.S. = Criminal Justice System; T.P. = Treatment Program

There were twenty-eight different narrative responses to this item. Twenty-two saw positive utilization of the results for policy planning, idea exchange, consensus development, etc.; three were uncertain of usefulness, and two were skeptical of its utility. There were fifty-four statements of personal-study objectives; these were distilled to twelve clusters, and are shown as Table 4.

Respondents were asked to comment on a list of standard definitions which had been prepared for use in the study. Twelve respondents commented on these definitions. As anticipated, there was little agreement with these definitions on the part of those who commented. Apart from giving each respondent a common base line, one of the reasons for including the list of definitions was to determine what interest would be aroused. In most of the social-service areas, and particularly in the field of drug abuse, there is much dissension, even among policymakers, regarding critical definitions (e.g., drug abuse, modality types). Clearly it is not sufficient to gloss over this issue continuously in the hopes that at some time in the future standard definitions will somehow be devised and agreed upon by a reasonable majority of those in the drug-abuse field. A Delphi study specifically related to the formulation of standard definitions is presently being planned.

Second Questionnaire

Preparation of the second questionnaire began shortly after the first completed Round One questionnaires had been received. It was decided that, because of the complexity and time required for completion of the transition matrix, this section would be deleted from the second questionnaire and included as part of a subsequent round. The second questionnaire and a summary of the first round results were disseminated in mid-May.

The second questionnaire included two sections:

1. *National Drug-Abuse Policy Objectives*. Respondents were asked to rate fifty-five objectives on the basis of feasibility and desirability, and to rate the importance of the key indicators associated with them.
2. *Policy Issue Statements*. Respondents were asked to re-rate the original twelve issue statements; rate the narrative comments associated with them; and rate the fifteen new issues suggested by respondents.

There is usually a decrease in response rates for the second round of a Delphi study, particularly those involving voluntary participation. For this reason, and because of the length of the Round Two questionnaire, we anticipated a response rate of approximately 45 percent to 50 percent. In fact, twenty-five respondents, 71%) completed the questionnaire; a most unusual and gratifying response rate, and one higher than that for Round One. (See Table 2.)

The absolute range of time to complete the second questionnaire was $1\frac{1}{2}$ to $8\frac{1}{2}$ hours; the interquartile range was $2\frac{1}{2}$ to five hours; the median time to complete was three hours. For the *policymakers'* subpanel, the median was $5\frac{1}{4}$ hours; for all others the median was three hours.

Table 4

RESPONDENTS' OBJECTIVES FOR PARTICIPATING IN THE STUDY

To test the Delphi technique; determine the value of the process in sharpening views in social policy fields or in bringing forth practical ideas or new insights.

To explore the limits of possible public policy formulation; to learn more about policy formulation.

To be involved in the formulation and development of drug abuse policy; influence policy through identification of critical policy distructions; to share in a process that may lead to wiser process than we now have in the drug abuse area; help develop policy view in a nonpolitical forum.

To see if any consensus is possible in drug abuse policy; see if the group can solve the problem or give direction.

To distill and synthesize the collective thinking of some of the best minds in drug abuse; obtain the benefit of the ideas of the others as a stimulus to my own thinking; to learn what the group knows about, and applies in responding to drug abuse.

To assess the extent to which my views in drug abuse coincide with or differ from those of my colleagues; check my own opinions against those of the group.

To clarify my own thinking pertaining to drug abuse policy; get a broader perspective.

To test ideas in a multidisciplinary quasianonymous environment.

To develop priorities for the organization or agency in which I work.

To make a useful contribution to knowledge; help solve a problem; provide ideas that may arise from my special knowledge and experience.

To gain personal enjoyment by responsive discussions.

To satisfy my sense of duty to participate in such studies as an active researcher and/or policymaker.

Third Questionnaire[2]

For this questionnaire, panelists were asked to respond to two series of questions:

1. *National Policy Objectives.* There were twenty-five objectives included in this section; these had to be re-rated because there was a broad distribution of voting responses, differences in voting between policy experts and policy nonexperts, or because original objectives had been combined or divided.
2. *Transition Matrix.* In this section, the transition factors suggested by respondents in the first questionnaire were further developed.

The policy issues were not included for consideration in this round in order to maintain expected time for completion to a reasonable level. The data from this third questionnaire, and information gathered during previous studies of this type, will be used as a basis for developing policy and program options in future rounds. The respondents were again sent two copies of the questionnaire, and an introduction and summary volume which included results of the previous questionnaire.

Results

Objectives

First Questionaire. In the first questionaire respondents were asked to list up to five national drug-abuse policy objectives (for a five-year time frame), and up to three key indicators for each of these. They listed a total of seventy-eight such objectives; from this list, fifty-five different objectives could be discerned. The 187 key indicators listed by the respondents were culled to 153.

Even after we distilled the original seventy eight objectives to fifty five, there still were similarities between them. Rather than risk a possible misinterpretation of the objectives as stated by the respondents, we decided not to distill the objectives any further. It is difficult to be precise in a Policy Delphi, particularly one in a field as complex as drug abuse; we have therefore emphasized development and creativity.

Second Questionaire. The respondents were asked to rate the fifty-five objectives derived from their responses to Round One Objectives Section. The two rating scales used were *feasibility* and *desirability*. They were also asked to rate the 153 indicators relating to each objective; in this case, the scale used was *importance*. These scales are shown in Table 5. The objectives were presented in arbitrarily defined categories (prevention, treatment, law enforcement, organization, e.g.) simply as a means of ease in completion; these categories had no other significance.

---

[2]First Round—National Drug-Abuse Policy Delphi.

Table 5

FEASIBILITY/PRACTICALITY SCALE

Scale Reference	Definitions
1. Definitely Feasible	Can be implemented No research and development work required (necessary technology is presently available) Definitely within available resources No major political roadblocks Will be acceptable to general public
2. Probably Feasible	Some indication this can be implemented Some research and development work required (existing technology needs to be expanded and/or adopted) Available resources would have to be supplemented Some political roadblocks Some indication this may be acceptable to the general public
3. May or May Not be Feasible	Contradictory evidence this can be implemented Indeterminable research and development effort needed (existing technology may be inadequate) Increase in available resources would be needed Political roadblocks Some indication this may not be acceptable to the general public
4. Probably Infeasible	Some indication this cannot be implemented Major research and development effort needed (existing technology is inadequate) Large scale increase in available resources would be needed Major political roadblocks Not acceptable to a large proportion of the general public
5. Definitely Infeasible	Cannot be implemented (unworkable) Basic research needed (no relevant technology exists, basic scientific knowledge lacking) Unprecedented allocation of resources would be needed Politically unacceptable Completely unacceptable to the general public

Table 5 (Continued)

DESIRABILITY/BENEFITS SCALE

Scale Reference	Definitions
1. Highly Desirable	Will have a positive effect and little or no negative effect Social benefits will far outweigh social costs Justifiable on its own merit Valued in and of itself
2. Desirable	Will have a positive effect with minimum negative effects Social benefits greater than social costs Justifiable in conjunction with other items Little value in and of itself
3. Neither Desirable nor Undesirable	Will have equal positive and negative effects Social benefits equals social costs May be justified in conjunction with other desirable or highly desirable items No value in and of itself
4. Undesirable	Will have a negative effect with little or no positive effect Social costs greater than social benefits May only be justified in conjunction with a highly desirable item Harmful in and of itself
5. Highly Undesirable	Will have major negative effect Social costs far outweigh any social benefit Not justifiable Extremely harmful in and of itself

Table 5 (Continued)

IMPORTANCE SCALE

Scale Reference	Definitions
1. Very Important	A most relevant point First order priority Has direct bearing on major issues Must be resolved, dealt with or treated
2. Important	Is relevant to the issue Second order priority Significant impact but not until other items are treated Does not have to be fully resolved
3. Moderately Important	May be relevant to the issue Third order priority May have impact May be a determining factor to major issue
4. Unimportant	Insignificantly relevant Low priority Has little impact Not a determining factor to major issue
5. Most Unimportant	No priority No relevance No measurable effect Should be dropped as an item to consider

The feasibility and desirability ratings of the objectives were analyzed so as to develop a summary list of objectives, and determine to what extent there was consensus or polarization.

Respondents' ratings on the feasibility and desirability of objectives were translated into group scores by summing the scale values and dividing the total by the number of ratings. This procedure, it should be noted, treats nominal scales as interval data. The feasibility and desirability scores were used to categorize objectives as follows:

Group Score	Feasibility	Desirability
Less than 1.80	Highly feasible	Highly desirable
Equal to or greater than 1.80 but less than 2.60	Feasible	Desirable
Equal to or greater than 2.60 but less than 3.40	May or may not be feasible	Neither desirable nor undesirable
Greater than 3.40 but less than or equal to 4.20	Probably infeasible	Undesirable
Greater than 4.20	Definitely infeasible	Highly undesirable

Objectives were first grouped on the basis of their feasibility and then sorted on the basis of their desirability. This produced the rating of objectives depicted in Table 6.

The twenty-five objectives which scored Highly Feasible and Highly Desirable, or Feasible and Highly Desirable, are shown as Tables 7 and 8, respectively.

No objectives were rated "Definitely Infeasible" and none was rated as either "Undesirable" or "Highly Undesirable." These results indicate that the majority (55%) of the objectives listed were rated at least "Feasible" and "Desirable." The following items or sets of items deserve special attention because of the distinctions in rating patterns. The feasibility of twenty-one objectives was indeterminable, either because there was polarization (with some respondents rating an objective feasible while others rated it infeasible); a broad distribution

Table 6

NATIONAL DRUG ABUSE POLICY OBJECTIVES: Summary of Feasibility and Desirability Ratings

FEASIBILITY \ DESIRABILITY	Highly Desirable	Desirable	Indeterminate Desirability	Undesirable	Highly Undesirable
Highly Feasible	5	0	0	0	0
Feasible	20	5	1	0	0
Indeterminate Feasibility	9	6	6	0	0
Infeasible	2	1	0	0	0
Highly Infeasible	0	0	0	0	0

Table 7

OBJECTIVES VOTED "HIGHLY FEASIBLE" AND "HIGHLY DESIRABLE"
(in decreasing order of desirability)

Objective	Feasibility Score	Desirability Score
To conduct research on treatment modalities and effectiveness.	1.50	1.14
To train in-line treatment personnel to enhance their skill in helping the drug dependent person.	1.77	1.27
To have available more adequate epidemiological estimates of a) prevalence, by drug type; b) incidence, by drug type and individual characteristics; c) discontinuance, by drug type.	1.73	1.41
To develop, validate and disseminate information on efficacious programs in vocational rehabilitation, and early intervention.	1.79	1.46
To increase research into brain chemistry and psycho-social correlates and clinical-social treatment.	1.70	1.70

Table 8

OBJECTIVES VOTED "FEASIBLE" AND "HIGHLY DESIRABLE"
(in decreasing order of desirability)

Objectives	Feasibility Score	Desirability Score
To increase the efficiency and quality of treatment for drugs and alcohol overuse, based on increased evaluative clinical research.	2.04	1.13
To measure effectiveness of different kinds of treatment/treatment programs for different types of users.	2.13	1.13
To reduce overdose deaths and damage.	2.26	1.17
To establish a flexible by comprehensive response system.	2.24	1.19
To minimize the adverse consequences of drug abuse.	2.38	1.23
To measure the extent of the problem by scientific sampling with rigorous respect for reliability of data.	2.04	1.25
To provide enough effective treatment centers for all forms of drug abuse so that no abusers need continue because he has been denied treatment according to his needs.	1.91	1.27
To evaluate training programs.	1.20	1.27
To determine priorities in terms of short- and long-term strategies for achieving our goals.	2.41	1.32
To improve basic progress in social science research pertaining to drug abuse.	2.00	1.35
To assign funding on the basis of priorities.	2.41	1.38
To create an explicit strategy which defines the problem and which includes strategic objectives in order of priority.	2.32	1.41
To delineate the gaps remaining to be filled in -- in knowledge and understanding, through further experience and research.	2.18	1.46
To reallocate prevention and treatment endeavors independently of particular drug fashions, i.e., emphasis on heroin and not on alcohol so that efforts are aimed at individuals suffering from adverse habits of use of any psychoactive drugs or combination of these.	2.29	1.50*
To make available more trained health professionals.	1.86	1.52
To develop adequate alternative models in prevention and phased intervention.	2.38	1.67
To establish means of transmitting scientific findings to the public.	2.22	1.70
To increase law enforcement pressure on illicit drugs sold in <u>large</u> quantities.	2.00	1.71
To incorporate drug treatment into standard health delivery systems.	2.14	1.77
To increase public awareness of the difference between drug use (responsible) and abuse (irresponsible).	2.46	1.79

* indicates that over one-fifth of respondents did not respond to this item.

Table 9

OBJECTIVES WHICH EXHIBITED POLARIZATION OR A BROAD DISTRIBUTION OF RESPONSES (Feasibility and/or Desirability)

Objective	Scale Value; Percent voting				
	1	2	3	4	5
Polarized or broad distribution on Feasibility					
To achieve improvements in public education which will enhance the quality of the individual's life, strengthen his preference for leisure-time activity not involving the use of drugs, and enhance his capacity for meeting personal problems without resorting to the non-medical use of dangerous drugs.	0	38	25	25	12
To develop a national public discourse and standards on use and abuse of drugs.	14	27	27	23	9
To reduce non-prescribed use of psychoactive drugs.	8	34	21	29	8
To reduce the number of persons engaged in the non-medical or recreational use of dangerous drugs.	0	33	22	33	12
To utilize outreach and coercive techniques to engage those not entering treatment on their own.					
To consider social action approaches, as alternatives to treatment, including concern with housing, employment, education, and counselling, in addition to other novel approaches.	23	23	23	26	5
To reduce/eliminate criminal penalties for personal use and possession of drugs currently defined as "illegal".	4	13	35	35	13
To establish the principle that whether or not an individual uses a mind altering substance is a matter of personal choice with minimum governmental interference.	9	26	13	30	22
To organize an effective, coordinated approach to research, prevention, treatment and rehabilitation among all appropriate Federal, State, local and private agencies in the health and social services areas, thus developing the mechanism for timely government responses.	23	18	27	27	5
To recognize that all substance abuse planning, programming, etc., should be administered together.	5	24	19	47	5
To minimize damage caused by government reaction to drug use.	0*	39*	22*	33*	6*

* Indicates that over one-fifth of respondents did not respond to this item.

Table 9 (Continued)

OBJECTIVES WHICH EXHIBITED POLARIZATION OR A BROAD DISTRIBUTION OF RESPONSES (Feasibility and/or Desirability)

Objective	Scale Value; Percent voting				
	1	2	3	4	5
Polarized or broad distribution on Desirability					
To utilize outreach and coercive techniques to engage those not entering treatment on their own.	32	18	9	23	18
To provide voluntary drug treatment available to anyone who wishes it, without government regulations.	35	5	23	14	23
To reduce legal-political-governmental control of treatment, including urines; dosage; age; identification requirements.	18	27	14	27	14
To reduce/eliminate criminal penalties for personal use and possession of drugs currently defined as "illegal".	35	13	17	13	22
To establish the principle that whether or not an individual uses a mind altering substance is a matter of personal choice with minimum governmental interference.	30	9	26	13	22
To increase med-social research -- hallucinogen for eco-systems and positive social values through use (Ritualistic, etc.)	36*	0*	36*	21*	7*
To develop a group with a primary interest in the problems people have with drugs.	32	14	27	9	18

* Indicates that over one-fifth of respondents did not respond to this item.

(with respondents voting approximately equally for four or more of the five scale values); or truly indeterminable (with the modal response being "May or may not be feasible"). In only eight of the twenty-one objectives which scored "May or may not be feasible" was this the modal response; in the case of eleven objectives the reason was that the voting was either polarized or broadly distributed, as Table 9 shows. All seven of the objectives which respondents scored "Neither desirable nor undesirable" were either polarized or of a broad distribution.

By reviewing the frequency distribution and the scale scores we were able to identify objectives in which there was a significant voting difference between those who rated themselves experts in national drug-abuse policy and those who did not. Table 10 lists these items. Some of the differences were in items that could be of major importance in the formulation of a national drug-abuse policy; most of the differences had to do with the feasibility of attaining objectives.

Table 10

OBJECTIVES WHICH EXHIBITED VOTING DIFFERENCES BETWEEN SELF-RATED POLICY EXPERTS AND NONEXPERTS (Feasibility and Desirability)

Objective	Policy expertise	Scale value; Percent voting 1	2	3	4	5	Feasibility Score	Desirability Score
<u>Differences on feasibility</u>								
To develop adequate alternative models in prevention and phased intervention.	Expert	9	27	55	9	0	2.64	2.00
	Non-Expert	15	69	8	0	8	2.15	1.39
To reduce the supply of illicit drugs available for abuse.	Expert	9	64	18	0	9	2.36	1.36
	Non-Expert	8	25	25	17	25	3.25	1.85
To reduce non-prescribed use of psychoactive drugs.	Expert	0	27	46	27	0	3.00	1.73
	Non-Expert	15	39	0	31	15	2.96	1.69
To reduce prescribed use of psychoactive drugs/diminish misuse by physicians.	Expert	27	9	55	9	0	2.46	2.18
	Non-Expert	32	23	15	15	15	2.62	1.69
To incorporate drug treatment into standard health delivery systems.	Expert	36	19	36	9	0	2.18	1.73
	Non-Expert	27	55	9	0	9	2.09	1.82

To establish an effective social rehabilitation system for drug abusers who have become desocialized.	Expert	0	30	40	20	10	3.10	1.20
	Non-Expert	46	9	36	0	9	2.18	1.09
To consider social action approaches, as alternatives to treatment, including concern with housing, employment, education, and counselling, in addition to other novel approaches.	Expert	0	20	0	80	0	3.60	1.80
	Non-Expert	25	25	34	8	8	2.50	1.75
To develop a group with a primary interest in the problems people have with drugs.	Expert	30	10	40	0	20	2.70	3.50
	Non-Expert	42	25	8	17	8	2.25	2.00
Differences on desirability								
To train in-line treatment personnel to enhance their skill in helping the drug dependent person.	Expert	40	60	0	0	0	2.00	1.60
	Non-Expert	100	0	0	0	0	1.58	1.00
To develop a group with a primary interest in the problems people have with drugs.	Expert	10	10	30	20	30	2.70	3.50
	Non-Expert	50	17	25	0	8	2.25	2.00

In five cases, the modal response of the policy experts was "May or may not be feasible," while nonexperts voted the same objective feasible. These related to major strategies such as "to develop adequate alternative models in prevention and phased intervention," to reduce prescribed use of psychoactive drugs/diminish misuse by physicians," and even "to incorporate drug treatment into standard health delivery systems." Since these objectives were all held to be at least desirable, and since one is unlikely to propose objectives one is unsure are achievable, bringing to light this additional information may broaden the policy options available to decision-makers. Alternately, it could be that the view of nonexperts in these cases is overly optimistic. In one case ("to consider social-action approaches as alternatives to treatment...") the modal response of the nonexpert was "May or may not be feasible"; the policy experts were sure it was "Probably feasible."

Policy experts and nonexperts differed on two objectives which represent a major effort in the present national drug-abuse prevention strategy. Policy experts scored "to reduce the supply of drugs available for abuse" as "Feasible," while nonexperts were less certain, scoring it "May or may not be feasible." The reverse was true of the objective "to establish an effective social-rehabilitation system for drug abusers who have become desocialized." Policy experts were not sure if this objective was attainable and scored it "May or may not be feasible"; nonexperts, on the other hand, scored it "Feasible."

Only one objective exhibited a voting difference between policy experts and nonexperts on both feasibility and desirability. This objective ("to develop a group with a primary interest in the problems people have with drugs") was scored "Undesirable" and "May or may not be feasible" by policy experts, but "Desirable" and "Feasible" by nonexperts.

Although there was a big difference in the desirability score between policy experts and nonexperts on the objective "to train in-line treatment personnel to enhance their skill in helping the drug-dependent person," this was mostly in agreement with 40 percent of experts and 100 percent of the nonexperts voting this objective as "Highly desirable."

Objectives that score "Probably infeasible" or "May or may not be feasible" (and this scale value was the modal response), were dropped from consideration, unless there was a significant difference in voting between policy expert and nonexpert.

Third Questionaire. In this questionaire, twenty-five objectives were listed, which required revoting (desirability and feasibility). Objectives are presented for revoting because of polarization on the part of the panel; because there was a broad distribution of voting responses; or because there were differences in voting between policy experts and policy nonexperts. In some cases, original objectives were combined or divided after respondents' comments had been reviewed; in this instance, voting was required on the newly developed objective. The remaining objectives will be held over until a later round. Consideration of the key indicators associated with objectives rated at least feasible and desirable will also be held over to a subsequent round.

Transition Model and Matrix

First Questionnaire. Social policy is the result of multiple interacting forces. Policy is often seen as being based on advocacy rather than derived from a careful analysis of empiric findings. Although policy may have to be developed even in the absence of information, a rational examination of the bases on which policies have been built is a fruitful way of providing the policymaker with insight to develop more appropriate policies.

In the case of drug abuse, the factors which cause people to pass through various states of drug dependence can be systematically examined. Such examination allows the policymaker to estimate the importance of specific variables, and the extent to which they are subject to his influence. A systematic examination of factors also allows any counterintuitive effect of the variables to be brought to light.

In this part of the study we hoped to elicit from respondents factors which control the rates of flow from general population through the various states of drug abuse. The simplified model shown in Fig. 1 developed for this purpose was intentionally simplified to allow for examination of the five transition states included in the matrix. More complex (and probably more realistic) models would have diverted attention from the question at hand.The five transitions in which we are particularly interested are:

- General Population To Potential User
- Potential User To Experimental User
- Experimental User To Occasional Abuser
- Occasional Abuser To Drug-Dependent Person
- Drug-Dependent Person To Formerly Drug-Dependent Person

Respondents were asked to list the factors which affected each of the transitions and state whether a specific factor increased (promoted) or decreased (inhibited) the rate of flow of individuals from one state to another.

The number of factors listed by respondents ranged from a low of two to a high of over forty; we identified a total of 128 distinct factors.

Using the criterion that a factor must register three votes for a single transition, twenty-five significant factors were identified from our total list of 128 factors. The twenty-five are shown in Table 11, the vote is shown by transition state. The table shows the number of respondents who thought that the factor increases the transition rate from one state to another, the number who said it decreases the rate, and a residue who did not indicate direction. Because the number of votes is small, and also because the interpretation of respondents' indication of direction was sometimes difficult, reference will be

Figure 1

DRUG INVOLVEMENT TRANSITION MODEL

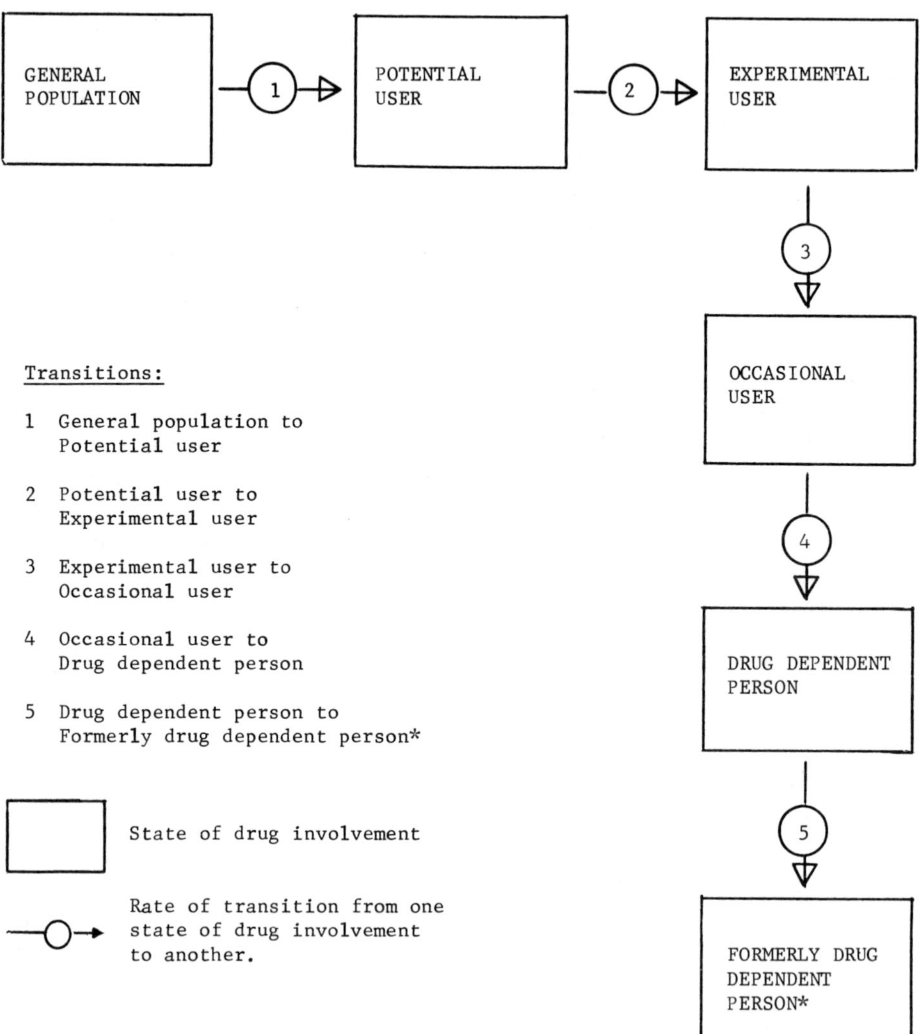

* one should not presume that this is a permanent state; it indicates that point at which, either through treatment or self-denial, one is no longer drug dependent.

Table 11

THE TWENTY-FIVE TRANSITION FACTORS RECEIVING AT LEAST THREE VOTES ON A SINGLE TRANSITION (presented by factor category)

Factor	Response	Transition #1(General Population to Potential User)	#2 (Potential User to Experimental User)	#3(Experimental User to Occasional User)	#4(Occasional User to Drug Dependent Person)	#5 (Drug Dependent Person to Former Dependent Person)	Total Impact Score and (Rank)
age	Total	4	5	4	3	2	18
	Inc	4	4	4	3	2	
	Dec	0	1	0	0	0	(8)
	EW	0	0	0	0	0	
family breakdown	Total	3	2	3	3	3	14
	Inc	2	2	2	2	2	
	Dec	1	0	1	1	1	(12=)
	EW	0	0	0	0	0	
peer pressure	Total	6	8	8	7	6	35
	Inc	5	7	7	5	2	
	Dec	0	0	0	1	3	(2)
	EW	1	1	1	1	1	
proportion of drug users in peer groups	Total	5	4	4	4	4	21
	Inc	5	4	4	4	0	
	Dec	0	0	0	0	4	(5)
	EW	0	0	0	0	0	
religious faith/ training	Total	2	3	2	1	2	10
	Inc	0	0	0	0	2	
	Dec	2	3	2	1	0	(20=)
	EW	0	0	0	0	0	
availability of effective stress-relieving alternatives to drug use/lack of boredom	Total	5	5	5	3	3	21
	Inc	3	3	2	1	2	
	Dec	2	2	3	2	1	(5)
	EW	0	0	0	0	0	
opportunity for meaningful social definition/ availability of adult roles for adolescents	Total	1	1	1	1	3	7
	Inc	1	0	0	0	2	
	Dec	0	1	1	1	1	(25)
	EW	0	0	0	0	0	
feeling of social inadequacy or alienation/ alienation from standards of adult-society	Total	3	3	3	3	1	13
	Inc	3	3	3	2	0	
	Dec	0	0	0	1	1	(16=)
	EW	0	0	0	0	0	
inability to resolve frustrating personal problems	Total	3	3	3	3	2	14
	Inc	2	2	2	2	0	
	Dec	0	0	0	0	1	(12=)
	EW	1	1	1	1	1	
personal stress/ situational crises/ increase in intensity of day-to-day pressures	Total	3	3	4	2	2	14
	Inc	2	2	3	2	2	
	Dec	0	0	0	0	0	(12=)
	EW	1	1	1	0	0	
protracted adolescence/ youth	Total	3	2	1	1	1	8
	Inc	3	2	1	1	0	
	Dec	0	0	0	0	1	(23=)
	EW	0	0	0	0	0	
poor school achievement/dropping out of school	Total	3	2	3	3	3	14
	Inc	2	1	2	2	0	
	Dec	0	0	0	0	2	(12=)
	EW	1	1	1	1	1	
meaningful and satisfying life role/ economic opportunity	Total	2	2	2	2	3	11
	Inc	1	1	2	2	1	
	Dec	1	1	0	0	2	(18=)
	EW	0	0	0	0	0	

Inc = Increase
Dec = Decrease
EW = Either Way

Table 11 (Continued)

THE TWENTY-FIVE TRANSITION FACTORS RECEIVING AT LEAST THREE VOTES ON A SINGLE TRANSITION (presented by factor category)

Factor	Response	Transition #1 (General Population to Potential User)	#2 (Potential User to Experimental User)	#3 (Experimental User to Occasional User)	#4 (Occasional User to Drug Dependent Person)	#5 (Drug Dependent Person to Former Dependent Person)	Total Impact Score and (Rank)
specific drug effects/ pharmacological properties of drugs of abuse	Total	2	2	4	4	3	15
	Inc	1	1	2	2	0	
	Dec	0	0	2	2	1	(11)
	EW	1	1	0	0	2	
availability of (illicit/euphoria producing/ anxiety reducing) drugs	Total	10	13	13	10	8	54
	Inc	9	12	12	10	0	
	Dec	0	0	0	0	8	(1)
	EW	1	1	1	0	0	
supply of drugs/ heroin	Total	2	4	3	2	9	20
	Inc	2	4	3	2	7	
	Dec	0	0	0	0	2	(7)
	EW	0	0	0	0	0	
advertising of drugs in the mass media especially on TV	Total	3	1	1	1	2	8
	Inc	2	0	0	0	1	
	Dec	0	0	0	0	0	(23=)
	EW	1	1	1	1	1	
dissemination of research results/reported research data (eg. effects of drug use)	Total	3	3	2	2	1	11
	Inc	1	1	0	0	0	
	Dec	1	2	2	2	1	(18=)
	EW	1	0	0	0	0	
drug abuse education programs	Total	6	7	5	2	4	24
	Inc	3	3	2	1	4	
	Dec	3	4	3	1	0	(4)
	EW	0	0	0	0	0	
availability of community based treatment/access to treatment centers	Total	0	0	2	5	9	16
	Inc	0	0	0	1	7	
	Dec	0	0	1	3	1	(10)
	EW	0	0	1	1	1	
severity of legal sanctions/laws against use/legal prohibition of drug use	Total	4	5	3	1	4	17
	Inc	1	1	1	1	4	
	Dec	3	4	2	0	0	(9)
	EW	0	0	0	0	0	
enforcement activities/ law enforcement pressure/application of legal sanctions	Total	4	5	5	5	6	25
	Inc	0	1	2	2	4	
	Dec	4	4	3	3	2	(3)
	EW	0	0	0	0	0	
public acceptance of legal drug taking behavior	Total	3	3	2	1	0	9
	Inc	2	2	1	1	0	
	Dec	0	1	0	0	0	(22=)
	EW	1	0	1	0	0	
economic social deprivation	Total	2	2	2	3	1	10
	Inc	2	2	2	2	0	
	Dec	0	0	0	1	1	(20)
	EW	0	0	0	0	0	
number of persons using the drug	Total	3	3	3	3	1	13
	Inc	3	3	3	3	0	
	Dec	0	0	0	0	1	(16=)
	EW	0	0	0	0	0	

Inc = Increase
Dec = Decrease
EW = Either Way

made only to the total number of votes indicating that a particular factor affected a given transition. It should be noted in passing, however, that for some factors respondents appeared to disagree on the direction in which a factor affected a transition rate. A wide range of factors made the listing; some of them were interwoven into the fabric of society, some were clearly interdependent, and in other cases any relationship between factors was less clearcut. The dominant factor was "the availability of drugs," which scored half again as many votes as the second factor, "peer pressure," which in turn scored almost half as many votes again as the third factor, "enforcement activities/law-enforcement pressure."

The top ten factors, arranged in rank order of voting across all transitions, together with the number of votes cast for the factor *not* affecting a particular transition are shown in Table 12.

It is important to note that the votes of "no effect" were obtained from respondents who completed the matrix *and who specifically recorded the fact that a particular transition was not affected by the factor they had listed.* This means other respondents may also not think a particular factor affects a transition, and so in order to balance the picture the number of nonvotes is also shown. Nevertheless, it is interesting to note that even for the most influential factor ("availability of drugs") some respondents specifically said it did not affect transition #1 (general population to potential user) and transitions #4 and #5 (occasional user to drug-dependent person, and drug-dependent person to formerly dependent person). The factor with most disagreement was "drug-abuse education" (all transitions). We presume that this item refers to drug education as it is practiced now, rather than what its potential impact might be if practiced "correctly." It is particularly noteworthy that no less than eight respondents specifically said that "availability of community-based treatment" did not affect the transition from general population to potential user, and from potential user to experimental user. This number of "no effect on " votes was twice as high as for any other factor, and particularly interesting in that not a single respondent stated that the factor actually affected these two transitions.

The wide range of impacting factors and the fact that some of them are closely related to the kind of society in which we live illustrates the problem of creating specific drug-abuse programs.

Second Questionnaire. The transition matrix was not considered in this round.

Third Questionnaire. In this questionnaire we examined the variables that influence the transition from one state of drug involvement to another. Respondents were asked to rate the twenty-five variables that were most frequently cited in response to the first questionaire; ratings included estimating the importance of each factor in controlling a particular rate of transition, and the extent to which these factors can be influenced by the national drug-abuse policymaker (manageability). These data can then be used by

Table 12

TOP TEN FACTORS ACROSS ALL TRANSITIONS

Factor	Response	Transition #1(General Population to Potential User)	#2 (Potential User to Experimental User)	#3(Experimental User to Occasional User)	#4 (Occasional User to Drug Dependent Person)	#5 (Drug Dependent Person to Former Dependent Person)
availability of illicit/ euphoria producing/anxiety reducing)drugs	Affect	10	13	13	10	8
	No Affect	3	0	0	2	4
	No Response	4	4	4	5	5
peer pressure	Affect	6	8	8	7	6
	No Affect	1	0	1	2	3
	No Response	10	9	8	8	8
enforcement activities/law enforcement pressure/ application of legal sanctions	Affect	4	5	5	5	6
	No Affect	2	1	1	1	0
	No Response	11	11	11	11	11
drug abuse education programs	Affect	6	7	5	2	4
	No Affect	2	1	3	6	4
	No Response	9	9	9	9	9

proportion of drug users in peer groups	Affect	5	4	4	4	4
	No Affect	0	0	1	1	1
	No Response	12	13	12	12	12
availability of effective stress-relieving alternatives to drug use/lack of boredom	Affect	5	5	5	3	3
	No Affect	1	1	1	2	3
	No Response	11	11	11	12	11
supply of drugs/heroin	Affect	2	4	3	2	9
	No Affect	1	0	0	1	0
	No Response	14	13	14	14	8
age	Affect	4	5	4	3	2
	No Affect	0	1	1	2	3
	No Response	13	11	12	12	12
severeity of legal sanctions/ laws against use/legal prohibition of drug use	Affect	4	5	3	1	4
	No Affect	1	0	2	4	1
	No Response	12	12	12	12	12
availability of community based treatment/access to treatment centers	Affect	0	0	2	5	9
	No Affect	8	8	5	3	0
	No Response	9	9	10	9	8

respondents to reformulate priorities for national policy, and develop programs from the objectives and policy-issue-statements sections of this study.

Respondents were also asked to rate the criticality of the five separate transitions from one state of drug involvement to another, in terms of the priority each transition should receive as part of the national drug-abuse policy. Respondents did this by allocating one hundred points over the five transitions so as to reflect their priorities. One of the intents of this question is to develop alternative methods of determining program-funding priorities.

Policy Issue Statements

First Questionnaire. For the first questionnaire, panelists were asked to respond to twelve should/should not statements, constructed on current policy issues in the drug-abuse field, and to indicate the importance of these statements. Respondents were also asked to list issues they thought were sufficiently important to be included in the study.

Pro and *contra* arguments for these issue statements were to have been elicited as part of the second questionnaire; in fact, the respondents used their first questionnaire comments regarding each of the statements to do just that. A total of 184 separate comments were made, and were fed back as part of the second questionnaire; respondents were asked to rate their importance.

A total of forty-six additional policy issues was suggested by respondents in the first questionnaire; once again, these seemed to exemplify the diverse and complex nature of this field. From this list we were able to formulate an additional sixteen policy issues in the form of should/should not statements.

Second Questionnaire. The issue statements section of the second questionnaire included a statistical summary of responses from the first questionnaire, and a list of the narrative comments made by respondents from each issue statement. The respondents were asked to revote the issue statements, and rate the narrative comments with regard to *importance*. In addition, thirteen of the issues posed by respondents themselves in the first questionnaire were fed back in the form of should/should not statements for initial voting. It should be noted that three of the original issue statements were rewritten for clarity, following the feedback from the respondent panel.

In preparation for this questionnaire, the analysis or the policy issue statements was confined to comparing the group's response in Round One and Round Two on those items which were common to both rounds (no major shifts were observed in the group's position); and to comparing the Round Two responses of self-rated policy experts to those of nonexperts. Both rating responses and importance ratings were compared.

Respondents were almost unanimous in thinking that "Treatment programs *should* offer treatment for more than one type of dependence within the same facility" and that "The federal government *should* allocate funds to community health centers so that these centers can offer treatment for drug abuse." In this round, substantially more respondents felt that the private insurance plans should be encouraged by the federal government to provide coverage for the treatment of drug dependence than in Round One. In Round One, 42 percent of respondents felt that marijuana should be legalized. This support dropped to 32 percent in Round Two, but 81 percent of respondents felt that "The personal use of marijuana *should* be decriminalized" (a new item).

Issues were classified according to their importance, determined by means of the group importance score. Table 13 lists all thirty-one issues in order of importance; the most important being "There should/should not be a national registry of drug-dependent persons. . . ." Also shown in Table 13 is a comparison of the way self-rated policy experts and nonexperts voted on the Round Two issue statements. Really major differences were observed on only three items. The greatest difference was in relation to the issue "Patients who enter treatment should/should not be required to remain in treatment for a minimum amount of time"; 70 percent of policy experts voted *should* while 77 percent of nonexperts voted *should not*. Only 56 percent of policy experts voted *should* compared to 96 percent of nonexperts on the issue "When an individual is referred for treatment in lieu of incarceration, he should/should not have the right to choose the treatment he prefers." On the other hand, 100 percent of policy experts voted that "Regulations *should* be passed by Congress prohibiting any and all alcohol and tobacco advertising via any media," compared to only 58 percent of nonexperts.

There were four issues that exhibited marked differences between policy experts and nonexperts in the importance of issues (see Table 13). Policy experts scored the importance of the issue "Treatment programs should/should not offer treatment for more than one type of dependence within the same facility" 2.60, while nonexperts thought it considerably more important, scoring it 1.67. A similar situation pertained to the issue "The manufacture and sale of cigarettes should/should not be outlawed." Although both policy experts and nonexperts considered the issue not to be particularly important, experts scored 3.40, while nonexperts scored 2.42.

Marked differences were also observed between self-rated policy experts and nonexperts in the importance scores of these issues:

- "All drug use or possession for personal use should/should not be decriminalized." (policy experts = 2.30, nonexperts = 1.50)
- "A greater proportion of the funds available for drug abuse should/should not be allocated to basis research." (Policy experts = 2.30, nonexperts = 1.55)
- "Persons who are identified as drug-dependent should/should not be required to receive treatment." (Policy experts = 2.40, nonexperts = 1.64)
- "Minimum mandatory prison sentences should/should not be imposed for certain drug-trafficking violations." (Policy experts = 1.90, nonexperts = 2.64)
- "The personal use of marijuana should/should not be legalized." (Policy experts = 2.60; nonexperts = 1.83)

Table 13

POLICY ISSUE STATEMENTS: RESPONSES AND IMPORTANCE RATINGS BY SELF-RATED POLICY EXPERTS AND NONEXPERTS (in decreasing order of group importance)

Policy Issue Statement	Percent voting "should" (1)		Importance Score (1)	
	Policy Experts	Policy Nonexperts	Policy Experts	Policy Nonexperts
Very Imporant				
There SHOULD/SHOULD NOT be a national registry of drug-dependent persons (including identifying information such as the social security number of the individuals).	0	17	1.56	1.50*
There SHOULD/SHOULD NOT be an explicit national policy regarding drug abuse.	80	100	1.70	1.40*
There SHOULD/SHOULD NOT be a maximum amount of time for an individual to be maintained on methadone.	20	31	1.90	1.33
Private health insurance plans such as Blue Cross/Blue Shield SHOULD/SHOULD NOT be encouraged by the Federal government to cover treatment of drug dependency in their insurance plans.	78	85	1.88*	1.46
Personal drug-dependence SHOULD/SHOULD NOT be a criminal offense.	20	8	1.50	1.75
When an individual is referred for treatment in lieu of incarceration, he SHOULD/SHOULD NOT have the right to choose the treatment he prefers.	56	92	1.90	1.50
Important				
The national policy SHOULD/SHOULD NOT be monitored by a non-political group.	89	91	1.89	1.80*
Personal use of marijuana SHOULD/SHOULD NOT be decriminalized.	78	83	1.78	1.92
All drug use or possession for personal use SHOULD/SHOULD NOT be decriminalized.	40	54	2.30	1.50
The Federal government SHOULD/SHOULD NOT allocate funds to community health centers so that these centers can offer treatment for drug abuse.	100	83	1.90	1.91*
A greater proportion of the funds available for drug abuse SHOULD/SHOULD NOT be allocated to basic research.	70	67	2.30	1.55*
Treatment programs SHOULD/SHOULD NOT offer treatment for more than one type of dependence within the same facility.	100	92	2.60	1.67
THE FEDERAL SPONSORING AGENCY/SINGLE STATE AGENCY should be responsible for evaluating treatment programs.	50*[2]	75*[2]	2.11	1.77
Persons who are identified as drug-dependent SHOULD/SHOULD NOT be required to receive treatment.	20	31	2.40	1.64*
Pre-trial detention SHOULD/SHOULD NOT be used in certain drug trafficking cases.	50	55	1.70	2.36*
Minimum mandatory prison sentences SHOULD/SHOULD NOT be imposed for certain drug trafficking violations.	60	55	1.90	2.64*
The present drug scheduling criteria ARE/ARE NOT adequate.	44[3]	40[3]	2.11	2.11*
The National Institute for Drug Abuse and the National Institute for Alcohol Abuse SHOULD/SHOULD NOT be combined to form a single agency.	70	77	1.50	1.75
The United States SHOULD/SHOULD NOT actively assist or influence other governments to cease production of crops such as opium, coca, or cannabis.	60	73	2.30	2.00*

(1) Percent of those responding
(2) Refers to the Federal Sponsors Agency response
(3) Refers to the "are" response
* Indicates that over one-fifth of respondents did not complete this item

Table 13 (Continued)

POLICY ISSUE STATEMENTS: RESPONSES AND IMPORTANCE RATINGS BY SELF-RATED POLICY EXPERTS AND NONEXPERTS (in decreasing order of group importance)

Policy Issue Statement	Percent voting "should" (1)		Importance Score (1)	
	Policy Experts	Policy Nonexperts	Policy Experts	Policy Nonexperts
Patients who enter treatment SHOULD/SHOULD NOT be required to remain in treatment for a minimum amount of time.	70	23	2.20	2.08
The personal use of marijuana SHOULD/SHOULD NOT be legalized.	33	31	2.60	1.83
Non-medical personnel who are employed as direct servicers in drug treatment programs SHOULD/SHOULD NOT be certified to work in that capacity in those programs.	80	91	2.10	2.33
A drug SHOULD/SHOULD NOT be subject to tight controls (production quotas, export-import regulations, prescription controls, etc.) if it has potential for abuse, irrespective of its historical abuse patterns.	67	82	2.22	2.27*
The Drug Enforcement Administration SHOULD/SHOULD NOT be dissolved (as part of a commitment to decriminalization of drug use and in an effort to decrease poor police practice and unnecessary government expense).	20	9	2.40	2.18*
Regulations SHOULD/SHOULD NOT be passed by Congress prohibiting any and all alcohol and tobacco advertising via any media.	100	58	2.30	2.17
The international treaty actions and commitments of the United States (Single Convention and Psychotropic Convention) SHOULD/SHOULD NOT be removed and our international stance vis-a-vis drug be radically altered.	44	17*	2.11	2.63*
Practicing physicians SHOULD/SHOULD NOT be required to attend continuing education in psychopharmacology or be required to pass periodic tests regarding psychopharmacology.	70	62	2.50	2.36*
A citizen-government-pharmaceutical council SHOULD/SHOULD NOT be formed to recommend guidelines to the pharmaceutical industry for its goals in psychoactive drug development and for its practices in sales to physicians.	80	80	2.50	2.55*
Licensed medical personnel who are employed in drug treatment programs SHOULD/SHOULD NOT be certified to work in those programs.	90	80	3.00	2.54
Moderately Important				
Special teams of Federal, State and local agents SHOULD/SHOULD NOT be formed to operate across municipal and state lines for the purpose of suppressing local drug peddling.	78	55	2.67	2.22*
Manufacture and sale of cigarettes SHOULD/SHOULD NOT be outlawed.	0	8	3.40	2.42

(1) Percent of those responding
* Indicates that over one-fifth of respondents did not complete this item

Preliminary Epilogue

Resource Requirements

One of the advantages of the Delphi technique as a tool in policy analysis is its minimal cost for maximum output. The costs for completion of a Delphi study such as this one can range from $15,000 to $40,000 for a nine- to twelve-month effort, depending upon staff and direct-cost expenditures required. For example, if the effort is included as part of ongoing staff assignments, then staff and space costs may not be directly chargeable; if computer services are available, then a sizable cost category is deleted. The amount of data which may be derived, and the opportunity afforded to facilitate a "discussion" of the issues by divergent experts in the field, render the technique unusually cost-effective.

Considerable time should be spent in coceptualization of the study design and development and pretesting of the questionnaire. In any area as complex and diffuse as drug abuse, the study-design team needs to allocate substantial effort to this phase of study development.

Applications

The relative success of this National Drug-Abuse Policy Delphi has resulted in considerable interest in utilization of the technique not only in the drug-abuse field, but in other social policy areas as well. The opportunities it affords for idea exchange among diverse professionals and interest groups; and the continuous flow of significant data for policy review are but two of the positive attributes of the method. The potential for its application is extensive; as this is the first study of its type, all of us who are interested in its future application can profit from the lessons learned from this effort.

The process of conceptualizing and analyzing policy options is supremely complex; it may be that the Delphi policy method will be a significant advance in the field of applied decision theory and policy analysis, as it relates to the social policy area in particular.

Prospectus

This phase of the study was completed in December 1974. During the succeeding rounds, the objectives will be further summarized; the policy issues on which there is significant divergent opinion will be explored; and policy options will be developed from the objectives, policy issues, and transition factors. An interactive conference will be held at the conclusion of the study; part of this conference will involve introduction of the computer conferencing technique to respondents.

We shall evaluate the present effort to determine to what extent the study objectives were met, and to what extent the respondents' objectives for participating in the study were met. The preliminary steps in this evaluation have already been taken: in the first questionnaire, respondents were asked to list their objectives for participating in the study, for example. At the conclusion of the study, the respondents will be asked to measure the degree to which their objectives have been reached and whether they might have developed other objectives during the course of the study. In addition, they will be asked to evaluate the study on the basis of questionnaire design, content, and other relevant areas. The results of this evaluation will be utilized in developing an ongoing interactive policy planning system which the author is presently designing, as well as other specific studies which are expected to stem from the present effort.

The evaluation of the impact of a study such as this is a much more complex problem, but one which we believe is ultimately of more significance. We have just begun to develop plans for a long-term evaluation of the study. This will include, for example, as assessment of the degree to which study results were reviewed and considered in the formulation of national drug-abuse policy.

Acknowledgments

I should like to acknowledge the extraordinary gratis assistance of Dr. Norman Dalkey and Dr. Murray Turoff; and that of Dr. Peter Goldschmidt and M. Alexander Stiffman of the Johns Hopkins University; without the support and dedicated efforts of these individuals, and the 39 respondents, the study would not be possible.

I should also like to thank the National Coordinating Council on Drug Education for sponsoring the continued development of the study design as originally conceived and implemented.

III. B. 4. A Delphi Evaluation of Agreement between Organizations

CHESTER G. JONES

Introduction

Delphi [1] is often used to combine and refine the opinions of a heterogeneous group of experts in order to establish a judgment based on a merging of the information collectively available to the experts. However, in this process it is possible to submerge differences of opinion and thus suppress the existence of uncertainty. In many situations it might be advisable to run separate Delphis using more homogeneous groups of experts in order to highlight areas of disagreement. This paper will report on an activity that did just this and point out several areas in which the types of responses obtained were fundamentally very different. In some cases these differences were quite unpredictable, and so, a highlighting of the variations greatly increased the information obtained. Running one Delphi using a subset of the experts from each group would probably not have illuminated some of the differences in opinion. The mere weight of pressure to move toward the median response [2] would have caused a joint Delphi to converge toward a middle position. In addition, the presence of disagreement is much more significant when large groups share similar positions. The traditional approach to Delphi generally results in the using of a small number of experts from any one area.

One concern that is often raised about the credibility of Delphi results is that individual experts may bias their responses so that they are overly favorable toward areas of personal interest. This is of particular concern when experts are asked to evaluate areas in which they are presently working and when the final Delphi results could impact the importance attached to these areas. In this paper results will be presented that indicate that no such bias occurred in the Delphis reported on. It appears that the particular groups of experts used were able to rise above the desire to protect personal interests.

Background

The United States Air Force presently maintains an official list of System Concept Options (SCOs) in order to indicate to the Air Force Laboratories potential future technology needs. This activity is primarily a means of com-

The Delphi Method: Techniques and Applications, Harold A. Linstone and Murray Turoff (eds.)
ISBN 0-201-04294-0; 0-201-04293-2

municating to the laboratory planners the thinking of Air Force System planners. However the number of potentially worthwhile systems possibilities, and thus the number of technology needs, exceed the resources available to fulfill all the possibilities and needs. Clearly the Air Force Laboratories needed a means of establishing priorities for the System Concept Options. Thus it was decided to undertake a program of Delphi evaluation. This program was run by the Deputy for Development Planning, Aeronautical Systems Division, and was limited to considerations of those SCOs that fell under the Deputy's jurisdiction. Thirty SCOs were evaluated. They covered a rather large spread in need for technological support as well as proposed mission use. Some concepts represented a rather straightforward extrapolation of present technology, while others would require substantial technology development programs. The missions represented included most of the areas of interest to the Air Force including many strategic and tactical possibilities as well as systems intended to meet support and training requirements.

It was decided to conduct separate Delphis utilizing personnel from various Air Force organizations, in order to determine how closely the organizational opinions agreed. In this way it was believed that not only would a basis for prioritizing the systems be obtained, but in addition, the results would help to indicate areas of communication problems between organizations. If organization viewpoints in a particular area differed greatly, there would appear to be a need for increased communication about the area.

Delphis were conducted within the following four USAF organizations: Deputy for Development Planning, Aeronautical Systems Division (ASD/XR); Air Force Avionics Laboratory (AFAL); Air Force Aero Propulsion Laboratory (AFAPL); Air Force Flight Dynamic Laboratory (AFFDL). The experts chosen were senior managerial and technical personnel (both civilian and military), and were selected so that representation of most if not all of the major departments within the organizations was present. A total of sixty-one experts took part in the evaluations which involved three rounds of questioning.

The above organizations are of two different types. The Deputy for Development Planning is a systems planning organization having responsibility for identifying promising aerodynamic system concepts and defining them to the point where development decisions can be made. It has no direct responsibility for research activities. The three laboratories are responsible for developing technologies in their assigned areas which will improve system capabilities. The Avionics Laboratory is concerned with electronic systems, the Aero Propulsion Laboratory with atmospheric engines, fuel, etc., and the Flight Dynamics Laboratory with aircraft structures, controls, aerodynamics, etc. Thus the four groups that were asked to evaluate the list of SCOs are quite different in their areas of expertise. In particular it should be emphasized that the laboratory groups were being asked to compare SCOs some of which required considerable support from their particular laboratory, others of which required little or no support. All of the participants were, however, senior Air Force personnel and were thus knowledgeable of activities at other Air Force Laboratories.

Results obtained for three of the questions used will be discussed in this paper.

Question 1. Please rank-order the SCO list of systems on the basis of where current Air Force Laboratory Programs will make the greatest contribution toward success of the system.

Question 4. Given that each system becomes a technological success, rank order the SCO list in terms of importance of each system to National Defense.

Question 5. Considering technology, timing, and system importance, rank-order the SCO list according to where you think the Air Force Laboratories can make the greatest contribution to National Defense.

Each of these questions involved a complete ranking of thirty items, which proved to be a trying but not impossible task. It should be noted that succeeding round changes in answers often required a large restructuring of the list. That is, a change in the answer or rank of one system generally changes the rank of other systems (however, the participants were allowed to use a limited number of ties if necessary and thus a few participants avoided this problem). This interrelation of answers tends to make convergence difficult, since disagreement in one area impacts other areas.

Convergence

One indication of the effect of a Delphi experiment is the amount of convergence caused by the iteration process, where convergence is a measure of how much more agreement is achieved on succeeding rounds as opposed to the first-round response. In this effort, one measure of convergence was the change in the spread between the lower and upper quartile values for a given question and a given SCO. In all of the Delphi experiments the spread between the lower and upper quartile values generally showed considerable reduction during the course of the efforts. However, as indicated in Fig. 1, the average amount of convergence varied considerably from group to group for some questions.

All of the groups achieved basically the same degree of convergence for Question 1. However, the convergence on Questions 4 and 5 follow significantly different patterns. In particular, the ASD/XR group achieved less convergence on Questions 4 and 5 than that achieved by the laboratory groups. The ASD/XR group was the only group primarily composed of system planners and so this group's failure to converge as well as the other groups on Question 4 would seem to be quite important. The ASD/XR group should be the best suited to serve as experts concerning Question 4. Thus, for Question 4 the greatest uncertainty is associated with the most expert group. If one Delphi had

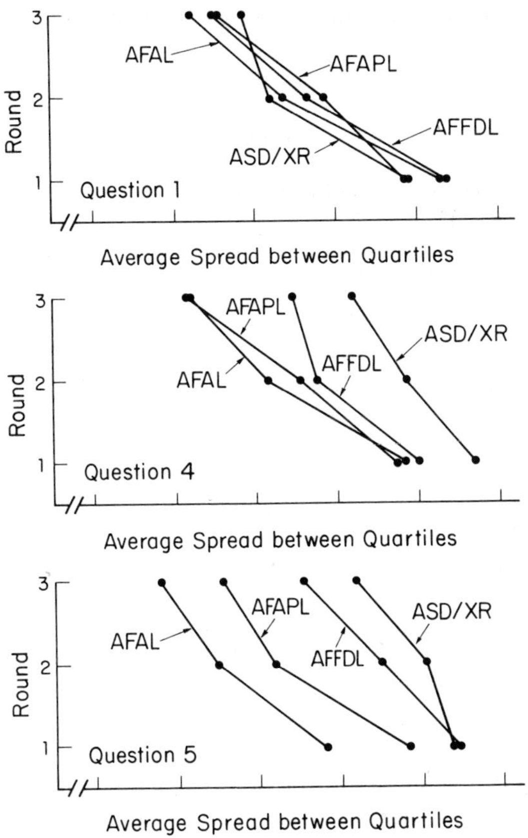

Fig. 1. Average interquartile spreads for questions 1, 4, 5.

been run combining experts from each of those groups, it appears possible that the greater convergence between the laboratory experts might have caused a considerably better overall convergence than that shown in the ASR/XR result. Thus the relatively less expert participants might have caused the creation of a false sense of expert agreement.

Correlation between Questions

In reviewing the results, it was obvious that some groups tended to give SCOs similar rankings for different questions, while other groups changed many of the SCOs rankings drastically from question to question. Table 1 shows the

Spearman rank correlation coefficient for each Delphi for each combination of questions.

Table 1

Spearman Rank Correlation Coefficient for Each Question Combination

QUESTIONS	ASD/XR	AFAL	AFAPL	AFFDL
Q1-Q4	+.295	+.788	+.571	+.448
Q1-Q5	+.315	+.863	+.904	+.698
Q4-Q5	+.746	+.925	+.579	+.844

Clearly the ASD/XR answers suggest a greater change in laboratory emphasis (as shown by the low correlation between Questions 1 and 4, and between Questions 1 and 5) than that indicated by the other three groups. The system planners thus indicated a greater need for laboratory redirection than the laboratory personnel. Again we have an area of disagreement that might be camouflaged had one combined Delphi been utilized.

It is interesting that the AFAPL results indicate the least correlation between Questions 4 and 5. Although it might seem that the answers to these questions should correlate closely, there are several possible reasons to explain lack of correlation:

(1) A system may be important but not need substantial laboratory support.

(2) The necessary laboratory support might best be supplied by non-Air Force Laboratories.

(3) A system might be important if technologically feasible, but the necessary technological developments might not be considered likely in the near future.

Thus there might be a logical explanation for this lack of correlation. However, the data are surprising enough to indicate the desirability of a more detailed review of the AFAPL results. A subsequent review of the AFAPL answers indicated that many of the comments used to justify the apparently inconsistent results did involve considerations such as those listed above. However this example shows the value of looking for correlation between answers, and then, highlighting comments that justify departures from expected correlation.

Bias by Time Period

Figure 2 shows the average evaluations for Question 5 when the SCOs are grouped according to date of estimated technological feasibility. Obviously the system planners (ASD/XR) with their more futuristic interests attach greater importance to the far-term, more advanced systems. This might be a result of the planners' greater awareness of the possible benefits these futuristic systems offer. However, a possible reason for the laboratory viewpoint might be a greater appreciation of the difficulty associated with solving the technological problems.

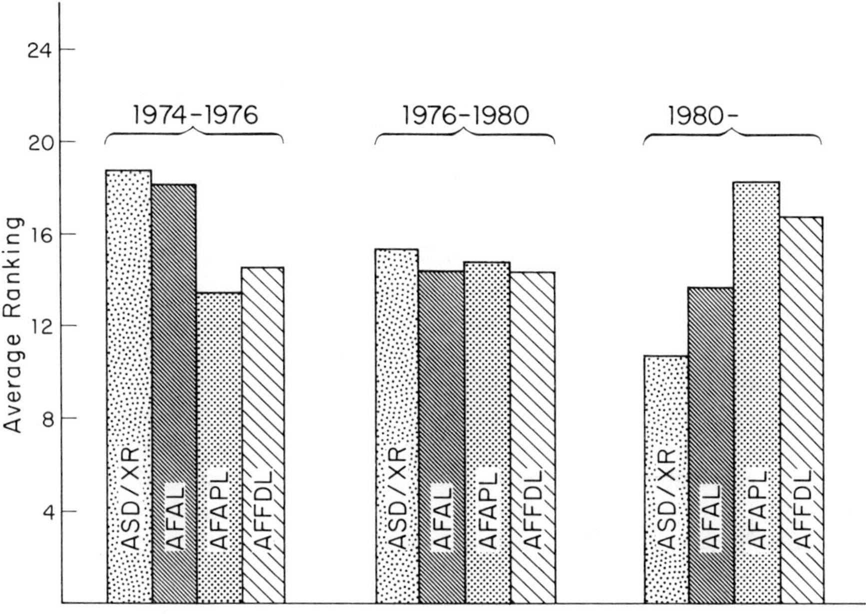

Fig. 2. Average responses for Question 5 when SCOs are grouped by year in which engineering development could start.

Again the results suggest the possibility of a communications gap. Both groups should benefit from an exposure to the reasoning that led to such diverse results. This type of exposure might best go beyond a Delphi-type exchange (which is generally limited in the amount of information transferred). Such a transfer of information is essential if the potential value of the SCO list is to be achieved. It is often not enough to establish priorities, unless all parties concerned accept and understand the logic that led to the priorities.

Laboratory Bias

There was some concern before the laboratory efforts were started that the results might tend to be biased. Although the laboratory participants were instructed to rank the laboratory efforts by the total efforts from all the Air Force Laboratories, it was hypothesized that the participants' greater knowledge about their own laboratory programs and the natural tendency to promote one's personal interests would lead to a bias in favor of their laboratory's efforts. In order to test this hypothesis, the rankings obtained on Question 5 for the SCOs that received crucial support from each of the laboratories were compared.

In mid-1972, each of the laboratories published reports that reviewed their Technology Planning Objectives (TPOs) and the relevance of each TPO to each SCO. The top relevancy category indicated a TPO that the laboratory felt was essential to a given SCO. Table 2 shows the average ranking given for Question 5 to the groups of SCOs having a top relevancy match with the various laboratory TPOs, respectively. The lowest number in each column indicates the organization placing the greatest emphasis on that laboratory's progress. Therefore, bias would be indicated if the lowest number in a given laboratory's column was on the row corresponding to that laboratory's Delphi. The Delphi conducted in Laboratory B gave poorer (larger numerically) rankings to SCOs that were felt to be essentially related to one or more of their TPOs than any of the other groups, while the Delphi conducted in Laboratory A gave neither the poorest nor the best rankings to SCOs that were felt to be essentially related to one or more of their TPOs. Although the Delphi conducted in Laboratory C did give the best (numerically lowest) ranking to SCOs considered to be essentially related to their TPOs, the average ranking is not too different from those obtained in the other Delphis.

Table 2

Average Answer to Question 5 for SCOs That Are Related to Programs of Particular Laboratories

DELPHI	Relevancy Match with		
	Laboratory A	Laboratory B	Laboratory C
ASD/XR	12.7	15.1	15.6
Laboratory A	12.0	17.6	15.9
Laboratory B	15.1	22.3	15.8
Laboratory C	11.6	20.0	14.9

Thus, the hypothesis that a given laboratory Delphi tends to indicate biased rankings for SCOs that receive crucial support from that laboratory's effort does not appear to be valid. The answers obtained from Question 5 do not indicate the presence of laboratory bias.

Summary

The results discussed in this paper indicate information that was obtained by comparing several Delphi experiments utilizing experts from different organizations that probably would not have been obtained had one Delphi been run utilizing a subgroup from each of the groups of experts. Clearly differing organization viewpoints were identified, despite the fact that all of the groups involved very senior Air Force personnel who shared access to a considerable common information base. That is, all of the organizations had detailed knowledge of many of the same programs.

A noticeable difference in the amount of convergence was observed where in one case the apparently more expert group showed the poorest convergence. Disagreement was also apparent concerning the question of whether or not the laboratory programs should be redirected (as well as the related question of whether laboratory efforts should be directed toward near-term or more futuristic technology needs).

Comparisons of results were also made to determine if the laboratory group gave answers that were biased to support their own program. This investigation failed to show the presence of any real bias. This finding is very encouraging, for it suggests that at least these groups of technical experts were able to place their professional ethics above the common desire to promote personal gain. Had this not been true, the worth of this activity would be greatly reduced.

References

1. Olaf Helmer, and Nicholas Rescher, "*On the Epistemology of the Inexact Sciences*," *Management Science*, **6**, No. 1 (October 1959).
2. Norman C. Dalkey, *The Delphi Method: An Experimental Study of Group Opinion*, The Rand Corporation, RM-5888-PR; 1969.

III. C. 1. Delphi Research in the Corporate Environment

LAWRENCE H. DAY

Introduction

The Delphi technique has become widely accepted in the past decade by a broad range of institutions, government departments, and policy research organizations ("think tanks"). These applications are described elsewhere in this book. The use of the Delphi approach in the corporate environment will be discussed in this section. Corporate utilization of Delphi is perhaps one of the least-known aspects of the technique's application. This is a result of corporations regarding the products of their Delphi exercises as proprietary and, hence, restricting their distribution or description in professional literature. A review of the long-term planning and futurist literature has revealed that few of the corporate efforts in this field have been documented in any detail.

The first part of this analysis will examine some uses of the Delphi technique in industry. This general review will be supplemented by an analysis of the application of the methodology in six Delphi studies conducted by the Business Planning group of Bell Canada. The Bell Canada experience will be followed by a description of some of the problems and issues that arise when using Delphi in the corporate environment. This review will conclude with some comments on the potential future of the Delphi technique in the business environment.

Examples of Corporate Delphi Research

Industrial Grouping or Professional Association Sponsorship

Delphi studies sponsored by corporations can be classified into three categories. The first category includes those studies sponsored by an industry association or a professional association.

These studies are usually of a broad nature and are concerned with projecting the future of an industry or perhaps even some broader societal field. The logistics of this application usually indicate that the study has to be contracted out to an independent consultant or research organization. While this type of application does not result in the day-to-day use of the methodology in business, the results of these Delphi studies are often exposed to a broad range

---

The Delphi Method: Techniques and Applications, Harold A. Linstone and Murray Turoff (eds.)
ISBN 0-201-04294-0; 0-201-04293-2

of high-level managers and executives in business. Thus, in this situation, corporations are the consumers of Delphi research rather than users of the technique itself.

Parsons & Williams Inc., an international consulting firm, conducted a Delphi study entitled "Forecast 1968—2000 of Computer Developments and Applications" in 1968.[1]. This study was undertaken for a conference on computer file organization held in Denmark sponsored by the International Federation of Information Processing Societies. This study examined future computer applications in business, the home, government, and institutions and projected the future of specific computing and technological developments [2]. Another recent Danish conference has also used a Delphi study as an input to panel discussion. This study, for a conference on long-term trends in personnel management called "Delphi 71–80", examined thirty-seven areas and predicted "how far society would be moved in certain directions by 1980" [3].

Sponsorship of Delphi studies by groups of firms are generally examinations of the future of an industry or an industrial segment. Current examples in this area include reviews of the cosmetics, recreation, and insurance markets. A "Delphi Panel on the Future of Leisure and Recreation" has been conducted by Social Engineering Technology Inc. (SET Inc.) [4]. This multiclient study was conducted by SET for a group of companies interested in future market opportunities in the recreational area that could develop through the impact of cultural change.

A Delphi study on the life insurance and other personal financial services markets is being conducted for two life insurance companies by the Canadian consultants Ducharme, Deom & Associes Ltee. The objective of this study is to "design a picture of the Life Insurance market and other personal financial services in the 1980's in terms of external environmental variables..." [5].

Individual Corporate Sponsorship

This second category of Delphi research is similar to the first. The grouping includes individual corporations who sponsor Delphi studies at research organizations on subjects of general or specific interest. The Institute for the Future (IFF) has conducted the largest number of these studies on this basis. In the case of IFF, the study results are in the public domain [6]. Several of these studies have been concerned with the impact of the computer/communications revolution:

(1) The Future of the Telephone Industry; sponsored by the American Telephone and Telegraph Co. (New York, N. Y.) [7].

(2) The Future of Newsprint; sponsored by MacMillan Bloedel Ltd. (Vancouver, B. C.) [8].

(3) On the Nature of Economic Losses Arising from Computer Based Systems in the Next Fifteen Years; sponsored by Skandia Insurance Co. (Stockholm, Sweden) [9].

Another IFF study sponsored by Owens-Corning Fiberglas Corp., examines "Some Prospects for Residential Housing by 1985" [10]. The research program by Owens-Corning also produced an IFF study on "Some Developments in Plastics and Competing Materials by 1985" [11].

An interesting IFF study sponsored by General Telephone and Electronics Corp., examines "Some Prospects for Social Change by 1985 and Their Impact on Time/Money Budgets" [12]. Hence the various corporate-sponsored studies at IFF fall into general research categories (e.g., GTE, Skandia, and Owens-Corning) and specific industry research studies (e.g., AT&T, and MacMillan Bloedel).

Delphi research has also been sponsored by corporations in other research organizations. The Danish study referenced above on personnel management was extended by the consultants (Management Training Division of the Danish Institute of Graduate Engineers) to three groups of employees from the printing firm CON-FORM [13]. The Pace Computing Corporation sponsored a study by marketing research consultants to determine the potential demand for its services [14].

The Delphi technique has recently come to the attention of other marketing research consultants. This should lead to an expansion of corporate sponsorship of these studies, as the market researchers will promote the technique with customers who might not normally become exposed to Delphi or other longer-term planning techniques. One Canadian market research organization has mentioned the technique in its periodic newsletter to clients [15]. As noted in the first category, this sponsorship leads to senior management exposure to the technique even though the corporations do not conduct the studies themselves.

Corporate In-House Delphi Research

This final category includes Delphi studies conducted by research or planning groups within the corporation itself. In this case, members of the corporation staff become very involved with Delphi as they must master the technique as well as use the study results. This category includes most proprietary uses of the studies and their results, and they usually have not been published or distributed widely.

The best-known example of corporate Delphi experience is that of TRW. While the TRW Delphi studies are unpublished and proprietary, a number of papers have been published by North and Pyke on the technique's use and selected study results [16]. TRW's modification of Delphi has been named PROBE. The initial study was started in 1965 and resulted in a fifty–page document containing a set of 401 forecasts published in June 1966 [17]. This has been refined in a second study called "Probe II." The reader is referred to the papers noted above for an elaboration of the TRW experience with Delphi.

A Delphi study was conducted in the U. K. by the Hercules Powder Co.

Ltd. on the future of the British Chemical Industry in the 1980s. This has been discussed in several articles by Parker of Hercules [18] and used as a case example in the book *Technological Forecasting* by Wills [19]. Other U. K. experiences with technological forecasting and Delphi were referenced in a recent article "Technological Forecasting in Six Major U. K. Companies" by Curill [20]. While he does not name specific companies, he notes that Delphi has been used by: a "Glass" Company, a "Consumer Goods" Company, two "Chemical Companies," and an "Electrical Engineering" Company and that this is one of the most popular techniques of those companies utilizing technological forecasting methodologies.

The medical field has been explored in a U. S. study by Smith, Kline and French, a major pharmaceutical manufacturer [21]. Three other large U. S. pharmaceutical companies are reported to have conducted studies as well [22].

Industries undergoing rapid change have been frequent targets of Delphi research. The merging computer and communications fields are an example of this phenomenon and a significant number of industrial studies have been conducted. IBM has conducted an internal study on future computer applications. ICL in England have also sponsored a Delphi study. In addition to sponsoring the IFF research, AT&T conducted a study, "Communication Needs of the Seventies and Eighties" (internal document) [23]. Bell Canada has undertaken six studies projecting technological and social trends in four main areas: education, medicine, business information systems, and "wired city" services (all proprietary) [24]. The Trans-Canada Telephone System conducted an internal Delphi study on future data service needs. British Columbia Telephone is conducting a Policy Delphi with senior managers.

Summary: Corporate Examples

The discussion of various forms and examples of corporate sponsorship of Delphi studies is not intended to be all-inclusive. It merely attempts to outline, on an international basis, a few of the known examples of the scope of Delphi usage in industry. The next discussion will center on our experiences in Bell Canada as one example of how corporate Delphi studies have been conducted.

Delphi and Bell Canada

Background

Bell Canada is an operating telecommunications company serving the provinces of Ontario and Quebec in Canada. In addition to offering voice, data, and visual telecommunications services, Bell Canada owns a large manufacturing subsidiary (Northern Electric) and an R & D subsidiary (Bell-Northern Research). There are also several other subsidiaries in the telephone, directory, and electronic components manufacturing fields. The Business Planning Group in Bell has the responsibility for identification of corporate opportunities (or threats) that will arise through changes in society and/or technology in the next decade or two.

The communications field is in the midst of rapid change which will have a significant impact on its intermediate and long term future. Highlights of these changes include:

- merging computer and communications technologies
- regulatory changes introducing new competitive elements
- emerging visual telecommunications markets
- perceived and projected social changes
- increasing costs of investment options

The Business Planning Group surveyed these various pressures in the late 1960s as it was developing a study plan to evaluate future trends in the visual and computer communications fields. There was a distinct lack of qualitative data on potential futures for these fields, especially in the Canadian environment. An examination of various potential technological forecasting techniques indicated that the Delphi technique would fill the perceived information gap.

Bell Canada Delphi Study Development

The individuals involved in designing, conducting and managing the Business Planning Delphi efforts have generally had a marketing background. This background includes both academic training and professional experience. These background factors were important determinants of the approach followed.

Initial steps relied upon the basic marketing approach of defining the "market segments" that will have the most important impact on future applications of visual and data communications. These segments were chosen after preliminary studies of potential segments and taking account of the time and resources available. The final choices were future applications in the educational, medical, information systems, and residential markets.

The basic philosophy in the studies was to examine the future of applications in these segments from a user point of view, not from the direction of technological imperatives. The initial questionnaires were prepared after extensive literature reviews of potential developments in each of the chosen areas. The approach in questionnaire design was to guide the discussions in some basic areas of interest in a segment rather than start with blank paper and ask the experts to suggest the most important areas of interest. Since the panelists were actively encouraged to suggest new questions or modifications to existing ones, the potential for significant study bias by the designers was low. This approach also helped reduce the number of rounds required for the studies and hence saved time for the participants and study managers alike.

Table 1

VALUE CHANGES IN NORTH AMERICAN SOCIETY 1970 – 2000					
	SIGNIFICANT INCREASE	SLIGHT INCREASE	NO CHANGE	SLIGHT DECREASE	SIGNIFICANT DECREASE
TRADITIONALISM					
HARD WORK AS A VIRTUE					
AUTHORITARIANISM					
MATERIALISM					
REWARDING WORK AS A VIRTUE					
INDIVIDUALISM					
INVOLVEMENT IN SOCIETY					
PARTICIPATION IN DECISION MAKING					
SELF EXPRESSION					
ACCEPTANCE OF CHANGE					

NOTE: The shaded areas above represent the median responses from the five Bell Canada Delphi studies noted in the footnotes. Shading over two areas indicates differences in opinions between the various panels.

Next, initial questionnaires were pretested with groups of readily available experts. This proved to be a very valuable step, as poorly worded questions or confusing questionnaire design were largely eliminated before the errors could be inflicted upon the Delphi panel. This step adds time to the study and may be somewhat ego deflating at times for the study managers; however, it pays good dividends in higher quality, less ambiguous results, and happier panelists.

Delphi Study Results—Education, Medicine, and Business

The initial studies in education, medicine, and business followed a similar format. The first part of the questionnaire asked the panelists to project their views on the long-term (thirty years) future of some basic North American values. The purpose in asking these questions was more to help the panelists get in a societal frame of mind when answering the rest of the questionnaire than to obtain the societal trend data itself. When the social trend views of the various groups of experts, as shown in Table 1, were compared after all of these studies were completed, it was interesting to note how similar the results were, considering the diverse background of the 165 individuals in the various panels (there was no interpanel communication during the studies).

Other areas of each study also explored nontechnological developments as well as the adoption of systems to serve various applications. Table 2 illustrates some of these nontechnical factors considered in the three studies [25].

Table 2

Nontechnological Factors Considered in the Bell Canada Delphi Studies

EDUCATION	MEDICINE	BUSINESS
1.Value Trends	1. Value Trends	1. Value Trends
2. Evolution in School Design	2. Trends in the Medical Profession	2. Changes in Business Procedures
3. Changing Role of the Teacher	3. Changes in the Medical Environment	3. Trends in Business Physical Environment

The Education Delphi examined potential adoption of three basic types of educational technologies: Computerized Library Systems (CLS), Computer Aided Instruction Systems (CAI), and Visual Display Systems (including IRTV—Instant Retrieval Television). The summary forecasts of the panel are shown below in Table 3 [26]. The projections on the use of terminals for input and output purposes for CAI and CLS are shown in Table 4 [27]. In both tables, threshold market penetration values of 20 percent and/or 55 percent were used for the technologies. This gives the panelists and readers some feeling for the scope of service acceptance in the markets under consideration.

The Medical Delphi explored acceptance of a number of developing medical technologies. These included: Multiphasic Screening, Computer-Assisted Diagnosis, Remote Physiological Monitoring, Computerized Medical Library Systems, and Terminal Usage. Table 5 illustrates some of the summary results from the medical study [28]. The format and adoption thresholds were similar to those in the Education Delphi.

The Business Information Processing Technology study examined trends in Management Information Systems, Mini and Small Computers, Terminals and Data Processing. Table 6 summarizes the median conclusions of the panel on 20 percent acceptance of various technologies both in business and in the home [29]. It should be noted that this Table, like the earlier ones, summarizes only the median statistical conclusions of the panels. Panelists were sometimes split into "schools of thought" on various issues. These opinion splits were often not reflected in graphic presentation of the results. In these cases the panelists were encouraged to debate their differences in writing through the rounds of the various studies. These differences are reflected in the reports along with supporting assumptions and comments of the panelists. It was found that the panelists' comments and their analyses were often very important modifications of the statistical projections shown in the above tables.

Design of the Future of Home Services Delphi

The three studies outlined above provided an important new source of input to the Business Planning group. However, one market area was still largely unresolved: the future of communications services in the residence market. Each of the above studies asked a few questions on home services but their combined answers still left a large information gap. Determining how to obtain the additional information reopened some important internal differences of opinion on the value of Delphi for this purpose.

The main issue revolved around the definition of what is an "expert" in the residential field. This question had developed in creating the earlier expert panels as well. In two cases (Education and Business) the question was whether selected industry specialists within the telecommunications industry were as knowledgeable as experts in the above fields when it came to projecting the future. This question was resolved by conducting two studies in the Education and Business fields. In both cases, independent panels of "internal" and "outside" experts were used.

The question in the residential study was whether housewives or researchers and planners were the best experts on the future adoption of communications services in the home. In this case the study was totally service- and *not*

Table 3

LIKELY TIMING OF ADOPTION OF TECHNOLOGICAL SYSTEMS

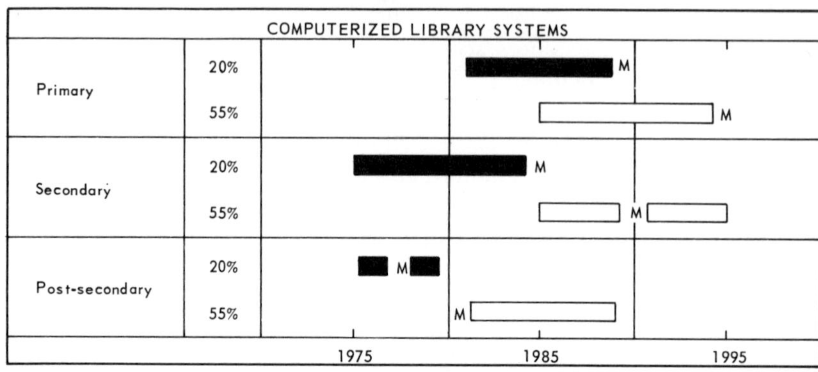

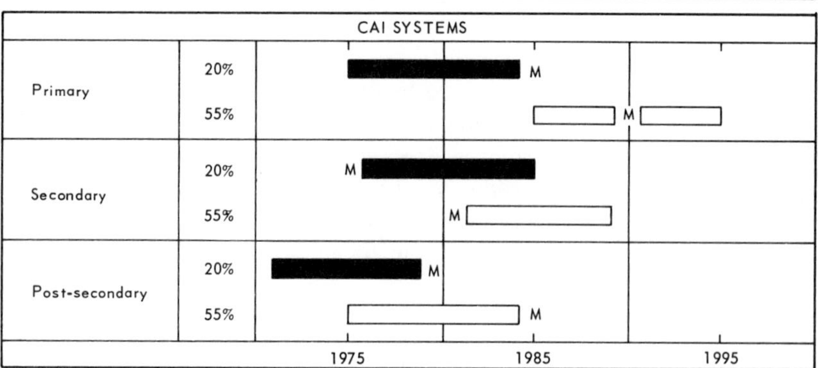

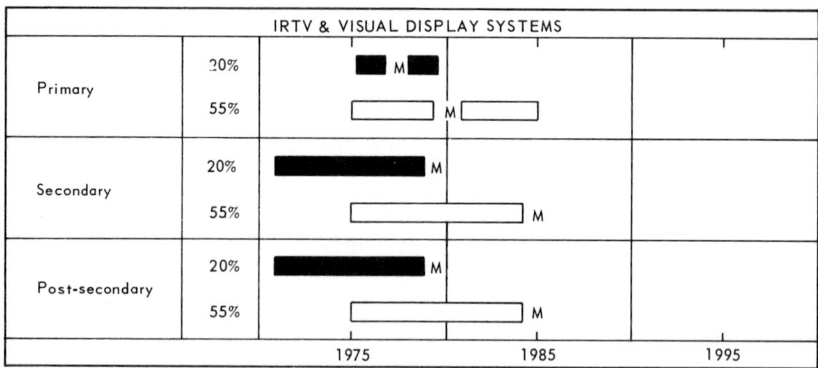

M MEDIAN

PERIOD OF SYSTEM REFINEMENT AND EARLY ADOPTION (INTER QUARTILE RANGE)

PERIOD OF EXTENSIVE ADOPTION (INTER QUARTILE RANGE)

SOURCE: DOYLE AND GOODWILL, EDUCATIONAL TECHNOLOGY, P. 64

NOTE: THE MEDIAN SYMBOL (M) ON THIS CHART AND THE FOLLOWING ONES ONLY INDICATES THE 5 YEAR TIME FRAME THAT THE MEDIAN LIES IN, NOT ITS EXACT LOCATION. THUS, THE MEDIAN AND THE FIRST OR THIRD QUARTILE MAY BE IN THE SAME 5 YEAR PERIOD IN SOME CASES.

Table 4

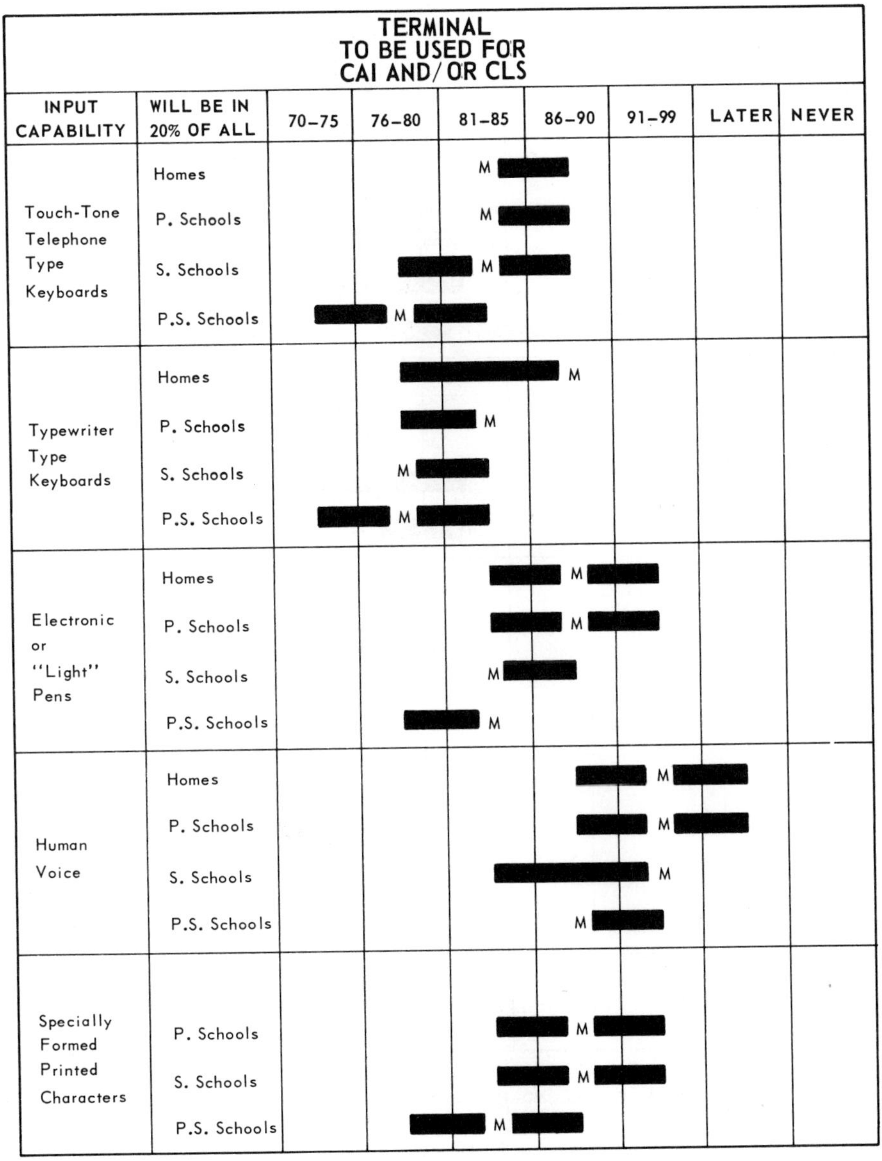

TERMINAL TO BE USED FOR CAI AND/OR CLS								
INPUT CAPABILITY	WILL BE IN 20% OF ALL	70–75	76–80	81–85	86–90	91–99	LATER	NEVER
Touch-Tone Telephone Type Keyboards	Homes			M ▬	▬			
	P. Schools			M ▬	▬			
	S. Schools		▬	▬ M ▬	▬			
	P.S. Schools	▬	▬ M ▬	▬				
Typewriter Type Keyboards	Homes		▬	▬	▬ M			
	P. Schools		▬	▬ M				
	S. Schools		M ▬	▬				
	P.S. Schools	▬	▬ M ▬	▬				
Electronic or "Light" Pens	Homes			▬	▬ M ▬	▬		
	P. Schools			▬	▬ M ▬	▬		
	S. Schools			M ▬	▬			
	P.S. Schools		▬	▬ M				
Human Voice	Homes				▬	▬ M ▬	▬	
	P. Schools				▬	▬ M ▬	▬	
	S. Schools			▬	▬	▬ M		
	P.S. Schools				M ▬	▬		
Specially Formed Printed Characters	P. Schools			▬	▬ M ▬	▬		
	S. Schools			▬	▬ M ▬	▬		
	P.S. Schools		▬	▬ M ▬	▬			

Table 4 (continued)

TERMINAL TO BE USED FOR CAI AND/OR CLS								
OUTPUT CAPABILITY	WILL BE IN 20% OF ALL	70–75	76–80	81–85	86–90	91–99	LATER	NEVER
Computer Voice Reply	Homes				M			
	P. Schools				M			
	S. Schools				M			
	P.S. Schools			M				
Printed Page	Homes				M			
	P. Schools			M				
	S. Schools			M				
	P.S. Schools		M					
Traditional Television Screen Display	Homes			M				
	P. Schools			M				
	S. Schools			M				
	P.S. Schools		M					
Large Flat Television Screen Display	Homes					M		
	P. Schools				M			
	S. Schools			M				
	P.S. Schools				M			
Audio Video Recorder Recording	Homes				M			
	P. Schools			M				
	S. Schools			M				
	P.S. Schools		M					
Picturephone Screen Display	Homes				M			

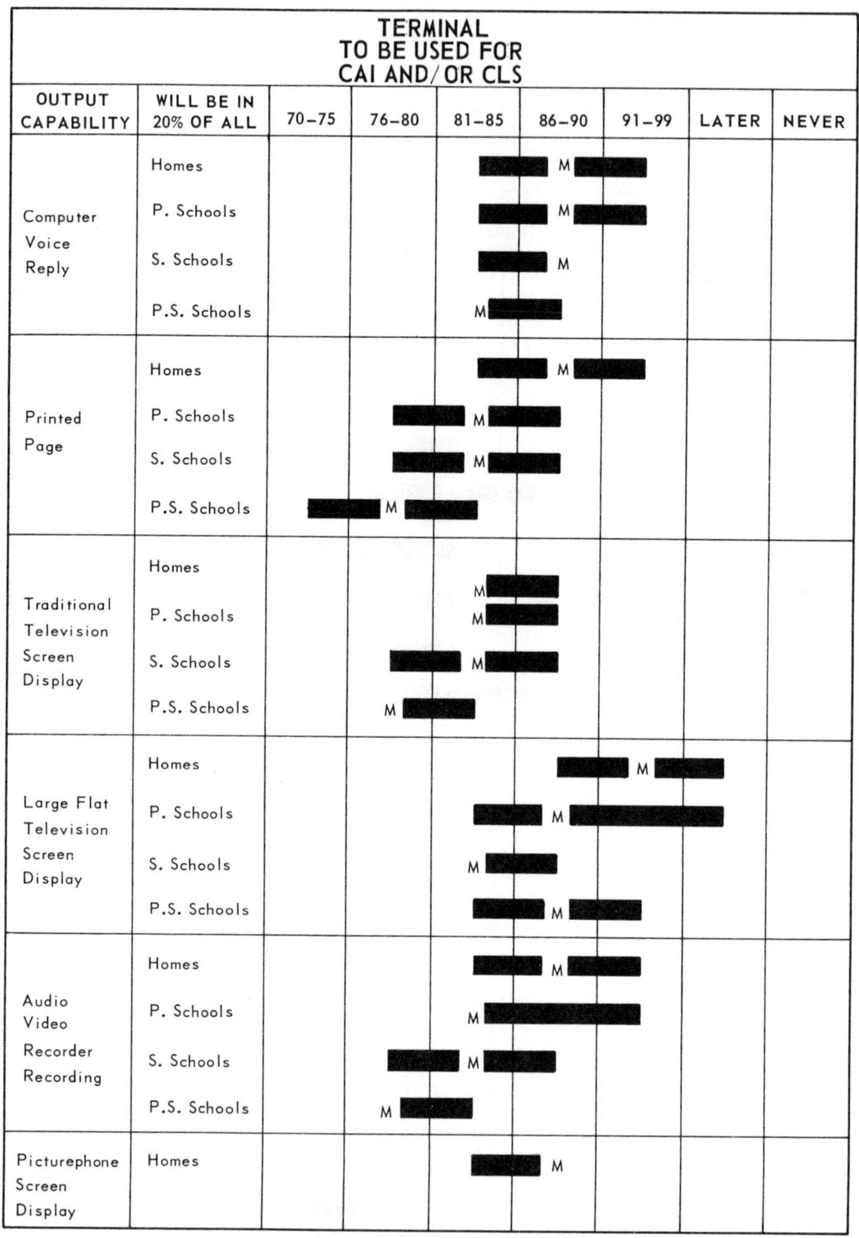

P. SCHOOLS = PRIMARY SCHOOLS
S. SCHOOLS = SECONDARY SCHOOLS
P.S. SCHOOLS = POST SECONDARY SCHOOLS

Source: Doyle and Goodwill, Educational Technology, P. 41–2

Table 5

PERIOD OF SYSTEM REFINEMENT AND EARLY ADOPTION OF TECHNOLOGICAL SYSTEMS (20% UTILIZATION)

INSTITUTION	TECHNOLOGICAL SYSTEM	70–75	76–80	81–85	86–90	91–99	LATER	NEVER
Physicians in Solo Practice	MS			M				
	CAD			M				
	CLS			M				
Physicians in Group Practice Who Share Facilities Only	MS		M					
	CAD			M				
	CLS			M				
Physicians in Group Practice Who Share Records & Incomes etc.	MS		M					
	CAD		M					
	CLS			M				
Acute General Hospitals	MS		M					
	CAD		M					
	CLS		M					
Occupational Medicine	MS		M					
	CAD			M				
	CLS			M				

MS– – MULTIPHASIC SCREENING
CAD– – COMPUTER ASSISTED DIAGNOSIS
CLS– – COMPUTERIZED LIBRARY SYSTEMS

Source: Doyle and Goodwill, Medical Technology, P. 60

Table 6
An Exploration of the Future in Business Information Processing Technology

DEVELOPMENTS*	1970–75	1976–80	1981–85
THE WORK LOCATION (WHERE WHITE-COLLAR AND CLERICAL EMPLOYEES WILL PERFORM THEIR JOB DUTIES)		OFFICE CENTER	
MANAGEMENT INFORMATION SYSTEMS		MIS FOR LARGE FIRMS	MIS FOR MEDIUM FIRMS INTEGRATED MIS
MINI AND SMALL COMPUTERS		FREE-STANDING COMPUTERS COMMUNICATION COMPUTERS DUAL-PURPOSE MACHINES	
TERMINALS		T-T TELEPHONE (BUSINESS) TYPEWRITER KEYBOARDS (BUSINESS) COMPUTER VOICE REPLY (BUSINESS) PRINTED PAGE (BUSINESS) TRADITIONAL TV SCREEN (BUSINESS) ACOUSTICALLY COUPLED TERMINALS (BUSINESS) PLUG-IN PORTABLE TERMINALS (BUSINESS)	T-T TELEPHONE (HOMES) PICTUREPHONE FOR FACE-TO-FACE COMMUNICATION (BUSINESS) TOUCH-SENSITIVE INPUT (BUSINESS) OCR (BUSINESS) 2-WAY LARGE SCREEN COLOR TV (BUSINESS) AUDIO-VIDEO RECORDERS (BUSINESS) PICTUREPHONE DISPLAY (BUSINESS)
DATA PROCESSING	MEDIUM-SIZE FIRMS IN MANUFACTURING AND SERVICE INDUSTRIES WILL BE UTILIZING D.P. FACILITIES	SMALL-SIZE FIRMS IN MANUFACTURING INDUSTRIES WILL BE UTILIZING D.P. FACILITIES	SMALL-SIZE FIRMS IN SERVICE INDUSTRIES WILL BE UTILIZING D.P. FACILITIES
TECHNOLOGY IN THE HOME			ONE-WAY (INCOMING) AUDIO-VISUAL COMMUNICATION

DEVELOPMENTS*	1986–90	1991–99	LATER
THE WORK LOCATION (WHERE WHITE-COLLAR AND CLERICAL EMPLOYEES WILL PERFORM THEIR JOB DUTIES)	NEIGHBORHOOD R.W.C. THE MOBILE WORKER	HOME REMOTE WORK CENTER	
MANAGEMENT INFORMATION SYSTEMS	MIS FOR SMALL FIRMS DISTRIBUTED MIS		
MINI AND SMALL COMPUTERS			
TERMINALS	LIGHT PENS (BUSINESS) HUMAN VOICE INPUT (BUSINESS) PICTUREPHONE FOR FACE-TO-FACE COMMUNICATION (HOMES) PICTUREPHONE FOR DATA INPUT (BUSINESS) HANDWRITTEN INPUT (BUSINESS) COMPUTER VOICE REPLY (HOMES) TRADITIONAL TV SCREEN (HOMES) AUDIO-VIDEO RECORDERS (HOMES) FULLY PORTABLE WIRELESS TERMINALS (BUSINESS)	PICTUREPHONE FOR DATA INPUT (HOMES) SCRIBBLEPHONE (HOMES) SCRIBBLEPHONE (BUSINESS) PRINTED PAGE OUTPUT (HOMES) LARGE FLAT COLOR TV DISPLAY (HOMES) LARGE FLAT COLOR TV DISPLAY (BUSINESS) PICTUREPHONE DISPLAY (HOMES)	TYPEWRITER KEYBOARDS (HOMES) LIGHT PENS (HOMES) HUMAN VOICE INPUT (HOMES) HANDWRITTEN INPUT (HOMES) TOUCH-SENSITIVE INPUT (HOMES) OCR (HOMES) 2-WAY LARGE SCREEN COLOR TV (HOMES)
DATA PROCESSING			
TECHNOLOGY IN THE HOME	TWO-WAY AUDIO-VISUAL COMMUNICATION		

*THE RESULTS INDICATE WHEN THE DEVELOPMENT WILL REACH A 20 PERCENT LEVEL OF ACCEPTANCE BY THE APPROPRIATE UNIVERSE. THE MEDIAN "EXPERT" RESULTS ARE PRESENTED.

technology-oriented. This issue was resolved by establishing two competing panels to forecast the future in this area. One panel consisted of housewives (experts through experience) and the other of experts through research or planning for "wired city" services. The study design and steps followed are shown in Table 7.

Table 7

Study Design: Future of Communications Services into the Home

1. Literature Search
2. Assemble Panels of "Experts" and Housewives
3. Design Draft Questionnaire
4. Pretest Questionnaire
5. Print and Distribute Revised Questionnaire (Identical to Both Groups)
6. Prepare Statistical Analysis of 1st-Round Answers
7. Prepare Analysis of Supporting Comments from Each Group
8. Design, Pretest, Print and Distribute 2nd-Round Questionnaire showing:
   a) 1st-Round Statistical Results from Each Group on One Page
   b) 1st-Round Supporting Comments from Each Group on Opposite Page
   c) Ask for Resolution of Answers within Each Panel
   d) Highlight Differences between Panels and Ask for Resolution
9. Prepare Final Analysis

The important steps that are different from normal Delphi studies are 7 and 8 in Table 7. The results should stimulate debates between the panels if this approach is going to derive the maximum benefits from both panels. Table 8 shows a typical two-page feedback and question set from the Home Communications Delphi [30]. The importance of obtaining feedback comments from the panelists is illustrated in the table.

This study examined future acceptance of electronic shopping from the home, remote banking, electronic home security services, and electronic programmed education in the home. The study also explored the future of ten types of information retrieval services that may be offered to homes. Table 9 illustrates some summary results of the study [31].

Summary: Bell Canada Delphi Studies

Business Planning efforts in the six studies outlined above have resulted in an important increase in the availability of qualitative data for planning purposes. Experience with the technique resulted in significant modifications from the original RAND approach, especially with the emphasis on analyzing the panelists' comments and establishing threshold levels of acceptance. The use of

Table 8

SHOP–FROM–HOME SERVICE

In predicting which types of products will be purchased through a Shop–from–Home system, the housewives and experts disagreed on a number of items. The summarized answers are presented below with the answers for the expected costs of such a service. Some typical comments are presented on the facing page. There appear to be significant differences in such items as produce, small and large appliances.

TYPE OF PRODUCT:

E / H	% YES	% NO	over 20% more	5–20% more	0–5% more	same	0–5% less	5–20% less	over 20% less
meat	E 48 / H 58	E 52 / H 42			E / H	E / H			
produce	*E 52 / H 75*	*E 48 / H 25*			E	E / H			
other perishables (dairy, bread, etc.)	E 100 / H 83	E 0 / H 17			E	H			
groc. dry goods	E 100 / H 96	E 0 / H 4			E	H	H		
clothing	E 58 / H 48	E 42 / H 52			E	H	E		
small appliances	*E 74 / H 96*	*E 26 / H 4*			E	H	E		
drugs and cosmetics	E 90 / H 75	E 10 / H 25			E	E / H			
large appliances	*E 47 / H 79*	*E 53 / H 21*			E	H			

NOTE: Shaded area represents significant differences between panels
E = Expert Panel: median or percentage response.
H = Housewife Panel: median or percentage response.

Do you wish to change any response? Could you comment on the differences between the two groups? Do you have any concluding comments

ROUND II COMMENTS

(Cont'd.)

Table 8 (continued)

SHOP-FROM-HOME (CONTINUED)

The panelists had the following generally favorable comments on the service and its expected costs.

EXPERTS

1. "Will succeeds only if it offers an economic advantage (savings on sales people, expensive store space, electricity and display, etc.)"
2. "Assumption – meat and produce supplied from a familiar supplier and good previous experience."
3. "Better price and product comparisons can allow the purchaser to save money as well as time."

HOUSEWIVES

1. "A useful service for those living in suburbs or remote rural areas whose access to cities and stores is difficult, also for elderly people."
2. "I don't think it should cost any more because it would not need a huge store or many salesclerks, and certainly shoplifting would be a thing of the past for anyone participating in the service."
3. "Small or large appliances (if brand name is known) present no problems for Shop-from-Home."

These comments reflect some of the typical reservations about the service:

EXPERTS

1. "Meat, produce – these items are bought by feeling, smelling, and seeing the specific items in question and comparing them to the others available in the display. This cannot be done remotely, hence the continued existance of meat markets, fruit stands, etc."
2. "Insuring quality will raise the price of perishables above attractive cost."
3. "People will expect large cost reductions for appliances in return for not being able to see the merchandise. . .Several visits to warehouses will be typical even with home shopping."

HOUSEWIVES

1. "For most housewives, shopping is a diversion and a break in a routine."
2. "Meat and produce – Individual specific quantities required; bult packaging not always desirable."
3. "The cost of these items in our daily living budget is expensive enough as it is; if it would cost more to buy these things at home, not many people would take advantage of it unless they were unable to get out of the house. . . Clothing and large appliances would be difficult with this service; with clothing you like to examine the fabric to see how well it is made; with large appliances you would want to discuss with the salesman the pro's-con's of the appliance."

Source: Bedford, _Questionnaire_, pp. 2-3

Table 9

SHOPPING FROM THE HOME

How will it be used?

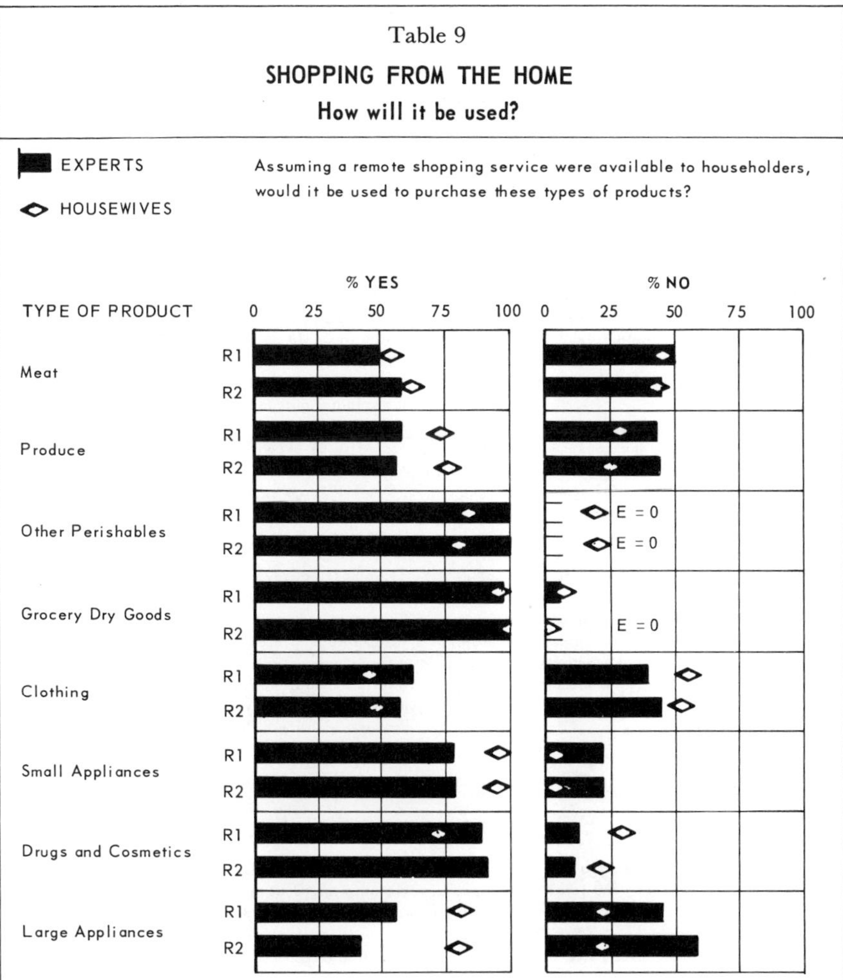

While there seemed to be general agreement between panels that grocery dry goods and certain perishables would be purchased through a remote shopping service, the housewives believed that a number of other products would also be purchased with the aid of this service. The comments reinforce this pattern ("that housewives would 'trust' the store . . . . but experts have a smaller trust"). At the same time however, the comments reflect the feeling that this service would eliminate a significant element in housewives' social activities. It seems that if remote shopping is to become widely accepted, housewives will need to have recourse to alternate modes of socializing.

There did not appear to be any major change in panelists' attitudes between rounds. The largest shift was the decrease in the number of experts expecting large appliances to be purchased from the home.

(Cont'd.)

Table 9 (continued)

COMMENTS

EXPERTS	HOUSEWIVES
ROUND ONE	
"Meat, produce – these items are bought by feeling, smelling, and seeing the specific items in question and comparing them to the others available in the display. This cannot be done remotely, hence the continued existence of meat markets, fruit stands, etc."	"A useful service for those living in suburbs or remote rural areas whose access to cities and stores is difficult, also for elderly people. Clothing would be difficult in my opinion as there would be no way of trying on and fitting. Endless returns and exchanges would defeat the original purpose of the shop-from-home service. Otherwise, most things could conveniently be bought from the home."
"In regard to meats, the consumer will always want hands-on experience. The butcher has a very poor public image. Consumers will want to see how much fat the steak has."	"Small or large appliances (if brand name is known) present no problems for Shop-from-Home."
"It's not a function of the product but rather how a given purchaser perceives the product. The gourmet thinks of food as an art form, and wants to be there. The cafeteria operator couldn't care less. Since many stores are our contemporary art museums, the process of shopping is quite complex."	"For most housewives, shopping is a diversion and a break in a routine."
	"Clothing and large appliances would be difficult with this service; with clothing you like to examine the fabric to see how well it is made; with large appliances you would want to discuss with the salesman the pro's-con's of the appliance."
ROUND TWO	
"I still don't accept the idea that housewives will buy meat, produce, and large appliances via shop-from-hone service. The first time the housewife gets burned using S-F-H (ie. a poor steak, unfresh meat, etc.), she will go back to using the store."	"Change my response for meat, produce and clothing to **NO** because one of the main reactions to the product is through a confrontation with it which results in the buying or rejecting of it."
"My '**NO**' responses were based on the belief that an intelligent housewife would wish to carefully choose meat and produce herself unless her butcher and grocer were intelligent people who knew her states and preferences well."	"It is rather amusing that housewives would 'trust' the store for meat and produce but the experts have smaller trust. Go with the housewives – we do most of the shopping."
"Experts tend towards **NO**, probably because they buy these products as men with a technical bent."	"I still don't like the concept of shopping on a larger scale from home. It strikes me that this system is geared to the larger chain stores which already offer catalogue service as a type of shop-from-home feature (with regard to clothing, appliances, etc.) But I wonder how such a service would affect smaller businesses and specialty stores which definitely couldn't operate such a service themselves."
"Experts seem to ascribe greater weight to first hand observation of product in buying decision than do housewives. Husbands may be more similar to experts in this propensity."	

Source: Bedford, Future of Communications Services in the Home, pp. 28-29.

Delphi to evaluate the marketability of services by users rather than predicting the median dates of potential technological development was also helpful. An analysis of completed studies has also revealed comparison information on the use of internal panels vs. external panels. Thus, Business Planning has learned much about the technique while obtaining useful information. This three-year intensive involvement with the technique has also given Business Planners a realistic view of some of the issues that arise when operating with Delphi in the corporate environment.

Delphi in the Corporate Environment

The issues discussed below are based upon Bell Canada experience and on discussions with individuals who have conducted similar studies in other corporations. These issues will probably be faced by any group in industry that launches a serious attempt to conduct professional quality research in this area.

Should Corporations Pay for Basic Delphi Research?

This, of course, is the fundamental question that must be answered. The emphasis here is on in depth research, since this will often result in a significant allocation of time and money resources in an area where immediate payoff is not clear to senior management. Other forms of business research (market research, operations research, economic research, etc.), have more precise goals and utilize more understandable techniques. The benefits of Delphi research will not be reiterated here, but the corporate planner has to recognize that this is one area not easily understood by busy executives.

This is part of the more basic question on the value of long-term planning in business. Generally, long-term planning has become an accepted part of business today. Delphi research is most needed in the long-term planning function where the conditions of uncertainty are the most evident.

Basic research of this nature is beginning to fall into the general area of corporate social responsibility. Many corporate decisions made today will have important secondary effects for decades to come. The rapid rise of interest in government, academia, and business in what is termed "Technology Assessment" [32] is one reason for considering this as a part of corporate social responsibility. Delphi study results can be used as corporate inputs to the development of technology assessment equations [33].

The North American telecommunications industry is generally privately owned but regulated by government agencies. Recent studies in the U. S. [34] and Canada [35] have noted and projected an accelerating trend on the sharing of planning data between the regulators and the corporations. While many industries are not formally regulated, none can escape the growing governmental and public scrutiny of the consequences of their actions and plans. Sharing of basic planning information such as Delphi study results can help develop a common assessment data base on both a corporate and a public basis.

The social/political reasons outlined above are especially important when evaluating the cost/benefit analysis of undertaking corporate Delphi research. However, this is a more obvious reason for doing this type of research: the results can be used to help make business decisions.

Using Delphi Results in Business

Many corporate Delphi studies conclude with the publication of a report to the panelists and management outlining the study findings. The problem of recognizing the value of this research develops if that is where the Delphi studies end. These basic Delphi reports are important as:

(1) Educational tools to inform senior managers of the panelists' views of potential futures or various areas of interest to the business.
(2) Trading documents with other planners and researchers.
(3) Environmental trend documents that can help technological planners in research labs.

The use of the Delphi results must then become more directed. One useful way of using the specific results is to regard them as a data base to be drawn upon when preparing corporate recommendations in specific topic areas. The Delphi forecasts should be combined with other relevant material (trend extrapolations, multiclient study results, market research data, etc.) in order to present a comprehensive estimate of the impact of a forthcoming decision [36]. These combinations may be in the form of cross-impact matrices, scenarios, market analyses, etc. The use of the Delphi data with other material helps create confidence in the overall package. It is rare that the Delphi results alone can help resolve an issue when preparing a recommendation. Of course, this approach is useful in the nonbusiness environment as well.

The Bell Canada Delphi study results are regarded as part of a data base. Each of the Delphi forecasts has been abstracted, key-word indexed, and stored in an on-line computerized information retrieval system. Other items in the data base are also stored in the same manner. These items may be forecasts from trend studies, material from other internal research, appropriate forecasts from studies available from government institutions, policy research institutes, corporations, etc. The data base is used in the creation of several types of Business Planning outputs (Note: the Delphi material is an *input*, not an output). These outputs include:

(1) Specific Service and Business Proposals designed to exploit identified opportunities.

(2) Environmental Outlook Reports that identify trends and potential future events which may impact on the company or a specific corporate function (i.e., marketing).

(3) Targeted Outputs designed to present selected material to various government commissions, task forces, as well as other research organizations.

(4) Subject Sourcebooks and Information Packages which combine all of the available information on a specific field of interest into an annotated document for use by other planners in the Bell Canada Group.

(5) Methodology Analyses that document what we have learned using a particular technological forecasting technique.

In all of the above cases, the Delphi research material has been combined, massaged, analyzed and placed in perspective vis-à-vis other future information.

Delphi research can also be used in obtaining certain types of information not usually available from normal marketing research activities. Statistical polling of consumers can only produce a limited base of attitudinal data. Feedback and interaction are not possible here. On the other hand, group depth interviews can run into many of the problems that the lack of anonymity produces. The modified version of Delphi used by Bedford enables the researcher to generate opinions and conflicts between potential consumer groups for new products or services. This controlled conflict with feedback produces valuable behavioral information that would not emerge using other techniques. This data can be used for product or service modification or redefinition of market opportunities.

Delphi research in business must be regarded as a means toward ends rather than as an interesting intellectual end in itself. Use of the technique as indicated above can result in an affirmative answer to the question: "Should corporations pay for basic Delphi research?"

Misusing Delphi Results in Business

Delphi study results can be used to advantage in the corporate environment. The reverse situation is also possible. One of the most common situations is for the results of the study to be viewed as representing a corporate position, policy, or forecast.

This is not the case for the results of the vast majority of Delphi studies which represent the combined and refined wisdom of the particular panel of experts on the study [37]. One of the recurring problems with the Bell Canada Delphi's has been the suggestion that the studies represent a corporate position even though this suggestion is explicitly refuted in the reports.

A related problem is the temptation of corporate public relations groups to distribute the studies as another P. R. tool. This can be especially problematic, since Delphi panelists are assured that their contributions are provided in confidence on a professional basis. The use of the study results in this manner could backlash on the study director, especially if he hoped to conduct future studies using panelists drawn from the same population. Of course, the value of the documents as trading vehicles would diminish as well if they were handled in this manner.

A further issue is related to the perceived precision of the study results. Many Delphi studies process the interim and final results using computers. This permits the presentation of statistical results that "appear" very precise to the casual observer or individuals accustomed to dealing with the results of economic and statistical research. The findings of Delphi studies are subject to more interpretation than are most research results. The planning group should try to ensure that others using the results as a data base are aware of the various strengths and weaknesses of the information.

In-House vs. Consultant-Conducted Studies

Another question that must be resolved is whether or not to conduct the study using in-house or consultative resources. This decision can be analyzed by considering the following factors.

(1) Single vs. Multiple Studies. There is a definite learning curve involved when conducting Delphi studies. Serious attempts utilizing the technique require an initial time and resource investment to learn how Delphi studies are effectively conducted. This investment will pay continuing dividends if a number of studies are planned. These rewards include the development of a more knowledgeable planning staff that fully understands the strengths and weaknesses of the data obtained from the studies. On the other hand, conducting a single study with a planning group unfamiliar with the use of the technique may be a costly venture that produces mediocre results.

(2) Study Sophistication. The corporation may be dealing with a subject matter that is changing rapidly and is very complex (e.g., computer technology). The firm may also want a large number of factors considered in the study. In this case, the use of outside consultants who have considerable experience in conducting large complex Delphi's (i.e., I.F.F.) may be more productive. The use of these consultants will also ensure that the best modification of the technique is applied to the company's problem. Experienced consultants are constantly learning more about the technique and are modifying it as a result of that experience. The time delays that occur before this experience is reflected in the professional literature may mean that the corporate researcher is using a somewhat less than optimum version of the technique.

(3) Proprietary Research. The problems involved with proprietory research that are discussed in the next section are also factors to consider when choosing between in-house and consultant conducted studies.

Proprietary Nature of Delphi in the Corporate Environment

One of the usual descriptors of corporate market research is that the results are considered proprietary. Many studies are conducted to further a competitive advantage. Corporate Delphi research is often conducted in this environment with similar objectives. In these instances the results of the studies are not designed for outside consumption. This creates problems if external expert panelists are used in the study. The usual contract with the panelist is a full or partial payment with a copy of the study results. This usually attracts high-caliber panelists who are interested in adding the study results to their own store of information. The presence of the report in turn results in dissemination of its contents to the panelist's professional colleagues, either by photocopying or by requests to the study director for additional copies. This process of information dissemination through "invisible colleges" usually means that proprietary studies are not too practical with external panelists.

One solution to this situation is to utilize in-house experts. Of course, this is practical only if there are a significant number of internal experts in the subject matter of interest. The penalty of using this approach is the loss of the independent outside viewpoint.

The use of mixed panels of in-house and external experts creates another potential problem. The in-house panelist may have access to confidential corporate market or technological research and use this in making and justifying his projections. The study director may have to edit this data out of the panelist feedback material unless the company is prepared to let the corporate information out to the external panelists. This situation can create some intellectual dissonance for the study director, since the secret data could help resolve specific questions under consideration by the panel. The best solution in this case may be to try in advance to avoid subject matter in the study where confidential company research is underway.

Conclusions

The preceding list of issues that must be considered when conducting corporate Delphi research is not exhaustive. The main purpose of this section was to examine some of the common or most important issues that the business planner must face when deciding whether or not, or how, to use Delphi research. As with many situations, a heavy application of common sense when planning Delphi research will avoid some of the potential problems outlined above.

Future Use of Delphi in Corporations

The near future should see continued rapid expansion of the Delphi technique in business. The methodology appears to be currently reaching the "faddish" stage. Many low-quality studies (which may be mislabeled "Delphi") will be conducted. This could result in a credibility gap with those trying to use the technique to its best advantage. If this credibility gap does occur, there may be a numerical decline in the number of studies conducted, but a general improvement in the overall quality of corporate Delphi research.

Widespread use of the methodology will result in continued rapid modification of the original RAND design. Mini-Delphi's will be used to develop specific forecasts or evaluate potential policy changes. The latter area will receive further attention with the continued development of interest in technology assessment. The use of on-line Delphi techniques will spread, especially as corporate management information systems and remote access terminals become widespread [38]. The availability of standard packages that permit any researcher with access to an on-line Delphi system to act as a study director will also encourage further use of the technique [39].

Delphi will become popular for certain types of market research studies. This will probably occur more as a result of the promotional activities of market research firms than from the conscious decision of corporate researchers or marketing academics. This opinion is held since there is little overlap between the current professional literature of the marketers and the long-term planners [40], whereas consultants are presently indicating interest in the technique.

In conclusion, Delphi has a healthy future in the corporate environment. This is a future for a whole *family* of Delphi-inspired techniques in a broad range of applications. Use of the term "Delphi" to describe a monolithic technique has rapidly become obsolete in this environment. This expanding family of techniques will be the property of the market researcher, market planner, policy planner, systems researcher, etc., as well as the long-term business planner.

References

1. *Forecast 1968-2000 of Computer Developments and Applications*, Parsons and Williams, Copenhagen, 1968.
2. Sample results are shown in "Computers in the Crystal Ball," *Science*, August 1969, p. 15; see also Joseph Martino, "What Computers May Do Tomorrow," *The Futurist*, Oct. 1969, pp. 134-135.
3. Ole Lachmann, "Personnel Administration in 1980: A Delphi Study," *Long Range Planning*, 5, No. 2 (June 1972) p. 21.
4. *Delphi Panel on the Future of Leisure and Recreation*, SET Inc., Los Angeles, 1972 (multiclient proprietary study).
5. Letter from Ducharme, Déom, and Associés Inc., August 18, 1972.
6. See the Articles of Incorporation; the Institute aim is "to make available without discrimination the results of such research and scientific advances to the public"; also quoted on inside front covers of many IFF reports.
7. Paul Baran and A. J. Lipinski, *The Future of the Telephone Industry*, R-20, Institute for the Future, Menlo Park, Sept. 1971.
8. Paul Baran, *The Future of Newsprint*, R-16, Institute for the Future, Menlo Park, December 1971.
9. J. R. Salancik, Theodore J. Gordon, and Neale Adams, *On the Nature of Economic Losses Arising from Computer-Based Services in the Next Fifteen Years*, R-23, Institute for the Future, Menlo Park, March 1972.

10. Selwyn Enzer, *Some Prospects for Residential Housing by 1985*, R-13, Institute for the Future, Menlo Park, January 1971.
11. Selwyn Enzer, following article (III C 2).
12. Selwyn Enzer, Dennis L. Little, and Frederick D. Luzer, *Some Prospects for Social Change by 1985 and Their Impact on Time/Money Budgets*, R-25, Institute for the Future, Menlo Park, March 1972.
13. Ole Lachmann, "Personnel Administration," p. 23.
14. Marvin A. Jolson and Gerald L. Rossow, "The Delphi Process in Marketing Decision Making," *Journal of Marketing Research* **8**, (November 1971). 444.
15. Contemporary Research Center, "Thinking about Tomorrow," *Printout*, April 1972.
16. See, for example: Harper Q. North, *Technological Forecasting in Industry*, NATO Defense Research Group Conference, National *Physical Laboratory*, England, Nov. 1968.
Harper Q. North and Donald L. Pyke, "'Probes' of the Technological Future," *Harvard Business Review,* May-June 1969.
Harper Q. North and Donald L. Pyke, "Technological Forecasting to Aid Research and Development Planning," *Research Management*, **12** (1969).
Harper Q. North and Donald L. Pyke, "Technological Forecasting in Planning for Company Growth," *IEEE Spectrum*, Jan. 1969.
Donald L. Pyke, "A Practical Approach to Delphi," *Technological Forecasting and Long Range Planning*, American Institute of Chemical Engineers, Nov. 1969.
Donald L. Pyke, *The Role of Probe in TRW's Planning, Technological Forecasting—An Academic Inquiry*, Graduate School of Business, University of Texas, April 1969.
onald L. Pyke, "Long Range Technological Forecasting: An Application of the Delphi Process," Fifth Summer Symposium ofEnergy Economic Division of ASEE, UCLA, June 21, 1968.
17. North and Pyke, "Technological Forecasting in Planning for Company Growth," *IEEE Spectrum*, Jan. 1969, p. 32.
18. See, for example: E. F. Parker, "Some Cexperience with the Application of the Delphi Method," *Chemistry and Industry*, Sept. 1969. p. 1317.
E. F. Parker, "British Chemical Industry in the 1980's—A Delphi Method profile," *Chemistry and Industry*, Jan. 1970, p. 138.
19. Gordon Wills *et al. Technological Forecasting*, pp. 194-203, 208-9 Penguin Books, London, 1972. Also discussed in Gordon Wills, "The Preparation and Deployment of Technological Forecasts," *Long Range Planning,* Mar. 1970, pp. 51, 53, 54.
20. D. I. Currill, "Technological Forecasting in Six Major U. K. Companies," *Long Range Planning*, March 1972, p. 72.
21. A. D. Bender, A. E. Strack, G. W. Ebright, and G. von Haunalter, "Delphi Study Examines Developments in Medicine," *Futures*, June 1969, p. 289.
22. George Teeling-Smith, "Medicine in the 1990's: Experience with a Delphi Forecast," *Long Range Planning*, June 1971, p. 69.
23. *Communications Needs of the Seventies and Eighties*, Bell System, April 1972 (internal document).
24. Frank J. Doyle and Daniel Z. Goodwill, *An Exploration of the Future in Educational Technology*, Business Planning, Bell Canada, Montreal, January 1971 (proprietary).
Frank J. Doyle and Daniel Z. Goodwill, *An Exploration of the Future in Medical Technology*, Business Planning, Bell Canada, Montreal, Oct. 1971 (proprietary).
Daniel Z. Goodwill, *An Exploration of the Future in Business Information Processing Technology*, Business Planning, Bell Canada, Montreal, Oct. 1971 (proprietary).
Daniel Z. Goodwill, *Perspectives D'Avenir des Techniques de l'Informatique,* Planification du Service Téléphonique, Business Planning, Bell Canada, Montreal, August 1971 (proprietary).
Michael T. Bedford, *The Future of Communication Services in the Home*, Business Planning, Bell Canada, Montreal, November 1972 (proprietary).
25. Sources: Doyle and Goodwill, *Educational Technology* and *Medical Technology*; Goodwill, *Business Technology*.
26. Doyle and Goodwill, *Educational Technology*, p. 64.
27. *Ibid* pp. 40-41.
28. Doyle and Goodwill, *Medical Technology*, p. 60.
29. Goodwill, *Business Technology*, p. 99.
30. Michael T. Bedford, *The Future of Communication Services in the Home*, Round II Questionnaire, Business Planning, Bell Canada, Montreal, Jan. 1972, pp. 2-3.

31. Bedford, *Communications in the Home*, Nov. 1972.
32. One recent indication of this interest was the formation of the "International Society for Technological Assessment."
33. Bell Canada has recently attempted a modest pilot assessment: Philip Feldman. *A Technology Assessment of Computer Assisted Instruction*, Business Planning, Bell Canada, Montreal, Sept. 1972.
34. Baran, *Future of the Telephone Industry*.
35. Canadian Computer/Communications Task Force, *Branching Out*, Vol. 1, Information Canada, Ottawa, May 1972.
36. The Bell Canada Business Planning approach has been described in J. Martino, "Technological Forecasting is Alive and Well in Industry." *The Futurist*, **IV**, No. 4 (Aug. 1972), pp. 167-68.
37. An exception to this is the Policy Delphi conducted within a corporation.
38. See Chapter VII. For information on on-line Delphi see: Murray Turoff, "Delphi Conferencing," *Technological Forecasting and Social Change*, **3** (1972) pp. 159-204.
    Murray Turoff, "Delphi and Its Potential Impact on Information Systems," *Fall Joint Computer Conference Proceedings 1971*, Vol. 39, Afips Press, 1971.
    A. J. Lipinski, H. M. Lipinski, R. N. Randolph, *Computer-Assisted Expert Interrogation—A Report on Current Methods Development*, *Technological Forecasting and Social Change* **5** (1973) pp. 3-18.
39. A. J. Lipinski *et al.* pp. 13ff.
40. One exception is the book by Wills, which takes a marketing approach when reviewing forecasting and Delphi studies.

Acknowledgments

Many of the current and past members of the Bell Canada Business Planning Group have been involved with the Bell Delphi studies outlined in this paper. Mike Bedford, Frank Doyle, and Dan Goodwill spent many months on the research, design, conduct, and management of those studies. Don Atkinson, Ogy Carss, Ken Hoyle, and Sinc Pritchard provided the necessary management support and maintained a belief that these efforts would produce useful results.

I would also like to thank Phil Feldman for his efforts in tracking down many of the references listed. In addition to some of the individuals mentioned above, Tony Ryan and Phil Weintraub provided many useful comments and suggestions on earlier drafts of this article. L.H.D

III. C. 2. Plastics and Competing Materials by 1985: A Delphi Forecasting Study

SELWYN ENZER

The application of Delphi to the identification and assessment of possible developments in plastics and competing materials[1] posed a severe challenge to the technique. Before launching into a discussion of this project it is worth considering the advantages offered by the technique for this application. Since the study was conducted with questionnaires transmitted through the mails, it permitted many widely separated people to participate without the difficulty of having them travel to be co-located at any specific time. It permitted the group to focus on what they regarded as major developments very quickly and discuss only those prospects in detail. Furthermore, because anonymity was employed, each participant was forced to judge the potential of each possibility on the basis of his knowledge and the supporting arguments presented. In other words, the tendency to judge those developments suggested by the most notable panelists were eliminated by virtue of anonymity.

This study was originally scheduled to be completed in three rounds of interrogation. However, as it evolved, only two rounds appeared necessary. This occurred by virtue of the high degree of specialization which appeared in the first-round responses and became even more evident in the second round.

The ability to tailor-make plastics for various applications, enhanced by growth in understanding of organic chemistry, alloying, reinforcing, etc., plus the responsiveness of the material itself, have led many researchers to believe that the types of plastics produced in the future will be determined more by what is desired (and pursued) than by what is possible. Thus in many ways this study was more an investigation of material needs and resource allocations than of technological possibilities.

The study focused upon possible combinations of material property[2] changes that are likely to affect widespread material usage. A prime difficulty encountered in this study arose from discussing yet unknown (and hence unnamed) materials. In general, it is easier to discuss improvements in the properties of steel, aluminum, concrete, boron, niobium, etc., than to discuss the prospects for development of, and properties of, material *X*, *Y*, or *Z*. Yet in many cases

---

[1]Selwyn Enzer, *Some Developments in Plastics and Competing Materials by 1985*, Report R-17, Institute for the Future (January 1971).

[2]The term "material property" included not only physical properties such as strength, density, toughness, and others, but also processability and cost.

The Delphi Method: Techniques and Applications, Harold A. Linstone and Murray Turoff (eds.)
ISBN 0-201-04294-0; 0-201-04293-2

this study had to do exactly that. As a result, it probably tended to focus more on changes in existing materials than it did on totally new materials.

Since the study focused on material property changes that may be realized in existing materials as well as new materials and their properties, the number of alternatives to be contemplated was vast. To address this challenge a matrix-type categorization of materials and properties was used as the point of departure. For this purpose a breakdown similar to that presented in "The Anatomy of Plastics," *Science and Technology* (F. W. Billmeyer and R. Ford), was used. This matrix of materials and properties was divided into five subcategories:

- Engineering Plastics
- General Purpose and Specialty Plastics
- Glass Fiber Reinforced Plastics
- Foamed Plastics
- Nonplastics

The panel was asked to: (1) review the materials and properties presented, indicating where they thought changes were likely to occur within the next fifteen years which would significantly affect the widespread use of that material; and (2) add and describe the anticipated properties of new materials which they thought were likely to evolve and gain widespread use by 1985. In both of these steps the panel was also asked to describe the new chemical, physical, or other technological developments that they believed would lead to the creation of the new material.

These inputs from the first Delphi round were used to prepare a three-part questionnaire for the final round of interrogation. These parts were: (1) a summary of the assessments of anticipated changes in existing material properties, indicating those selected for more detailed investigation; (2) a listing of both plastic and nonplastic materials with the nature of the anticipated major changes described (those respondents who had anticipated these changes were asked to estimate the new material properties they expected would exist by 1985 and to estimate the 1985 annual consumption by application); and (3) a list of new materials anticipated by 1985 and a description of their properties (those respondents who had anticipated these items were asked to estimate the properties and consumption patterns they expected for these by 1985). All of these parts were open-ended in that any of the respondents could still add additional items or comment on any item.

Anticipated Changes in Properties of Existing Materials

The Delphi panel was presented with descriptions of the major uses, properties, and proprietary qualities of 37 plastics and 16 nonplastics all currently in widespread use. These 37 plastics are presented in Table 1. As indicated earlier, they were asked to: (1) identify likely changes in the properties of these materials which would significantly affect their widespread use by 1985; and (2) identify new materials (in each of the categories shown) which are likely to be developed and would be in widespread use by 1985.

Table 1

Existing Plastics

Engineering Plastics:	Glass Fiber Reinforced Plastics:
ABS	ABS
Acetal	Epoxy
Fluorocarbons	Nylon
Nylon	Polyester
Phenoxy	Phenolics
Polycarbonate	Polycarbonate
Polyimide	Polystyrene
High Density Polyethylene	Polypropylene
Polypropylene	San
Polysulfone	Polyethylene
Urethane	
Poly (Phenylene Oxide)	

General Purpose & Specialty Plastics:	Foamed Plastics:
Acrylics	Polyethylene
Cellulosics	Polystyrene
Cast Epoxy	Polyurethane (low density)
Ionomer	PVC
Melamines & Ureas	Polyurethane (high density)
Phenolics	
Low Density Polyethylene	
Polystyrene	
Vinyls (PVC)	
San	

The format used for this portion of the assessment is shown in Fig. 1. This figure is divided into four columns. Column 1 lists the material and its typical uses. Column 2 describes the properties of that material which are the key to its current widespread use. Column 3 is divided into subcolumns which contain specific material properties and the current performance ratings of each

EXISTING GENERAL PURPOSE & SPECIALTY PLASTICS

1. MATERIAL & Typical uses	2. Proprietary qualities (relative to typical uses): (+) Assets (-) Liabilities	3. Property: Price	Processability	Tensile strength	Stiffness	Impact strength	Hardness	Useful temperature range	Chemical resis.	Weather resistance	Water resistance	Flammability	4. Aspects of changes in material properties:
ACRYLICS Windows; fiber optics; building panels; lighting; tubing	(+) optical clarity; weather resistance (-) abrasion resist.	2	3	2	3	1	2	1	2	3	2	1	
CELLULOSICS Packaging; film; toys; telephones; instrument glass	(+) tough; clear (-) abrasion resist.	2	3	2	2	2	3	2	1	2	1	1	
CAST EPOXY Printed circuits; potting compounds	(+) strong & flexible	2	2	3	2	1	3	3	2	2	1	1	
IONOMER Molded housewares; toys; extruded tubing; sheeting; packaging	(+) transparent, tough & flexible; chemical resistance (-) strength; temp. range	2	3	1	1	3	1	1	2	2	2	1	
MELAMINES & UREAS Dishes; wood laminates; appliance cabinets; electrical devices	(+) appearance (finishability); surface hardness (-) impact strength; temp. range	2	2	2	3	1	3	2	2	2	2	2	
PHENOLICS Appliance cabinets & parts; bonding resins; electrical parts	(+) cost; strong, hard, rigid; abrasion resist. (-) chemical resist.	3	2	2	3	1	3	2	1	2	2	1	
LOW DENSITY POLYETHYLENE Dishes; bottles; pipes; tubing; film packaging	(+) flexible; cost (-) strength; weatherability; flammability	3	3	1	1	3	1	2	3	1	3	1	

3	Outstanding in property indicated; among the best performers available.
2	Acceptable performance in this property; still suitable in most cases.
1	Not acceptable if indicated property is important to intended use.

Fig. 1. Typical questionnaire for eliciting changes in existing plastics.

material relative to the others in that category. This rating is indicated with a "1," "2," or "3" in accordance with the code noted at the bottom of the figure.

The results of this assessment were presented to the panel in the format presented in Fig. 2. Shown in the subcolumns of Column 3 are the changes anticipated by the panel. These changes are noted, using the code presented in the upper-right-hand corner of the figure. Column 4 contains the panel's comments. These comments and all other changes suggested by the panel are in italics.

Items noted as being "included in Package No. 2" were reassessed by the panel in the second round in greater detail. The comments received from the panel regarding these materials were presented in greater detail in a subsequent section of the questionnaire.

In the course of this assessment several variations in the original list of existing plastics were made by the panel. As a result, the existing materials that were investigated further in round two are presented in Table 2.

The questionnaire format used for this interrogation and typical results are presented in Fig. 3. As before, the information presented to the panel is in roman type, the additions are in italics.

Table 2

Existing Plastics for More Detailed Considerations

Engineering Plastics:	Glass Fiber Reinforced Plastics:
ABS	ABS
PVF	Epoxy
Nylon	Nylon
Polyimides	Polyester (Molding Compounds)
High Density Polyethylene	
Polypropylene	
Polysulfone	
General Purpose & Specialty Plastics:	**Foam Plastics—Rigid:**
Acrylics	Polystyrene Foams
Epoxy	Low Density Polyurethane Foam
Ionomer	Variable (High Overall) Density Integral Skin Urethane Foam
Phenolics	
	Foam Plastics—Flexible:
	PVC Foam
	Variable Density, Integral Skin Urethane Foam

EXISTING GENERAL PURPOSE & SPECIALTY PLASTICS

1. MATERIAL & Typical uses	2. Proprietary qualities (relative to typical uses): (A) Assets (L) Liabilities	3. Property: Price	Processability	Tensile strength	Stiffness	Impact strength	Hardness	Useful temperature range	Chemical resis.	Weather resistance	Water resistance	Flammability	4. Comments
ACRYLICS Windows; fiber optics; building panels; lighting; tubing	(A) optical clarity; weather resistance (L) abrasion resist.	2 +	3	2	3	1 ++	2 +	1 ++	2	3	2	1 +++	*(Included in package No. 2).*
CELLULOSICS Packaging; film; toys; telephones; instrument glass	(A) tough; clear (L) abrasion resist.	2	3	2	2	2	3	2 (−)	1	2	1	1 +	●*Not competitive with low cost plastics, e.g., vinyls.*
CAST EPOXY Printed circuits; potting compounds	(A) strong & flexible	2 ++	2	3	2	1 +++	3	3	2 +	2	1 ++	1 +++	*(Included in package No. 2).*
IONOMER Molded housewares; toys; extruded tubing; sheeting; packaging	(A) transparent, tough & flexible; chemical resistance (L) strength; temp. range	2 ++	3	1 +	1 +	3	1 ++	1 ++	2	2	2	1 ++	*(Included in package No. 2).*
MELAMINES & UREAS Dishes; wood laminates; appliance cabinets; electrical devices	(A) appearance (finishability); surface hardness (L) impact strength; temp. range	2	2	2	3	1 ++	3	2	2	2	2	2	
PHENOLICS Appliance cabinets & parts; bonding resins; electrical parts	(A) cost; strong, hard, rigid; abrasion resist. (L) chemical resist.	3	2 ++	2 +	3	1 ++	3	2 +	1 +++	2	2	1 +++	*(Included in package No. 2).*
LOW DENSITY POLYETHYLENE Dishes; bottles; pipes; tubing; film packaging	(A) flexible; cost (L) strength; weatherability; flammability	3	3	1 +	1	3	1	2	3	1 ++	3	1 +++	●*Compounding can improve weather resistance.*

○ Degradations encircled indicate panel disagreements with the original rating not a forecast.

CODE

Property changes anticipated by the panel, likely to affect the widespread use by 1985, are noted in *italics* as follows:

Direction of expected change / % of the panel	40 or more	20 to 40	less than 20	None
Improvement	+++	++	+	
Degradation	−−−	−−	−	

3 — Outstanding in property indicated; among the best performers available.

2 — Acceptable performance in this property; still suitable in most cases.

1 — Not acceptable if indicated property is important to intended use.

Fig. 2. Typical feedback of results of initial estimates of changes in existing materials.

ENGINEERING PLASTICS

MATERIALS FOR WHICH THE PANEL ANTICIPATES A SIGNIFICANT INCREASE IN USE BY 1985			COMMENTS BY THE PANEL	AGREE	NO CONSENSUS	DISAGREE	APPLICATIONS		
KEY PROPERTIES IN WHICH IMPROVEMENTS ARE CONSIDERED LIKELY BY THE PANEL	CURRENT PROPERTY VALUE	ESTIMATE PROPERTY VALUE LIKELY BY 1985					MARKET BREAKDOWN (MAJOR USES)	VOLUME-MILLION LB/YR CURRENT	VOLUME-MILLION LB/YR ESTIMATE 1985
1. ABS			- Reduction in depolymerization will improve flammability.		x		TOTAL VOLUME	508	*1500*
Price, ($/lb.)	.28-.44	*.20-.30*	- Alloying and blending will improve temperature range and flammability	x			Automotive	80	*250*
Maximum Service Temperature *(D 648) °F	180-245	*250*	- Improved shaping of disperse particles as well as processing techniques will orient particles.	x			Major Appliances	60	*150*
Flammability, in./min.*(D 265)	1.0-2.0	*0.65-2.0*					Pipe & Fittings	60	*200*
Impact Strength Notched Izod, ft. lb./in. *(D 256) @ 73°F	2.0-10.0	*2.5-11.*	- Reduction in compound prices; scale and competition lowering prices.	x			Bus. Machines, Phones	40	*90*
@ 40°F	.8-3.5	*.9- 5.0*	- Availability of composite forms as sheet and ability to fabricate in inexpensive equipment will make this material competitive with glass-polyester.	x			Recreational Vehicles	45	*150*
Deflection temperature, °F *(D 648) @ 264 psi	214-244	*214-280*	- Platability makes this material increasingly attractive for automobile parts, this will be especially important if low temperature strength and crack resistance can be improved.	x			Luggage	27	*65*
@ 66 psi	215-250	*215-280*	- *Properties will be very dependent upon filler.*				Other:	196	*350*
			- *ABS will include chemical and cross-linked materials.*				*Interior panels & sheeting*		

*ASTM Test Method

Fig. 3. Typical results of final assessment of important changes in existing plastics.

The comments received from the panel are presented in Column 2 of this figure. Because these generally referred to the reasons why the material property changes were anticipated, the panel was asked to indicate whether or not they agreed or disagreed with each statement. The results of this assessment are also shown in Column 2. Those items presented in italics in this column were added in round two and hence were not assessed by the entire panel.

Column 3 presents the current major markets and their annual volume usage. Shown in italics are new markets suggested by the panel and their estimated 1985 usage.

In that portion of the investigation concerned with nonplastics, many new material developments were suggested, but only a few of these were regarded as threats to the growth of plastics. This can be seen in the following general comments received from the panel.

- The main competition between plastics and aluminum will occur in the construction field, particularly in residential housing and light industrial buildings. New developments in aluminum will hurt plastics in the applications which are primarily structural. On balance, however, these developments will affect the use of other metals more than plastics.
- In general, plastics will continue to replace iron and steel in some applications. This will be significant to the plastics industry; however, it will be a relatively small change to the steel industry. Any development which brings steel closer to "one-step" finishing, with improved environmental resistance, will be important in this regard, since it will blunt some of the basic advantages that plastics have over steel, allowing the use of "conventional" technology and existing capital equipment. Such developments will bring steel and plastics closer to a straight-cost competition. However, these developments must be realized before potential markets have switched from steel to plastics to maintain the continuity of technology and equipment.
- Developments in concrete appear more likely to enhance the demand for plastics than to replace or be replaced by them. Developments in wood and plywood are more likely to be in combination with plastics and hence are apt to increase the demand for such materials. However, unlike the concretes, wood will increasingly be replaced by plastics, particularly in furniture and siding.

Other Materials Suggested by the Panel as Likely to Become Important by 1985

In addition to the changes suggested in the existing materials, other materials (some already in existence) were suggested by the panel as prospects for widespread use by 1985. These are presented in Table 3.

These materials were submitted for consideration by the entire panel in the final round of reestimation. Format and typical results are presented in Fig. 4.

Table 3

Other Materials Suggested by the Panel as Likely to Become Important by 1985

Engineering Plastics:	Other Fiber Reinforcements and Reinforced Plastics:
Polybutadiene (High 1, 2 Content)	Boron Fibers
Polyethyleneterephthalate	Graphite Fibers
Polyphenylene Oxide Derivatives	Fiber Strengthened Oxides
New Thermoplastic	Aluminum Oxide Fiber & Whisker Composites
New Tougher Plastics	Boron/Epoxy
	Boron/Polyimide
	Graphite/Epoxy
	Graphite/Polyimide
General Purpose and Specialty Plastics:	**Foamed Plastics:**
PVC—Polypropylene Copolymers	Phenolic Foams
Ethylene—Polar Copolymer	Vinyl Foam
Acrylic—PVC	Polyolefins (Ethylene, Propylene, etc.)
New Polyolefins	Isocyanurate—Urethane
New Thermosetting Resins	Silicone Foams
Completely Nonburning Organic	Special Hi-Temp Foams
Semiorganic & Inorganic	Structural Foams
	Foamed Thermoplastics
	Injection Molded Urethane Foams
Glass Fiber Reinforced Plastics:	**Miscellaneous:**
PVC	Silicate Glasses & Polymers
Polyimides (or Amidemides)	Titanium Alloys
Polysulfones	Cermets
Polyurethane	
Vinyl Ether	
Thermoplastic Polyester	
Thermoplastic Sheet	
New Thermoplastic Resin	

As seen, this is similar to the format presented earlier. One notable difference is in Columns 5 and 6, which contain estiamtes of the likelihood of these materials being in widespread use by 1985 and the annual production estimated by that time. These estimates should be treated with even greater care than those presented earlier, since they often represent the comments of as few as two or three respondents who were familiar with the development.

OTHER ENGINEERING PLASTICS

1. MATERIAL & Typical uses (suggested by the panel)	2. Proprietary qualities (relative to typical uses): (+) Assets (−) Liabilities	3. Anticipated 1985 Properties: Price	Processability	Tensile Strength	Stiffness	Impact Strength	Hardness	Useful temperature range	Chemical resist.	Weather resist.	Water resistance	Flammability	4. Panel Comments	5. Likelihood of being in widespread use by 1985	6. Potential Volume by 1985 (Million lbs/year)
POLYBUTADIENE (HIGH 1, 2 CONTENT) *Very high performance electrical applications, thermoset molding compound, excellent high performance laminates, Asd laminate*	(+) low density, high rigidity, excellent electrical properties, clarity, high heat distortion (−) processing	2	2	2	3	2	2	2	2-3	2	2-3	2	- These products, which are just developing, will make great progress as availability and costs improve. - *Compared to thermosets these polymers are easily processed and have outstanding high temperature rigidity and low density.*	*80%*	*110*
POLYETHYLENETEREPHTHALATE (saturated polyesters) Large volume in fiber applications	(+) chemical resistance, mechanical properties (−) processing (especially unfilled)	2	2	3	3	2	2	2	2	2	2	2	- Fiber filling will upgrade properties significantly. - *Friction fatigue and bearing props of Nylon without cold flow or creep.*	*65%*	*75*
POLYPHENYLENE OXIDE DERIVATIVES		*1-2*											- Technology exists to make a family of these polymers and polymer blends. *(Yes, but this is true for every family and is not an adequate criterion for expansion.)*	*65%*	*750*

3	Outstanding in property indicated; among the best performers available	2	Acceptable performance in this property; still suitable in most cases.	1	**Not acceptable if indicated property is important to intended use.**

Fig. 4 Typical results of assessment of other materials added by the panel

As before, the roman type represents information obtained from the panel in round one and hence presented to the entire group in round two. The entries in italics represent the results received in round two.

Overall Forecasts of U.S. Plastic Production

Estimates of future plastic production, both in total and in several major subcategories, were also made by the panel. To assist in making these forecasts, graphic data describing the production of (1) all plastics, (2) foamed plastics (total and flexible), and (3) fiber glass reinforced plastics, for the U.S. since 1950 were presented to the panel. Each respondent was asked to extrapolate his estimate of these trends out to 1985. These estimates were collated to display the spread of opinion among the respondents.

Also presented to the panel was the distribution of the current production of major markets. The respondents were asked to estimate the 1985 distribution of production among these markets, and to add others they thought likely to become important by 1985.

In the opinion of the panel, the growth of the plastics industry in the United States will continue at a rate about equal to its current pace. This is seen in the forecast presented in Fig. 5.[3] As seen, the median estimate of the panel suggests that 50 billion pounds of plastic will be produced in the U.S. by 1985, and half of the panel's estimates ranged between 41 and 75 billion pounds for 1985.

Beyond the production estimates themselves, the shape of the trend curves for the median and the upper and lower quartiles appear to suggest a wide divergence of opinion regarding the saturation level of plastic production. The median and lower quartile estimates indicate that the rate of plastic production will peak by around 1980. The upper quartile, on the other hand, suggests that the growth rate in this time period will still be increasing, with no reversal of this trend by 1985.

As can also be seen in Fig. 5, all of the current markets are major growth candidates. The major markets listed currently consume approximately 40% of the plastic production and the 1985 median estimates indicate that these markets will represent about 40% of the 1985 median forecast of plastic production.

Figure 6 presents comparable estimates for foamed plastics production. These estimates were made for both total foamed plastic production and the subcategory of flexible foamed plastics. The difference between these estimates represents, of course, rigid foamed plastics. Foamed plastics, which are presently produced at the rate of approximately 1 billion pounds per year, are

[3]In this and all similar figures illustrating the panelists' forecasts, solid lines represent statistics, dashed lines the median forecast, and shaded areas the interquartile range.

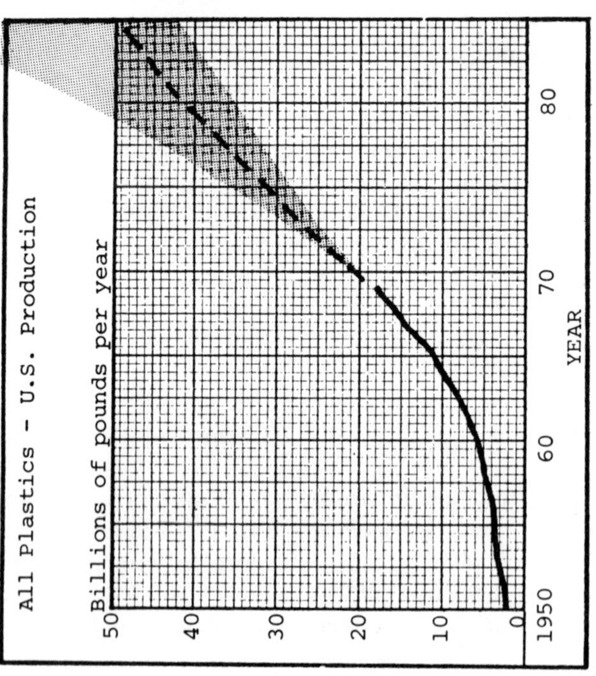

MARKETS	U.S. PRODUCTION (MILLIONS OF LB./YR.) CURRENT	BY 1985
Appliances	476	1,000
Construction	2,359	6,000
Consumer Products	1,245	3,500
Furniture	567	1,600
Packaging	2,892	5,500
Transportation	835	2,500

Fig. 5. Total U.S. plastic production.

MARKETS	U.S. PRODUCTION (MILLIONS OF LB/YR.)	
	CURRENT	BY 1985
Appliances	58	130
Construction	206	1,050
Furniture	270	825
Packaging	170	475
Transportation	200	575
Other	90	440

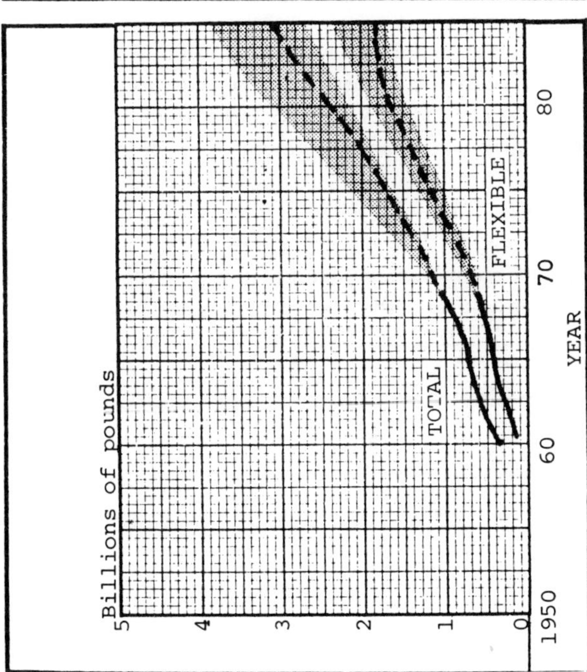

Fig. 6. Total U.S. production—foamed plastics.

expected to reach a production rate of between 2.7 and 3.8 billion pounds per year by 1985. The bulk of this growth is anticipated to result from the construction, furniture, packaging, and transportation markets.

Here again the shapes of the forecasts are quite revealing. These indicate that the largest portion of the growth of foamed plastics in the 1970–80 time period is expected to occur in the flexible foams. However, the panel estimates a leveling off of this growth after 1980, despite an increased rate of growth for foams in general. This indicates that the growth of rigid foams is likely to be slow until 1980 but is expected to increase more rapidly thereafter.

Figure 7 presents the panel's estimates for the growth of fiber glass reinforced plastics. As seen, the spread of opinion here is quite large, but even the conservative group, as indicated by the lower quartile curve, suggests a tripling of this production by 1985. Fiber glass reinforced plastics, which are presently produced at a rate slightly in excess of 1 billion pounds per year, are expected to reach a production rate of between 3.2 and 6.1 billion pounds per year. The major growth markets for this material are expected to be construction; marine products; transportation; and pipes, ducts, and tanks. Additionally, a significant growth in the use of fiber glass reinforced thermoplastics is anticipated. Presently only 6% of all fiber glass reinforced plastics are thermoplastics; by 1985 this figure is expected to reach 35%.

Interestingly, the median values for the numerical estimates of the market distribution for fiber glass reinforced plastics are considerably less than the median of the graphic estimate. However, since several of the respondents estimated only selected markets, consistency among these forecasts need not occur.

Along with these forecasts, comments were also elicited from the panelists. These comments are presented in Fig. 8. In general, these comments suggest that the growth of plastic production is related more to the nature of the products likely to be in demand and the natural environment (resources and pollution) than it is to technological progress per se.

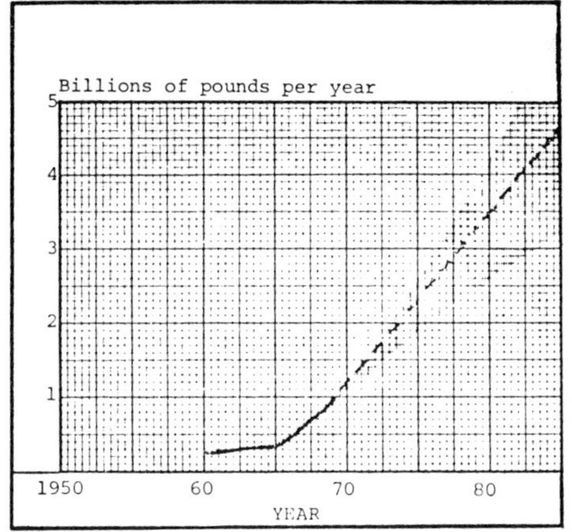

MARKETS	U.S. PRODUCTION (MILLIONS OF LB./YR.)	
	CURRENT	BY 1985
Appliances	30	120
Construction	108	600
Consumer Products	60	200
Marine Products	270	600
Transportation	200	975
Agriculture	34	100
Aerospace, aircraft	42	140
Electrical	82	225
Pipes, ducts & tanks	105	500
Miscellaneous	12	65
FGRP market distribution by plastic		
Type:		
Thermoplastics	6%	35%
Thermosets	94%	65%

Fig. 7. Total U.S. production—fiber glass reinforced plastics.

- PLASTICS, ESPECIALLY REINFORCED PLASTICS, WILL REPLACE WOOD AND METALS IN MANY APPLICATIONS.
- AS NATURAL RESOURCES CONTINUE TO BE DEPLETED, PLASTICS OF ALL KINDS BECOME MORE IMPORTANT.
- IMPROVEMENTS IN PROCESSABILITY MAKE PLASTICS INCREASINGLY MORE ECONOMICAL THAN TRADITIONAL ENGINEERING MATERIALS.
- PLASTIC USE WILL INCREASE BEYOND ITS NORMAL GROWTH BECAUSE OF ITS INCREASED USE IN: 1) ALL PLASTIC APPLIANCES, 2) MODULAR HOME CONSTRUCTION, AND 3) TRANSPORTATION EQUIPMENT.
- AIR POLLUTION WILL FORCE THE AUTO INDUSTRY TO ELECTRIC CARS AND HENCE ALL-PLASTIC BODIES.
- CONSUMERS WILL DEMAND, AND GET LONGER LIFE IN APPLIANCES, HENCE FUTURE PRODUCTION WILL NOT KEEP UP WITH GNP INCREASES, AND PLASTIC GROWTH RATE WILL DECREASE.

Fig. 8. Panel comments concerning future plastic consumption.

III. C. 3. A Delphi on the Future of the Steel and Ferroalloy Industries*

NANCY H. GOLDSTEIN

Introduction

In the spring and fall of 1970 a Delphi on the U. S. ferroalloy industry was conducted by the National Materials Advisory Board (NMAB) of the National Academies of Science and Engineering. The Board, concerned about a possible shortage of certain critical and strategic materials within the next decade or two, turned to the Delphi as a means of assessing the implications of technological change on usage trends of ferroalloys. The trends brought out by the Delphi could serve as a long-range planning guide for policy issues affecting the use of ferroalloys in steel making and certain other alloy production. This article will discuss the format of the Delphi, the selection of respondents, the manpower required to carry out the exercise, and the round-by-round method of conducting the Delphi. The article will then present a comparison of the Delphi exercise with a conventional panel study[1] which was conducted simultaneously with the Delphi exercise and conclude with some advice to prospective Delphi designers.

Form of the Delphi. The Steel and Ferroalloy Delphi included three rounds. The questions and exercises presented in each round were divided into three sections: Section I, Steel; Section II, Alloys; and Section III, Key Developments. Sections I and II generally presented trend lines for extension by the respondents and the assumptions underlying these extensions. Section III indicated future developments thought by the respondents to have a potential role in the steel and/or ferroalloy industry in the next two decades. More detailed descriptions of these three sections will be given in the round-by-round discussions which follow.

Selection of Respondents. The original Delphi respondents were not chosen randomly but were carefully selected from all sectors of the industry, government, the universities, institutes, and trade publications. Members of the NMAB Panel on Ferroalloys submitted suggestions for respondents and the

*The full report on this exercise is available from the National Technical Information Service, Springfield, Va., as "A Delphi Exploration of the U. S. Ferroalloy and Steel Industries," by Nancy H. Goldstein and Murray Turoff, NMAB-277, July 1971.

[1]Available as "Trends in the Use of Ferroalloys by the Steel Industry of the United States," NMAB-276, July 1971, by the Panel on Ferroalloys of the NMAB.

The Delphi Method: Techniques and Applications, Harold A. Linstone and Murray Turoff (eds.)
ISBN 0-201-04294-0; 0-201-04293-2

panel as a whole discussed each suggestion. One hundred names were chosen for the initial Delphi round. These one hundred potential respondents received a letter inviting them to participate, a card to return to the panel indicating their preference for participating or not, and a copy of the first-round questionnaire of the Delphi. Of the one hundred potential respondents, forty-two returned the card stating that they wished to participate and thirty-three actually responded to the first round. Response to the exercise was voluntary and no compensation was provided. This resulted in a much higher percentage response from industry-associated respondents who, as members of planning staffs, could consider the effort part of their job function. A much lower percentage response occurred from university people who probably considered this request as an uncompensated consulting effort.

The summary below shows the makeup of the final respondent group and the number of respondents replying to rounds one and two, one and three, and one, two, and three:[2]

OCCUPATIONS OF DELPHI RESPONDENTS

Occupation	Number of Respondents
Ferroalloy Producer	6
Nonferrous Alloy Producer	1
Specialty Metals	2
Powder Metals	2
Specialty Steel Producer	4
Steel Producer	4
Polymers	3
Institutes	4
University	4
Government	1
Technical Journal	2
Consultant	1
	34

PATTERN OF RESPONSES TO THE ROUNDS

Rounds	Number of respondents
1 and 2	3
1 and 3	2
2 and 3	3
1, 2, and 3	28

[2]No respondent replied to fewer than two rounds.

Manpower. The manpower required for this Delphi included two full-time professionals—a senior professional and his assistant—and intermittent temporary clerical and secretarial help. The exercise was conducted in cycles: one to two months waiting for the results of the previous round and making preparations for the handling of these results, and one to two months actually handling the results and preparing them in a form suitable for the next round. The requirements for secretarial help were also cyclical: little or no help was required during the waiting period but two full-time secretaries were required during the week-long rush period when the results had been tabulated and were being typed up for inclusion in the next round.

The Delphi ran for three rounds. Each round will be discussed below in terms of the design of that round and the handling of the results. It is, however, somewhat artificial to separate the design of a new round from the handling of the results of the previous round, since the form of the new round determines the method of handling and presenting the old round.

Tables 1 and 2 summarize (1) the effort involved in designing, monitoring, and analyzing the Delphi, (2) the contributions by the respondents, and (3) the flow of information in the Delphi rounds. While the clerical effort is broken out separately, a significant portion of this was actually done by the professionals involved. The availability of clerical-type support was a random process that did not always conform to requirements. One key element of both clerical and secretarial support is the benefit of having the same individuals to aid on every round, since there is a learning curve on the explicit procedures to be followed.

Round One

Design. Round one was divided into three sections. The first section entitled "Steel," presented graphs covering various aspects of the steel industry—total steel shipments, ratio of shipments to production, etc. A trend line, usually running from 1960–69, was shown on the graph and the respondents were asked to extend the line through to 1985. Three questions were associated with each graph:

(1) How reliable did the respondent consider his graph extension to be?

(2) What key developments (i.e., his assumptions) did the respondent assume in making his extension?

(3) What other developments (i.e., his uncertainties) might result in major revisions in the extension?

A flow chart of the steelmaking process was also presented at the end of Section I. Figures were given for 1969, and the respondent was asked to supply the corresponding figures for 1980.

Section II, entitled "Alloys," presented a number of graphs in the same manner as Section I. The graphs of Section II were concerned with aspects of the ferroalloy industry—U. S. consumption of chromium, tungsten, etc., and

Table 1

Table of Analysis Effort

Units: Weeks or Man-weeks	Elapsed Time	Senior Professional	Professional	Clerical Tabulation and Curve Extrapolation	Secretarial (Typing)
Pre-Round One	4 weeks	4 man weeks	4 man weeks*	—	1 man week*
During Round One	6 weeks	1 man week	1 man week	—	—
Pre-Round Two	6 1/2 weeks	6 man weeks	—	8 man weeks	2 man weeks
During Round Two	7 weeks	1 man week	3 man weeks	—	—
Pre-Round Three	4 1/2 weeks	1 man week	2 man weeks	5 1/2 man weeks	1 man week
During Round Three	8 weeks	1 man week	1 man week	1 man week	1 man week
Pre-Final Report	10 weeks	4 man weeks	2 man weeks	1/2 man week	2 man weeks
TOTALS	46 weeks	18 man weeks	13 man weeks	15 man weeks	7 man weeks

*Man week refers to time spent by male or female.

Table of Respondents' Contributions

	Number		Individual Effort (average man hours)	Number of Curves Extrapolated	Number of Comments Added	Number of Comments Evaluated	Number of Key Developments Evaluated
	Sent	Rec'd					
Round One	42	33	11.7	33	0	0	0
Round Two	52	34	7.0	36	401	401	36
Round Three	38	33	5.5	3	235	266	40
TOTALS			24.2	72		667	76

exports and imports of these materials. The exercise was separated into a Steel Section and a Ferroalloy Section because as majority of the expertise in the respondent group broke down into specialists in these two areas. While most respondents had something to contribute to both sections, it was clear, when the results came in, that a given individual usually focused most of his effort on one of the two sections.

Section III, entitled "Suggested Additional Variables and Key Developments," offered blank graph sheets and blank key development tables for respondents wishing to add to the items presented in Sections I and II. Figure 1 indicates the format.

The selection of categories to be presented in Sections I and II was made by the questionnaire designers, with suggestions and assestance from the Panel on Ferroalloys.

Handling the Results. Round one was mailed to forty-two respondents on June 16, 1970; thirty-three responded to this round.

Table 2

FLOW OF INFORMATION IN THE DELPHI ROUNDS*

	ROUND 1		ROUND 2		ROUND 3
HISTORICAL TREND LINE					
	CURVE PROJECTIONS				
		50% BOUNDARIES			
			RE-PROJECTION		
				50% BOUNDARIES	
		NEW CURVES ADDED			
			PROJECTIONS		
				50% BOUNDARIES	
					RE-PROJECTION
	ASSUMPTIONS UNCERTAINTIES				
		ASSUMPTIONS			
			VALIDITY JUDGMENTS		
				AGREEMENT AVERAGED DISAGREEMENTS	
					RE-EVALUATED
		KEY DEVELOPMENTS			
			PROBABILITY BY 1975 AND 1980 IMPACT ON INDUSTRY (QUANTITATIVE, QUALITATIVE)		
					RE-EVALUATION
FLOW OF STEELMAKING CHART					
	FLOW ESTIMATES AND MODEL CHANGES (1969 AND 1980)				
			RE-ESTIMATION		
				INPUT-OUTPUT MODEL	
					ESTIMATES (1969 AND 1980)

*Information presented within the boundaries was generated by the respondents. Information presented between the boundaries was provided to the respondents by the monitors.

1. Potential Development	2. Likelihood of Occurrence by 1980						Impact on U.S. Steel Industry if Development were to Occur				4. Nature of Impacts (add if you wish)
	Very (1.–.8) Probable	Probable (.8–.6)	Either Way (.6–.4)	Improbable (.4–.2)	Very (.2–0) Improbable	No Judgment	Strong	Moderate	Slight or None	No Judgment	
2.17 Development of an economical process for the recovery and utilization of Titanium scrap	32	37 A	11	16		5	16	16	53 A	16	Enhance Ti competitive position; Steel companies in Ti field will alter product balance; Ti scrap already being used in steel production; Reduce Ti price; Nonferrous metal prices have tremendous impact on steel demand (i.e., substitution prone) particularly in construction
2.18 Development of an economical process allowing a major improvement in Titanium workability	6	44 A	28	11	6	6	6	24 A	53	18	Same as 2.17
2.19 More than 20% of U.S. Manganese requirements met by ocean floor mining		22	17 A	39	13	9	11	5	63 A	21	Increased availability of Co and Ni; Implies higher cost for Mn and reduced consumption as a result
2.20 U.S. low-grade Manganese ores become economical for meeting 20% or more of U.S. requirements		20	25 A	25	15	15	6	6	59 A	29	Four times cost of imported material; Implies disruption of ocean transport or unavailability of foreign sources

Fig. 1. Format for Key Developments.

(A denotes average)

There were two principal elements to handling the results of round one. First, a determination was made, for each graph, as to the location of upper and lower limits of the extensions which would include 50 percent of the responses to that graph.

The second element involved the gathering and synthesizing of comments presented under Questions 2 and 3 of the graph sheets. The comments of all the respondents were collected, and each comment was then studied and determined to be either a forecasting assumption, an economic and international consideration, a key development, or a comment to be associated with that particular graph page. The comments were grouped accordingly, and the final product was retyped for inclusion on the second round. It was quite apparent that in many instances one respondent's assumption was another's uncertainty. The frequency with which a topic was brought up influenced the judgment on its choice as a key development for round two.

The process of collecting and editing the large number of comments obtained on the first round represented the largest single task in the exercise, in terms of both clerical time and professional judgment. Assumptions from all respondents were initially xeroxed, cut out, and taped on large sheets. Each sheet represented different topics or curves. On these large sheets duplications were crossed out and editing of assumptions to produce shorter wordings took place. This conglomeration was then retyped once and put through a final polishing, editing, and reordering before the final typing for the second-round questionnaire. The process of putting each set of assumptions through a two-stage editing process allowed each professional to check the other's work. It is noticeable to a certain extent that the availability of the xerox machine is a key feature in making large-scale Delphis possible via paper-and-pencil approaches. This is particularly true where one is handling a large volume of textual comments on the part of the respondents.

Section III in round one, "Suggested Additional Variables and Key Developments," produced few responses from the participants. Some of the responses in this section became key developments for Section III of round two, some became assumptions for round two, and the remainder were dropped from the exercise. Three additional curves suggested by the respondents were prepared for inclusion in round two.

The flow chart appended to Section I in the first round also received scant response (i.e., about ten respondents). The responses that were supplied were averaged for each entry in the flow chart and standard deviations were provided through the use of a simple computer program.

Round Two

Design. The questionnaire sent to the respondents in round two was patterned after the results described in the handling of round one.

The new Section I, "Steel," contained thirty-six "Forecasting Assumptions," thirty-five "Economic and International Considerations," and all the graphs contained in round one (Section I) with their associated reasons. The respondent was asked to associate a validity score with each forecasting assumption and economic and international consideration presented. The scores were based on the validity codes which are shown in Fig. 2.

For all graphs the original trend line was presented and the 50 percent confidence limits were indicated. The respondent was asked to reestimate his previous extension, after viewing the 50 percent limits, and to identify his estimate as reliable, as good as anyone's, or risky. Each graph also contained associated reasons given by the respondents for increasing or decreasing the graph extension. In Section I, a total of 116 reasons associated with the graphs were presented for evaluation. The respondent was asked to assign a rank of 1 to 6 to each reason given according to the validity scale. He was also invited to show additional reasons if he wished. A sample of a typical round-two question is presented in Fig. 3. This is exactly the same basic form as used in round one.

At the end of Section I, the flow charts were again presented. Two charts, for 1969 and 1980, showed the means and standard deviations for each chart entry and also showed some additional boxes and paths not included in round one. These modifications were suggested by the respondents and incorporated by the senior professional. The respondent was asked to circle the estimate presented if he agreed with it, or to cross it out and provide a new figure on the blank charts provided if he disagreed. The absence of either action was considered to be a No Judgment vote. It was a surprise to the designers that almost all the respondents to the flow chart chose to modify it, since this was not an action suggested in the instructions.

Section II, "Alloys," was similar to Section I in design. The category of forecasting assumptions included sixty-nine assumptions about individual alloys studied, as well as seventeen general assumptions. The respondents were again asked to assign each statement a validity score of from 1 to 6. There was no category of Economic and International Considerations, but all the graphs from Section II of the first round were included with the 50 percent confidence limits and 128 associated forecasting reasons. The respondents were asked to provide the same information requested in Section I.

Section II was entitled "Key Developments and Added Curves." Under Key Developments, thirty-six items were presented for scoring by the respondents. For each item, the respondent was asked to evaluate the likelihood of occurrence by 1975 on a scale of 1 to 6 and to indicate, on a scale of 1 to 4, the impact on the steel industry if the development were to occur. Figure 1 provides a sample of the form utilized for key developments.

In addition to scoring the developments, the respondents were asked to describe the nature of impacts they had characterized as strong or moderate.

Three new curves, supplied by respondents to round one, were also included in this section. The respondents were asked to handle these new curves in the same way they treated the original curves in round one (Sections I and II).

Numeric Scale	
1	CERTAIN (Average of 1 to 1.5) • Low risk of being wrong. • Decision based upon this will not be wrong because of this "fact." • Most inferences drawn from this will be true.
2	RELIABLE (Average of 1.6 to 2.5) • Some risk of being wrong. • Willingness to make a decision based upon this. • Assuming this to be true but recognizing some chance of error. • Some incorrect inferences can be drawn.
3	NOT DETERMINABLE (at this time) (Average of 2.6 to 3.5) • The information or knowledge to evaluate the validity of this assertion is not available to *anyone*—expert or decision-maker.
4	RISKY (Average of 3.6 to 4.5) • Substantial risk of being wrong. • Not willing to make a decision based upon this alone. • Many incorrect inferences can be drawn. • The converse, if it exists, is possibly RELIABLE.
5	UNRELIABLE (Average of 4.6 to 5) • Great risk of being wrong. • Worthless as a decision basis. • The converse, if it exists, is possibly CERTAIN.
6	NOT PERTINENT (Used to eliminate some assumptions from exercise) • Even if the assertion is CERTAIN or UNRELIABLE it has no significance for the basic issue. • It cannot affect the variable under question an observable amount.
blank	NO JUDGMENT • No knowledge to judge this item, but the appropriate individual (expert, decisionmaker) should be able to provide an evaluation I would respect.

Fig. 2. Validity or confidence scale.

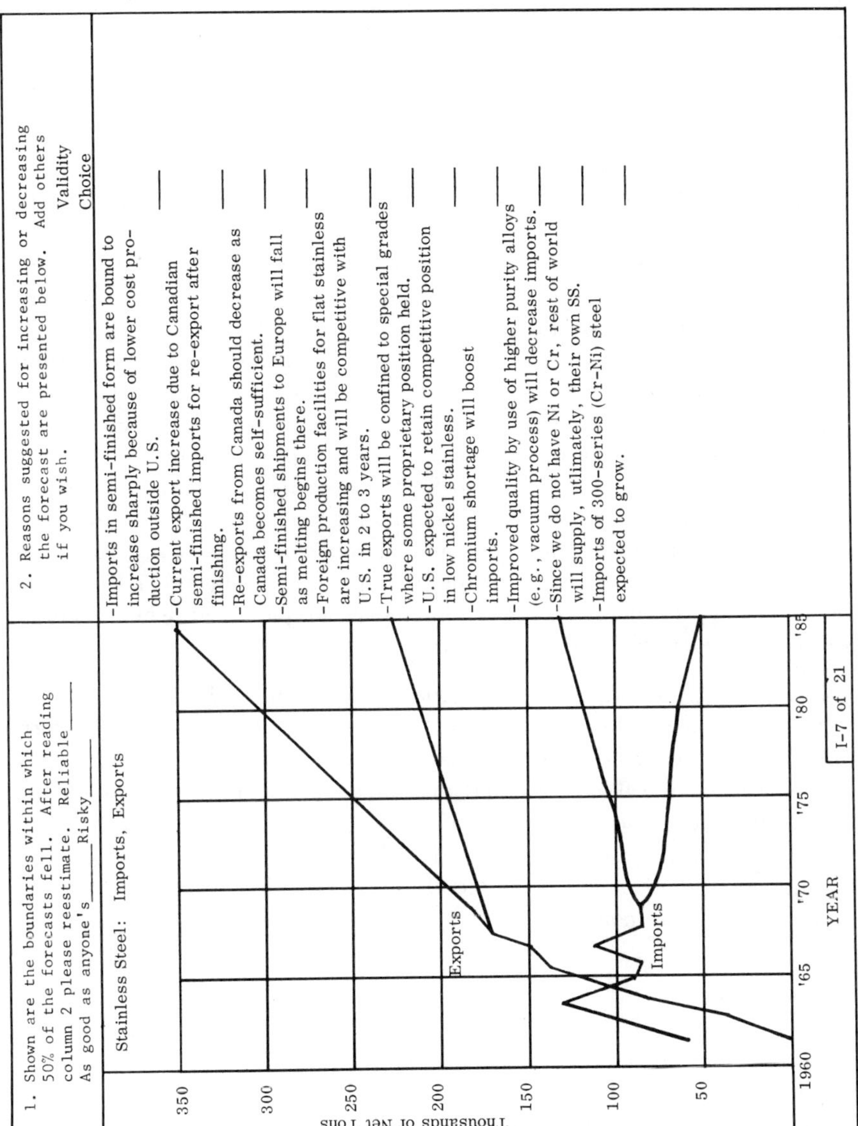

1. Shown are the boundaries within which 50% of the forecasts fell. After reading column 2 please reestimate. Reliable____ As good as anyone's____ Risky____

Stainless Steel: Imports, Exports

2. Reasons suggested for increasing or decreasing the forecast are presented below. Add others if you wish.

	Validity Choice
-Imports in semi-finished form are bound to increase sharply because of lower cost production outside U.S.	____
-Current export increase due to Canadian semi-finished imports for re-export after finishing.	____
-Re-exports from Canada should decrease as Canada becomes self-sufficient.	____
-Semi-finished shipments to Europe will fall as melting begins there.	____
-Foreign production facilities for flat stainless are increasing and will be competitive with U.S. in 2 to 3 years.	____
-True exports will be confined to special grades where some proprietary position held.	____
-U.S. expected to retain competitive position in low nickel stainless.	____
-Chromium shortage will boost imports.	____
-Improved quality by use of higher purity alloys (e.g., vacuum process) will decrease imports.	____
-Since we do not have Ni or Cr, rest of world will supply, utlimately, their own SS.	____
-Imports of 300-series (Cr-Ni) steel expected to grow.	____

Fig. 3. Form for Trend Extrapolations.

Handling the Results. Round two was sent to fifty-two potential respondents; thirty-four replies were received. Several respondents, representing the polymer industry, were added during this round. They had not been represented in round one and were introduced for the purpose of addressing specific issues on the substitution of plastics which had been generated in round one.

The results of round two required three separate types of handling: (1) new 50 percent confidence limits were supplied for the graphs; (2) verbal comments associated with the assumptions, the graphs, and the Key Development section were collected and considered; and (3) the numerical results, i.e., validity choices, Key Development scores, and flow chart inputs, were collected and tabulated.

There were several steps necessary in handling the large amount of data that was generated by the results of round two. It was first determined that several statistical calculations on the data would be desirable—specifically, the mean and standard deviation of the validity choices for each statement, a distribution showing the percentage of responses falling under each of the scores from 1 to 6 for each of the statements, and a matrix comparing the distribution, by numbers, of the different occupation categories represented by the respondents with the range of scores from 1 to 6. These occupation categories included: primary steel producers, research institutions, steel producers, ferroalloy producers, research institutions, government, and the universities. A computer program was written to carry out these computations. This breakdown allowed us to observe if there were any differences in judgment which may have reflected differences in affiliation of the respondents.

Examples of the statistical presentations are shown in Fig. 4.

The flow chart included in Section I received very little additional information in round two. The scant information received was averaged into previous information on the flow chart and presented in a summary for round three. Due to the differences of opinion among the respondents on how actually to model the flow of steel-making materials, it was felt by the designers that this one question could have constituted the total Delphi exercise among a select smaller group of respondents.

Round Three

Design. Round three again consisted of three major sections: Section I: A Summary of Round Two, Section I, Steel; Section II: A Summary of Round Two, Section II, Alloys; and Section III: Key Developments and Added Curves.

Sections I and II were provided for summary purposes and required no further input by the respondents. For each statement, the mean and standard deviation as calculated from round two results were shown. Several new

Question Number	Mean	Standard Deviation
1	3.4	1.0
2	2.6	0.9
3	3.0	1.2

Question Number	Validity Choice						Number Responding
	1	2	3	4	5	6	
1	.25		.25	.25		.25	25
2		.33	.33		.33		16
3	.20	.20		.20	.20	.20	19

Question 1

Occupation Category	Validity Choice					
	1	2	3	4	5	6
1 Steel	3	2	1	2	0	1
2 Ferroalloy	1	4	0	0	0	0
3 R&D	3	1	0	1	0	0
4 Government	0	0	1	3	2	0

Fig. 4. Examples of statistical presentations.

statements were added in Sections I and II; a few new developments were added to Section III to be assigned a validity score by the respondents. The new 50 percent confidence limits, taken from the results of round two, were also presented in a final form for each graph.

Section III, the only section to be returned by the respondents, contained all the new assumptions and all previously evaluated assumptions which exhibited a large standard deviation (i.e., disagreement). It also called for a reevaluation of all key developments.

The first portion of Section III indicated the percentage distribution of the scores from round two on the likelihood and impact of potential key developments. The estimated average score for each development was indicated and a summary of all verbal comments associated with each development was given. The respondent was asked again to give his preference on the likelihood and impact of each potential development.

The second portion of Section III presented the three curves shown in this section of round two and included a number of reasons given by the respondents of round two for their curve extensions. The respondent was asked to reestimate the curves after reading the associated past reasons and to rate the reliability of his estimation. He was also asked to vote on the reasons given for each curve using the validity scale from 1 to 6 described earlier.

The third portion of Section III contained all assumptions from Sections I and II which exhibited a considerable degree of disagreement. This category generally, although not exclusively, included assumptions with a standard deviation of 1.3 or greater. The respondent was asked to reevaluate his previous validity choice and submit a new score. Several new assumptions were also added and the respondent was requested to provide a validity choice for these new assumptions.

In the final portion of Section III a new chart was introduced showing percentage breakdowns of inputs and outputs for three major steel processes. The figures were supplied for 1969 and a blank sheet was provided for 1980. The respondent was asked to fill in the sheet for 1980 and to change any 1969 figures with which he disagreed. A space was provided for an explanation of any disagreements with 1969 figures. The results of this chart were to be considered a summary response, as the monitors did not plan to feed back the responses for changes.

The monitor also surveyed briefly the attitudes of the respondents toward the Delphi approach by asking the rspondents a number of questions, e.g., was the time spent in participating in Delphi well used; what organizations should sponsor an exercise of this type on a regular basis, etc.

Handling the Results. Round three was sent to thirty-eight respondents on December 10, 1970. Thirty-three respondents actually replied to Round 3.

A computer program provided the means, standard deviations, percentage distributions, and industry category matrices for all key developments, assumptions to be reevaluated, and new assumptions. The percentage distributions were then examined by the senior professional. If 20 percent or more of the vote fell into the "not pertinent" category (a validity score of 6), the items were dropped from the exercise. Eight items were dropped for this reason. The remaining items were regrouped so that every assumption was associated with a curve. The assumptions and reasons for each curve were then reordered according to their mean validity scores.

The final report was then prepared for the National Materials Advisory Board of the National Academies of Science and Engineering.

Comparison of Delphi and Panel Studies

A separate panel appointed by the National Materials Advisory Board approached the same problem considered by the Delphi exercise in a more "conventional" vein. The conventional study was carried out simultaneously with the Delphi, but the results were not compared until both exercises had been completed.

In the panel approach, individual members of the Panel on Ferroalloys reviewed portions of the problem with which they were most familiar. The Panel Report, NMAB-276, "Trends in the Use of Ferroalloys by the Steel Industry of the United States" consists of chapters, each of which is logical, comprehensive, and definitive with respect to its topic. While appropriate caveats exist, its forecasts are precise. The recommendations and conclusions therein represent the unanimous agreement of the panel and no areas of disagreement are spelled out. The result is typical of a competent panel (or committee) activity. Based upon the expertise of the carefully selected participants, the report is a reliable and comprehensive account of known information and of projections based on this information and on current research and development. In contrast, this Delphi was designed to complement the panel report. The planned approach was to provide an opportunity to indicate uncertainties or disagreements about the subject and to evaluate quantitatively the degree of uncertainty which exists within a large group of experts. The Delphi product attempts to present an awareness of the areas which are subject to differences of view and to highlight the topics which appear to concern the respondent group. The Delphi provides a group evaluation of every statement advanced by the respondents who, presumably, express their beliefs. Although the results of this exercise include a number of statements which were rated uncertain, risky, or unreliable by the whole group, this variation does not imply that one dissenter from the group will be incorrect in retrospect. The group view has a higher probability of being correct than the view of any one individual. However, in the past, developments that significantly affected industries were often unforeseen by most of the involved experts. Therefore, the reader is cautioned not to extrapolate blindly from the group judgments exhibited in a Delphi to assumed facts.

In this case, the Delphi exercise is a literal exploration of the minds of experts in the steel and ferroalloy industries regarding their views on individual items. This exploration allowed a broader coverage of the subject area than was possible in the panel report.The presentation of the Delphi results allows the reader to compare easily his judgments with those of the group. No attempt is made to arrive at conclusions or recommendations, or to present a definitive view as was done in the panel activity.

Where the panel and Delphi activities overlap, there is considerable agreement in their forecasts. Figure 5 compares the consumption in steel of a number of alloys, as predicted by the panel and the Delphi, and some of the qualitative features of the two methodologies.

Other comparisons could be made between information presented in the panel report and the Delphi predictions. For example, NMAB-276 projected that carbon steel shipments would increase by 22 percent in the next ten years;

Quantitative Comparison

	Predicted Ferroalloy Consumption in Steels and Superalloys in 1980 (In Short Tons of Container Element)	
	Panel Report NMAB-276	Delphi
Chromium	319,260	250,000–303,000
Cobalt	2,732	3,000–4,000
Columbium	1,977	1,300–1,850
Manganese	1,011,235	1,100,000–1,250,000
Molybdenum	21,540	17,400–21,000
Nickel	124,200	90,000–115,000
Tungsten	1,850	1,550–2,600
Vanadium	5,796	5,000–6,200

Qualitative Comparison

Type of Activity	Delphi	Committee
Form of Information	Specific Comments	General Discussion
Weighting of Information Provided	Most Rated on Reliability Scale	None
Disagreement among Committee Members or Respondents	Indicated in Reliability Score	Eliminated
Presentation of Back Ground Information	Only as Randomly Generated by Respondents	Thorough and Systematic
Recommendations	Not Specifically Stated	Consensus Recommendations Indicated
Range of Information Provided	Broad, Reflecting Wide Interest of Respondents	Limited to Specific Committee Subject Area

Fig. 5. Comparison of the panel and Delphi approaches.

the increase projected on the Delphi graph for the next decade was 22 to 26 percent over the current figure. Also, the panel report stated that High Strength Low Alloy Steel (HSLA) is the fastest growing segment of the steel industry; the Delphi results were that HSLA is one of the two fastest growing segments of the industry.

As was mentioned earlier, the panel report did not indicate the areas of disagreement. In the Delphi category, "not determinable" with respect to validity reflected either the inability of the entire group to determine the validity of an assumption or the averaging of opposing judgments on the validity of a given assumption. Of the 135 statements that fell into this classification, seventy-three reflected an actual disagreement among the respondents by exhibiting a high standard deviation.

The following assumptions exemplify those that fell into the "not determinable" category of the Delphi as a result of disagreement:

- Cobalt-iron base tool steels will be marketed.
- Continuing nickel shortages will establish permanent substitution.
- New techniques will allow significantly greater flexibility of substitution among alloying elements based upon price changes.
- Critical shortages of nickel will reoccur.
- No important new use for cobalt.
- Present and projected investment in ocean studies is too large to exclude development of economical offshore mining except in short term (i.e., next five years).
- Alloy steels will increase in nickel content.
- Tungsten content of carbides will decrease because of cost.
- Full alloy shipments will parallel automotive production.
- Low-cost method of preparing high-purity iron powder will be developed by 1975.
- More and cheaper scrap will result from urban waste recycling.
- Shortages of natural gas will be a primary limiting factor in the expansion or modernization of the steel industry.
- Electron-beam refining will grow significantly.

An additional fifty-seven assumptions were rejected by the Delphi respondents as either risky or unreliable. One must reflect that while each of the 667 assumptions were suggested by at least one expert in the Delphi respondent group, approximately two hundred of these, or 30 percent, were considered less than reliable by the group as a whole. Furthermore, there is often considerable value to decisionmakers in observing the nature of rejected assumptions. For this reason the final Delphi report listed all the evaluated assumptions pertaining to any curve in the order of decreasing validity so one may observe the complete span of the topics covered. Probably the most significant difference between the Delphi and committee approaches is the itemization of what the Delphi group could not agree on or what they rejected. Usually the psychological process in a committee of experts tends to eliminate these categories of information from the final report.

In summary, the two reports did not always cover the same subject matter. However, when they did touch upon the same subject matter, the results were generally compatible.

Advice to Future Delphi Monitors and Designers

The monitor's experience with the Steel and Ferroalloy Delphi gives rise to a number of observations and advice to those planning to monitor Delphis in the future.

(1) When presenting statements for a vote, or synthesizing the respondents' suggestions, be alert for ambivalent wording. Two separate statements may appear as one, leading to confusion as to what should be voted upon. Vague wording or easily misinterpreted wording may also lead to confusion.

(2) When editing respondents' comments for clarity, try to preserve the intent of the originator. When editing from round to round, avoid changing a statement so that it has one meaning in round one and another in round two.

(3) Lay out the expected processing of the data throughout all the rounds of the Delphi before you finalize the design. You may, by circumstance, be forced later to modify the procedure, but the process of planning ahead will usually turn up any gross problems in your initial questionnaire design and its impact on following rounds.

(4) Design the handling of your data so that each response can be processed (or punched for processing) as it comes in. Thus you will not have a frantic rush to analyze all the responses at once when the last tardy return comes in.

(5) Keep track of how different subgroups in your respondent group vote on specific items. This can be very useful in analyzing the results and will occasionally produce situations where you wish to let the respondent group know that polarizations or differences based upon background exist.

(6) If you are covering a number of fields of expertise, make sure that each field is adequately represented in your group.

(7) It should be mandatory that at least two professionals work on monitoring any one Delphi exercise, particularly when the abstracting of comments is a notable portion of the exercise. With two individuals one can always review what the other has done.

(8) Pretest your questionnaire on any willing guinea pigs you can find outside your respondent or monitor group. If you have a sponsor, it is useful to go over the design of each round with some of his people before finalizing it.

IV. Evaluation

A. Introduction

HAROLD A. LINSTONE and MURRAY TUROFF

Skeptics from the allegedly "hard" sciences have at times considered Delphi an unscientific method of inquiry. Of course, the same attitude is often encountered in the use of subjective probability (even in the face of considerable mathematical theory developed to support the concept). The basic reason in each case is the subjective, intuitive nature of the input.

Yet Delphi is by no means unordered and unsystematic. Even in the Gordon-Helmer landmark Rand study of 1964, an analysis of certain aspects of the process itself was included.[1] The authors observed two trends: (1) For most event statements the final-round interquartile range is smaller than the initial-round range. In other words, convergence of responses is more common than divergence over a number of rounds. (2) Uncertainty increases as the median forecast date of the event moves further into the future. Near-term forecasts have a smaller interquartile range than distant forecasts.

It was also observed in all early forecasting Delphis that a point of diminishing returns is reached after a few rounds. Most commonly, three rounds proved sufficient to attain stability in the responses; further rounds tended to show very little change and excessive repetition was unacceptable to participants. (Obviously this tendency should not unduly constrain the design of Policy Delphis or computerized conferencing which have objectives other than forecasting.)

We shall briefly review here some of the systematic evaluations made in recent years.

Dispersion as a Function of Remoteness of Estimate

Martino has analyzed over forty published and unpublished Delphi forecasts.[2] For every event the panel's median forecast dates (measured from the year of the exercise) and the dispersion were determined. A regression analysis was performed and the statistical significance presented in terms of the probability that the regression coefficient would be smaller than the value actually obtained if there were no trend in the data.

The results are quite clear-cut. The remoteness of the forecast date and the

---

[1]T. J. Gordon and O. Helmer, "Report on a Long Range Forecasting Study," Rand Paper P-2982, Santa Monica, California, Rand Corporation, September 1964.

[2]J. P. Martino, "The Precision of Delphi Estimates," *Technological Forecasting* 1, No. 3 (1970), pp. 293-99.

The Delphi Method: Techniques and Applications, Harold A. Linstone and Murray Turoff (eds.)
ISBN 0-201-04294-0; 0-201-04293-2

degree of dispersion are definitely related. The regression coefficient is in nearly all cases highly significant for a single panel addressing a related set of events. However, there is no consistent relation among different panels or within a panel when addressing unrelated events.

Martino also finds that the dispersion is not sensitive to the procedure used: in cases where only a single best estimate year is requested the result is similar to that where 10 percent, 50 percent, and 90 percent likelihood dates are stipulated.[3]

Distribution of Responses

Dalkey has analyzed the first-round responses by panels asked to respond to almanac-type questions, i.e., those with known numerical answers.[4] All responses to each question are standardized by subtracting the mean value and dividing by the standard deviation for that question. The resulting distribution of "standardized deviates" shows an excellent fit to a lognormal distribution. Martino has applied the same techniques to the TRW Probe II Delphi using 10 percent, 50 percent, and 90 percent likelihood dates for 1500 events.[5] Again there is a very good fit to a lognormal distribution.

Optimism-Pessimism Consistency

Another interesting analysis on the TRW Probe II data was undertaken by Martino to ascertain whether a panelist tends to have a consistently optimistic or pessimistic bias to his responses.[6] With each respondent providing 10 percent, 50 percent, and 90 percent likelihood dates, three standardized deviates can be computed for each individual and a given event. Taking the means over all events of the standardized deviates for a given individual and likelihood, we find an interesting pattern. Most panelists are consistently optimistic or pessimistic with respect to the three likelihoods, i.e., there are relatively few cases where, say, the 10 percent likelihood is optimistic while the 50 percent and 90 percent likelihoods are pessimistic. Considering the totality of events the individual panelist tends to be biased optimistically or pessimistically with moderate consistency. However, the amount of the bias is not very great; an optimistic panelist is pessimistic in some of his responses and vice versa. In other words, each participant exhibits a standard deviation which is comparable to, or greater than, his mean.

---

[3]In the first case the interquartile range of best estimates was used, in the second case the 10 percent to 90 percent span was taken.

[4]N. C. Dalkey, "An Experimental Study of Group Opinion," Rand RM-5888-PR, Rand Corporation, Santa Monica, California, March 1969.

[5]J. P. Martino, "The Lognormality of Delphi Estimates," *Technological Forecasting* **1**, No. 4 (1970), pp. 355-58.

[6]J. P. Martino, "The Optimism/Pessimism Consistency of Delphi Panelists," *Technological Forecasting and Social Change* **2**, No. 2 (1970), pp. 221-24.

Accuracy of Forecasts

An apparent indicator of the value of Delphi as a forecasting tool is its accuracy. Since the method was widely publicized only ten years ago, it is difficult to have sufficient hindsight perspective to evaluate its success by this measure. In any event caution is in order. The most accurate forecast is not necessarily the most useful one. Forecasts are at times most effective if they are self-fulfilling or self-defeating. The Forrester-Meadows World Dynamics model has been sponsored by the Club of Rome in the hope that it will act as an early warning system and prove to be a poor forecast. Delphi may be viewed similarly in terms of effectiveness.

We should also observe that long-range forecasts tend to be pessimistic and short-range forecasts optimistic. In the long term no solution is apparent; in the near term the solution is obvious but the difficulties of system synthesis and implementation are underestimated.[7] Thus in 1920 commercial use of nuclear energy seemed far away. By 1949 the achievement appeared reasonable and in 1964 General Electric estimated that fast breeder reactors should be available in 1970.[8] Today the estimate has moved out to the 1980s. The same pattern has been followed by the supersonic transport aircraft. Buschmann has formulated this behavior as a hypothesis and proposed an investigation in greater depth.[9] If this pattern is normal, forecasts should be adjusted accordingly, e.g., forecasts more than, say, ten years in the future brought closer in time and forecasts nearer than ten years moved out. Subsequently Robert Ament made a comparison between a 1969 Delphi study on scientific and technological developments and the 1964 Gordon-Helmer Rand study.[10] Focusing on those items forecast in both studies, he found that all items originally predicted to occur in years before 1980[11] were later shifted further into the future, i.e., the original year seemed optimistic by 1969. On the other hand, two-thirds of the items originally forecast to occur after 1980 were placed in 1969 at a date earlier than that estimated in the 1964 study. Thus we find evidence here, too, of Buschmann's suggested bias.

Grabbe and Pyke have undertaken an analysis of Delphi forecasts of information-processing technology and applications.[12] Forecast events whose occurrence could be verified cover the time period 1968 to 1972. Although six

---

[7]The large cost overruns on advanced technology aerospace and electronics projects are evidence of this trend (see Chapter III, A).

[8]E. Jantsch, "Technological Forecasting in Perspective," OECD, Paris, 1967, p. 106.

[9]R. Buschmann, "Balanced Grand-Scale Forecasting," *Technological Forecasting* **1** (1969), p. 221.

[10]R. H. Ament, "Comparison of Delphi Forecasting Studies in 1964 and 1969," *FUTURES*, March 1970, p. 43.

[11]T. J. Gordon and H. R. Ament, "Forecasts of Some Technological and Scientific Developments and Their Societal Consequences," *IFF Report R-6*, September 1969.

[12]E. M. Grabbe and D. L. Pyke, "An Evaluation of the Forecasting of Information Processing Technology and Applications," *Technological Forecasting and Social Change* **4**, No. 2 (1972), p. 143.

different Delphi studies were used, eighty-two out of ninety forecasts covering this period were taken from one study: the U.S. Navy Technological Forecast Project. The results appear to contradict the hypothesis that near-term forecasts tend to be optimistic. In this case information-processing advances forecast four to five years in the future occur sooner than expected by the panelists who were drawn largely from government laboratories. There is, of course, the possibility that these laboratories are not as close to the leading edge of technology in this field as industrial and university research and development groups. Alternatively, the meaning of "availability" of a technological application may be interpreted differently by the laboratory forecasters and by the authors of this article.

Delphi Statements

The statements which comprise the elements of a Delphi exercise inevitably reflect the cultural attitudes, subjective bias, and knowledge of those who formulate them. This was recognized by Gordon and Helmer a decade ago and led them to commence the first round with "blank" questionnaires. Every student knows that multiple-choice examinations require insight into the instructor's mode of thought as well as the substance of the questions. Misinterpretations of the given statements can arise in both superior and inferior students. Grabbe and Pyke present examples of good and poor Delphi statements.[13] Statements may be too concise, leading to excessive variations in interpretation, or too lengthy, requiring the assimilation of too many elements. Consequently, we would expect a constraint on the number of words leading to the widest agreement in interpretation. Salancik, Wenger, and Helfer have probed this question more deeply.[14] They use an information theory measure (bits) of the amount of information derivable from a distribution of responses to a Delphi statement to measure consensus and the number of words needed to describe an event as a measure of its complexity. The study uses a computer development and application Delphi study as a test case. The authors find a distinct relation between number of words used and amount of information obtained, i.e., agreement in forecast dates. Low and high numbers of words yield low consensus with medium-statement lengths producing the highest consensus. In the particular case considered, twenty to twenty-five words form the peak in the distribution. This study also finds that the more familiar respondents are with a specific computer application, the fewer words are needed to attain agreement. If many words are used, less information results as

[13] *Ibid.*

[14] J. R. Salancik, W. Wenger, and E. Helfer, "The Construction of Delphi Event Statements," *Technological Forecasting and Social Change* 3, No. 1 (1971), pp. 65-73.

to the occurrence of a familiar event. On the other hand, a longer-word description raises the consensus level for unfamiliar events.

A corresponding pattern is found when expert respondents are compared to nonexperts. The latter develop increasing consensus with longer-event descriptions. The experts, however, come to very high consensus with moderate-statement lengths (higher than the greatest nonexpert consensus) but fall to a very low level of agreement with long statements. Apparently the addition of words brings on an effect somewhat similar to that of disputations by Talmudic scholars about minutiae.

Basis for Respondents' Intuitive Forecast

Salancik has examined the hypothesis that the panelists in a forecasting Delphi assimilate input on *feasibility*, *benefits*, and *potential costs* of an event in an additive fashion to estimate its probable date of occurrence.[15] The subject of the test is again a panel forecast of computer applications. Separate coding of participants' reasons for their chosen dates in the three categories enables the author to make a regression analysis. The second-round median date is made a linear function of the number of positive and negative statements in each of the three categories. He finds that the multiple regression strongly supports the hypothesis. The more feasible, beneficial, or economically viable a concept is judged, the earlier it is forecast to occur. The three categories contribute about equally to the regression.

In a second study independent assessments of feasibility and benefits are rated for twenty computer applications and then combined to form the basis for a rank ordering. This ordering is then compared to the Delphi panelists' responses. Again the correlation supports the suggested model of Delphi input assimilation. This paper adds another beam of support to the idea that Delphi is a systematic and meaningful process of judgment synthesis.

Self-Rating of Experts

Dalkey, Brown, and Cochran tackle another aspect of Delphi: the expertise of the respondents.[16] With a given group we might consider two ways of improving its accuracy: iterating the responses and selecting a more expert subgroup. The latter process implies an ability to identify such a subgroup (e.g., by self-rating) and a potential degradation in accuracy due to the reduced group

---

[15] J. R. Salancik, "Assimilation of Aggregated Inputs into Delphi Forecasts: A Regression Analysis," *Technological Forecasting and Social Change* **5**, No. 3 (1973), pp. 243-48.

[16] N. Dalkey, B. Brown, and S. Cochran, "Use of Self-Ratings to Improve Group Estimates," *Technological Forecasting* **1**, No. 3 (1970), pp. 283-91.

size. The authors stipulate a minimum subgroup size to counteract this degradation and they force a clear separation in self-ratings of low- and high-expertise subgroups. The experiments were carried out by the authors using 282 university students and verifiable almanac-type questions. The conclusions: (1) self-rating is a meaningful basis for identification of expertise, and (2) selection of expert subgroups improves the accuracy to a somewhat greater degree than does feedback or iteration.

One must raise the question whether an experiment based on almanac type questions serves as an adequate basis for a conclusion about the validity of self-ratings of expertise for forecasting Delphis. While the lognormality behavior exhibited a similar pattern for factual (almanac-type) and forecasting cases, this similarity might not carry over for self-ratings.

And there are other fascinating unanswered questions. Why do women rate themselves consistently lower than men? Should only the expert subgroup results be fed back to the larger group in the iteration process? How do age, education, and cultural background condition the response of individuals?

The four articles in this chapter provide us with further evaluations of the process. When we use Delphi to draw forth collective expert judgments, we are actually making two substitutions: (1) expert judgment for direct knowledge, and (2) a group for an individual. In the first article, Dalkey strives to develop some mathematically rigorous underpinnings, i.e., a start toward a theory of group estimation. It quickly becomes evident that we still have much to learn about this process. Dalkey emphasizes the concept of "realism," or "track record," to describe the expert's estimation skill and the theory of errors for the group. But the final verdict on their applicability is by no means in.

Scheibe, Skutsch, and Schofer report on several highly instructive findings based on research in the application of Delphi to the derivation of explicit goals and objectives. Analysis of a Delphi goal-formulation experiment for urban systems planning yielded the following important results:

(1) The three-interval scaling methods used—simple ranking, a rating scale, and pair comparisons—give essentially equivalent scales. The rating scale is found to be most comfortable to use by the participants.

(2) Respondents are sensitive to feedback of the scores from the whole group and tend to move (at least temporarily) toward the perceived consensus.

(3) There is only a modest tendency for the degree of confidence of an individual with respect to a single answer to be reflected in movement toward the center of opinion, i.e., less confident members exhibit a somewhat larger movement in the second round.

(4) Stability of the distribution of the group's response along the interval scale over successive rounds is a more significant measure for developing a stopping criterion than degree of convergence. The authors propose a specific stability measure.

Next, Mulgrave and Ducanis discuss an experiment which focuses on the behavior of the dogmatic individual in successive Delphi rounds. Surprisingly, the high-dogmatism group exhibits significantly more changes than the low-dogmatism group. It is the authors' belief that the dogmatic individual looks to authority for support of his view. In the absence of a clearly defined authority, he views the median of the group response as a surrogate.

There clearly exists the possibility of an unnatural overconsensus. Conformists may "capitulate" to group pressures temporarily, on paper. It would be interesting to compare the behavior of such psychological types in a Delphi with that in a conventional committee.

Finally, Brockhoff examines a series of hypotheses on the performance of forecasting groups using the Delphi technique and face-to-face discussions in a Lockean context. He focuses on short-range forecasting and small homogeneous groups. Staff members of local banks trained in economics are queried about data concerning financial questions, banking, stock quotations, and foreign trade. Groups vary in size from eleven to four participants (the latter below the size considered minimal by Dalkey, Brown, and Cochran[17]). The Delphi process uses an interactive computer program for structuring the dialogue as well as computing intermediate and final results. The correlation of self-rating of expertise with individual or group performance, the relation between information exchange and group performance, and the relevance of almanac-type fact-finding questions for short-term forecasting analysis are among the questions examined. One may speculate whether the dogmatism aspect raised by Mulgrave and Ducanis plays a significant role in groups of the type used in Brockhoff's experiments.

For the reader the thrust of this chapter is that, to develop proper guidelines for its use, we can and should subject Delphi to systematic study and evaluation in the same way as has been the case with other techniques of analysis and communication. Much still needs to be learned!

---

[17] *Ibid.*

IV. B. Toward a Theory of Group Estimation *

NORMAN C. DALKEY

Introduction

The term "Delphi" has been extended in recent years to cover a wide variety of types of group interaction. Many of these are exemplified in the present volume. It is difficult to find clear common features for this rather fuzzy set. Some characteristics that appear to be more or less general are: (1) the exericse involves a group; (2) the goal of the exercise is information; i.e., the exercise is an inquiry; (3) the information being sought is uncertain in the minds of the group; (4) some preformulated systematic procedure is followed in obtaining the group output.

This vague characterization at least rules out group therapy sessions (not inquiries), team design of state-of-the-art equipment (subject matter not uncertain), brainstorming (procedure not systematic), and opinion polls (responses are not treated as judgments, but as self-reports). However, the characterization is not sufficiently sharp to permit general conclusions, e.g., concerning the effectiveness of types of aggregation procedures.

Rather than trying to deal with this wide range of activities, the present essay is restricted to a narrow subset. The subject to be examined is *group estimation*—the use of a group of knowledgeable individuals to arrive at an estimate of an uncertain quantity. The quantity will be assumed to be a physical entity—a date, a cost, a probability of an event, a performance level of an untested piece of equipment, and the like.

Another kind of estimation, namely, the identification and assessment of value structures (goals, objectives, etc.) has been studied to some extent, and a relevant exercise is described in Chapter VI. Owing to the difficulty of specifying objective criteria for the performance of a group on this task, it is not considered in the present paper.

To specify the group estimation process a little more sharply, we consider a group $I=\{I_i\}$ of individuals, an event space $E=\{E_j\}$ where E can be either discrete or continuous, and a response space $R=\{R_{ij}\}$ which consists of an estimate for each event by each member of the group. In addition, there is an external process $P=\{P(E_j)\}$, which determines the alternatives in E which will occur. Depending on the problem, P can either be a δ-function on E—i.e., a specification of which event will occur—or a probability distribution P_j on the

*Research reported herein was conducted under Contract Number F30602-72-C-0429 with the Advanced Research Projects Agency, Department of Defense.

The Delphi Method: Techniques and Applications, Harold A. Linstone and Murray Turoff (eds.)
ISBN 0-201-04294-0; 0-201-04293-2

event space. In general P_j is unknown. For some formulations of the group estimation process, it is necessary to refer to the a priori probability of an event. This is not the same as the external process, but rather, is (in the present context) the probability that is ascribed to an event without knowing the individual or group estimates. This a priori probability will be designated by $U = \{U(E_j)\}$.

In many cases the R_{ij} are simply selections from E. The weatherman says, "It will rain tomorrow"—a selection from the two-event space *rain tomorrow* and *no rain tomorrow*. The long-range technological forecaster says, "Controlled nuclear fusion will be demonstrated as feasible by 1983"—selection of a single date out of a continuum. In these cases the R_{ij} can be considered as 0's and 1's, 1 for the selected event and 0 for the others. Usually, the 0's are left implicit. More complex selections can be dealt with—"It will either rain or snow tomorrow." "Controlled nuclear fusion will be demonstrated in the interval 1980–1985"—by allowing several 1's and interpreting these as an or-combination. Selections can also be considered as special cases of probability distributions over the event space. In the case of probability estimates, the R_{ij} can be probability assignments for discrete alternatives, or continuous distributions for continuous quantities.

A kind of estimate which is sometimes used in applied exercises, but which is not directly expressible in terms of elementary event spaces, is the estimation of the functional relationship between two or more variables (e.g., the extrapolation of a trend). Such an estimate can be included in the present formalism if the relationship is sufficiently well known beforehand so that all that is required is specification of some parameters (e.g., estimating the slope of a linear trend). Although of major practical importance, estimates of complex functional relationships have received little laboratory or theoretical treatment. In particular, there has been no attempt to develop a scoring technique for measuring the excellence of such estimates.

In addition to the group I, event space E, and response space R, a Delphi exercise involves a process $G = G[I, E, R]$ which produces a group response G_j for each event E_j in the event space. Square brackets are used rather than parentheses in the expression for G to emphasize the fact that generally the group estimation process cannot be expressed as a simple functional relationship. The process may involve, for example, discussion among members of the group, other kinds of communication, iteration of judgments with complex selection rules on what is to be iterated, and so on.

One other piece of conceptual apparatus is needed, namely, the notion of *score*, or measure of performance. Development of scoring techniques has been slow in Delphi practice, probably because in most applied studies the requisite data for measuring performance either is unavailable, or would require waiting a decade or so. But in addition, the variety of subject matters, the diversity of motivations for applied studies, and the obscuring effect of the radical uncertainty associated with topics like long-range forecasting of social and technological events have inhibited the attempt to find precise measures of performance.

In the present paper, emphasis will be put on measures related to the accuracy of estimates. There is a large family of such measures, depending on the form of the estimate, and depending on the interests of the user of the estimate. For this essay, measures will be restricted to what might be called scientific criteria, i.e., criteria which do not include potential economic benefits to the user (or potential costs in terms of experts' fees, etc.) or potential benefits in facilitating group action.

For simple selections out of discrete event spaces a right/wrong measure is usually sufficient, for example, crediting the estimate with a 1 or 0 depending on whether it is correct or incorrect. However, as in the related area of performance testing in psychology, the right/wrong measure is usually augmented by computing a score—total number right, or proportion right, or right-minus-wrong, etc.—over a set of estimates.

For simple selections out of continuous spaces (point estimates), a distance measure is commonly employed, for example, difference between the estimate and the true answer. However, if such measures are to be combined into a score over a set of estimates, some normalizing procedure must be employed to effect comparability among the responses. One normalizing procedure for always positive quantities such as dates, size of objects, probabilities, and the like, is the log error, defined as

$$\text{Error} = \log\left|\frac{R_i}{T}\right|,$$

where T is the true answer and R_i is the individual response. The vertical bars denote the absolute value (neglecting sign). Dividing by T equates proportional errors, and taking the logarithm uniformizes under- and over-estimates. Comparable scoring techniques have not been worked out for quantities with an inherent zero, i.e., quantities admitting both positive and negative answers. Such quantities are rare in applied exercises. Whether this is because that type of quantity is inessential to the subject matter or whether it is due to avoidance by practitioners is hard to say.

For probability estimates, some form of probabilistic scoring system appears to be the best measure available. The theory of probabilistic scoring systems is under rapid development. It is usually pursued within the ambit of subjective probability theories, where the primary property sought is a reward system which motivates the estimator to be honest, i.e., to report his "true" belief.

This requirement can be expressed as the condition that the expected score of the estimator should be a maximum when he reports his true belief. If $q = \{q_j\}$ is the set of probabilities representing the actual beliefs of the estimator on event space $\{E_j\}$, $R = \{R_j\}$ is his set of reported probabilities, and $S_j(R)$ is the reward he receives if event E_j occurs, then the honesty condition can be written in the form

$$\sum_j q_j S_j(q) \geqslant \sum_j q_j S_j(R). \tag{1}$$

The expression on the left of the inequality is the individual's subjective expectation if he reports his actual belief; the expression on the right is his expectation if he reports something else.

Formula (1) defines a family of scoring (reward) systems often referred to as "reproducing scoring systems" to indicate that they motivate the estimator to reproduce his actual belief.

It is not difficult to show that the theory of such scoring systems does not depend on the interpretation of q as subjective belief; it is equally meaningful if q is interpreted as the objective probability distribution P on E. With this interpretation the estimator is being rewarded for being as accurate as possible —his objective expectation is maximized when he reports the correct probability distribution.

This is not the place to elaborate on such scoring systems (see [1], [2], [3]). Although (1) leads to a family of reward functions, it is sufficient for the purposes of this essay to select one. The logarithmic scoring system

$$S_j(R) = A \log R_j + B \tag{2}$$

has a number of desirable features. It is the only scoring system that depends solely on the estimate for the event which occurs. The expected score of the estimator is precisely the negative entropy, in the Shannon sense [4], of his forecast. It has the small practical difficulty that if the estimator is unfortunate enough to ascribe 0 probability to the alternative that occurs, his score is negatively infinite. This can usually be handled by a suitable truncation for very small probabilities.

Within this restricted framework, the Delphi design "problem" can be expressed as finding processes G which maximize the expected score of the group response. This is not a well-defined problem in this form, since the expectation may be dependent on the physical process being estimated, as well as on the group-judgment process. There are two ways to skirt this issue. One is to attempt to find G's which have some optimality property independent of the physical process. The other route is to assume that knowledge of the physical process can be replaced by knowledge about the estimators, i.e., knowledge concerning their estimation skill. The next section will deal with the second possibility.

There are two basic assumptions which underlie Delphi inquiries: (a) In situations of uncertainty (incomplete information or inadequate theories) expert judgment can be used as a surrogate for direct knowledge. I sometimes call this the "one head is better than none" rule. (b) In a wide variety of situations of uncertainty, a group judgment (amalgamating the judgments of a group of experts) is preferable to the judgment of a typical member of the group, the "*n* heads are better than one" rule.

The second assumption is more closely associated with Delphi than the first, which has more general application in decision analysis. These two assumptions do not, of course, exhaust all the factors that enter into the use of Delphi techniques. They do appear to be fundamental, however, and most of the remaining discussion in this paper will be concerned with one or the other of the two.

Individual Estimation

Using the expert as a surrogate for direct knowledge poses no problems as long as the expert can furnish a high-confidence estimate based on firm knowledge of his own. Issues arise when existing data or theories are insufficient to support a high-confidence estimate. Under these circumstances, for example, different experts are likely to give different answers to the same questions.

Extensive "everyday experience" and what limited experimental data exist on the subject strongly support the assumption that knowledgeable individuals can make useful estimates based on incomplete information. This general assumption, then, is hardly in doubt. What is in doubt is the degree of accuracy of specific estimates. What is needed is a theory of estimation that would enable the assignment of a figure of merit to individual estimates on the basis of readily available indices.

An interesting attempt to sidestep this desideratum is to devise methods of rewarding experts so that they will be motivated to follow certain rules of rational estimation. One approach to the theory of probabilistic scoring systems described in the introduction is based on this strategem [5].

The outlines of such a theory of estimation have been delineated in the literature of decision analysis; but it is difficult to disentangle from an attendant conceptualization of a *prescriptive* theory of decisionmaking, or as sometimes characterized, the theory of rational decisionmaking. In the following I will try to do some disentangling, but the subject is complex and *is* and *ought* may still intermingle more than one might wish.

In looking over the literature on decision analysis, there appear to be about six desirable features of estimation that have been identified. The number is not sharp, since there are overlaps between the notions and some semantic difficulties plague the classification. The six desiderata are *honesty, accuracy, definiteness, realism, certainty*, and *freedom from bias*.

Honesty is a clear enough notion. In most cases of estimation, the individual has a fairly distinct perception of his "actual belief," or put another way, he has a relatively clear perception whether his reported estimate matches his actual belief. This is not always the case. In situations with ambiguous contexts, such as the group-pressure situations created by Asch [6], some individuals appear to lose the distinction. The reason for wanting honest reports from estimators is also clear. Theoretically, any report, honest or not, is valuable if the user is aware of potential distortions and can adjust for them. But normally such information is lacking.

Accuracy is also a fairly straightforward notion, and is measured by the score in most cases. It becomes somewhat cloudy in the case of probability estimates for single events, where an individual can make a good score by chance. In this case, the average score over a sequence of events is more diagnostic. But the notion of accuracy then becomes mixed with the notion of realism. Given the meaningfulness of the term, the desirability of accuracy is clear.

Definiteness measures the degree of sharpness of the estimate. In the case of probabilities on discrete event spaces, it refers to the degree to which the probabilities approach 0 or 1 and can be measured by $\Sigma_{j=1}^{m} R_j^2$. In the case of probability distributions on continuous quantities, it can be measured by the variance or the dispersion. In the case of selections, the comparable notion is "refinement." For discrete event spaces, one report is a refinement of another if it is logically included in the second.

The reason for desiring definiteness is less clear than for accuracy or honesty. "Risk aversion" is a well-known phenomenon in economic theory, but "risk preference" has also been postulated by some analysts [7]. In the case of discrete alternatives, the attractiveness of a report that ascribes a probability close to 1 to some alternative, and probability close to 0 to the others is intuitively "understandable." There is a general feeling that probabilistic estimates close to 0 or 1 are both harder to make, and more excellent when made, than "wishy-washy" estimates in the neighborhood of $\frac{1}{2}$. There is also the feeling that an individual who makes a prediction with a probability of .8 (and it turns out correct) knows more about the phenomenon being predicted than someone who predicts a similar event with probability .6.

All of this is a little difficult to pin down. In the experiments of Girshick, *et al.* [8], there was almost no correlation between a measure of definiteness and the accuracy of the estimates. Part of the problem here appears to be an overlap between the notion of definiteness and uncertainty, which is discussed below. At all events, there appears to be little doubt that definiteness is considered a virtue.

Realism refers to the extent that an individual's estimates are confirmed by events. It is thus closely related to accuracy. However, accuracy refers to a single estimate, whereas realism refers to a set of estimates generated by an individual. Other terms used for this notion are *calibration* [9], *precision* [10], *track record.*

Because the notion of realism is central to the first principle of Delphi stated in the introduction, namely, the substitution of expert judgment for direct knowledge, it warrants somewhat extensive discussion.

In the case of probability judgments, it is possible in theory to take a sequence of estimates from a single estimator, all with the same estimated probability, and count the number of times the estimate was confirmed. Presumably, if the estimator is using the notion of probability correctly, the relative frequency of successes in that sequence should be approximately equal to the estimated probability. Given enough data of this sort for a wide range of different estimates, it is possible in theory to generate a realism curve for each individual, as illustrated in Fig. 1.

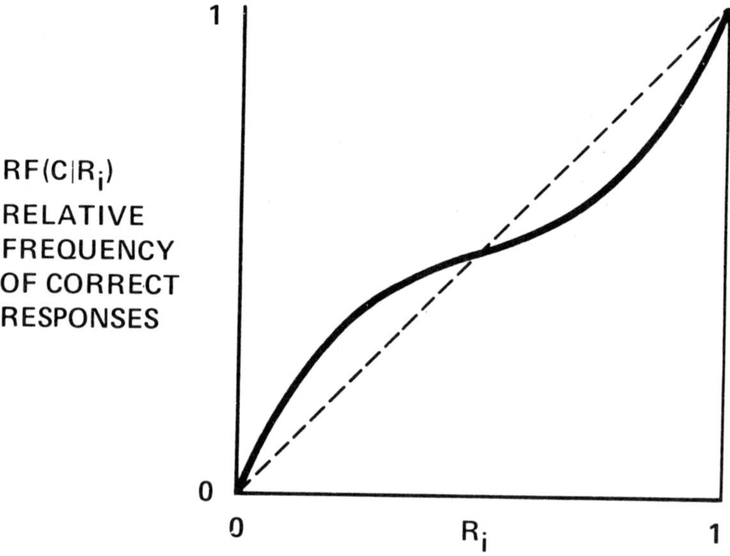

Fig. 1. Typical realism curve.

In Fig. 1 the relative frequency with which an estimate of probability R_i is verified, $RF(C|R_i)$ ("C" for "correct"), is plotted against the estimate. Realism can be defined as the degree to which the $RF(C|R_i)$ curve approximates the theoretically fully realistic curve, namely the dashed line in Fig. 1, where $RF(C|R_i)=R_i$. Figure 1 illustrates a typical realism curve where probabilities greater than $\frac{1}{2}$ are "overestimated" and probabilities less than $\frac{1}{2}$ are underestimated [11].

Various quantities can be used to measure the overall realism of an estimator. $\int_0^1 (RF(C|R_i)-R_i)^2 \; D(R_i)$ where $D(R_i)$ is the distribution of the estimator's reports R_i—roughly the relative frequency with which he uses the various reports R_i—is a reasonable measure. However, for most applications of the concept, it is the realism curve itself which is of interest.

If such a curve were available for a given individual, it could be used directly to obtain the probability of a given event, based on his report. In particular, if the individual were fully realistic, the desired probability would be R_i. At first sight, it might appear that one individual, given his realism curve, is all that is needed to obtain a desired estimate, since the curve furnishes an "objective" translation of his reports into probabilities. However, for any one specific estimate, the reports of several individuals typically differ, and in any case the realism curve is not, by itself, a measure of the expertness or knowledgeability of the individual. In particular, the frequency with which the individual reports relatively high probabilities has to be taken into account.

As a first approximation, the knowledgeability K_i of individual i can be measured by

$$K_i = \int_0^1 S(R_i)D(R_i),$$

where $S(R_i)$ is the probabilistic score awarded to each report R_i and $D(R_i)$ is, as before, the distribution of the reports R_i.

It is easy to verify two properties of K_i: (a) K_i is heavily influenced by the degree of realism of the estimator. For a given distribution of estimates, $D(R_i)$, K_i is a maximum when the individual is fully realistic. (b) K_i is also influenced by the average definiteness of the estimator. The higher the definiteness (e.g., measured by $\int R_i^2\, D(R_i)$), the higher the expected score.

Theoretically, one might pick the individual with the highest K rating and use him exclusively. There are two caveats against this procedure. On a given question, the individual with the highest average K may not furnish the best response; and, more in the spirit of Delphi, if realism curves are available for a set of individuals, then it is sometimes feasible to derive a group report which will have a larger average score than the average score of any individual—in short, the K measure for the group can be higher than the K measure for any individual.

As far as the first principle—substitution of expert judgment for knowledge—is concerned, the question whether realism curves exist for each individual is a crucial one. Detailed realism curves have not been derived for the types of subject matter and the type of expert desired for applied studies. In fact, detailed track records for any type of subject matter are hard to come by. Basic questions are: Is there a stable realism curve for the individual for relevant subject matters? How general is the curve—i.e., is it applicable to a wide range of subject matters? How subject is the curve to training, to use of reward systems like the probabilistic score, to contextual effects such as the group pressure effect in the Asch experiments?

Certainty is a notion that is well known in the theory of economic decisionmaking. It has not played a role in the study of estimation to the same extent. In the case of economic decisionmaking, the distinction has been made between *risk* (situations that are probabilistic, but the probabilities are known) and *uncertainty* (situations where the probabilities are not known) [12]. Many analysts appear to believe that in the area of estimation this distinction breaks down—uncertainty is sufficiently coded by reported probabilities. However, the distinction appears to be just as applicable to estimation as to any other area where probabilities are relevant. Consider, for example, the situation of two coins, where an individual is asked to estimate the probability of *heads*. Coin A is a common kind of coin where the individual has flipped it several times. In this case, he might say that the probability of heads is $\frac{1}{2}$ with a high degree of confidence. Coin B, let's say, is an exotic object with an unconventional shape, and the individual has not flipped it at all. In the case of coin B he might also estimate a probability of $\frac{1}{2}$ for heads, but he would be highly uncertain whether that is the actual probability. Probability $\frac{1}{2}$, then, cannot express the uncertainty attached to the estimate for the second coin.

A closer approximation to the notion of uncertainty can be obtained by considering a distribution on the probabilities. For example, the individual might estimate that the probability of the familiar coin has a tight distribution around $\frac{1}{2}$, whereas the distribution for the unfamiliar coin is flat, as in Fig. 2. The independent variable is labeled q to indicate that it is the individual's belief, and not necessarily his report, which is being graphed.

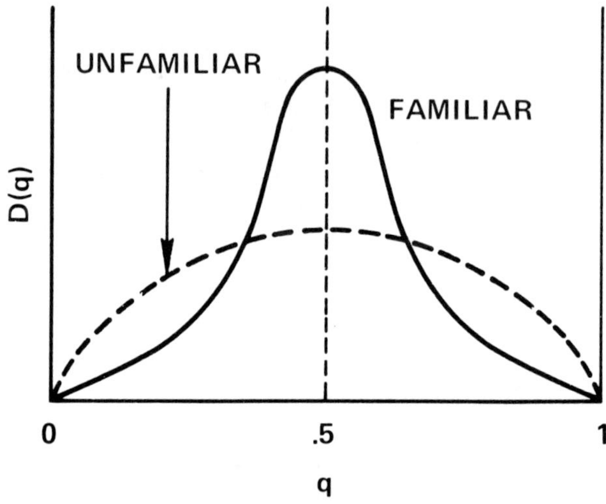

Fig. 2. Uncertainty represented as a higher-level distribution.

The use of a higher-level distribution is only an approximation to the notion of uncertainty, since the distribution itself might be uncertain, or, in more familiar language, the distribution may be "unknown." The use of additional levels has been suggested, but for practical reasons seems highly unappealing.

The problem of representing uncertainty in precise terms is closely related to past attempts to translate lack of information into probabilities by means of principles such as the "law of insufficient reason," or the rule of equal ignorance. These have invariably lead to paradoxes [13].

Using the idea of the dispersion of a second-level distribution as an approximate measure of uncertainty, there is some interaction between the notions of realism, definiteness, and certainty. It is not possible for a set of estimates to be simultaneously realistic, definite, and uncertain. Assuming that the individual will give as his first level report R_i the mean of his second-level distribution, then as R_i approaches 1 or as R_i approaches 0, the standard deviation of the distribution $D(q)$ approaches 0. Figure 3 illustrates this coupling for $R_i = .9$. If the individual is realistic and estimates a probability of .9 for a given event, then the standard deviation of his higher-level distribution for that estimate must be small.

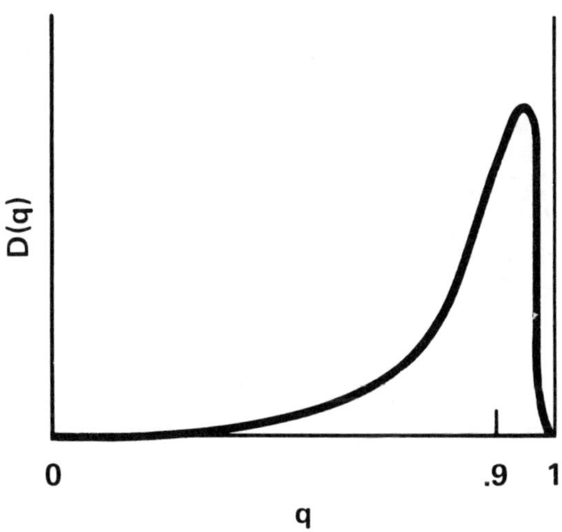

Fig. 3. Illustration of coupling between certainty and definiteness.

Unfortunately, the coupling applies only to the extremes of the 0 to 1 interval. At $q = \frac{1}{2}$, D can be about anything, and the estimator still be realistic. If the average probabilistic score for an estimate with a second-level distribution D is computed, the average score is influenced only by the mean, and otherwise is independent of D. Thus an average probabilistic score does not reflect uncertainty. It appears that something like the variance of the score will have to be included if certainty is to be reflected in a score.

At the present, the only "visible" index of certainty is the self-rating—i.e., a judgment by the individual of his competence or knowledgeability concerning the estimate. This has turned out to be a significant index for rating group estimates [14]; it is not so effective for individual estimates. Due to the lack of a theoretical definition of the self-rating, it has not been possible to include it in a formal theory of aggregation. However, the self-rating has proved to be valuable for selecting more accurate subgroups [15].

Bias is a term that has many shades of meaning in statistics and probability. I am using the term to refer to the fact that there may be subclasses of events for which $RF(C|R_i)$ may be quite different from the average relative frequency expressed by the realism curve. Of course, for this to be of interest, the subclasses involved must be identifiable by some means other than the relative frequency. It is always possible after the fact to select a subset of events for which an individual has estimated the probability R_i which has any $RF(C|R_i)$.

In the theory of test construction, e.g., for achievement tests or intelligence tests, it is common to assume an underlying scale of difficulty for the questions, where difficulty is defined as the probability that a random member of the target population can answer the question correctly [16]. This probability will range from 1 for very easy questions to 0 for very hard questions, as illustrated by the solid curve in Fig. 4. From the standpoint of the present discussion, the significant fact is that when a class of questions is identified as belonging to the very difficult group in a sample of the population, that property carries over to other members of the population—in short the property of being very difficult is relatively well defined.

At some point in the scale of difficulty, labeled *d* in Fig. 4, a typical member of the population could increase his score by abandoning the attempt to "answer" the question and simply flipping a coin (assuming that it is a true/false or yes/no type of question). Put another way, from point *d* on, the individual becomes a counterpredictor—you would be better off to disbelieve his answers.

Contrasted with this notion of difficulty is the notion that underlies theories of subjective probability that, as the individual's amount of information or skill declines, the probability of a correct estimate declines to 50 percent as illustrated by the dashed curve in Fig. 4. Ironically, it is the probabilistic notion that influences most scoring schemes, which assume that the testee can achieve 50 percent correct by "guessing," and hence the score is computed by subtracting the number of wrong answers from the number right. By definition, for the more difficult items, the testee cannot score 50 percent by "guessing" unless that means literally tossing a coin and not trusting his "best guess."

If it turns out that "difficult" questions in the applied area have this property, even for experts, then the first principle does not hold for this class. Although there are no good data on this subject, there does not appear to be a good reason why what holds for achievement and intelligence tests should not also hold for "real life" estimates. Almost by definition, the area of most interest in applications is the area of difficult questions. If so, assuming that the set of counterpredictive questions can be identified before the fact, then a good fair coin would be better than an expert. It is common in experimental design to use randomization techniques to rule out potential biases. There is no logical reason why randomization should not be equally potent in ruling out bias in the case of estimation.

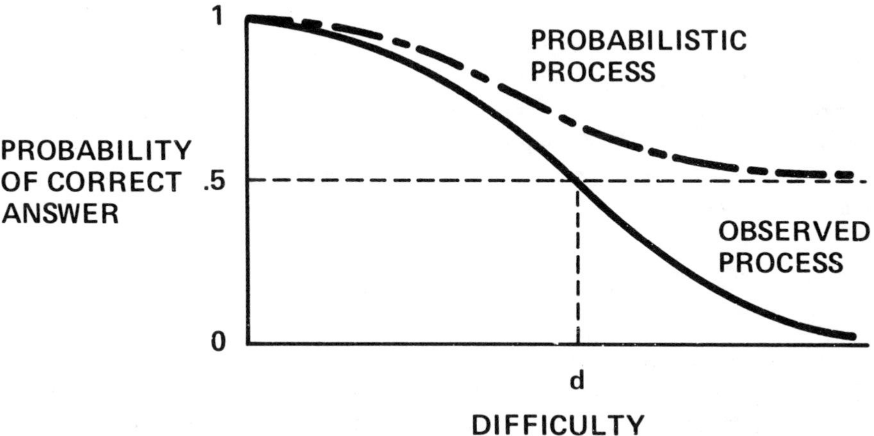

Fig. 4. Scale of difficulty in test construction.

The four notions, honesty, accuracy, definiteness, and precision, are all tied together by probabilistic scoring systems. In fact, a reproducing scoring system rewards the estimator for all four. As pointed out in the introduction, condition (1) defines the same family of scoring systems whether q is interpreted as subjective belief, or as objective probability. Thus, the scoring system rewards the estimator for both honesty and accuracy. In addition, the condition leads to the result that $\Sigma_j q_j S_j(q)$ is convex in q. This convex function of q can be considered as a measure of the dispersion of q; and in fact, three of the better-known scoring systems define three of the better-known measures of dispersion. Thus, if S_j is the quadratic scoring system

$$S_j(R) = 2R_j - \sum_j R_j^2,$$

then $\Sigma_j q_j S_j(q) = \Sigma_j q_j^2$ which is a measure of variance. If S_j is the spherical scoring system

$$S_j(R)=\frac{R_j}{\sqrt{\sum_j R_j^2}},$$

then $\Sigma_j q_j S_j(q)=\sqrt{\Sigma_j q_j^2}$, a measure similar to the standard deviation. Finally, for the logarithmic scoring system, $\Sigma_j q_j S_j(q)=\Sigma_j q_j \ln q_j$, which is the negative of the Shannon entropy, another measure of definiteness.

Realism enters in a more diffuse fashion. In general, the probabilistic score for a single event is not very diagnostic, since the individual may have obtained a high (or low) score by chance. Thus, as for most scoring systems, an average (or total) score over a large set of questions is the usual basis for evaluation. But over a large set of questions, the average score is determined by the realism curve of the individual, in conjunction with the relative frequency with which he makes reports of a given probability. In general, if the estimator is not realistic, he will lose

$$\int_0^1 (RF(C|R_i)S(RF(C|R_i)) - RF(C|R_i)S(R_i))D(R_i).$$

As pointed out above, the probabilistic score does not include a penalty for uncertainty, nor does it include a penalty for bias, except where bias shows up in the realism curve. The latter case is simply the one where, for whatever reason, the individual is faced with a stream of questions in which the number of questions biased in a given direction is greater than the number biased in the opposite direction.

To sum up this rather lengthy section: The postulate that, in situations of uncertainty, it is feasible to substitute expert judgment for direct knowledge is grounded in a number of empirical hypotheses concerning the estimation process. These assumptions are, primarily, that experts are approximately realistic in the sense defined above, that the realism curve is stable over a relatively wide range of questions (freedom from bias), and that knowledgeability is a stable property of the expert. At the moment, these are hypotheses, not well-demonstrated generalizations.

Theoretical Approaches to Aggregation

Assuming that, for a given set of questions, we can accept the postulate that expert judgment is the "best information obtainable," there remains the question how the judgments of a group of experts should be amalgamated. In the present section, three approaches to this issue are discussed. The discussion is limited to elementary forms of aggregation, where the theory consists of a mathematical rule for deriving a group response from a set of individual responses; thus, an *elementary* group estimation process can be defined as a function, $G = G(E, I, R)$.

Theory of Errors

This approach interprets the set of judgments of a group of experts as being similar to the set of readings taken with an instrument subject to random error. It seems most appropriate when applied to point estimates of a continuous quantity, but formally at least, can be applied to any type of estimate. In analogy with the theory of errors for physical measurements, a statistical measure of central tendency is considered to be the best estimate of the quantity. Some measure of dispersion is taken to represent a confidence interval about the central value.

Relevant aspects of the individual estimation process such as skill or amount of information of the expert, are interpreted as features of the "theory of the instrument."

This point of view appears to be most popular in the Soviet Union [17]; however, a rough though unexpressed version of this approach underlies much of the statistical analysis accompanying many applied Delphi studies. To my knowledge, this approach has not been developed in a coherent theory, but rather, has been employed as an informal "interpretation"—i.e., as a useful analogy.

The theory-of-errors approach has the advantages of simplicity, and similarity with well-known procedures in physical measurement theory. Much of the empirical data which have been collected with almanac and short-range prediction studies is compatible with the analogy. Thus, the distribution of estimates tends to follow a common form, namely the lognormal [18]. If the random errors postulated in the analogy are assumed to combine multiplicatively (rather than additively as in the more common Gaussian theory), then a lognormal distribution would be expected.

The geometric mean of the responses is more accurate than the average response; or more precisely, the error of the geometric mean is smaller than the average error. Since the median is equal to the geometric mean for a lognormal distribution [19], the median is a reasonable surrogate, and has been the most widely used statistic in applied studies for the representative group response.

The error of the median is, on the average, a linear function of the standard deviation [20], which would be predicted by the theory of errors. The large bias observed experimentally (bias = error/standard deviation) is on the average a constant, which again would be compatible with the assumption that experts perform like biased instruments.

Although the analogy looks fairly good, there are several open questions that prevent the approach from being a well-defined theory. There does not exist at present a "theory of the instrument" which accounts for either the observed degree of accuracy of individual estimates or for the large biases observed in experimental data. Perhaps more serious, there is no theory of errors which accounts for the presumed multiplicative combination of errors—especially since the "errors" are exemplified by judgments from different respondents.

Despite this lack of firm theoretical underpinnings, the theory-of-errors approach appears to fit the accumulated data for point estimates more fully than any other approach.

In addition, the measures of central tendency "recommended by" the theory of errors have the desirable feature that the advantage of the group response over the individual response can be demonstrated irrespective of the nature of the physical process being estimated. So far as I know, this is the only theoretical approach that has this property.

To make the demonstration useful in later sections, a somewhat more sophisticated version of the theory will be dealt with than is necessary just to display the "group effect."

Consider a set of individual estimates R_{ij} on an event space E_j, where the R_{ij} are probabilities, i.e., $\Sigma_j R_{ij}=1$. We assume there is a physical process that determines objective probabilities $P=\{P_j\}$ for the event space, but P is unknown. Consider a group process G which takes the geometric mean of the individual estimates as the best estimate of the probability for each event. However, the geometric means will not be a probability, and must be normalized. This is accomplished by setting

$$G_j=\frac{\left(\prod_{i=1}^{n} R_{ij}\right)^{\frac{1}{n}}}{\sum_{j=1}^{m}\left(\prod_{i=1}^{n} R_{ij}\right)^{\frac{1}{n}}}. \tag{2}$$

We can now ask how the expected probabilistic score of the group will compare with the average expected score of the individual members of the group. It is convenient to use the abbreviation C for the reciprocal of the normalizing term

$$C=\frac{1}{\sum_{j=1}^{m}\left(\prod_{i=1}^{n} R_{ij}\right)^{\frac{1}{n}}}.$$

Using the logarithmic scoring system and setting the constants $A=1$, $B=0$, we have

$$S_j(G)=\log\left(C\left(\prod_{i=1}^{n}R_{ij}\right)^{\frac{1}{n}}\right) \tag{3}$$

$$=\frac{1}{n}\sum_{i=1}^{n}\log R_{ij}+\log C. \tag{4}$$

Taking the expected score,

$$\sum_{j=1}^{m}P_jS_j(G)=\sum_{j=1}^{m}P_j\frac{1}{n}\sum_{i=1}^{n}\log R_{ij}+\log C, \tag{5}$$

and rearranging terms, where $\bar{S}\,(G)$ denotes the expected score of the group, i.e., $\bar{S}\,(G)=\sum_{j=1}^{m}P_jS_j(G)$

$$\bar{S}(G)=\frac{1}{n}\sum_{i=1}^{n}\sum_{j=1}^{m}P_j\log R_{ij}+\log C, \tag{6}$$

log C appears outside the summation, because, as a constant, $\sum_{j=1}^{m}P_j\log C=\log C$. The expression $\sum_{j=1}^{m}P_j\log R_{ij}$ is just the expected score $\bar{S}\,(R_i)$ of individual i, and the expression on the right of (6) excluding log C is the *average* individual expected score, which we can abbreviate as $\hat{S}\,(R)$. Thus

$$\bar{S}(G)=\hat{S}\,(R)+\log C. \tag{7}$$

Since C is greater than 1, log C is positive, and the expected group score is greater than the average expected individual score by the amount log C. C depends only on the individual responses R_{ij} and not on the specific events E or the objective probabilities P.

Formula (7) exemplifies a large variety of similar results that can be obtained by using different statistics as the aggregation rule and different scoring rules.[1]

Probabilistic Approach

Theoretically, joint realism curves similar to the individual realism curve of Fig. 1 can be generated, given enough data. In this case, the relative frequency $RF(C|R)$ of correct estimates would be tabulated for the joint space of

[1] Brown [21] derives a similar result for continuous distributions, the quadratic scoring system, and the mean as the group aggregation function.

responses R for a group. Such a joint realism curve would be an empirical aggregation procedure. $RF(C|R)$ would define the group probability judgment as a function of R.

Although possible in theory (keeping in mind all the caveats that were raised with respect to individual realism curves), in practice generating joint realism curves for even a small group would be an enormous enterprise. It is conceivable that a small group of meteorologists, predicting the probability of rain for a given locality many thousands of times, might cover a wide enough region of the R space to furnish stable statistics. However, for the vast majority of types of question where group estimation is desired, individual realism curves are difficult to come by; group realism curves appear to be out of the question for the present.

One possible simplification at this point could be made if general rules concerning the interdependence of individual estimates on various types of estimation tasks could be ascertained. In such a case, the joint realism curves could be calculated from individual realism curves. Although very iffy at this point, it is conceivable that a much smaller body of data could enable the testing of various hypotheses concerning dependence. In any case, by developing the mathematical relationships involved, it is possible to pursue some theoretical comparisons of probabilistic aggregation with other types of aggregation.

In the following, the convention will be used that whenever the name of a set of events occurs in a probability expression, it denotes the assertion of the joint occurrence of the members of the set. For example, if X is a set of events, $X=\{X_i\}$, then $P(X)=P(X_1 \cdot X_2 \ldots X_n)$, where the period indicates "and." In addition, to reduce the number of subscripts, when a particular event out of a set of events is referred to, the capital letter of the index of that event will be used to refer to the occurrence of the event. Thus $P(X_j)$ will be written $P(J)$.

The degree of dependence among a set of events X is measured by the departure of the joint probability of the set from the product of the separate probabilities of the events. Letting D_X denote the degree of dependence within the set X, we have the definition

$$D_X = \frac{P(X)}{\prod_{i=1}^{n} P(X_i)}. \tag{8}$$

This notion is usually introduced by taking into account dependence among subsets of X as well as the more global notion defined by (8). However, for generating a probabilistic aggregation function, interactions among subsets can be ignored, proving we maintain a set fixed throughout any given computation.

A useful extension of the notion of dependence is that of dependence with respect to a particular event, say $E_j = J$.

$$D_X^J = \frac{P(X|J)}{\prod_{i=1}^{n} P(X_i|J)}. \tag{9}$$

From the rule of the product, we have

$$D_X^J = \frac{P(J \cdot X)P(J)^n}{P(J)\prod_{i=1}^{n} (J \cdot X_i)}. \tag{10}$$

The probability we want to compute is $P(J|R)$; that is, we want to know the probability of an event E_j given that the group reports R. Again, from the rule of the product, we have

$$P(J|R) = \frac{P(R \cdot J)}{P(R)}.$$

Substituting R for X in (10) and multiplying the top and bottom of the right-hand side by $P(R)/\prod_{i=1}^{m} P(R_i)$, and rearranging, gives

$$P(J|R) = \frac{D_R^J \prod_{i=1}^{n} P(J|R_i)}{D_R U(J)^{n-1}}. \tag{11}$$

Formula (11) presents the computation of the joint probability in terms of the individual reports, the dependency terms, and the "a priori" probability $U(J)$. The $P(J|R_i)$ can be derived from individual realism curves. In case the estimators are all fully realistic, then $P(J|R_i) = R_i$. $U(J)$ is the probability of the event J based on whatever information is available without knowing R.[2]

The ratio D_R^J / D_R measures the extent to which the event J influences the dependence among the estimates. If the estimates are independent "a priori," $D_R = 1$. However, the fact that estimators do not interact (anonymity) or make separate estimates, does not guarantee that their estimates are independent.

[2]All of the formulations in this subsection are presumed to be appropriate for some context of information. This context could be included in the formalism, e.g., as an additional term in the reference class for all relative probabilities, or as a reference class for "absolute" probabilities. For example, if the context is labeled W, $U(J)$ would be written $P(J|W)$, $P(J|R)$ would be written $P(J|R \cdot W)$. However, since W would be constant throughout, and ubiquitous in each probability expression, it is omitted for notational simplicity.

They could have read the same book the day before. The event related dependence D_R^J is even more difficult to derive from readily available information concerning the group.

If there is reason to believe that a particular group is completely independent in their estimates, and in addition each member is completely realistic, (11) reduces to

$$P(J|R) = \frac{\prod_{i=1}^{n} R_i}{U(J)^{n-1}}. \tag{12}$$

The simplicity of (12) is rather misleading; it depends on several strong assumptions. (11) on the other hand, is exact, but contains terms which are difficult to evaluate.

An exact expression for $P(J|R)$ can be obtained which does not involve D_R by noting that

$$P(J|R) = \frac{P(J|R)}{P(J|R) + P(\bar{J}|R)}.$$

Substituting for $P(J|R)$ on the right-hand side from (11) and the corresponding expression for $P(\bar{J}|R)$ ($\bar{J}$ denotes the complement of J or "not-J") and dividing top and bottom by $D_J^R / U(J)^{n-1} D_R$ we obtain

$$P(J|R) = \frac{\prod_{i=1}^{n} P(J|R_i)}{\prod_{i=1}^{n} P(J|R_i) + D \prod_{i=1}^{n} (1 - P(J|R_i))}, \tag{13}$$

where

$$D = \frac{D_R^{\bar{J}}}{D_R^J} \left(\frac{U(J)}{U(\bar{J})} \right)^{n-1}$$

If the estimators are all fully realistic and fully independent, and the a priori probability $= \frac{1}{2}$, (13) reduces to

$$P(J|R) = \frac{\prod_{i=1}^{n} R_i}{\prod_{i=1}^{n} R_i + \prod_{i=1}^{n} (1 - R_i)}. \tag{14}$$

To complete this set of estimation formulae, if there are several alternatives in E, and it is desired to compute the group estimate for each alternative from the individual estimates for each alternative, (13) generalizes to

$$P(E_k|R) = \frac{\prod_{i=1}^{n} P(E_k|R_i)}{\sum_{j=1}^{m} D_{jk} \prod_{i=1}^{n} P(E_j|R_i)}, \tag{15}$$

where

$$D_{jk} = \frac{D_R^{E_j}}{D_R^{E_k}} \left(\frac{U(E_k)}{U(E_j)} \right)^{n-1}.$$

(14) is similar to a formula that can be derived using the theorem of Bayes [22]. Perhaps the major difference is that (14) makes the "working" set of estimates the $P(E_j|R_i)$ which can be obtained directly from realism curves, whereas the corresponding formula derived from the theorem of Bayes involves as working estimates $P(R_i|E_j)$ which are not directly obtainable from realism curves. Of course, in the strict sense, the two formulae have to be equivalent, and the $P(R_i|E_j)$ are contained implicitly in the dependency terms. Without some technique for estimating the dependency terms separately from the estimates themselves, not much is gained by computing the group estimate with (14).

Historically, the "a priori" probabilities $U(J)$ have posed a number of conceptual and data problems to the extent that several analysts, e.g., R. A. Fisher [23], have preferred to eliminate them entirely and work only with the likelihood ratios—in the case of (14), the ratios

$$\frac{\prod_{i=1}^{n} R(E_j|R_i)}{\prod_{i=1}^{n} P(E_k|R_i)}.$$

This approach appears to be less defensible in the present case, where the a priori probabilities enter in a strong fashion, namely with the $n-1$ power.

For a rather restricted set of situations, a priori probabilities are fairly well defined, and data exist for specifying them. A good example is the case of weather forecasting, where climatological data form a good base for a priori probabilities. Similar data exist for trend forecasting, where simple extrapolation models are a reasonable source for a priori probabilities. However, in many situations where expert judgment is desired, whatever prior information exists is in a miscellaneous form unsuited for computing probabilities. In fact, it is in part for precisely this reason that experts are needed to "integrate" the miscellaneous information.

Some additional light can be thrown on the role of a priori probabilities as well as the dependency terms by looking at the expected probabilistic score. In the case of the theory-of-errors approach, it was possible to derive the result that, independent of the objective probability distribution P, the expected probabilistic score of the group estimate is higher than the average expected score of individual members of the group. This result is not generally true for probabilistic aggregation.

Since probabilistic aggregation depends upon knowing the a priori probabilities, a useful way to proceed is to define a *net* score obtained by subtracting the score that would be obtained by simply announcing the a priori probability. Letting $S^*(G)$ denote the expected net score of the group and $S^*(R_i)$ the expected net score of individual i, and $S(E)$ the score that would be obtained if $\{U(E_j)\}$ were the report, $S^*(G)=S(G)-S(E)$ and $S^*(R_i)=S(R_i)-S(E)$. The net score measures the extent to which the group estimate is better (or worse) than the a priori estimate. This appears to be a reasonable formulation, since presumably the group has added nothing if its score is no better (or is worse) than what could be obtained without it.

Many formulations of probabilistic scores include a similar consideration when they are "normalized." This is equivalent to subtracting a score for the case of equally distributed probabilities over the alternatives. Thus the score for an individual is normalized by setting $S^*(R_i)=S(R_i)-S(Q)$ where $Q_j=1/m$ and m is the number of alternatives. In effect this is assuming that the a priori probabilities are equal.

Computing the expected group net score from (11) we have

$$\sum_{j=1}^{m} P_j G_j - S(E) = \sum_{j=1}^{m} P_j \ln\left(\frac{D_R^J \prod_{i=1}^{n} P(J|R_i)}{(J)^{n-1} D_R} \right) - S(E) \tag{16}$$

$$= -(n-1)\sum_{j=1}^{m} P_j \ln U(J) + \sum_{j=1}^{m} P_j \sum_{i=1}^{n} \ln P(J|R_i)$$

$$+ \sum_{j=1}^{m} P_j \ln D_R^J - \ln D_R - S(E) \tag{17}$$

$$= -nS(E) + n\hat{S}(R) + \sum_{j=1}^{m} P_j \ln D_R^J - \ln D_R, \tag{18}$$

whence $S^*(G)=nS^*(R)+$Expectation of dependency terms.

If the average net score of the individual members is positive (i.e., the average member of the group does better than the a priori estimate), then the group score will be n times as good, providing the dependency terms are small or positive. On the other hand, if the average net score of the individual members is negative, then the group will be n times as bad, still assuming the dependency terms small. Since the logarithm of D_R will be negative if $D_R<1$, (18) shows that the most favorable situation is not independence where $D_R=1$, $\ln D_R=0$, but rather, the case of negative dependence, i.e., the case where it is less likely that the group will respond with R than would be expected from their independent frequencies of use of R_i.

The role of the event-related dependency term $\sum_{j=1}^{n} P_j \ln D_R^J$ is somewhat more complex. In general, it is desirable that D_R^J be greater than one for those alternatives where the objective probability P_j is high. This favorable condition would be expected if the individuals are skilled estimators, but cannot be guaranteed on logical grounds alone.

One of the more significant features of the probabilistic approach is that under favorable conditions the group response can be more accurate than any member of the group. For example, if the experts are fully realistic, agree completely on a given estimate, are independent, and finally, if it is assumed that the a priori probabilities are equal (the classic case of complete prior ignorance), then formula (14) becomes

$$P(J|R)=\frac{p^n}{p^n+(1-p)^n}, \qquad (19)$$

where p is the common estimate, and n is the number of members of the group. If $p>.5$, then $P(J|R)$ rapidly approaches 1 as n increases. For example, if $p=2/3$ and n is 5, then $P(J|R)=32/33$. If the theory-of-errors approach were being employed, the group estimate would be $2/3$ for any size group.

In this respect, it seems fair to label the probabilistic approach "risky" as compared with the theory-of-errors approach. Under favorable conditions the former can produce group estimates that are much more accurate than the individual members of the group; under less favorable conditions, it can produce answers which are much worse than any member of the group.

Axiomatic Approach

A somewhat different way to develop a theory of group estimation is to postulate a set of desired characteristics for an aggregation method and determine the process or family of processes delimited by the postulates. This approach has not been exploited up to now in Delphi research. The major reason has been the large number of nonformal procedures associated with an applied Delphi exercise—formulation of a questionnaire, selection of a panel of experts, "interpretation of results," and the like. However, if the aggregation process is defined formally as in the two preceding subsections, where questionnaire design is interpreted as defining the event space E, and panel selection is reduced to defining the response space R, then the axiomatic approach becomes feasible.

Considering the group estimation process as a function $G=G(E,I,R)$, various properties of this function appear "reasonable" at first glance. Some of the more evident of these are:

(A) *Unanimity*. If the group is in complete agreement, then the group estimate is equal to the common individual estimate; i.e., if $R_{ij}=R_{kj}$ for all i and k, then $G(R)=R$.

(B) *Monotony*. If R and R' are such that $R_{ij} \geqslant R'_{ij}$ for all i, then $G_j(R) \geqslant G_j(R')$. If R and G are defined as real numbers then they fulfill the usual ordering axioms, and condition B implies condition A.

(C) *Nonconventionality*. G is not independent of the individual estimates; i.e., $G(R) \neq G(S)$ for every possible R and S.

(D) *Responsiveness*. G is responsive to each of the individual estimates; i.e., $G(R) \neq G'(T)$, where T is a proper subvector of R.

(E) *Preservation of Probability Rules*. If G is an aggregation function which maps a set of individual probability estimates onto a probability, then G preserves the rules of probability. For example, if $T_{ij}=R_{ij}S_{ij}$ for all i and j (as would be the case if R_{ij} is the estimated probability of E_j and S_{ij} is the estimated relative probability of an event E'_j given that E_j occurs) then

$$G(T_j)=G(R_j)G(S_j).$$

This set of conditions will be displayed more fully below.

All of these conditions have a fairly strong intuitive appeal. However, intuition appears to be a poor guide here. The first four postulates are fulfilled by any of the usual averaging techniques. But A, which is perhaps the most apparently reasonable of them all, is not fulfilled by the probabilistic aggregation techniques discussed in the previous subsection. It was pointed out there that one of the more intriguing possibilities with probabilistic aggregation is that the group estimate may be higher (or lower, depending on the interaction terms) than any individual estimate.

It can be shown that there is no function that fulfills all five of the postulates; in fact, there is no function that fulfills D and E. The proof of this impossibility theorem is given elsewhere [24]; it will only be sketched here.

Three basic properties of probabilities are (a) normalization, if p is a probability, $0 \leqslant p \leqslant 1$; ($b$) complementation, $P(J)+P(\bar{J})=1$; and (c) multiplicative conjunction, i.e., $P(J_1 \cdot J_2)=P(J_1)P(J_2/J_1)$. The last is sometimes taken as a postulate, sometimes is derived from other assumptions.

If the individual members of a group are consistent, their probability judgments will fulfill these three conditions. It would appear reasonable to require that a group estimate also fulfill the conditions, consistently with the individual judgments. In addition, condition D, above, appears reasonable. This leads to the four postulates:

P1. $0 \leqslant G(R) \leqslant 1$.
P2. $G(1-R)=1-G(R)$.
P3. $G(R \cdot S)=G(R)G(S)$.
P4. $G(R) \neq G'(S)$, where S is a subvector of R (condition D).

Here, $R \cdot S$ is the inner product of the two vectors R and S, i.e.,

$$R \cdot S=(R_1 S_1, R_2 S_2, \ldots, R_n S_n).$$

P1-P3 have the consequence that G is both multiplicative and additive. The multiplicative property comes directly from P3, and the additive property—i.e., $P(R+S)=P(R)+P(S)$—is derived by using the other postulates. For functions of a single variable, there is only one which is both multiplicative and additive, namely the identity function $f(x)=x$. There is no corresponding identity function for functions of several variables except the degenerative function, $G(R)=G'(R_i)=R_i$, which violates P4.

This result may seem a little upsetting at first glance. It states that probability estimates arrived at by aggregating a set of individual probability estimates cannot be manipulated as if they were direct estimates of a probability. However, there are many ways to react to an impossibility theorem. One is panic. There is the story that the logician Frege died of a heart attack shortly after he was notified by Bertrand Russell of the antinomy of the class of all classes that do not contain themselves. There was some such reaction after the more recent discovery of an impossibility theorem in the area of group preferences by Kenneth Arrow [25]. However, a quite different, and more pragmatic reaction is represented by the final disposition of the case of 0. In the 17th century, there was long controversy on the issue whether 0 could be treated as a number. Strictly speaking there is an impossibility theorem to the effect that 0 cannot be a number. As everyone knows, division by 0 can lead to contradictions. The resolution was a calm admonition, "Treat 0 as a number, but don't divide by it."

In this spirit, formulation of group probability estimates has many desirable properties. It would be a pity to forbid them because of a mere impossibility theorem. Rather, the reasonable attitude would appear to be to use group probability estimates, but at the same time not to perform manipulations with the group aggregation function which can lead to inconsistencies.

Coda

The preceding has taken a rather narrow look at some of the basic aspects of group estimation. Many significant features, such as interaction via discussion or formal feedback, the role of additional information "fed-in" to the group, the differences between open-ended and prescribed questions, and the like, have not been considered. In addition, the role of a Delphi exercise within a broader decisionmaking process has not been assessed. What has been attempted, albeit not quite with the full neatness of a well-rounded formal theory, is the analysis of some of the basic building blocks of group estimation.

To summarize briefly: The outlines of a theory of estimation have been sketched, based on an objective definition of estimation skill—the realism curve or track record of an expert. Several approaches to methods of aggregation of individual reports into a group report have been discussed. At the moment, insufficient empirical data exist to answer several crucial questions concerning both individual and group estimation. For individual estimation, the question is open whether the realism curve is well defined and sufficiently stable so that it can be used to generate probabilities. For groups, the degree of dependency of expert estimates, and the efficacy of various techniques such as anonymity and random selection of experts in reducing dependency have not been studied.

By and large it appears that two broad attitudes can be taken toward the aggregation process. One attitude, which can be labeled conservative, assumes that expert judgment is relatively erratic and plagued with random error. Under this assumption, the theory-of-errors approach looks most appealing. At least, it offers the comfort of the theorem that the error of the group will be less than the average error of the individuals. The other attitude is that experts can be calibrated and, via training and computational assists, can attain a reasonable degree of realism. In this case it would be worthwhile to look for ways to obtain a priori probabilities and estimate the degree of dependency so that the more powerful probabilistic aggregation techniques can be used.

At the moment I am inclined to take the conservative attitude because of the gaping holes in our knowledge of the estimation process. On the other hand, the desirability of filling these gaps with extensive empirical investigations seems evident.

References

1. John McCarthy, "Measures of the Value of Information," *Proc. Nat. Acad. of Sci.* **42** (September 15, 1956), pp. 654–55.
2. Thomas Brown, "Probabilistic Forecasts and Reproducing Scoring Systems," The Rand Corporation, RM-6299-ARPA, July 1970.
3. L. J. Savage, "Elicitation of Personal Probabilities and Expectations," *J. Amer. Stat. Assoc.* **66** (December 1971), pp. 783–801.
4. E. E. Shannon and W. Weaver, *The Mathematical Theory of Communication*, University of Illinois Press, Urbana, 1949.
5. Savage, *op. cit.*
6. S. E. Asch, "Effects of Group Pressure upon the Modification and Distortion of Judgments," in E. E. Maccoby, T. M. Newcomb, and E. L. Hartley (eds.), *Readings in Social Psychology*, Henry Holt, New York, 1958, pp. 174–83.
7. C. H. Coombs, "A Review of the Mathematical Psychology of Risk," presented at the Conference on Subjective Optimality, University of Michigan, Ann Arbor, August 1972.
8. M. Girshick, A. Kaplan, and A. Skogstad, "The Prediction of Social and Technological Events," *Public Opinion Quarterly*, Spring 1950, pp. 93–110.

9. G. A. S. Stael von Holstein, "Assessment and Evaluation of Subjective Probability Distributions," Economic Research Institute, Stockholm School of Economics, 1970.
10. Girshick, *et al., op. cit.*
11. W. Edwards, "The Theory of Decision Making," *Psychol. Bulletin* 5 (1954), pp. 380–417.
12. F. Knight, *Risk, Uncertainty and Profit*, Houghton Mifflin, Boston, 1921.
13. Hans Reichenbach, *The Theory of Probability*, University of California Press, Berkeley, 1949, Section 68.
14. N. Dalkey, "Experimental Study of Group Opinion," *Futures* 1 (September 1969), pp. 408–26.
15. N. Dalkey, B. Brown, and S. Cochran, "The Use of Self-Ratings to Improve Group Estimates," *Technological Forecasting* 1 (1970) pp. 283–92.
16. J. P. Guilford, *Psychometric Methods*, McGraw-Hill, New York, 1936, pp. 426ff.
17. N. Moiseev, "The Present State of Futures Research in the Soviet Union," in *Trends in Mathematical Modeling*, Nigel Hawkes (ed.), Springer-Verlag, Berlin, 1973.
18. N. Dalkey, "Experimental Study of Group Opinion," *op. cit.*
19. J. Aitchison and J. A. C. Brown, *The Lognormal Distribution*, University of Cambridge Press, Cambridge, Eng., 1957.
20. N. Dalkey, "Experimental Study of Group Opinion," *op. cit.*
21. T. Brown, "An Experiment in Probabilistic Forecasting," The Rand Corporation, R-944-ARPA, March 1973.
22. Peter A. Morris, *Bayesian Expert Resolution*, doctoral dissertation, Stanford University, Stanford, California, 1971.
23. R. A. Fisher, "On the Mathematical Foundations of Theoretical Statistics," *Philos. Trans. Roy. Soc.*, London, Series A, Vol. 222, 1922.
24. N. Dalkey, "An Impossibility Theorem for Group Probability Functions," The Rand Corporation, P-4862, June 1972.
25. K. J. Arrow, *Social Choice and Individual Values*, John Wiley and Sons, New York, 1951.

IV. C. Experiments in Delphi Methodology*

M. SCHEIBE, M. SKUTSCH, and J. SCHOFER

Introduction

The emphasis in the Delphi literature to date has been on results rather than on methodology and evaluation of design features. The other articles in this chapter do address the latter aspects. Still, quite a number of issues remain unsolved, particularly those concerned with the details of the internal structure of the Delphi. For example, the way in which subjective evaluation is measured may affect the final output of the Delphi. A number of variables enter here. Ostrom and Upshaw [1] have noted that the range of the scale provided has a marked effect on judgment. Persons playing the role of judges who estimated themselves as "relatively harsh" assigned average "sentences" of four years to "criminals" when presented with a one-to-five-year scale, and twenty-one years when presented with a 1 to 25-year scale. The difficulties involved with the selection of a suitable scale range can be solved by the employment of an *abstract* scale rather than one representing, for example, hard dollars or years. An abstract scale allows relative measures to be made. Abstract scales are particularly suited to the measurement of values, as for example in the development of goal weights to represent relative priorities for goal attainment.

A number of psychological scaling techniques which result in abstract scales are available. This study reports on the comparison of several scaling techniques which were tested in the context of an experimental Goals Delphi.

Another issue is that of the effects of feedback input, which form the sole means of internal group communications in the Delphi process. It is important to the design of Goals Delphis to determine the nature and strength of the feedback influence. In the experiment reported below, the impact of feedback was identified by providing participants with modified feedback data. The resulting shifts of opinion were then used as measures of feedback effectiveness.

Methods for the measurement of consensus are also considered and a redefinition of the endpoint of a Delphi is offered. Instead of consensus, the *stability* of group opinion is measured. This allows much more information to be derived from the Delphi, and in particular, preserves opinion distributions that achieve a multimodal consensus.

---

*This study was supported by the Urban Systems Engineering Center, Northwestern University, NSF Grant GU-3851.

The Delphi Method: Techniques and Applications, Harold A. Linstone and Murray Turoff (eds.)
ISBN 0-201-04294-0; 0-201-04293-2

A number of Delphi studies have used high/low self-ratings of participant confidence. Evidence of the value of such confidence ratings in improving the results of the Delphi is somewhat limited, except under certain conditions of group composition [4]. In this study, the use of high/low self-confidence ratings is again evaluated, and the influence of a number of other personal descriptive variables is tested.

Other design features include the application of short turn-around times using a computerized system for supporting inter-round analysis of the Delphi data. Although Turoff (Chapter V, C) has used a more complex, interactive computer system for this purpose, a simpler program is used here merely to accelerate accounting tasks.

Description of Procedure

The objective of this experimental Delphi study was the development and weighting of a hierarchy of goals and objectives for use in evaluating a number of hypothetical transportation facility alternatives. In the terminology suggested by Wachs and Schofer [2], *goals* are long-term horizonal aims derived directly from unwritten community values; *objectives* are specific, directional, measurable ends which relate directly to goals. Previous experiments by Rutherford, *et al.* [3] had indicated that Goals Delphis should be initiated by the development of objectives rather than goals, for the tendency toward upward drift in generality can be minimized if the Delphi participants are first asked to work at the more specific level. The development of goals, once the objectives have been defined, can be accomplished with much greater ease than can the reverse process.

The flow chart of Fig. 1 illustrates graphically the process by which the Goals Delphi and the design experiments were carried out. First, the initial list of objectives was generated. The process administrators presented the hypothetical transportation situation to the participants by means of a verbal description (Appendix I of this article), and a map (Appendix II of this article). The participants were then given a set of five blank 3 × 5 cards and asked to list no more than five objectives which they felt were applicable in the hypothetical situation. In all, seventy-seven objectives were submitted.

To derive goals from the list of objectives, and to eliminate the overlaps between them, a grouping procedure was followed. The process administrators first rejected those objectives that were exact duplicates of others, and assigned the remaining ones to sets. Each set represented objectives tending toward a common goal. Nine major goals were established, and these were "named" appropriately. Statements which were not strictly objectives were left out of the grouping process. The complete list of goals and objectives is given in Table 1. These goals were then returned to the participants for their evaluation. They were given the opportunity to add new objectives, and several were received and incorporated into the goal set.

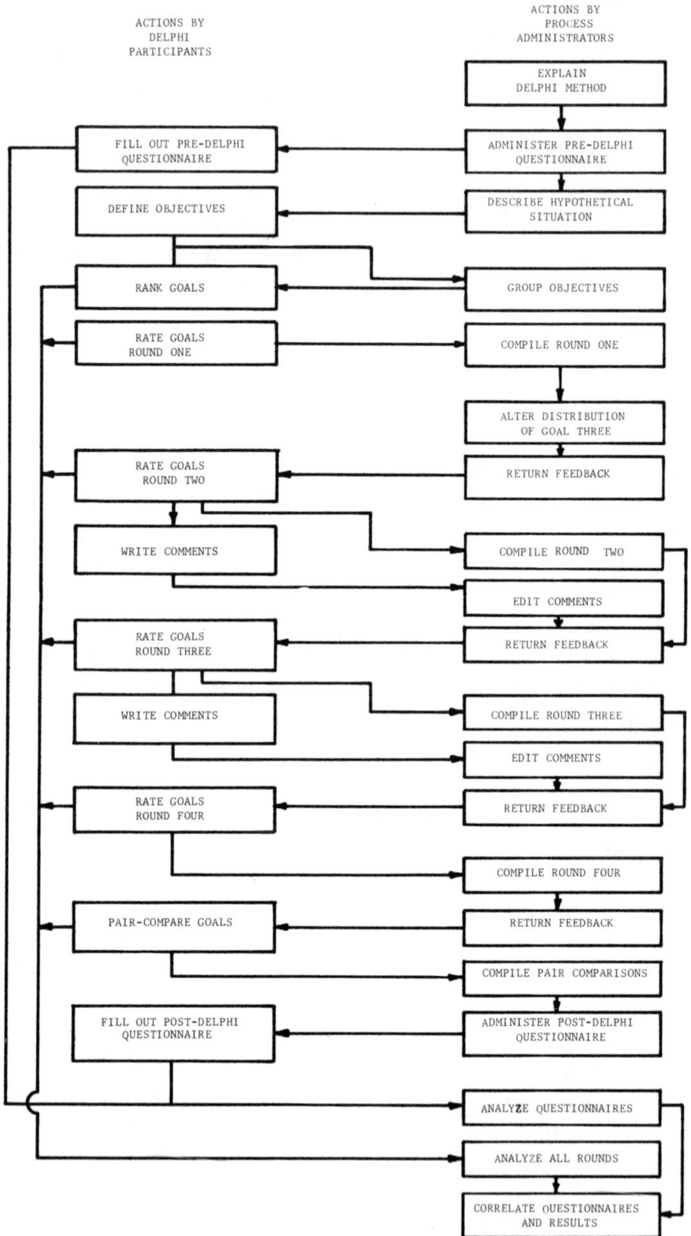

Fig. 1. The Goals Delphi experimental design.

Table 1: Goals and Objectives Suggested for the Hypothetical Transportation Scenario

GOAL I:

Minimize the adverse environmental impacts of the system.

Objectives:

Minimize air, noise, and water pollution.

Minimize negative illumination and vibration effects.

GOAL II:

Preserve the recreational and social environment.

Objectives:

Provide compensation for parkland removed.

Minimize the amount of parkland taken.

Minimize the demolition of historic buildings.

Minimize the amount of non-urban land required for the new facility.

Minimize the adverse social consequences of the transportation facility location by providing adequate compensation to families and businesses (and employees).

Minimize the amount of urban land required for the new facility.

Minimize the number of residences relocated.

Minimize the disruption of existing neighborhoods, people and businesses.

GOAL III:

Minimize operating and construction costs.

Objective:

Minimize operating and construction costs.

GOAL IV:

Maximize mobility and accessibility for the urban area residents.

Objectives:

Minimize travel time.

Minimize monetary travel cost.

Reduce congestion

Locate the facility to increase mobility and accessibility in the area.

Provide sufficient mobility for <u>all</u> members of society.

Table 1 (cont.)

GOAL V:

Design the facility to operate as safely as possible.

Objective:

Design the facility to operate as safely as possible.

GOAL VI:

Coordinate the transportation system with land use development.

Objectives:

Maximize the flexibility of the facility so as not to hinder possible further urban development in the area.

Encourage the desired land use development pattern as stated in the land use plan.

GOAL VII:

Design the facility so that its aesthetic appeal is in harmony with the environment.

Objectives:

Use transport facilities to highlight the character of the city to increase awareness, interest, and participation by the facility users.

Design the facility to be visually pleasing to the surrounding community.

GOAL VIII:

Maximize positive economic benefits for the entire area.

Objective:

Maximize positive economic benefits for the entire area.

GOAL IX:

Minimize the adverse impacts occurring during construction.

Objective:

Minimize the adverse impacts occurring during construction.

Following the development of the goals hierarchy, a decision was made (largely because of time constraints) to concentrate attention at the goal level. The objectives, therefore, were not included in the weighting procedure. Objectives, however, were at all times appended to the goals related to each, so that participants would always be aware of the specific meaning of each goal.

Participants were first asked to do a simple ranking of goals. As discussed elsewhere in this paper, one of the purposes of the Delphi was to compare different scaling methods. Participants were therefore asked to follow the ranking with a rating analysis. Nine-point Likert scales were used (0 = unimportant, 9 = very important). This type of scale was felt to be easily understood by the participants. In addition, when the ends are anchored adjectively, as in semantic differential scales, this scale is commonly found to have interval properties. Using the computer program developed for this study, the results of this first round were analyzed. The program was processed using a remote terminal; goal weights served as inputs, and histograms and various distributional statistics were produced as outputs. Frequency distributions of scores for each goal prepared by the computer were presented to the participants, along with the mean for each distribution.

Participants were asked to once again rate each goal on a nine-point scale, using the information from the previous round as feedback. In addition, those participants whose score on any goal was significantly distant from the group mean value for that goal were asked to write a few words explaining the reasons behind their positions. These statements were edited and returned in the next round. This procedure continued for a total of four rounds. The results are given in Fig. 2 and 3, which show the histograms produced in the first and the final rounds.

After the fourth weighting round, the participants were asked to perform a pair-comparison rating of all the goals. This was done to compare this scaling method with the nine-point rating scale and the ranking methods.

The initial development of the goals was accomplished during one two-hour class period. The rank ordering of the goals and the first three weighting rounds were conducted in a second two-hour period two days after the first. The fourth round took place an additional five days later, while the pair comparisons were made a week after the first weighting round.

An Experiment on the Effect of Feedback

There has been quite a bit written about the uses of feedback in the Delphi technique. Most of this, such as the work of Dalkey, Brown, and Cochran [4] at Rand, has concentrated on the effects of different types of feedback, such as written statements and various statistical measures. The effects of this feedback, particularly in the almanac-type Delphis, have been measured by comparing

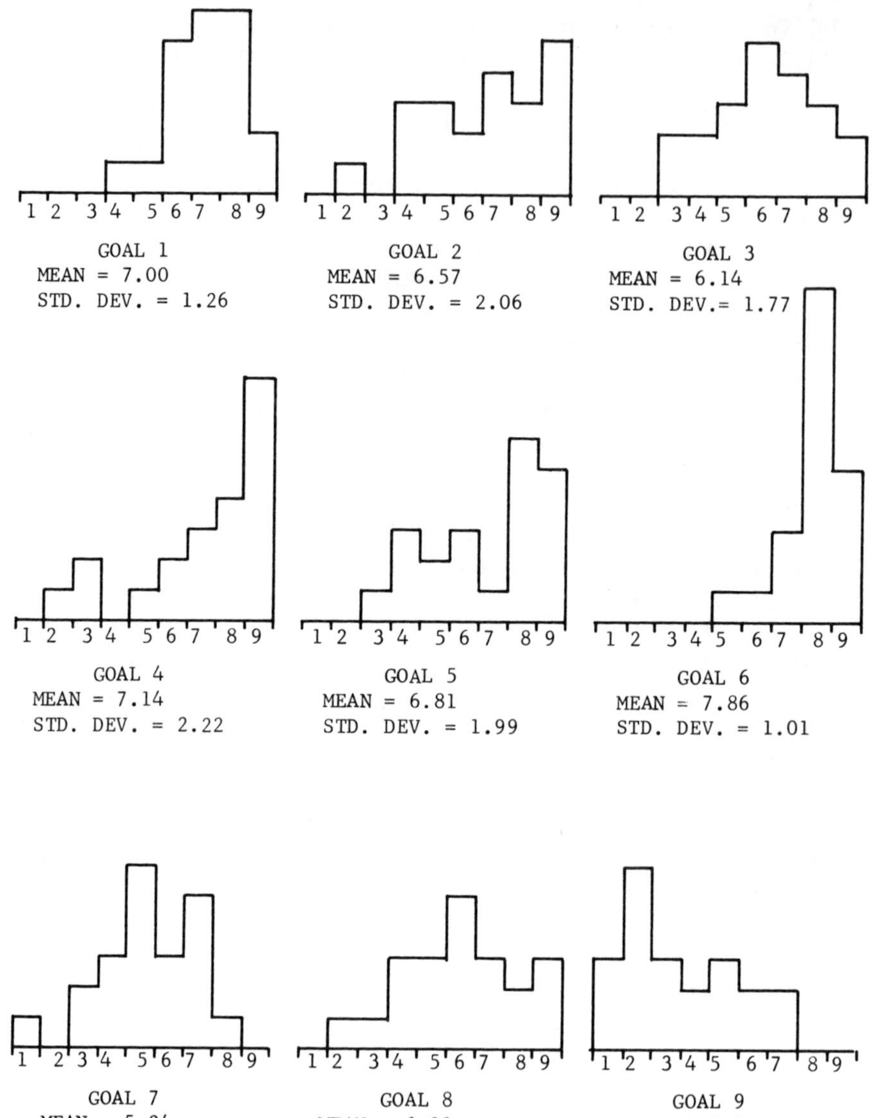

Fig. 2. Results of round one rating analysis.

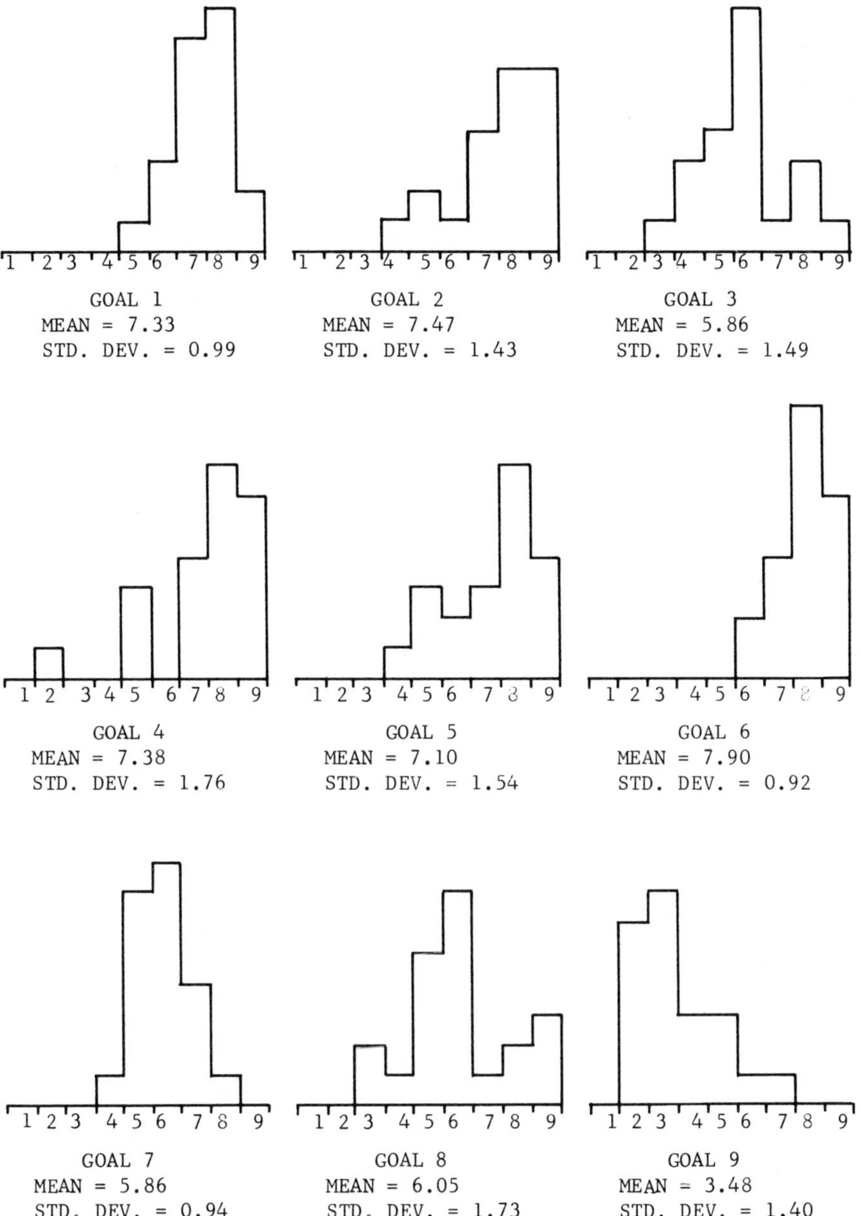

Fig. 3. Results of round four rating analysis.

the accuracy of the opinions of a group given a certain feedback with that of a group given different feedback or no feedback at all. In Policy and Value Delphis the effect of feedback is evaluated by measuring the degree of consensus which is reached and the speed with which it is reached.

There seems to be very little in the literature, however, which examines the round-by-round effect of feedback or investigates the manner in which the feedback affects the distribution of scores in a particular round. In this study, it was decided to investigate these aspects of feedback, since the kind and amount of feedback used in the Delphi may be an important variable in its results. A greater understanding of the impacts of feedback might lead to better Delphi design. The method employed was to provide participants with false feedback data, and then to observe the effect of this on the distribution of priority-weight scores.

Two types of feedback were used in this study. The first was a graphical representation of the distribution of scores together with a listing of the mean of the distribution. In addition, in later rounds edited, anonymous comments by the participants concerning the importance of the various goals were distributed. During the experiment on feedback, one goal was chosen and the distribution was altered by the administrators so as to change markedly the position of the mean. Since this was done after the first weighting round, no written feedback accompanied the altered distribution. The goal chosen for this test was Number 3 ("Minimize the operating and construction costs"). This goal was chosen because it appeared to have a good consensus after the first iteration. In addition, it was judged to be substantively important. It was felt that most participants would be very surprised by the altered distribution.

The second-round distribution showed that the feedback had had an effect, since a number of persons shifted their positions away from the true mean. By the third round, the distribution was once again similar to what it had been in the first round, although the distribution was shifted slightly to the lower end of the scale and remained that way permanently, showing residual effects of the gerrymandering. Figure 4 shows the actual distributions and the altered feedback used.

In attempting to explain the reasons behind these changes, the following hypothesis is offered. Upon seeing the first round of feedback information, the respondents had three options: they could ignore the feedback and keep their votes constant; they could rebel against the feedback and move their votes to the right, in the interest of moving the group mean closer to their true desire; or they could acknowledge the feedback and move their votes nearer the false mean. If they had followed either of the first two options, it would indicate that the feedback was not effective in changing individual attitudes. That the third option was in fact taken, however, indicates that the feedback did have an effect on the participants.

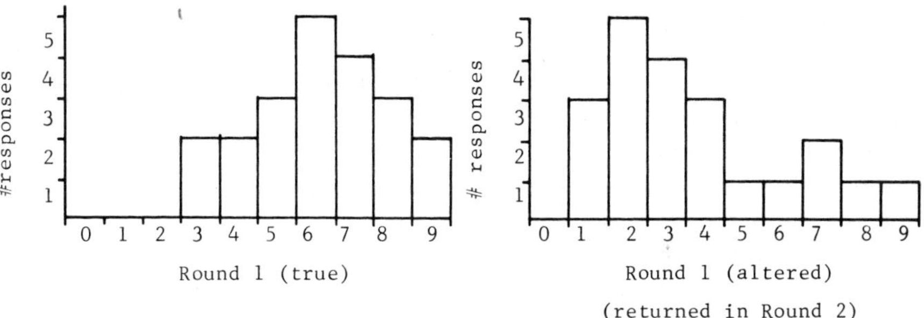

responses
Round 2

responses
Round 3

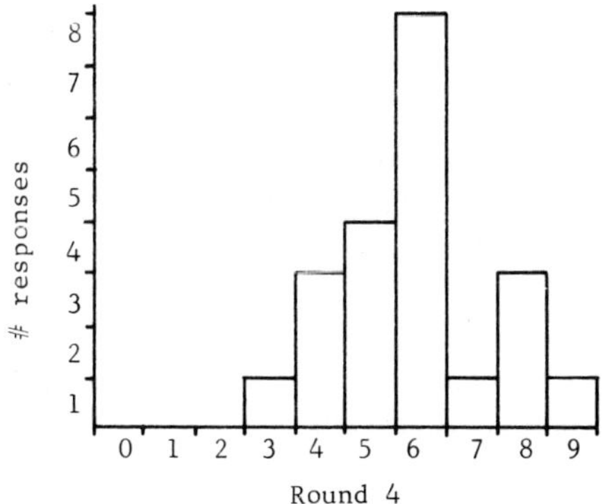

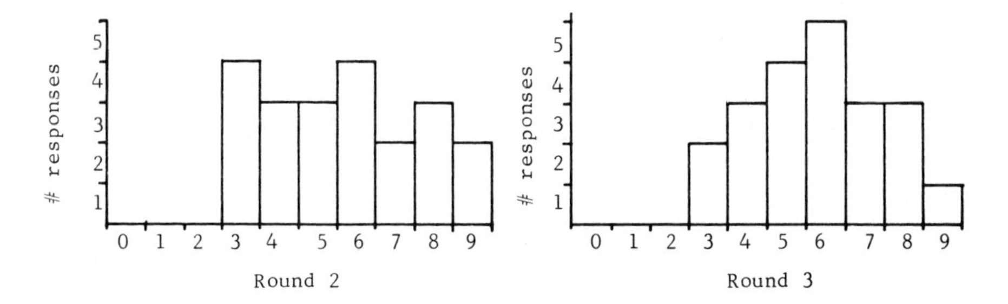

Fig. 4. Distributions generated for Goal 3.

The third round, as a result of the feedback of the second round, also shows the effect of feedback. The second-round distribution showed that participants were attempting to increase the priority for Goal 3, although with respect to their true initial opinions, they were actually decreasing this priority. It seems likely that many respondents, upon seeing this, felt that the group was moving closer to their original position and they decided to return to their first-round vote, since it no longer appeared that this position would be far distant from the mean value of the group.

This experiment suggests that the respondents are, in fact, sensitive to the feedback of distributions of scores from the group as a whole. These results seem to indicate that most repondents are both interested in the opinions of the other members of the group and desirous of moving closer to the perceived consensus.

Comparison of Scaling Techniques

With the exception of Dalkey and Rourkc [5], there is little discussion in the literature of the different methods of scaling which could be used in a Delphi. The two most common methods which are used are simple ranking and a Likert-type rating scale. Even when these methods are used, there have seldom been attempts to ensure that the scales developed are, in fact, interval scales.

The necessity of having an interval scale is seldom emphasized in Delphis. There is the suspicion that on some occasions the scales derived are ordinal scales. An ordinal scale merely shows the rank order of terms on the scale, and no statement can be made concerning the distance between quantities.

Presumably, the primary reasons for using a Delphi, especially when comparing policies or measuring values, include the determination of not only which policies are considered most important, but also the degree to which each policy is preferred over the other possibilities. In order to assure that this can be determined, an interval scale must be obtained.

In this study, three methods which usually yield interval scales were tested. These methods were simple ranking, a rating-scale method, and pair comparisons. The purpose in trying three scales was to determine if all three methods yielded approximately equivalent interval scales. If this is found to be the case, then in future designs any one of these scales could be used. In this situation, it would probably be wisest to choose that scaling method which was considered easiest to perform by the participants in the Delphi. In this study it was found that the rating-scale method was considered by the participants as the most comfortable to perform. The limitation of the pair-comparison method is that it is time consuming. For example, to apply this method to a set of ten objectives, each participant must make forty-five judgments. The ranking method is fairly easy for a small number of goals, but becomes increasingly difficult as the number of goals increases, for it essentially requires the participant to order the entire list of items in his mind. In addition, many participants felt uncomfortable performing this method because they were prevented from giving two goals an equal ranking (i.e., forced ranking). While this dilemma might possibly have encouraged more thought concerning underlying priorities, it was felt that the frustration caused had a negative effect on the end result. The rating-scale method was found to be quick, easy to comprehend, and psychologically comforting. The participant's task is easy, since he must rate only one item at a time. The problem that remained was to determine whether such a scaling procedure would yield a scale with interval properties.

In this experiment it was found that each of the three methods yielded somewhat different scales. Using the Law of Comparative Judgment [6], scale values for each goal for each round were derived. These values were then translated onto a scale from one to nine. Graphical representations of these scales are given in Fig. 5. Because of the presence of feedback, the four rating rounds are not independent. Each one depends on those previous to it. The scales derived in each successive round should not be identical, for if the scale remains constant from round to round, the justification for using an iterative approach is lost. In addition, because of the order in which the scales were developed, the ranking scale can only be compared with the first-round rating scale and the pair-comparison scale can only be compared with the fourth-round rating scale. Because of four rounds of feedback between them, the ranking scale and pair-comparison scale should be compared only cautiously, and should not be expected to be identical

The interscale comparison shown in Fig. 5 is not especially encouraging. The pair-comparison method is known to produce interval scales, and the similarity in results of this approach and the round-four rating results is not strong. The scales produced by ranking and round-one rating are, however, not too different from each other. It is possible to interpret the progression of rating scales from round one to round four as a movement in the direction of the pair-comparison scale. This experiment did not pursue further weighting rounds, but, as discussed below, major changes in weights beyond round four do not seem likely. In addition, later pair-comparison responses might differ from that shown in Fig. 5. Given the complexity of the pair-comparison method for participants, however, it may not be unreasonable to accept cautiously the results of simple rating methods as fair approximations to an interval scale.

The Effect of Personal Variables on Participant Behavior

Dalkey, Brown, and others have considered and used the confidence of participants in their responses to reach more accurate estimates of quantitative phenomena in Delphi exercises. Working with almanac-type data not available to the participants, they found that by selecting for inclusion in the feedback only those responses considered "highly confident" by their proponents, a

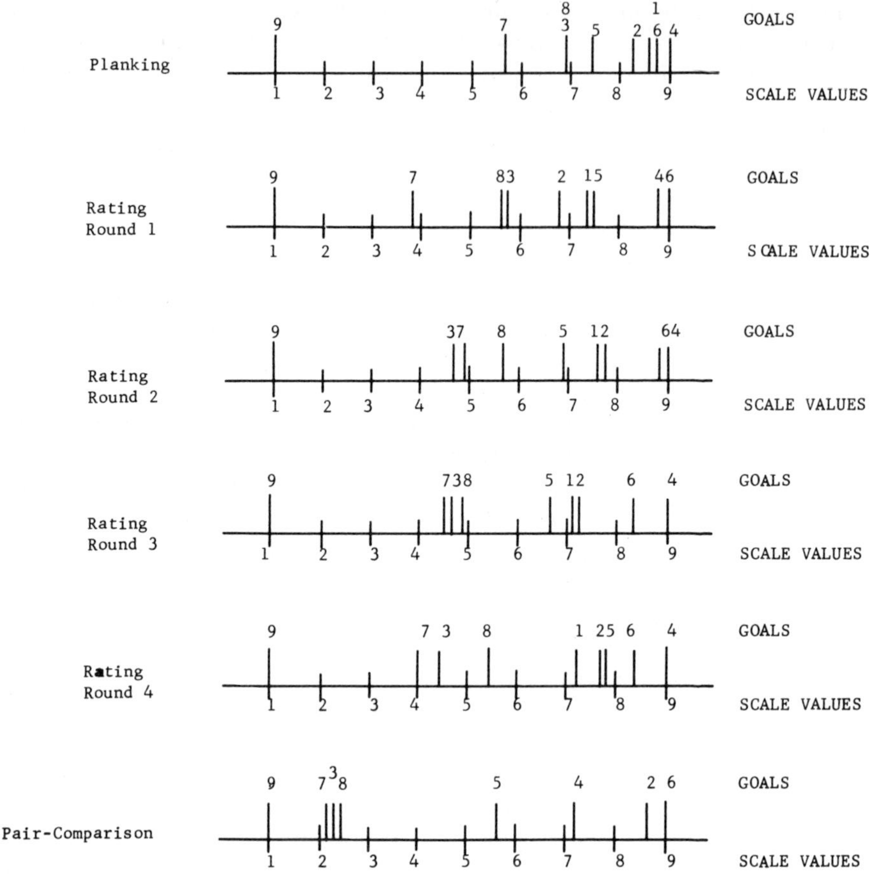

Fig. 5. Comparisons of goal weights on various categorical judgment scales.

slightly superior result was achieved [7]. Later, they found that in situations in which relative confidence was measured and in which the "highly confident" group was reasonably large, a definitely superior result could be expected [4].

Studies in the psychology of small groups, however, indicate that highly confident persons should be less influenced by group pressure than those with less confidence, and therefore it would be expected that highly confident individuals move less toward consensus than do others in the Delphi context. Later, Dalkey *et al.* [4] showed that "over consensus" may occur, and the ratio of average error to standard deviation may actually increase, if consensus is forced too quickly. In order to reach some greater understanding of theory and observation, therefore, several simple hypotheses were tested. It was felt that confidence might involve more than simply confidence in individual answers, and therefore a selection of variables representing various aspects of personal confidence were sought, as well as high/low confidence in each response. Appendix III of this article shows the questionnaire issued to all participants before the Delphi. The variables measured were as follows:

Pre-Delphi Survey Question Number	Independent Variable
1	A. Perceived academic standing relative to other participants
2	B. At-oneness with other participants
3	C. Familiarity with other participants
4	D. Confidence of own capacity in the subject area of the Delphi
5 and 6	E. Experience in the field
7	F. Expectation of the Delphi process (confidence in Delphi)

Each of these confidence variables was then correlated against dependent variables describing the amount of movement actually made by each participant toward the center of the distribution. It was, of course, not possible in this value-judgment Delphi to test accuracy as well, as was done in Dalkey's experiments.

The dependent variables were summed for every participant over each of his nine responses, and were as follows:

(a) Total amount of change from round one to round four
(b) Total number of monotonic (as opposed to oscillating) changes made
(c) Amount of change made in second round
(d) Number of responses exactly on the mode in round three
(e) Number of responses within three places of the mode in round three

It was hypothesized that high confidence would be associated with small amounts of total change, monotonic rather than oscillating change, and low confidence with a high degree of change in round two and a high conformity to the consensus in round three.

The response with regard to confidence in individual answers (measured by high/low ratings of confidence on the Delphi answer sheets) was correlated significantly, negatively, but not very highly with the amount of change in the second round, although not with the total change between rounds one and four.

Percent of "highly confident" answers, however, was cross-correlated positively with perceived academic status, although this was not significantly connected with either movement variable. Amount of change in the second round was also just correlated (positively) with the "at-oneness" variable, although there was no relationship at all between "at-oneness" and percent highly confident. These represented the only significant correlations found.

The evidence for the effect of confidence on the tendency to converge is somewhat sketchy. The only conclusions that can be drawn from the experience is that the initial surprise on being confronted with some distribution of group opinion may to some extent cause the less confident members who believe that they associate with the rest of the group to move toward the center of opinion, but that this tendency is certainly not an overwhelming one.

At the end of the Delphi a second questionnaire (Appendix IV of this article) was used to determine whether the kind of feedback provided had any conscious effect on movement in the Delphi. The variables were as follows:

Post-Delphi Survey Question Number	Independent Variable
1	G. Satisfaction with the results
2,3 and 4	H. Success of feedback in the learning process
5 and 6	I. Frustration with communication levels
7	J. Optimism for Delphi
8	K. Feeling of being rushed

These variables were correlated with the same dependent variables. Both optimism for the future of Delphi and satisfaction with the process correlated significantly and positively with the number of monotonic changes made, perhaps indicating that those people who were not caused to change their opinion radically were in better spirits after the Delphi than the others. However, the success-of-feedback variable was strongly and negatively correlated with the propensity to conform to the mode in round three. In other words, those who did conform to the visible majority had difficulty in giving and taking ideas from the feedback. This is interesting in that it indicates the different kinds of feedback that may affect people in different ways. The tendency to converge strongly has elsewhere been shown by Schofer and Skutsch [8] to be due to emphasis in the visible consensus and on the need to create consensus. Satisfaction with the process was also negatively correlated with the conformity variable. (Satisfaction and agreement with feedback were strongly cross-correlated.) Clearly, the people who were strongly conforming were not happy with the Delphi at all. The question of what is cause and what is effect, however, remains to be answered. Yet one might speculate that, especially in a value-oriented Delphi, the group pressure from some forms of feedback can be overly strong, forcing participants to take positions which they find uncomfortable. While compromise may be uncomfortable in any situation, the real danger here is that participants may leave the process without really compromising their feelings at all. That is, perhaps the anonymity of the Delphi itself may have encouraged participants to capitulate, but only on paper. They may later hold to their original views, and, if the results of the Goals Delphi are used to develop programs to meet their needs, participants might ultimately be quite dissatisfied with the results.

A cautionary note is relevant at this point. Another study by Skutsch [9] has shown that the form of the feedback itself influences consensus development. Despite the fact that participants in this experiment were encouraged to report their verbal rationale for their positions, the rapidity with which the process was carried out tended to discourage such responses. As a result, histograms of value weights formed the bulk of the feedback. It is just this kind of limited, "hard" feedback which tends to force what might be an irrational consensus, one which might be only temporary.

Opinion Stability as a Method of Consensus Measurement

In most Delphis, consensus is assumed to have been achieved when a certain percentage of the votes fall within a prescribed range—for example, when the interquartile range is no larger than two units on a ten-unit scale. Measures of this sort do not take full advantage of the information available in the distributions. For example, a bimodal distribution may occur which will not be registered as a consensus, but indicates an important and apparently insoluble cleft of opinion. Less dramatically, the distribution may flatten out and not reach any strongly peaked shape at all. The results of the Delphi are no less important for this, however. Indeed, considering that there is a strong natural tendency in the Delphi for opinion to centralize, resistance in the form of unconsensual distributions should be viewed with special interest.

A measure which takes into account such variations from the norm is one that measures not consensus as such, but *stability* of the respondents' vote distribution curve over successive rounds of the Delphi. Because the interest lies in the opinion of the group rather than in that of individuals, this method is preferable to one that would measure the amount of change in each individual's vote between rounds.

To compare the distributions of opinion between rounds, the histograms may be subtracted columnwise and the absolute value of the result taken. In Table 2 this approach is applied to the weight histograms reported for Goal 5 in the first three rounds of the Delphi. Columnwise subtraction between the first and second, and the second and third, rounds gives the results shown in Table 2. The absolute values of the differences between histograms are aggregated to from total units of change; but since any one participant's change of opinion is reflected in the histogram differences by two units of change, net person-changes must be computed by dividing total units of change by two. Finally, the percentage change is determined by dividing net changes by the number of participants. Clearly, in the example shown in Table 2, the distribution of value weights for Goal 5 became more stable between rounds one and three.

The question of what represents a reasonable cut-off point at which the response may be said to be unchanged, and therefore finally in its stable position, poses some problems, however. Since there is no underlying statistical theory in what has so far been proposed, no true statistical level may be set, as might, for example, be possible with a statistical change in variance test.[1]

Empirical examination of the responses in the Delphi, however, showed that at any point in time a certain amount of oscillatory movement and change within the group is inevitable. This might be conceptualized as a sort of underlying error function, a type of internal system noise. What is needed is a "confidence" measure which allows the distinction to be drawn between this kind of movement and strong group movements that represent real changing opinion. Such an estimate has tentatively been made from studies of observed probability of movement.

Leaving aside objective 3 (for reasons made apparent earlier), the propensity of the individual to alter his score as a function of distance from the center point was measured. This was done by calculating the proportion of respondents at each scale distance from the mode that moved toward the mode between rounds. The results, displayed in Fig. 6, show a strong tendency for increased amounts of movement with distance from the center point. They also show that a percentage change is to be expected among respondents who are already dead on the mode itself. The amount of movement at the mode (about 15%) has therefore been taken to represent the base of oscillatory movement to be expected, and this is supported by the fact that the amount of change at the centroid does not alter appreciably between rounds.

Using the 15% change level to represent a state of equilibrium, any two distributions that show marginal changes of less than 15% may be said to have reached stability; any successive distributions with more than 15% change should be included in later rounds of the Delphi, since they have not come to the equilibrium position.

[1]Conventional variance tests were found to be unsuited to the case of change in histogram shape in this context. Most rely on independent samples; none is strong enough to pick up small changes in shape, and none robust enough to deal with non-normal distributions.

Table 2: Example of Stability Measurement Computations for Goal 5

Rating	1	2	3	4	5	6	7	8	9
Absolute difference in number selecting rating, rounds 1-2 (a)	0	0	0	2	2	2	4	0	2
Total units of change (b)	12								
Net person-changes (c)	6								
Number of participants	21								
Percent change (d)	28.6%								

Rating	1	2	3	4	5	6	7	8	9
Absolute difference in number selecting rating, rounds 2-3 (a)	0	0	1	0	1	1	0	0	1
Total units of change (b)	4								
Net person-changes (c)	2								
Number of participants	21								
Percent change (d)	9.5%								

(a) These numbers are the absolute differences in the histograms for the two successive rounds.

(b) These numbers are the sums of the absolute differences in the histograms.

(c) Net changes are total units of change divided by 2.

(d) Percent change is net change divided by the number of participants.

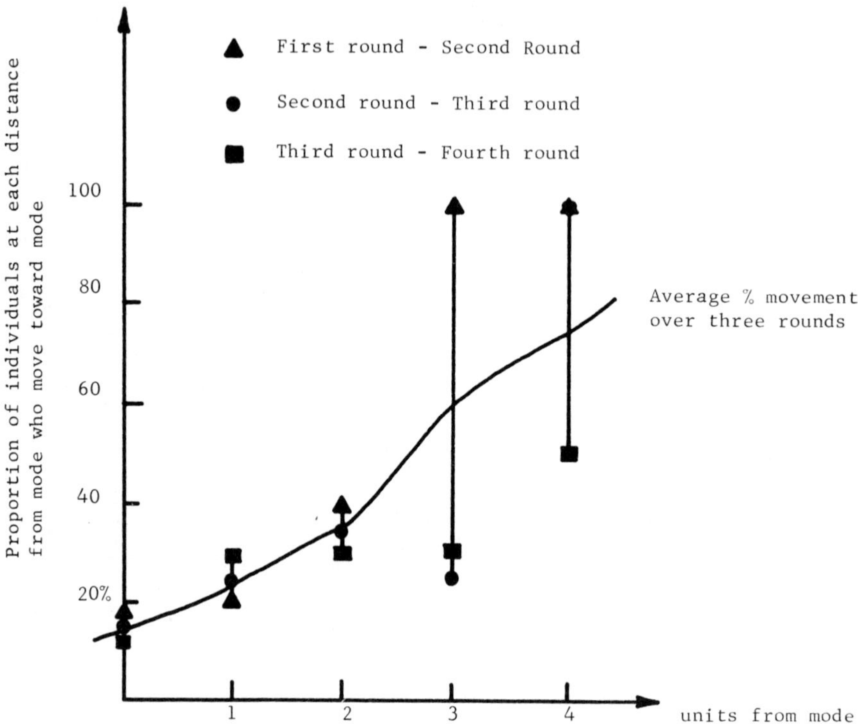

Fig. 6. Proportion of respondents shifting their positions, as a function of distance from the mode.

The results for all nine goals included in this experimental Delphi using this analysis are shown in Table 3. From these data, there can be no doubt as to the general tendency toward stabilization. Only one goal, 7, had not reached a stable position by the end of the third round, although 3, 8, and 9 were all only just stable.

In general, this method seems to have a number of advantages. Firstly, it allows the use of more of the information contained in the distributions. There are applications in which, at the end of the Delphi process, the entire distribution may be used, as for example in linear-weighting evaluation models where goal-weight distributions are treated stochastically, such as that by Goodman [10]. In addition, this stability measure is relatively simple to calculate, and has much greater power and validity than parametric tests of variance.

Perhaps most important, one of the original objectives of Delphi was the identification of areas of difference as well as areas of agreement within the

Table 3: Results of Stability Analysis

Goal	Amount change Rounds 1-2	Amount change Rounds 2-3
1	13% stable	10%
2	17%	10% stable
3	38%	14% just stable
4	10% stable	10%
5	26%	10% stable
6	2% stable	2%
7	24%	33% increasingly unstable
8	22%	14% just stable
9	24%	14% just stable

participating group. Use of this stability measure to develop a stopping criterion preserves any well-defined disagreements which may exist. To the organizer of a Goals Delphi, this information can be especially useful.

Delphi Service Program

In order to make several iterations possible in the space of a very short time period, a computer time-sharing terminal was used to process the results of this Delphi experiment. Unlike the systems described by Price (see Chapter VII, B), the program used in this Delphi was an accounting device only; verbal feedback was compiled and read to participants by the organizers.

In this application, histograms produced by the computer terminal were copied by hand onto an overhead projector transparency to provide immediate feedback to participants, who themselves determined their positions in the distributions relative to the group. It is anticipated that, for future experiments, computer-generated histograms will be produced in multiple copies, one of which will be provided to each participant.

This type of computer support, oriented toward the use of a single terminal for all participants, may be especially desirable for Goals Delphi applications, where, because of the lay nature of the respondents, it seems especially desirable to keep all of those involved in a single room, and to maintain a relatively high rate of progress throughout the survey.

Closure

The potential applicability of the Delphi method to goal formulation and priority determination for public systems is very great. Yet, because the detailed characteristics of the design of the process can have important effects on the nature of the outcomes, it will be important to tailor the Goals Delphi to the problems at hand. The structuring of internal characteristics which are appropriate to a Goals Delphi should be based on a rather complete understanding of the linkages between form and function in the Delphi environment. While considerable experience must be gained before Delphi can be offered as a routine goal-formulation process, this discussion has suggested some structural and process features relevant to this important application of the Delphi method.

Appendix I: Hypothetical Decision Scenario

The following transportation-facility-location problem is offered as an appropriate context for developing local-scale transportation planning objectives. Within this context, there is a need to establish an objective set, and to evaluate *quantitatively* several alternative plans in the context of the objectives.

A two-mile transportation link is proposed in an urban area. It is to run from the Central Business District (CBD) to new, developing suburbs to the north. This area is presently served by a four-lane boulevard with an average daily traffic (ADT) of forty thousand vehicles, and by a four-lane street with an ADT of twenty thousand vehicles. This street, however, circles an historical area by means of four 90° turns, and traffic must travel this section at twenty-five mph. The southeast corner of the historical area comes within five hundred yards of the edge of a lake, and the main street presently is only one block from the lake at this point. A tollway also passes the suburb and proceeds in a southeasterly direction. The tollway passes within one mile of the CBD, with its alignment located in a ravine. The elevation of the ravine is such that to build a connector to the tollway from the CBD would require a great deal of earthwork, and even with this the grade would be about 3%.

The alignment of the boulevard is such that it begins in the CBD, proceeding northwesterly through a low-income area to a large park, where it turns and continues in a northeasterly direction through a middle-income residential area to the suburbs.

The four-lane street heads due north from the CBD, passes through an industrial park, and then makes four sharp turns around the historical area and proceeds directly into the suburbs.

Citizen opposition can be expected in four areas. Public opinion has long been against any changes in the historical area. A citizen group can be expected to form opposing an alignment through the park. One can also be expected to form opposing removal of houses in the middle-income areas north and east of the park. Problems can also be expected if the alignment goes through the low-income area, requiring relocation of some households.

Appendix II: Map of Hypothetical Transportation Scenario

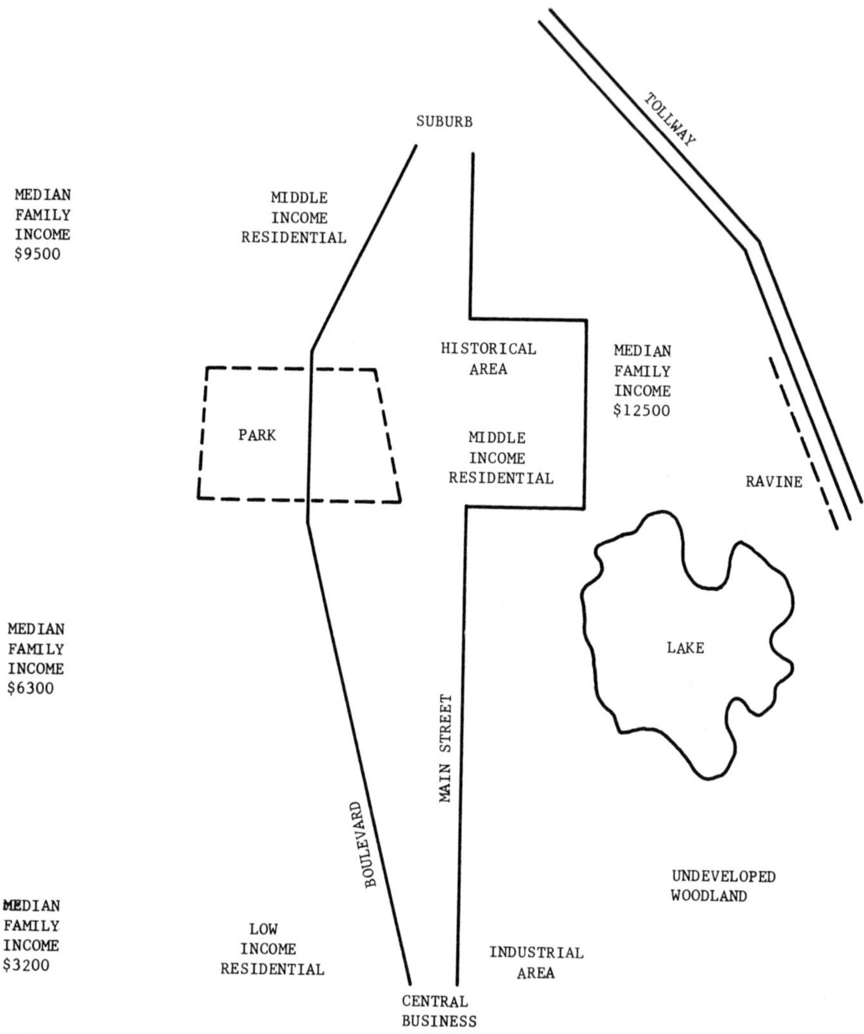

Appendix III: Pre-Delphi Self-Rating Form

Code Number________________

PRE-DELPHI SURVEY

1.	As a transportation planner, in a class of transportation planners, my skills in planning would put me about here, relative to the others.	Very Highly Skilled — No Skill at all 1 2 3 4 5 6 7
2.	I think my ideas are, in essence, in agreement with the rest of the class.	Yes, absolutely — No, not at all 1 2 3 4 5 6 7
3.	I know most of the people in the class very well.	Yes, pretty much — No, none at all 1 2 3 4 5 6 7
4.	I have some definite ideas about what the goals in transportation planning are and should be.	Yes, lots — No, none 1 2 3 4 5 6 7
5.	I have been in transportation for longer than most of the other people here.	Yes — No 1 2 3 4 5 6 7
6.	I have a lot of experience in planning outside of school.	Yes — No 1 2 3 4 5 6 7
7.	I am anticipating that the Delphi is going to be a good thing for goal setting.	Yes, I think it will be — No, I think it may be a waste of time 1 2 3 4 5 6 7

Appendix IV: Post-Delphi Evaluation Form

Code Number________________

POST-DELPHI SURVEY

Statement	1	2	3	4	5	6	7	Opposite
1. I feel satisfied with with the results in general.	1	2	3	4	5	6	7	I'm not really happy with the results at all.
2. I learned ideas from the feedback.	1	2	3	4	5	6	7	I didn't learn a thing from the feedback.
3. In general, I agreed with the ideas in the feedback.	1	2	3	4	5	6	7	I disagreed with everything in the feedback.
4. I could express my ideas OK this way.	1	2	3	4	5	6	7	I couldn't really write what I wanted to say.
5. I feel as if I really wanted to talk to people.	1	2	3	4	5	6	7	I didn't feel the need to talk at all.
6. I have a feeling people didn't understand or think about my reasons.	1	2	3	4	5	6	7	I think people understood my reasons pretty well.
7. I think the Delphi could be operational in goal setting more generally.	1	2	3	4	5	6	7	I don't think it could be operational, really.
8. I think it went too fast.	1	2	3	4	5	6	7	I think it went too slowly.

References

1. T. H. Ostrom and H. S. Upshaw, "Psychological Perspective and Attitude Change," in A. G. Greenwald *et al.*, *Psychological Foundations of Attitudes*, Academic Press, New York, 1968.
2. M. Wachs and J. L. Schofer, "Abstract Values and Concrete Highways," *Traffic Quarterly*, **23**, No. 1 (1969).
3. G. S. Rutherford, J. L. Schofer, M. Wachs, and M. Skutsch, "Goal Formulation for Socio-Technical Systems," *Journal of the Urban Planning and Development Division, Proceedings of the American Society of Civil Engineers*, forthcoming.
4. N. C. Dalkey, B. Brown, and S. Cochran, "The Delphi Method IV: Effect of Percentile Feedback and Feed-in of Relevant Facts," Rand Corporation, RM-6118-PR, 1970.
5. N. S. Dalkey and D. L. Rourke, "Experimental Assessment of Delphi Procedures with Group Value Judgment," Rand Corporation, R612-ARPA, 1971.
6. W. S. Torgerson, *Theory and Methods of Scaling*, Wiley, New York, 1958.
7. N. C. Dalkey, "The Delphi Method: An Experimental Study of Group Opinion," Rand Corporation, RM-5888-PR, 1969.
8. M. Skutsch and J. L. Schofer, "Goals Delphis for Urban Planning: Concepts in their Designs," *Socio-Economic Planning Sciences*, **7**, pp. 305-313, 1973.
9. M. Skutsch, "Goals and Goal Setting: A Delphi Approach," unpublished master's thesis, Department of Industrial Engineering and Management Science, Northwestern University, 1972.
10. K. M. Goodman, "A Computerized Investigation and Sensitivity Analysis of a Linear Weighting Scheme for the Evaluation of Alternatvies for Complex Urban Systems," unpublished master's thesis, Department of Civil Engineering, Northwestern University, 1971.

IV. D. Propensity to Change Responses in a Delphi Round as a Function of Dogmatism

NORMAN W. MULGRAVE and ALEX J. DUCANIS

Since one of the assumptions of Delphi is that it "reduces the influence of certain psychological factors such as ... the unwillingness to abandon publicly expressed opinions, and the bandwagon effect of majority opinion" (Helmer, 1966), it would seem of interest to examine the effect of personality upon an individual's performance during several Delphi rounds. Specifically, the question can be raised concerning the willingness of a more dogmatic individual to change his answer in a Delphi round (whether he is an expert or a nonexpert). Since dogmatic thinking is characterized by resistance to change (Rokeach, 1960), it might be posited that the dogmatic individual would be less likely to change his position when confronted by the opinions of others. It might be further presumed that the type of question asked, i.e., those upon which the individual could be considered either more or less expert, might also affect performance of highly dogmatic as opposed to less dogmatic individuals. It was therefore predicted that the number of changes made by a high dogmatism group (D_H) would be less than a low dogmatism group (D_L), and that the (D_H) group would change less on questions on which they might be considered expert than on questions on which they would be considered less expert.

Method

The subjects for the study were ninety-eight graduate students enrolled in a class in Educational Psychology, most of whom were school teachers.

Berger's (1967) revision of Rokeach's Dogmatism Scale (the FCD Scale) was administered on the first day of class. Subsequently the class was used as a Delphi panel and asked to make certain estimates. Four types of questions were utilized. Ten questions defined the subjects as nonexpert, such as the number of farms in the United States. Ten other questions concerning class size, teachers' salaries, length of the school year, and similar items defined the subjects as experts.

Eighteen other questions were value-oriented items. Subjects were to respond in terms of what certain values in the United States *will* be in 1980, and also

The Delphi Method: Techniques and Applications, Harold A. Linstone and Murray Turoff (eds.)
ISBN 0-201-04294-0; 0-201-04293-2

what they *should* be. The latter set made each person a fully qualified "expert."

With these question categories as a base, it was possible to use the questions to define the respondents as expert or nonexpert.

The Delphi procedure was continued through three rounds. During rounds two and three each subject was given the group median, interquartile range and his own response to the previous round for each item. He was asked to "review [his] projection on the basis of the information provided" and to change his answer if he wished to do so.

Results

The D_H and D_L groups were identified as those scoring in the upper or lower 27 percent of the class on the FCD Scale (Berger, 1967). Second- and third-round changes were tabulated for those who were both inside and outside the interquartile range for each of the four sets of questions. The results of that tabulation are shown in Tables 1 and 2.

Table 1
Proportion of Individuals Changing Answers
Round Two

Type of Item	Percent Changing High Dogmatism		Percent Changing Low Dogmatism	
	Inside Range	Outside Range	Inside Range	Outside Range
Almanac	29.5	76.2	13.6	63.1
"School" Questions	20.3	64.7	6.1	54.0
Predictions of Values	13.4	74.8	4.5	50.8
What Values Should Be	12.8	71.9	7.2	59.3

Table 2
Proportion of Individuals Changing Answers
Round Three

Type of Item	Percent Changing High Dogmatism		Percent Changing Low Dogmatism	
	Inside Range	Outside Range	Inside Range	Outside Range
Almanac	6.9	29.2	6.8	14.5
"School" Questions	4.1	18.6	3.4	14.8
Predictions of Values	8.9	19.7	1.9	6.7
What Values Should Be	9.2	15.8	1.9	6.0

A significant difference was found between the groups in the number of times they changed their answers. For round two the value for chi-square was computed as 18.48 with 7 degrees of freedom. This value is significant at the .01 level. The corresponding value for the third round was 14.78, which is significant at the .05 level.

Discussion

It would seem that personality characteristics of the individual involved in the Delphi panel have some effect upon his propensity to change. Of interest as well was the finding that the High Dogmatism group exhibits significantly more changes. Thus the prediction that they would be less likely to change is not upheld. A possible explanation for this may be that if the dogmatic individual looks to authority for his support, then in the absence of any clearly defined authority the dogmatic individual would tend to seek the support of whatever authority seems present. In this case authority would be the median of the group response.

The second prediction that the High Dogmatism group would change less on questions where they could be considered expert, i.e., "school questions" and "what values should be" than on questions where they could not be considered expert, i.e., "almanac questions" and "what values will be," was upheld on the second round (chi-square 6.622 with one degree of freedom) but not on the final round. There were no significant differences on either round for the Low Dogmatism group.

These results seem to indicate that the High Dogmatism group is less likely to change an answer to a question on which they consider themselves expert than one on which they consider themselves less expert, but that in the presence of some "perceived" authority such as the group median, High Dogmatism groups will exhibit more change than Low Dogmatism groups.

References

V. F. Berger, "Effects of Repression and Acquiescence Response Set on Scales of Authoritarianism and Dogmatism," unpublished master's thesis, University of Oklahoma, 1967.

N. C. Dalkey, "The Delphi Method: An Experimental Study of Group Opinion," Rand Corporation, 1969.

O. Helmer, "The Use of the Delphi Technique in Problems of Educational Innovations," The Rand Corporation, 1966, P-3499.

O. Helmer and N. Rescher, "On the Epistemology of the Inexact Sciences," Rand Corporation, 1960, R-353.

J. Martino, "The Consistency of Delphi Forecasts," *The Futurist* 4 (1970), p. 63.

M. Rokeach, *The Open and Closed Mind*, Basic Books, New York, 1960.

IV. E. The Performance of Forecasting Groups in Computer Dialogue and Face-to-face Discussion

KLAUS BROCKHOFF*

The Problem

Advances in mathematical and statistical techniques, the availability of efficient computers as well as ideas and attempts to utilize certain organizational structures in the compilation of expertise are basic elements for an intensified discussion of the problems of forecasting specific future developments and events. As J. Wild has shown, those conditional forecasts which are derived from models by certain statistical techniques are based as much on empirical knowledge as on ad-hoc extrapolations, projections, or expert opinions.[1] However, the isolation of independent variables from the surrounding conditions and the intra-personal process of information processing often do not become clearly visible in the latter. Thus there is a danger that uncontrolled, or uncontrollably, misinterpretations and false judgments may occur.

We do not want to infringe upon the controversy on the superiority of "forecasts" as compared with "projections," which is being carried out in the theory of science[2] as well as on an empirical-pragmatic[3] level. It has by no means been settled for the forecasts. This seems particularly true when the comparison is drawn on the basis of a benefit-cost relationship.[4] If only for this reason we are interested in the question whether or not the utilization of the empirical knowledge of groups of experts in the derivation of statements about future developments or events can be improved upon by organizational arrangements. Improvement is meant as an increase in the accuracy of these statements. This is one reason for the development of the Delphi method.[5]

---

*In collaboration with D. Kaerger and H. Rehder.

[1] J. Wild, "Probleme der theoretischen Deduktion von Prognosen," *Zeitschrift für die gesamte Staatswissenschaft (ZfgS)* **126** (1970), pp. 553–75.

[2] *Ibid.* and E. v. Knorring, "Probleme der theoretischen Deduktion von Prognosen," *ZfgS* **128** (1972), pp. 145–48; J. Wild, "Zur prinzipiellen Überlegenheit theoretisch deduzierter Prognosen," *ZfgS* **128** (1972), pp. 149–55.

[3] E.g., R. M. Copeland, R. J. Marioni, "Executives' Forecasts of Earnings per Share versus Forecasts of Naïve Models," *Journal of Business* **45** (1972), pp. 497–512, and the literature quoted there.

[4] Thus the suggestion in H. A. Simon, D. W. Smithburg, V. A. Thompson, *Public Administration*, New York, 1961, p. 493.

[5] O. Helmer, N. Rescher, "On the Epistemology of the Inexact Sciences," *Management Science* **6** (1959), pp. 25–52.

The Delphi Method: Techniques and Applications, Harold A. Linstone and Murray Turoff (eds.)
ISBN 0-201-04294-0; 0-201-04293-2

The conditions of group performance have been investigated in thousands of studies. Only few studies have been devoted to the question whether the peculiar organizational structure of the Delphi group[6] leads to higher group performance than the face-to-face discussion in a group in which the each-to-all pattern of the communication system can be activated.[7] Beyond this the unwieldy, nonhomogeneous, and inaccurate definition of types of tasks by which the performance of groups is judged[8] and thus the classification of concrete formulations of the question make it very difficult to derive statements as to the particular capacity of groups in forecasting.[9]

A large proportion of the statements as to the superiority of certain forms of group organization compared with others was obtained by observing group performance in solving certain kinds of problems and by assuming that the results would apply to tasks which appeared comparable. Thus references to the ability of groups to forecast particular future events were judged on the basis of their performance in responding to almanac-type questions. Martino has demonstrated that the answers to almanac-type questions observe the same type of distribution as the answers to forecasting questions.[10] However, it has not been investigated whether the parameters of the distribution vary significantly from one type of task to the other under conditions which are comparable otherwise. Thus it appears desirable to reconsider the original assumption.

In the following report we try to investigate some of these questions experimentally.

2. *Initial Hypotheses on the Performance of Forecasting Groups*

2.1 Group Performance and Group Size

2.1.1 Measurement of Variables It has been shown in various studies that the performance of a group may depend on its size.[11]

---

[6]Cf. below, section 2.3.1.

[7]On the restriction of the each-to-all pattern cf. M. E. Shaw, "Some Effects of Varying Amounts of Information Exclusively Possessed by a Group Member upon His Behavior in the Group," *Journal of General Psychology* **68** (1963), pp. 71–79.

[8]For the differentiation of types of tasks cf., e.g., M. E. Shaw, *Group Dynamics: The Psychology of Small Group Behavior*, New York, 1971, pp. 59, 403ff.; A. P. Hare, *Handbook of Small Group Research*, New York, 1962, pp. 246ff; C. G. Morris,"Task Effects on Group Interaction," *Journal of Personality and Social Psychology* **4**, (1966), pp. 545–54. For problems of definition also R. Ziegler, *Kommunikationsstruktur und Leistung sozialer Systeme*, Meisenheim a. Glan, 1968, pp. 96ff.

[9]The limited choice of type of tasks and the strict assumptions on experimental group problem solving are most criticized. H. Franke, *Gruppenproblemlösen, Problemlösen in oder durch Gruppen? Problem und Entscheidung*, Heft 7, München, 1972, pp. 1–36, here p. 26 *et seq.*

[10]J. P. Martino, "The Lognormality of Delphi Estimates," *Technological Forecasting*, **1**, (1970) pp. 355–358.

[11]Cf., e.g., J. D. Steiner, "Models for Inferring Relationships between Group Size and Potential Group Productivity," *Behavioral Science* **11** (1966) pp. 273–83; F. Frank and L. R. Anderson, "Effects of Task and Group Size upon Group Productivity and Member Satisfaction," *Sociometry* **34** (1971), pp. 135–49.

The *group size* is determined by the number of members of a group. This measure refers solely to formal criteria. Thus a person who does not contribute to the activity of the group, either because of his own reticence or because of a formal system of communication which does not accept his contributions, is still considered a member of the group.

Group performance can be "synthesized" by a statistical aggregation of individual performances.[12] However, such groups lack the essential characteristic of communication which exists in natural groups. In the following we will report entirely about experiments with natural groups.[13]

It may seem natural to study group performance of groups with a considerable number of members. However, in the experiments to follow we have deliberately concentrated on small groups of four to eleven people. One reason for this is that very many small and medium-sized organizations are applying Delphi. They can call in only small groups of experts. Even so they may wonder about their performance, and how to measure it. With regard to the possible objection that the results from the observation of small groups may be subject to considerable "noise," we may say that basic to most evaluations is the use of the median of individual responses. The median, however, is not sensitive to large dispersions, even if they are one-sided. On the other hand, it goes without question that it would be desirable to repeat our experiments in order to check on the reliability of the results.

Very different things can be understood by the "performance" of a group. The tripel, number of pieces of information exchanged, time needed for solving a problem and number of mistakes, can be considered a "classical" yardstick of performance. Ziegler ascribes the origin of this tripel to a paper written by Bavelas in 1950[14]. As Barnard's definition of performance—"the accomplishment of the recognized objectives of cooperative action"[15]—makes clear, however, this classical tripel is not compulsory. Indeed, it generally remains unclear whether performance refers to the goals (recognized objectives) of the members of the group, to those of the group, or to those presented to the group.

The task given to the groups is to find an answer A to a question which deviates as little as possible from the answer A' which can be verified now or in the future. Increasing performance then means that $|A - A'|$ approaches 0. In order to make comparisons between different questions or different groups a

[12]For a chronology of the publications on statistically "synthesized" performance, cf. J. Lorge, D. Fox, J. Davitz, and M. Brenner, "A Survey of Studies Contrasting the Quality of Group Performance and Individual Performance, 1920–1957," *Psychological Bulletin* 55 (1958), pp. 337–72, here pp. 367f.

[13]Here natural group does not mean only a natural group with an each-to-all pattern of the communication system. With this I depart from the narrower definition in my earlier publication. See K. Brockhoff, "Zur Erfassung der Ungewissheit bei der Planung von Forschungsprojekten (zugleich ein Ansatz zur Bildung optimaler Gutachtergruppen)," in H. Hax, *Entscheidung bei unsicheren Erwartungen*, Köln, 1970, pp. 159–88, here pp. 167f.

[14]R. Ziegler, *op. cit.* p. 18; see also p. 55.

[15]C. J. Barnard, *The Functions of the Executive*, Cambridge, Mass., 1962, p. 55.

standardization is necessary. Thus the relative deviation of an estimate from the correct answer is used:

$$\frac{|A-A'|}{A'}.$$

We call this expression the "error." If the error refers to a person, we speak of an individual error. If the error refers to group performance, we speak of a group error. If A is the median of the estimates of all members of a group, we speak of a median group error (MGE).

The "mean group error" as used by Dalkey[16] is not identical with the MGE as given here. The basic difference is that Dalkey uses the logarithm of the quotient A/A' in order to test hypotheses about the distribution of his "mean group error." The distribution is of secondary importance for our present considerations regarding performance. For this reason we do not use logarithms here.

The MGE is used here directly as a measure of performance. It is not put into relation to the expenditures made for its derivation.

Further measures of group performance which are mentioned are the ability of the group to survive in a changing environment, its satisfaction, and the habitual change of its members.[17] We do not intend to study group performance in such a broad context (although we have unsystematically collected remarks on member satisfaction).

2.1.2 The Relationship of Performance to Group Size The study of hypotheses about a relationship between group size and group performance occupies a prominent position in small-group research. A brief survey of the diversity of empirical results was compiled by Türk.[18] A uniform result cannot truly be expected, because the individual studies were carried out under different conditions (types of tasks, performance measures, etc.). With respect to forecasting it has been hypothesized that the mean group error decreases with increasing group size.[19] It should be taken into account, however, that this statement has been formulated only for synthetic groups, i.e., a statistical aggregation of individual judgments, and with reference to the performance of the group in answering fact-finding questions.[20]

---

[16]Cf. N. C. Dalkey, "The Delphi Method: An Experimental Study of Group Opinion," Rand Corp., RM 5888 PR, 1969. Also H. Albach, "Informationsgewinnung durch strukturierte Gruppenbefragung—Die Delphi-Methode", *Zeitschrift für Betriebswirtschaft* (*ZfB*), Suppl. **40**, Yr. 1970, pp. 11–26, here p. 20.

[17]Summarized, e.g., by M. Deutsch, "Group Behavior," in D. L. Sills (ed., *International Encyclopedia of the Social Sciences* **6**, New York, 1968, pp. 265–75, here p. 274.

[18]K. Türk, "Gruppenentscheidungen. Sozialpsychologische Aspekte der Organization kollektiver Entscheidungsprozesse," *ZfB* **43**. (1973), pp. 295–322, here p. 302.

[19]N. C. Dalkey, *op. cit.* pp. 9f.

[20]"These were questions where the experimenters knew the answer but the subjects did not": N. C. Dalkey, *op. cit.*, p. 10, fn.

The assumption of a decreasing error with increasing group size is based on the probability model of search.[21] From this model one deduces the possibility of compensating individual errors by calculating the mean for the group.

In natural groups, however, the rigid conditions on which alone the statements of the probability model are valid cannot always be fulfilled. Since, however, a consistent system of other factors that influence group performance as well (as the direction of their influence cannot be given), we may formulate:

Hypothesis G: With increasing group size, the group performance increases ceteris paribus.

It becomes clear in section 2.5 to what extent the restriction *ceteris paribus* can be repealed in our experiments.

2.2 Group Performance and Expertise

2.2.1 The Measuring of Expertise H. A. Simon represents "expertise" as a possible basis of authority or as a form of authority.[22] By expertise we mean expert knowledge upon which professional authority can be founded. This expert knowledge can be "proven"[23] by demonstration or by recourse to confirmation through third parties. A "proof" by recourse to third parties can hardly confirm more than a refutable conjecture as to the expertise of a person. If influencing variables of group performance are being sought, the experiments which make these variables visible can hardly contain an analyzable test of expertise at the same time. It is necessary to measure expertise as an independent variable by some other means.

One could proceed by testing which persons demonstrate expert knowledge in solving fact-finding questions. When such persons have been found, they can be engaged in forecasting. This takes for granted that the answering of both types of questions can be considered to be identical types of tasks.[24] Until now, however, no empirically tested statement to this effect exists.

---

[21]Cf., e. g., P. R. Hofstätter, *Gruppendynamik. Die Kritik der Massenpsychologie*, 11th ed., Reinbek, 1970, pp. 35ff., 160ff.

[22]H. A. Simon, *Administrative Behavior*, 3rd ed., New York and London, 1965, p. 76.

[23]F. Landwehrmann, "Autorität," in E. Grochla (ed.), *Handwörterbuch der Organisation*, Stuttgart, 1969, col. 269–73, here col. 270, refers to H. Hartmann, *Funktionale Autorität*, Stuttgart, 1964.

[24]For a procedure oriented thus, cf. M. A. Jolson, G. L. Rossow, "The Delphi Process in Marketing Decision Making," *Journal of Marketing Research*, **8** (1971), pp. 443–48. Another procedure, based on the solution of test questions and a test of the understanding of professional terminology, is described by A. J. Lipinski, H. M. Lipinski, R. H. Randolph, "Computer-Assisted Expert Interrogation: A Report on Current Methods Development," *Technological Forecasting and Social Change* **5** (1973), pp. 3–18, here pp. 9f. (The same in S. Winkler [ed.], *Computer Communications, Impact and Implications*, New York, 1973, pp. 147–54. The authors also test the "quality of respondents' comments" [presumably on factual questions], the degree of attention and the degree of optimism [with the aid of a price list for old phonograph records] and inquire from this a rank order of expertise.)

Further, one could consider whether expertise ratings by third parties can be used. However, it may be of interest to maintain the anonymity of all persons who may possibly participate in a forecasting group.[25] Another problem that arises is what criteria should be used in choosing those persons who are to judge the expertise of others.

Thus there remains the possibility of determining expertise by self-rating. For this purpose an ordinal scale is generally used, from which one value can be chosen to indicate expertise. We worked with a scale of real numbers graded from 1 to 5, in which low numbers must be used to express a low degree of expertise while high numbers may be used to express a high degree of expertise. Such a determination of expertise is employed already in some forecasting groups by their management.[26] These results of individual forecasts are weighted according to the self-ratings when a group judgment is derived.

Measurements of expertise which are obtained for individuals should also permit statements as to the expertise of the total group. Since self-ratings are measured on an ordinal scale it is not permissible to form an arithmetic mean of all the self-ratings of the members of the group. We therefore characterize the expertise of a group by the median of the individual self-ratings.

Whether such self-ratings have a high positive correlation with ratings by third parties has not yet been studied in realistic situations. For this reason it remains an open question whether corresponding confirmatory results of psychological tests[27] can be applied to real situations, which generally are not free of conflicting interests.

Even if no significant positive correlation exists between ratings by third parties and self-ratings concerning expert knowledge, it is not determined which of the ratings is more correct. Thus in the approach used here, which involves self-ratings, the question remains whether the participants in the experiment rate themselves correctly when compared with an (inapplicable) objective standard.

2.2.2. The Relationship between Expertise and Group Performance We assume that groups with high self-ratings of expertise perform better than groups whose members rate themselves as less qualified. With this assumption we follow Dalkey, Brown, and Cochran[28]. Their results must, however, be examined with care insofar as they were obtained from answering fact-finding questions. Furthermore, the subjects were able to compare all questions to one another

[25] Cf. Section 2.3.1.

[26] D. L. Pyke, "A Practical Aporoach to Delphi: TRW's Probe II," *Futures* 2 (1970), pp. 143–52; H. P. North, D. L. Pyke, "Probes of the Technological Future," *Harvard Business Review 3 (1969), pp. 68–76; A. J. Lipinski, L. M. Lipinski, R. H. Randolph, op. cit.*, pp. 11ff.

[27] M. A. Wallach, N. Kogan, J. Bem, "Group Influence on Individual Risk Taking," *Journal of Abnormal and Social Psychology* **65** (1962), pp. 75–86, here p. 83.

[28] N. C. Dalkey, B. Brown, S. Cochran, "The Use of Self-Ratings to Improve Group Estimates," *Technological Forecasting*, **1** (1970) pp. 283–292.

before rating their expertise with respect to each question. In many applications it is not possible to present all questions at once. We shall therefore follow a different procedure by presenting tasks in a sequential manner. Even so, we assume that the basic relationship is still valid. Thus, we arrive at

Hypothesis E: With increasing expertise, group performance increases ceteris paribus.

2.3 Group Performance and Communication System

2.3.1 The Characteristics of Face-to-Face Discussion Groups vs. Delphi Groups The question whether decentralized or centralized organizations exhibit higher performance is another of the classical questions in organization research. The attempt to set up a universally applicable rule of organization to answer this question had to be dropped because little by little conditions become known on which first the one organizational structure and then the other seemed more advantageous for accomplishing specific tasks. Fundamentally, it is assumed in these studies that all organizational structures considered are capable of accomplishing certain tasks. A formal organization is characterized essentially by its communication system and the distribution of competence among its members.[29] It is further assumed that fact-finding questions as well as forecasting questions can be answered more accurately by groups than by individuals, if the (expected) error[30] is taken as the measure of accuracy.

Fact-finding questions and forecasting questions can be discussed in natural groups with an each-to-all pattern of communication. If it is desired to have a group judgment, this task can be left up to the group, or a rule for aggregating the group judgment from the individual judgments of the participants can be given. Depending on the situation, the use of such a rule can be restricted to the case where the group does not agree on a group judgment within a given period of time.[31]

Particularly Carzo's results indicate that in natural groups in which communication is not limited, solutions to complex tasks are reached rapidly, with few errors, and to the satisfaction of the group members.[32] One objection to this is that this form of organization may also produce dysfunctional effects.

A first dysfunctionality may arise as certain group members consciously or unconsciously influence the group result to a greater degree than their expertise

[29]This is made particularly clear by H. Albach, "Organisation, betriebliche," in *Handwörterbuch d. Sozialwiss.* **8**, Stuttgart, Tübingen, Göttingen, 1961, pp. 111–17.

[30]Cf. H. Schöllhammer, "Die Delphi-Methode als betriebliches Prognose-und Planungsverfahren," *Zeitschrift für betriebswirtschaftliche Forschung (ZfbF)*, N. S., 22nd Yr. (1970), pp. 128–37, here particularly fn. 8.

[31]Cf. the experiments by B. Contini, "The Value of Time in Bargaining Negotiations—Some Empirical Evidence," *American Economic Review* **58** (1968), pp. 374–93.

[32]Cf. R. Carzo, Jr., "Some Effects of Organization Structure on Group Effectiveness," *Administrative Science Quarterly* **7** (1963), pp. 393–424.

warrants.[33] A further dysfunctionality arises if the exchange of information is interfered with by "noise."[34] A more far-reaching possibility for the occurrence of dysfunctional effects is that the transmission of information necessary for task accomplishment from some group members to others is blocked by somebody interested[35] or that the transmission of information from some group members in the time period given for accomplishing the task is altogether impossible. In the latter case the participation of the individual group members in the exchange of information may be independent of the degree of expertise. Then participants with a high degree of expertise cannot necessarily influence group performance.

One can summarize that possible dysfunctionalities in a natural group with an each-to-all pattern of communication result from utilizing the communication system and the system of competence in a manner which does not correlate positively with the degree of expertise. Assuming that these effects often interfere with group performance, rules should be set up which reduce the effects of dysfunctionality or prevent their appearance. A set of such rules was suggested and introduced by Helmer and Dalkey[36] and given the name "Delphi." In spite of diverse variations in procedure, the applications known to date have as their primary objective: "...the establishment of a meaningful group communication structure."[37]

2.3.2. The Relationships According to Dalkey's studies, Delphi groups demonstrate a certain, though not significant superiority when compared with certain other groups in solving fact-finding questions.[38] We apply this perception to our

Hypothesis D: The performance of Delphi groups is ceteris paribus higher than the performance of natural groups with an each-to-all pattern of communication.

The diverse arrangements of Delphi experiments make it necessary to investigate some of their special features as well. Of particular importance is the question whether the performance of Delphi groups is the same in each round, or whether it increases with increasing number of rounds, at least up to

[33]This very abridged presentation must be viewed on the basis of the entire discussion concerning the question: which conditions promote behavioral conformity in individuals in groups; cf. A. P. Hare, *Handbook ..., op. cit.*, chapters, 2, 13 (there in reference to status rivalry); L. Festinger, E. Aronson, "The Arousal and Reduction of Dissonance in Social Contexts," in D. Cartwright, A. Zander (eds.), *Group Dynamics: Research and Theory*, Evanston, Ill., 1960, pp. 214–31.

[34]A. P. Hare, *op. cit.*, chapter 10, pp. 272ff.

[35]On this broad field, see H. H. Kelley, J. W. Thibaut, "Group Problem Solving," in G. Lindzey, E. Aronson (eds.), *Handbook of Social Psychology* **4**, Reading, Mass., 1968, pp. 1–101, here pp. 6ff., 26ff.

[36]N. C. Dalkey, O. Helmer, "An Experimental Application of the Delphi Method to the Use of Experts," *Management Science* **9** (1963), pp. 458–67.

[37]M. Turoff, "Delphi and its Potential Impact on Information Systems," *AFIPS Conference Proceedings*, Fall Joint Computer Conference (Fall 1971), **39**, pp. 317–26, here p. 317.

[38]N. C. Dalkey, *op. cit. passim*, particularly p. 22.

the fourth round.[39] This leads to

Hypothesis R′: The performance of Delphi groups increases ceteris paribus with increasing number of rounds.

We test this hypothesis here for up to five rounds. It cannot be expected that hypothesis R' is valid for an unlimited number of rounds, since growing dissatisfaction of the participants and increasing time requirements make it seem senseless to continue the consultations indefinitely. Therefore, we modify hypothesis R' to

Hypothesis R: The performance of the Delphi groups increases ceteris paribus with increasing number of rounds only at first. Finally, the increase in performance can be reduced and inverted. The question in which round the "performance inversion" begins must be answered empirically.

Finally, Helmer and Dalkey[40] showed an interest in the observation that the variance of the responses around the median decreases with increasing number of rounds. The reduction of variance is not in itself a criterion for increased performance. One must view this observation on the basis of hypothesis R: together with it, variance reduction gains importance in the sense that it means increasing certainty and accuracy of the answers.[41] We therefore also test this question by investigating

Hypothesis V: The variance of answers around the median decreases ceteris paribus with increasing number of rounds.

Two statistical measures of variance are at our disposal for testing this hypothesis: average quartile difference and average variance from the median. The latter measure offers certain advantages for a comparison between groups of different sizes. For this reason it is given preference here.

2.4 Group Performance and Type of Task: Fact-finding Questions and Forecasting

Only a few of the generally available results of forecasts by Delphi groups can be tested against reality, because they mainly refer to events which are

[39]Cf. also J. B. Martino, "An Experiment with the Delphi Procedure for Long-Range Forecasting," *IEEE Trans. on Engineering Management* 15 (1965), pp. 138–44.

[40]O. Helmer, *Social Technology*, New York and London, 1966, pp. 101ff.; N. C. Dalkey, *op. cit.*, p. 20.

[41]They are based according to the Delphi method, on renewed intrapersonal conflict solution and problem solving, after being provided with additional data. On the interpersonal process which is to be eliminated here, cf. J. Hall and M. S. Williams, "A Comparison of Decision-making Performances in Established and Ad Hoc Groups," *Journal of Personality and Social Psychology* 3 (1966), pp. 214–22. On the practical organization of the elimination of dysfunctionality and pressure to conform, cf. 3.2, below.

expected to take place in the distant future. For this reason, the comparison of performance of face-to-face discussion groups and Delphi groups is made by observing tasks which appear similar.[42] The problem of solving fact-finding or almanac-type questions is assumed to be similar to forecasting. The answers of such questions are, as a matter of principle, unknown to the participants but known to the experimenters. These two applications of Delphi, its use in forecasting and its simulation with only subjectively unknown bits of knowledge, exist as yet side-by-side without comparison. Since the complexity of a task is an important determinant of group performance, but the criteria for determining tasks of varying degrees of complexity are not clear enough, we want to test directly

Hypothesis F: The performance of a group in answering fact-finding questions is ceteris paribus equal to that in forecasting.

The hypothesis is deduced from the assumption that both types of tasks exhibit the same degree of complexity. The tests should be carried out separately for face-to-face groups and Delphi groups of the same size. If the hypothesis is refuted, many of the statements about Delphi-groups which we rederived using fact-finding questions cannot be maintained. In order to test hypothesis F, we chose only facts referring to events that had occurred, on the average, six months before the experiments. The forecasts, on the other hand, refer mainly to a period of time which did not exceed six months after the tests were carried out.

2.5 The Relationship between the Hypotheses

The repeated use of *ceteris paribus* in the hypotheses leads us to assume that they are in fact related to each other. It seems senseless to describe the great variety of possible combinations. The rejection of certain hypotheses can reduce the test program considerably as it leaves only a small number of relationships which need to be tested. Hypothesis E, e.g., can be tested for a given group size, a given type of question, a given type of group, and, in Delphi groups, with regard to the results of any round. No matter what the result is, it is irrelevant for the further experiments, if expertise should be distributed equally in the different groups. Since the distribution of the actual expertise becomes known from the results of the experiments, it is advisable to bring about the necessary clarifications at first.

The situation is similar when discussion groups and Delphi groups are compared with each other. If hypothesis R is not tested for the latter, it cannot

[42] N. C. Dalkey, *op. cit.*, pp. 9f. Dalkey cites (p. 23) a paper by Campbell, in which similar methods evidently were used to the ones planned here. The original publication was not available.

be determined from which round the results should be taken if compared with the performance of the face-to-face discussion groups.

The presentation of our results in Section 4 is organized according to such reasoning.

3. *The Experiments*

3.1 Participants, Group Formation, Place, and Time

All experiments were carried out as part of a lab. course listed in the University of Kiel catalogue. It was planned to have "students" and practitioners work in separate groups and to compare the results. However, since we could not give credits for the course, too few students registered to be able to form even *one* small group.

Practitioners were designated and chosen from the permanent staffs of the local banks with the assistance of the bank managers. Bank employees were chosen because hardly any other line of business is represented in the area by enough individual organizations with personnel trained in economics and with relatively uniform fields of business. At the same time, the size of the participating organizations is generally so large that persons who perform specialized functions (long-term credits, short-term credits, investment brokerage, etc.) and who differ with respect to the lengths of their employment could be chosen. This seemed desirable in order that definite differences could show up in the self-ratings of expertise in reference to the individual tasks.

The thirty-two participants were randomly assigned to four groups, having five, seven, nine, and eleven participants, respectively. At the face-to-face discussions, however, registered participants were absent for various personal reasons, so that the groups had four, seven, eight, and ten participants only.

The experiments began with an introductory lecture about forecasting methods and an exercise in the use of the displays which were used in the experiments with the Delphi groups. After that, each subject participated in a session in which the members were organized as a Delphi group. At a later session a face-to-face discussion took place. After the experiments were concluded, an opportunity for criticism was given. Eight months later, the results were communicated. We have tried to motivate our participants to cooperate well in the preparatory lecture and demonstration. Besides, we offered book prizes for outstanding performances with regard to different types of questions and the two basic group structures.

The experiments were conducted in May and July 1973. With three exceptions they were scheduled for Thursdays to conform with late closing time of banks. The Delphi groups worked in the computer center of the University of Kiel; the face-to-face discussions were carried out in a library room.

3.2 The Organization of the Delphi Groups and the Face-to-Face Groups

The Delphi groups were set up so that the participants received all information from the experimenters on a computer-generated display.[43]

The participants in a group were not supposed to establish immediate contact with each other. They responded to all questions by writing an alpha-numeric text in their normal language. The responses can be divided into three classes: (1) responses that were known only to the experimenters; (2) responses which, after the responses of all participants had been received by the experimenter, became objects of computing procedures, the results of which were made known to all participants; (3) responses which were recorded and made known to all participants without any changes. The first class includes the name of the participant and the degree of expertise that he expresses with regard to each question. This is handled differently in the face-to-face discussions groups. A response of the second category is an individual estimate. After the computation of the median of the responses of the group members, this figure is made known to all participants. The third category includes all arguments for divergent opinions of those whose responses lay outside of the lower or upper quartiles.

Computer communication has been praised as a means to enable experts to communicate with each other even though they are separated from each other by large distances. In the real world this could mean savings in travel expenses and in the efforts expended in coordinating dates for groups of experts. Beyond this it is of importance for the experiments that the computation of quartiles, and the preparation, distribution, collection, reproduction, and renewed distribution of questionnaires do not have to be carried out by hand during the sessions. This gives one the chance to shorten the experiments considerably.[44]

The entire exchange of information between the participants as well as between the experimenter and the participants during the sessions, with the exception of certain recurrent standard formulations, was stored on tape as a record of the experiments.

Figure 1 shows part of a record. A separate data file, an abstract from the records, is kept on tape. It serves as the data base for the diverse computations.

The abstract from the record shows the beginning of a session of the Delphi group with seven participants. Vertical lines on the left edge of the text signal those portions of the texts which appear in the same form on the display of each participant. In section 1 the names of the participants are given to the experimenter. Section 2 contains the first fact-finding question for the group. In

---

[43]The programming was done by D. Kaerger, who will report separately on problems that arose herewith. The program had to be in FORTRAN IV. It was run on a PDP 10.

[44]For a brief discussion of the advantages and disadvantages of "computer communication" compared with direct communication, cf. A. I. Lipinski, H. M. Lipinski, R. H. Randolph, *op. cit.*, particularly pp. 11–12.

```
GEBEN SIE BITTE IHREN NAMEN AN

BRIDSTRUP
CARL MOELLER
BOLTE                                                                 1
JOERG BISKUP
THOMAS M. RIECKEN
GLOCKNER
KAEMPFER

FUER WELCHEN BETRAG KAUFTEN DIE KREDITINSTITUTE
DER BRD 1972 FESTVERZINSLICHE WERTPAPIERE                             2
IN MRD.DM
FAKTEN... KAUF VON FESTVERZINSLICHEN WERTPAPIEREN
DURCH KREDITINSTITUTE IN DEN JAHREN 1967 BIS 1969
JEWEILS ZWISCHEN 1Ø UND 15 MRD.                                       3
BRUTTOABSATZ FESTVERZINSLICHER WERTPAPIERE 1971
3Ø,3 MRD. DM
GEBEN SIE IHREN EXPERTENGRAD--ZIFFER 1. BIS 5.-- ZU DIESER FRAGE AN..

EXPERTENGRAD  FRAGE   1 TEILNEHMER  3        3
EXPERTENGRAD  FRAGE   1 TEILNEHMER  4        2
EXPERTENGRAD  FRAGE   1 TEILNEHMER  5        1                        4
EXPERTENGRAD  FRAGE   1 TEILNEHMER  6        1
EXPERTENGRAD  FRAGE   1 TEILNEHMER  7        3
EXPERTENGRAD  FRAGE   1 TEILNEHMER  8        4
EXPERTENGRAD  FRAGE   1 TEILNEHMER  9        4

GEBEN SIE IHRE SCHAETZUNG FUER RUNDE   1 AN..

SCHAETZUNG  1  FRAGE   1   TEILNEHMER 3          5Ø.ØØ
SCHAETZUNG  1  FRAGE   1   TEILNEHMER 4          45.ØØ
SCHAETZUNG  1  FRAGE   1   TEILNEHMER 5          35.ØØ                5
SCHAETZUNG  1  FRAGE   1   TEILNEHMER 6          5Ø.ØØ
SCHAETZUNG  1  FRAGE   1   TEILNEHMER 7          48.ØØ
SCHAETZUNG  1  FRAGE   1   TEILNEHMER 8          4Ø.ØØ
SCHAETZUNG  1  FRAGE   1   TEILNEHMER 9          21.5Ø

ABWEICHUNGSFRAGE AN TEILNEHMER   9   GESTELLT
 TEILNEHMER   9 -INFORMATION..
KEINE
ABWEICHUNGSFRAGE AN TEILNEHMER   6   GESTELLT                         6
 TEILNEHMER   6 -INFORMATION..
1972 WAR EIN BESONDERS GUTES JAHR FUER DEN ABSATZ VON WERTPAPIEREN
FUER WELCHEN BETRAG KAUFTEN DIE KREDITINSTITUTE
DER BRD 1972 FESTVERZINSLICHE WERTPAPIERE                             7
IN MRD.DM
```

Fig. 1. Abstract from a record

```
FAKTEN... KAUF VON FESTVERZINSLICHEN WERTPAPIEREN
                                                                       8
DURCH KREDITINSTITUTE IN DEN JAHREN 1967 BIS 1969
JEWEILS ZWISCHEN 1Ø UND 15 MRD.
BRUTTOABSATZ FESTVERZINSLICHER WERTPAPIERE 1971
3Ø,3 MRD. DM
ZUSATZINFORMATION..   MITTELWERT =      45.ØØ
KEINE INFORMATION
1972 WAR EIN BESONDERS GUTES JAHR FUER DEN ABSATZ VON WERTPAPIEREN     9

GEBEN SIE IHRE SCHAETZUNG FUER RUNDE  2 AN..
SCHAETZUNG  2  FRAGE   1   TEILNEHMER 3            5Ø.ØØ
SCHAETZUNG  2  FRAGE   1   TEILNEHMER 4            45.ØØ
SCHAETZUNG  2  FRAGE   1   TEILNEHMER 5            35.ØØ
                                                                       1Ø
SCHAETZUNG  2  FRAGE   1   TEILNEHMER 6            5Ø.ØØ
SCHAETZUNG  2  FRAGE   1   TEILNEHMER 7            48.ØØ
SCHAETZUNG  2  FRAGE   1   TEILNEHMER 8            35.ØØ
SCHAETZUNG  2  FRAGE   1   TEILNEHMER 9            42.3Ø
ABWEICHUNGSFRAGE AN TEILNEHMER   5   GESTELLT
 TEILNEHMER   5 -INFORMATION..
```

FIGURE 1 : ABSTRACT FROM A RECORD

Fig. 1. (continued)

section 3 you see information which is considered relevant for the judgment of the problem and which is given to each participant. Additional information of this sort is not given to participants when forecasting questions are asked. This difference is justified by the assumption that the subjects need some information to refresh their memories with respect to judging "facts" which are about six months old. It is expected that they do not need help in evaluating present-day facts as a basis for their forecasts.

In the following section, 4, participants are asked to give their degree of expertise. It is given only once for each question. In section 5, we enter the response portion of the first round.

The numbering of the participants in sections 4 and 5 serves only as an internal identification. It begins with "3", because "1" and "2" are reserved for the experimenter and the tape on which we store the record.

After the estimates have been made (in section 5) the 0.25, 0.5, and 0.75 quartiles are calculated. The participants whose estimates lie outside the 0.25

and 0.75 quartiles are shown the quartile values which they exceeded or fell short of, and are asked to give reasons for this divergence, in case they think they possess particular additional information. The text of the questions, which is repeated at this place, is not put out in the record shown here. In section 6, data necessary for the analysis are recorded.

The second round begins with section 7. First, the question is repeated, then the relevant information. In section 9, additional data are given, namely, group responses from the previous round and any additional information which was collected in section 6 of the previous round. This information is presented in the following order: first, additional information from those participants whose estimates fell short of the lower quartile; then, information from those whose estimates exceeded the upper quartile. In the present case, there is only one item of additional information. The original text accompanying this information, which does not refer to its sources, is not reproduced here.

Beginning with section 10, the process which was described for section 5 and the succeeding sections is repeated. Five rounds are carried out for each question. This scope was chosen in compliance with the observation that after the fourth round generally the results do not improve (see section 2.3.2). An additional round is added here to test this statement.

After the five rounds are completed the next question is asked. It is a forecasting question. The two types of questions are asked alternately so that possible effects of learning or fatigue do not influence only one type of question.

In face-to-face discussion groups, the group members are asked to introduce themselves to each other by their name, field of employment, official position, and the number of years spent in banking. The idea of this was to provide each participant with a basis for judging the experience of the discussion partners in the following discussions. To what extent this information was taken into account in the formation of the group judgments could not be registered explicitly. Furthermore, the participants were asked to specify their degree of expertise for each question on a record. They noted their personal estimates for each question before any discussion took place. A discussion of the problem was expected to follow and a unanimous group estimate was demanded. A discussion leader was not appointed.

3.3 The Questions

The fact-finding questions and the forecasting questions refer to finance, banking, stock quotations, and foreign trade. We could choose from a list of ninety fact-finding questions and thirty forecasting questions that were kept on a separate tape. The questions were chosen at random from this stock. Each question of the two different types had the same chance of being chosen. However, no question appeared twice in a group.

All questions refer to items which are reported in the monthly statistics of the German Federal Bank (Deutsche Bundesbank), the daily stock market quotations of the Frankfurt Stock Exchange, and the market reports of the big banks. Only very few of the fact-finding questions refer to facts which are reported in the foreign trade statistics.

In all cases the correct responses can be verified objectively at the time of the experiments or at a later date. In the opening lecture it was called to the attention of the participants that, for example, the questions about certain past or future interest rates did not refer to the rates of the respective local institutions, which may depend largely on effects of local competition. Rather, they refer to the rates which are listed as averages in the statistics of the Federal Bank.

In five cases the wording of the questions was unclear to the participants.[45] This resulted partly from an inexact formulation on our part and partly from imperfect knowledge of the definitions as used by the Federal Bank on the part of the participants. In the face-to-face discussions, clarifications could be made immediately. In two cases in the Delphi groups, the "correct answer" in the sense in which it was understood by the participants was carried on rather than the answer to the original formulation of the question. Further cases of general misunderstanding did not become evident.

In a final discussion many of the participants expressed the feeling that the fact-finding questions were rather irrelevant and annoying: all the facts could be looked up with no trouble. This point of view was not expressed with regard to the forecasting questions. It should be recorded, however, that both sets of questions refer to the same objects, although at different points in time. (Thus, for example, one question asks for the price of a share of RWE common stocks six months before the experiments and another for the same quotation six months afterward).

3.4 The Volume of the Tests

It was originally planned that each group of different size and organization should be asked to give ten forecasts and to answer ten fact-finding questions. This plan was carried out with one exception. The Delphi group with eleven participants was able to handle only eight questions of each kind.

The time spent on the discussions amounts to between 140 and 200 minutes. In the Delphi rounds between 200 and 240 minutes of connect time were spent per participant.[46] (This length of time does not correspond to the CPU time,

---

[45]On the significance of the wording of questions and its influence on the level of estimates, cf. J. R. Salancik, W. Wenger, and E. Helfer, "The Construction of Delphi Event Statements," *Technological Forecasting and Social Change* 3 No. 1 (1971), pp. 65–73.

[46]With 20 questions, that is 10 to 12 minutes; with 16 questions, 12 to 15 minutes. Lipinski, Lipinski, and Randolph, *op. cit.*, report 15 minutes per question, on comparable hardware.

however.[47] The following points can be considered as possible explanations for the greater length of time spent on the Delphi rounds: (1) Participants write more slowly than they speak. (2) Communications between the experimenter and the participants takes place "sequentially," i.e., if participants j and k are involved, $j < k$, the message of participant k to the experimenter or back can be exchanged only after the same kind of message has been exchanged between participant j and the experimenter. This pattern of sequential communication is determined by the available technology. (3) Communication among the participants and between the participants and the experimenter can take place only during the periods of time in which the computer, which operates in a time-sharing mode, is available for the job. Although the CPU was not busy with batch operation during the experiments, the demand for memory space for other jobs which were also initiated at remote terminals was noticeable during the experiments. The participants considered such delays very disturbing. Finally let us point out that since the available teletype terminals type more slowly than the displays, preliminary experiments indicated that the former were not suitable for the experiments.

The operating system allowed for the connection of 13 displays at one time. This determined a possible maximum of group size. However, as each participant was supposed to join each type of group only once, we have limited maximum size to 11. The minimum size of groups was determined by the consideration that we wanted to have two clearly determinable participants whose estimates lay outside the quartile values. This can be achieved with five people. The fact that we had one man less in the discussion group did not interfere with this principle.

4. The Results[48]

4.1 The Distribution of Expertise

We want to investigate whether the expertise of the group members is distributed evenly or unevenly in the groups. It can be assumed that the distribution is even, because the participants and the questions were assigned to the groups at random.

However, the preliminary question whether expertise is rated differently with regard to fact-finding questions and forecasts must be clarified. Only if the

---

[47] D. Kaerger will contribute to the further analysis. See also, Institute for the Future, "Development of a Computer-Based System to Improve Interaction among Experts," First Annual Report to the National Service Foundation, 8/1/73, p. 6, Table 2. The relationship of CPU time to connect time varies from 1:110 to 1:135.

[48] The tests quoted in the following are described, e.g., by S. Siegel, *Non-Parametric Statistics for the Behavioral Sciences*, New York, and London, 1956; G. A. Lienert, *Verteilungsfreie Methoden in der Biostatistik*, Meisenheim a. Glan, 1962.

hypothesis of an uneven distribution is rejected can a comparison between the groups be made with aggregated data from fact-finding questions and forecasts. Therefore, we first test "auxiliary hypothesis 1": *expertise is rated differently in each group with regard to fact-finding questions and forecasts.*

The comparison of the quartiles given in Table 1 reveals different results in only five out of twenty-four cases. However, the narrow limits of a scale from 1 to 5 does not allow this result to appear sufficiently reliable.[49] We therefore compare the differences between the cumulative relative frequencies of the expertise ratings within each group for fact-finding questions and forecasts. The Kolmogorov-Smirnov Test shows no significant difference of the distributions on the 5 percent level.

Table 1
Quartiles of Expertise in Fact-Finding Questions and Forecasting Questions

		Quartiles					
		Lower Quartile		Median		Upper Quartile	
Type of Group	Group Size	Fact-finding questions	Forecasting questions	Fact-finding questions	Forecasting questions	Fact-finding questions	Forecasting questions
Face-to-Face Discussion Group	4	1	1	1	2	2	3
	7	2	2	2	2	3	3
	8	1	1	2	2	2	3
	10	1	1	2	1	2	2
Delphi Group	5	1	1	2	2	3	3
	7	2	2	2	3	3	3
	9	1	1	2	2	3	3
	11	1	1	2	2	3	3

Thus, auxiliary hypothesis 1 is rejected. We can use the entire set of data to investigate auxiliary hypothesis 2. It says that *between the groups no differences occur in the relative frequencies with which the different scale values of expertise are chosen.*

Table 2 shows the relative frequencies of the distributions of expertise.

---

[49]This conjecture is based on general reflections and empirical results on the correct construction of a scale. On the inferiority of the scale with five divisions compared with scales with more graduations in estimates of decisionmaking groups, see G. Huber and A. Delbecq, "Guidelines for Combining the Judgments of Individual Group Members in Decision Conferences," manuscript, TIMS-meeting, Detroit, 1971, pp. 5ff. It could not be investigated whether these statements can be applied to our results, because of the difference in the task and because American subjects may be less familiar with a scale using five divisions than are Germans.

Table 2
Distribution of Expertise (%)

Type of Group	Group Size	Degree of Expertise				
		1	2	3	4	5
Face-to-Face Discussion Groups	4	44	32	13	5	6
	7	21	38	34	6	1
	8	30	44	19	4	3
	10	45	33	15	6	1
Delphi Groups	5	39	32	16	9	4
	7	15	35	37	9	4
	9	29	28	25	13	5
	11	39	30	18	9	4

Within each type of group comparable results appear, as expected.[50] The distribution of expertise is in both cases indistinguishable between the smallest and the largest groups; the distribution varies greatly between the group with seven participants and the largest and smallest groups, respectively (cf. Table 3).

Table 3
Maximum Absolute Difference between Cumulative Distributions of the Relative Frequencies of the Degree of Expertise between Pairs of Groups (in %)

Face-to-Face Discussion Groups				Delphi Groups			
Group Size	7	8	10	Group Size	7	9	11
4	23	14	5	5	24	14	2
7		15	24	7		14	24
8			15	9			12

[50]A comparison with the groups with the same rank of size within the type of group shows no significant differences at $p \geqslant 5$ percent, neither with the Kolmogorov-Smirnov Test nor with the sign test. The latter test was carried out in compliance with the possible objection that the samples were interrelated.

An investigation of the observations for all groups of a given type leads us to reject auxiliary hypothesis 2. If one were to formulate auxiliary hypothesis 2 for a pairwise comparison between groups it would be refuted in all cases except when the smallest group is compared with the largest group or the second largest group respectively (Kolmogorov-Smirnov Test on the 5 percent level). A consideration of the data in Table 2 now gains increased importance. The significant differences are due to the fact that in the middle-sized groups, and particularly in the groups with seven participants, the ratings of expertise seem higher than in the smallest or largest groups. It is obvious, and is not examined more closely here, that the lower degrees of expertise are chosen much more frequently than the higher degrees. Whether this is an illustration of the often assumed pyramid of qualifications and abilities, or whether it only reflects a fear of using the higher values on the scale, cannot be determined definitively. The second assumption is supported by the observation that in the Delphi groups, where greater anonymity is guaranteed, 9.6 percent of the self-ratings fall into the categories four and five, whereas in the face-to-face discussion groups, where the self-ratings were occasionally asked for directly by other members of a group, only 5.5 percent fall into these categories. The results of the next section contribute to the extension of these reflections.

4.2 The Significance of the Self-Ratings

A first indication as to the validity of hypothesis E can be gained by observing individual errors and self-ratings. We test whether the lowest individual errors in each group and in relation to each question are attained by those persons who rate their expertise highest. Individual errors (see section 2.1.1) are taken as absolute values. In the Delphi groups we can determine this error in every round. We restrict ourselves here to the first and the last round. In the face-to-face discussion groups the only data for judging individual errors is from the questionnaires filled out before entering the discussion of each question. We test auxiliary hypothesis 3:

The distributions of the highest self-ratings with regard to each question and the self-ratings of those who attain the highest level of performance (i.e., the lowest individual error taken as an absolute value) coincide.

Auxiliary hypothesis 3 is tested separately for fact-finding questions and forecasting questions in face-to-face discussion groups and Delphi groups. In the latter case it is also tested separately for the first and the last rounds. In all cases, auxiliary hypothesis 3 is refuted at a high level of significance in a χ^2 test (0.01). Thus, one must assume that in the situations investigated here, self-ratings with regard to expertise do not give enough information as to which persons actually possess expertise. For this reason either the ability to give a self-rating which corresponds to actual expertise must be studied more closely and, if possible, promoted or other methods of determining expertise before the beginning of the questioning must be tested for their effectiveness. The Institute for the Future emphasizes the latter problem[51] in its recent studies.

Auxiliary hypothesis 3 was based on individual performance. Hypothesis *E*, on the contrary, refers to the performance of the entire group. We attempt to operationalize this viewpoint by testing the rank correlation[52] between group performance with respect to each question and the average expertise of the group with respect to the same question, separately for each type of group, each group size, and each of the two types of questions. The skew distribution of the degrees of expertise already lets us expect that important information cannot be gained from such a test. Indeed, we do not find any significant rank correlation coefficient in the relationships tested for fact-finding questions. A classification of the data in a 2×2 contingency table according to the criteria: low and high degree of expertise *vs.* upper and lower half of the scale of the rank figures for group performance, does not lead to significant relationships. With regard to forecasts, significant relationships (at the 5 percent level) show up in Delphi groups with five and nine participants in the third as well as in the fifth round. However, as is shown in Table 6, these groups are not noted for particularly good overall results. In this respect the correlation seems unimportant.

Since the conclusion that objectively existent expertise is not an essential factor of individual or group performance contradicts the definition of expertise and thus is not tolerable as an explanation of the results, it can only be concluded that the self-rating of expertise by practitioners for tasks of the present type does not coincide with their objective expertise.

After the experiments were concluded, this unsatisfactory result led to the question whether explanations can be found for the choice of the rank figures of expertise by the individuals. We attempt to explain the behavior of our subjects by the following auxiliary hypothesis 4:

The degree of expertise with regard to a question is related to the number of years that a subject spent in banking. Furthermore, it is higher whenever the subject matter of the question coincides with one of the fields handled during the years in the profession.

The necessary data were collected by questionnaire. Answers from up to twenty-eight participants were available. In the analyses a high positive correlation showed up between age and number of years in the profession, so that separate hypotheses for these two variables were not tested. Further, a positive correlation showed up between the number of fields one had experience in (a list of possible fields was presented which could, however, be supplemented)

---

[51]Cf. Institute for the Future, "Development of a Computer-Based System to Improve Interaction among Experts," *op. cit.*

[52]Spearman rank correlation with correction for ties.

and the number of years in the profession. Thus the two components of hypothesis 4 can no longer serve as mutually independent variables for explaining the degree of expertise. Therefore, we tested the hypothesis in a simplified form, once for the influence of the number of years in the profession and once for the influence of the fields of employment on the choice of the degree of expertise. In neither case, a Kolmogorov-Smirnov Test refutes the hypothesis (0.001 level).

Thus an observation that was gained in the face-to-face discussion groups is confirmed. Once in a while during the discussions the number of years of "banking experience" was brought into play to decide points of controversy, obviously with the view that this is a criterion for measuring expertise objectively. The same is true for the present field of employment of the persons in question. Obviously, however, these criteria for judging on the expertise are not sufficient in the light of very special questions. It would probably be better to consider, for example, the regular observation of special sections of the bank statistics.

4.3 The Performance of the Delphi Groups

We formulated two hypotheses regarding the performance of the Delphi groups. We first test hypothesis *V*. To do so, we determine how frequently the measure of variance for the last round is smaller than that for the first round (cf. Table 4).

Hypothesis *V* cannot be refuted. When up to five rounds are carried out in Delphi groups a reduction of variance of the estimates takes place.

A closer examination shows that it cannot be rejected that variance reduction appears with equal frequency in all groups, and that variance reduction occurs independent of the type of question (chi-square test, 5 percent level).

Table 4
Frequency of Variance Reduction in Delphi Groups, Round Five compared with Round One (% of all possible cases)

Group Size	Fact-Finding Questions	Forecasting Questions
5	100	80
7	90	80
9	90	100
11	100	100

Hypothesis *R* will be tested now for judging performance. In order to represent the performance of a group for a certain type of question, individual performances must be aggregated. Since the individual performances are "index numbers" only the geometric mean can be chosen for this. At first we turn to the fact-finding questions. If considered individually, it becomes evident that the lowest and the highest median group error (taken as an absolute value) lie with approximately equal frequency in the first two rounds (cf. Table 5). In case of identical figures for the observed variable, its first appearance was considered.

Table 5
Relative Frequency of the Lowest (and Highest) Median Group Error by Number of Rounds, Group Size, and Type of Question

Group Size	Round (Fact-Finding Questions)				
	1	2	3	4	5
5	0.70(0.40)	0.20(0.40)	0.10(0.20)	0.0(0.0)	0.0(0.0)
7	0.40(0.70)	0.30(0.20)	0.10(0.0)	0.10(0.10)	0.10(0.0)
9	0.40(0.80)	0.30(0.0)	0.30(0.0)	0.0(0.20)	0.0(0.0)
11	0.635(0.40)	0.125(0.20)	0.125(0.10)	0.0(0.10)	0.125(0.0)

Group Size	Round (Forecasting Questions)				
	1	2	3	4	5
5	0.50(0.70)	0.40(0.20)	0.0(0.10)	0.0(0.0)	0.10(0.10)
7	0.70(0.60)	0.10(0.20)	0.0(0.20)	0.10(0.0)	0.10(0.0)
9	0.70(0.70)	0.30(0.10)	0.0(0.0)	0.0(0.10)	0.0(0.10)
11	0.375(0.865)	0.50(0.0)	0.0(0.125)	0.0(0.0)	0.125(0.0)

The median values for each group, which are easily read from Table 5, all lie in the first or second round. If we consider the lowest median group errors (taken absolutely), we observe that round two evidently is of greater importance concerning forecasting questions than concerning fact-finding questions, whereas rounds one and three are of much greater importance for fact-finding questions than for forecasting questions. On the other hand, the highest median group errors (taken absolutely) are much more heavily concentrated in the first round for forecasts than for fact-finding questions.

This analysis does not, however, take into consideration the degree to which the medians deviate from the correct values. This may be evaluated by looking at the geometric mean of the individual errors for each round, each group, and each type of question. For it could be possible that good results deteriorate not at all or only very little, while poor results improve greatly with an increasing number of rounds.

Data to judge on this question are presented in Table 6.

Table 6
Geometric Mean of the Individual Errors (taken as absolute values)

	Round (Fact-Finding Questions)				
Group Size	1	2	3	4	5
5	0.20	0.19	0.18	0.23	0.27
7	0.13	0.09	0.07	0.10	0.11
9	0.22	0.14	0.09	0.17	0.16
11	0.24	0.29	0.23	0.28	0.22

	Round (Forecasting Questions)				
Group Size	1	2	3	4	5
5	0.28	0.27	0.28	0.35	0.35
7	0.44	0.42	0.43	0.36	0.36
9	0.25	0.20	0.16	0.16	0.16
11	0.12	0.09	0.10	0.10	0.10

Data are analyzed by Friedman's two-way analysis of variance by ranks.

A significant difference for the entire set of data for fact-finding questions barely fails to be demonstrated at the 10 percent level ($\chi_r^2 = 7.6$ compared with the tabulated value of 7.78). The computed value of χ_r^2 for forecasting questions is considerably lower (5.75) and fails the 20 percent level. The phenomenon that for fact-finding questions in all groups except the one with eleven participants the highest performance is attained in the third round does not influence the tests significantly. This is even less so for forecasting questions, where the best performance is achieved twice in the second round and otherwise in the fourth or fifth rounds.

If one assumes that people get bored with answering the same questions over and over again and if one therefore cuts out the results of the last round, we observe a minor increase in the test statistics. The calculated value for fact-finding questions exceeds the significance level of 10 percent. For forecasts the increase is so slight ($\chi_r^2 = 5.77$) that no further consequences can be drawn from it. Taking everything together, our results seem to indicate that it is not reasonable to extend the number of rounds in Delphi groups beyond the third round.

This would support hypothesis *R* while it refutes hypothesis *R′*. Since in all cases investigated it is not assumed that the self-rating of expertise varies with the number of rounds, the result cannot be tested further in this respect.

The observations of group performance are not supported by the results of the best individual performance. In no case, i.e., neither when all rounds are considered nor when the number of rounds taken into consideration is limited, can a significant difference in the best individual performance be observed which would vary with the number of rounds. The best individual performance is very often maintained over consecutive rounds. So, if one would have objective criteria by which one could pick real experts, one might expect a greater stability of their judgments as compared with that of all members of our present groups.

4.4 The Size of the Groups

We test hypothesis *G* directly, separately for face-to-face discussion groups and Delphi groups. We do not correct the hypothesis to include the distribution of expertise (as determined subjectively by self-ratings) between groups, as it has practically a random influence.

When the data in Table 6 are analyzed by the rows, significant differences (on the 0.1 percent level) show up in Friedman's analysis of variance by ranks ($\chi_r^2 = 12.84$ for fact-finding questions and $\chi_r^2 = 15$ for forecasting questions). It is clear that the group performance in all rounds may be rank ordered as follows:

Fact-finding questions: Group with seven participants on top, followed by the groups with nine, five, and eleven participants.

Forecasting questions: Group with eleven participants on top, followed by the groups with nine, five, and seven participants.

The groups of seven and eleven reverse their rank order of performance in fact-finding questions as compared with forecasting questions.

This observation cannot be explained by varying self-ratings of expertise, as can be read from the refutation of the hypothesis concerning a different distribution of expertise in fact-finding questions and forecasting questions (see section 4.1).

If one considers the best individual performances directly, the result for the groups is confirmed, except for a shift in the rank orders of the groups with five and eleven participants for fact-finding questions, and of the groups with seven and five participants for forecasting questions. The individual estimates of the participants can thus be considered to be one factor which influences the result.

The quality of the estimates could be determined by the frequency of information exchange between the participants.[53] The frequency of information exchange is shown in Table 7.

Table 7
Frequency of Information Exchange between the Participants in Delphi Groups

	Absolute Number of Information Exchanges				% of All Possible Exchanges of Information			
	After rounds				After rounds			
Group Size	1	2	3	4	1	2	3	4
5	7	9	4	2	17.5	22.5	10.0	5.0
7	10	13	11	15	25.0	32.5	26.5	37.5
9	43	33	17	20	53.8	41.3	21.3	25.0
11	26	17	11	11	40.7	26.5	17.2	17.2

We have aggregated data for fact-finding questions and forecasting questions as we have found that the frequency of information exchange does not vary significantly with the type of question (binomial test, 5 percent level).

We find that the absolute frequency of information exchange between the participants varies significantly between the individual groups ($\chi_r^2 = 19.8$ is significant on the 5 percent level). Nevertheless the group with nine participants clearly leads the sequence, followed by those groups with eleven, seven, and five participants. If the frequency of information exchange is related to the number of possibilities for information exchange (which depends on the number of questions as well as on the number of participants outside of the quartiles, of course) a significant difference between the groups can likewise be determined ($\chi_r^2 = 15.0$). In this case only the groups with seven and eleven participants change their positions in the rank order as compared with the rank order of absolute frequencies of information exchange. One could assume that

[53]See I. Lorge *et al.*, "Solutions by Teams and Individuals to a Field Problem at Different Levels of Reliability," *Journal of Educational Psychology* **46** (1955), pp. 17–24.

the participants themselves, as keepers of information, channel varying amounts of information into the group. If so, it should be possible to find a rank correlation between group performance and the frequency of information exchange with regard to each question within each group. It turns out, however, that a significant result (5 percent level) can be identified only for the group with nine participants.

Low, and in part negative, values for the rank correlation, particularly in the group with eleven participants, can probably be explained by the fact that the opportunities for information exchange were used to transmit signs of impatience toward the end of each session. These were not considered to contribute to group performance in the stipulated sense, and thus were not counted in selecting the data of Table 7.

In the face-to-face discussions group performance is distributed differently (see Table 8). For the fact-finding questions we discover the following:

If performance is measured again by a median group error which is composed of individual estimates given before entering discussion, we find that the group with seven participants attains the highest level of performance. The groups with ten, four, and eight participants follow in that order. It should be noted that these data are only approximately comparable to those for round one in Table 7 because the initial data used here are given before the start of the exchange of information. If we take the geometric mean of the absolute values of the differences between the group estimate and the correct values, the group with ten participants appears at the top of the scale of performance. The groups with seven, four, and eight participants follow.

Let us now turn to the forecasting questions.

As before, a corresponding rank order of group performance cannot be

Table 8

Geometric Mean of the Median of Individual Relative Errors (taken as absolute values) before Discussion and the Relative Error of the Group Estimate in Face-to-Face Discussion Groups

Group Size	Fact-Finding Questions		Forecasting Questions	
	Geometric Mean Individual Errors	Group Errors	Geometric Mean Individual Errors	Group Errors
4	0.171	0.163	0.179	0.184
7	0.116	0.154	0.103	0.147
8	0.257	0.220	0.072	0.121
10	0.141	0.139	0.187	0.212

demonstrated for the two types of group structure. Here the group with eight participants leads, and the lowest level of performance is attained by the group with ten participants (see Table 8).

When these results are interpreted it should be noted that the fact that in several cases one member of the Delphi groups did not participate in the face-to-face discussion groups does not affect the results of the latter in a uniform fashion.

Moreover, the direct observation of the group activity suggests that the number of members present in a discussion group cannot be considered the decisive variable. It is more important to consider the number of members of a face-to-face discussion group who actively take part in the discussion, as the nonparticipants do not influence the group judgment in any way. This situation occurs in our face-to-face discussion groups. According to our observations, in the group with eight participants, one member of the group only very rarely took part in the discussions. In the group with ten members two persons did not express any opinions during the entire experiment. When the group average was determined by voting, they conformed to the majority opinion, which was apparent. Another person only rarely joined in the consultations. Thus the "active" part of the group is almost always reduced to seven persons.[54]

Since in the two smaller groups the group itself required that each member participate on each question, there the "active" group corresponds in size with the entire group.[55] Thus our observation material is reduced practically to "active" discussion groups with four, seven, and possibly eight participants.

After these reservations as to the interpretation of the results, it is not surprising that the rank order of performances, if partitioned with respect to fact-finding questions and forecasting questions, as well as with respect to both types of group structures, does not generally exhibit comparable results in groups of varying size. Results coincide in the smallest group only. This, however, cannot be considered significant.

4.5 Hypothesis *D*

The problems of drawing a comparison between the results from the Delphi groups and the face-to-face discussion groups have already been mentioned. Besides, one has to find a generally accepted criterion on which to base the

[54]Unfortunately, only short notices about the impressions of the author after each session of the face-to-face discussion groups exist as a proof of these statements. Video tapes of the discussions do not exist; they would certainly have contributed to the interpretation of the results.

[55]Our observations coincide with the statements of Alter *et al.*, who confirm clique formation with disturbing internal discussion, a higher absolute proportion of inactive group members, and poorer agreements with dominant participants as dysfunctional effects in large groups in brainstorming sessions. Cf. U. Alter, H. Geschka, H. Schlicksupp, "Methoden und Organisation der Ideenfindung," report on a group project of Battelle Institute, Frankfurt (July 1972), Methodological Appendix, p. 20.

comparison. Dalkey's statements are based on the number of cases of superior performance. However, the interest in group performance can be based with at least the same right on the amount of the errors that occur. We take up this latter point.

The geometric average of all group performances for fact-finding questions in face-to-face discussion groups (0.167) is higher than the corresponding value for the third round of Delphi groups, which is as low as 0.127. However, the result is reversed for forecasts. Here the corresponding value of 0.209 for the third round in Delphi groups is higher than the result of 0.162 for the face-to-face discussion groups. Thus an unequivocal relationship cannot be established. The performance of the only group with identical size under different organizational structures also differs greatly.

It is further noteworthy that with regard to the forecasts, the discussion in the discussion groups shows no progress in performance if the results are compared with the geometric means of individual errors before discussion. On the other hand, the result of the discussion in all the Delphi groups is that the mean error is reduced up to the third round. Furthermore, we find that the performance level of the face-to-face discussion groups is approximately equal for fact-finding questions and for forecasts, whereas the performance level of the Delphi groups with regard to forecasts is much lower as compared with fact-finding questions.

4.6 Hypothesis *F*

The preceding statements have already made it clear that different results are definitely produced when group performance with regard to fact-finding questions is differentiated from group performance with regard to forecasts (cf. Tables 6, 8). A general confirmation or refutation of hypothesis *F* on the basis of group performance is not possible. We therefore formulate and test the auxiliary hypothesis 5 in addition to what we have found until now. It says:

There exists a positive relationship between the rank order of performance of individuals in each group with regard to fact-finding questions and forecasts. Performance is defined as the geometric mean of the individual errors (taken as absolute values) in answering fact-finding questions and forecasting questions.

The auxiliary hypothesis is refuted in each of the groups. With increasing group size we calculate rank correlations of -0.300, -0.391, $+0.357$ and -0.150. In the majority of cases not even the sign corresponds to the expectations of the auxiliary hypothesis.

5. *Summary*

Although one should not overestimate the results from the very few experiments presented here, they do lead to doubts as to the efficiency of the Delphi method. Of course, it must be admitted that the attempt to use the Delphi method for short-term forecasting is a comparatively tough test, for it was originally designed for long-range forecasting. At another occasion (sales estimates) it was also found that the errors of short-term forecasts can be very much higher than those of long-term forecasts.[56] If one assumes that the results of the forecasts could be interpreted as an attempt to estimate an unknown status quo at the time when the experiments took place as well, then they should be corrected by the average value of the relative difference between the status quo and the realization of each topic. These corrections vary between 0.042 and 0.098 in the individual groups, according to the choice of questions. Their application does not alter the rank order of the results as compiled in Tables 6 and 8. Therefore we may refer directly to the text with our summary:

(1) It cannot be discerned that fact-finding questions are suitable test material for recognizing expertise or appropriate organizational structures for forecasting groups.

(2) A general positive relationship between group size and group performance cannot be recognized.

(3) In face-to-face discussion groups the measure of the group size must be determined by the number of active participants. Appropriate precautions should be developed.

(4) Variance reduction almost always occurs in Delphi groups between the first and the fifth rounds, but the best results are as a rule already known in the third round. Further rounds may impair the results.

(5) Self-ratings of expertise show a positive relationship to the performance of the persons questioned in only two of four Delphi groups. They tend to be lower in face-to-face discussion groups than in the Delphi groups, and are determined substantially by the extent of professional experience rather than being set with regard to the questions in case. It is important to employ and develop better methods for the determination of expertise.

(6) A direct comparison of Delphi groups and face-to-face discussion groups was not possible because several participants dropped out. However, the results, if separated for fact-finding questions and forecasts, do not point in one direction.

(7) Proponents of the Delphi method will point out that our subjects, being banking experts, are better able to express themselves in a face-to-face discussion than on a display, even if its use has been explained to them. This should apply particularly when the space for exchanging information among participants is limited. The fear of making mistakes in the operation of the display could lead to exaggerated caution. However, if one agrees with this argument, the first point of this summary has to be explained also, as the results are not uniform in this respect. Anyhow, it appears to be important to avoid a "new

[56] J. Berthel, D. Moews, *Information und Planung in industriellen Unternehmungen*, Berlin, 1970, pp. 158 ff. (with data from fifteen firms).

Taylorism," on which some *mis*-concepts are founded.[57] However, it must be granted that the originators of the Delphi method did not say that it has to be operated as a computer dialogue.

(8) Only in the Delphi group with the greatest exchange of information did we observe a positive relationship to group performance. The results indicate that in small Delphi groups more opportunities for information exchange should be given. However, it probably must be tested whether the information given by the participants does coincide with what others would want to know, i.e., whether it adds to their knowledge.[58] How can the "confusion effect" in the majority of the discussions, which is recognized when the "reference values" mentioned there are compared (cf. Table 8), be explained without this distinction and the assumption of a difference between information supplied and information demanded?

(9) It must be admitted that in our strongly discipline-oriented group there has been relatively little opportunity for improving estimate by sharing information as compared to interdisciplinary groups concerned with other tasks. However, this affects all our groups in the same way. This criticism would be valid if it could be demonstrated that different groups react differently to these two types of tasks.

---

[57]See W. Kirsch, "Auf dem Wege zu einem neuen Taylorismus?" in H. R. Hansen, M. P. Wahl (eds.), *Probleme bein Aufbau betrieblicher Informationssysteme*, Müchen, 1973, pp. 338–48.

[58]E. Witte (ed.), *Das Informationsverhalten in Entscheidungsprozessen*, Tübingen, 1972, especially, pp. 44ff.; R. Bronner, E. Witte, B. R. Wossidlo, "Betriebswirtschaftliche Experimente zum Informations Verhalten in Entscheidungsprozessen," in E. Witte, *op. cit.*, pp. 186ff.

V. Cross-Impact Analysis

V. A. Introduction

HAROLD A. LINSTONE and MURRAY TUROFF

The process of summarizing the views, judgments, and opinions of a group of individuals leads to a certain degree of quantification in most Delphi exercises. Certainly the field of statistics offers a host of techniques for analyzing and summarizing the responses to Delphi questions. In addition, the nature of certain common question types used in the Delphi process lends itself to the incorporation of underlying models for the judgmental process that have been developed in such fields as psychology, decision theory, and operations research. In fact, the art of Delphi design has proceeded to a point where the body of literature has adapted many concepts and techniques from these other areas (see references on "Related Work" in Appendix) to specific application for the Delphi process.

In this and the following chapter, the reader is introduced to a few of the more significant attempts in this area of specialized techniques. The last words on the "best" methods to handle certain problems are far from being written. Therefore, a number of the papers in these sections present differing approaches, philosophies, or views on the handling of the same application.

One very common problem in Delphi is to get at the underlying relationships among possible future events. The existence of interrelationships is the reason for the complexity of many biological and social systems and for the counterintuitive nature of their behavior. Most individuals are simply unable to follow the impact of one change through the system; they assume independence of its various parts. One common approach, known as "cross impact," aims to alleviate this difficulty and probe the effect of interactions among elements of a system. The papers in this chapter represent three differing views or approaches to the cross-impact problem, and there are more to be found in the literature referenced in these articles. Our choice of these particular papers rests on both the completeness of the discussion of their approaches and their occurrence as recent work in the field.

The first paper dealing with cross impact, by Dalkey, utilizes the laws of probability calculus and rests on a fundamental assumption of the correctness of the Bayes theorem for analyzing the consistency of human judgment. In a sense this paper represents an approach arising out of current practices in the area of decision theory.

The second paper on cross impact, by Turoff, rejects the Bayes-type

The Delphi Method: Techniques and Applications, Harold A. Linstone and Murray Turoff (eds.)
ISBN 0-201-04294-0; 0-201-04293-2

approach and seeks to establish an interaction model analogous to a quantum mechanical interaction model from the field of physics. It is quite clear these two papers are based upon opposing assumptions. In a broader context this difference of view is a result of our lack of understanding of the human perception and thought process. However, in the use of the cross-impact technique to supply relative weightings of relationships, it would probably be safe to say that either approach will give about the same relative rankings of impacts. If the reader desires, on the other hand, to lay bets on the actual probability estimates obtained, then he will have to first answer for himself such questions as whether people are, or should be, Bayesian decisionmakers.

The third paper on cross impact, by Kane, formulates a dynamic time dependent model for the problem. This approach rests on control-theory concepts and bears a similarity, with respect to what the user sees, to the system dynamics type analysis (e.g., Forrester-Meadows). This third procedure defines the cross-impact problem in such a manner that it cannot, as an approach, be directly compared with the first two papers. In any case, the problem it solves is of interest and applicable to a large number of common situations. Its simplicity makes it a particularly useful pedagogical device. It has proven to be an excellent means to introduce laymen to complex systems, since the technique exhibits many of the characteristics which makes their behavior appear counterintuitive.

We can anticipate much more work in cross-impact analysis as we search for ways to include the more subtle interactions which so far have eluded our methodological capabilities.

V. B. An Elementary Cross-Impact Model

NORMAN C. DALKEY*

Abstract

Cross-impact analysis is a method for revising estimated probabilities of future events in terms of estimated interactions among those events. This Report presents an elementary cross-impact model where the cross-impacts are formulated as relative probabilities. Conditions are derived for the consistency of the matrix of relative probabilities of *n* events. An extension also provides a necessary condition for the vector of absolute probabilities to be consistent with the relative probability matrix. An averaging technique is formulated for resolving inconsistencies in the matrix, and a nearest-point computation derived for resolving inconsistencies between the set of absolute probabilities and the matrix.

Although elementary, the present model clarifies some of the conceptual problems associated with cross-impact analysis, and supplies a relatively sound basis for revising probability estimates in the limited case where interactions can be approximated by relative probabilities.

I. Introduction

One of the more promising new tools for long-range forecasting is cross-impact analysis. The general notion was first suggested by Helmer and Gordon with the game "Futures" created for the Kaiser Corporation. Cross-impact analysis has now been expanded and applied to a number of forecasting areas by Gordon and others at The Institute for the Future [1]. The motivation for cross-impact analysis arises from a basic aspect of long-range forecasting. There are usually strong interactions among a set of potential technological events or among a set of potential social developments. In assessing the likelihood that any given event or development will occur, the interactions with other events are clearly relevant. However, the number of first-order potential interactions increases as the square of the number of events. Even if a matrix describing the interactions is available—say from estimates furnished by a panel of experts—the task of thinking through the implications rapidly gets out of hand. Some computational aid is required to take account of the large number of interdependencies.

Gordon and others at The Institute for the Future have developed two major approaches to the computational program.[1] Both approaches involve (1) preliminary estimates of the absolute probabilities (i.e., the probabilities of the individual events), (2) estimates of the interdependencies in terms of a cross-impact matrix, (3) a Monte Carlo sampling of chains of events in which the probability of an event in the chain is modified by the cross-impact of the previously occurring event in the chain, and (4) reestimation of the absolute probability of each event in terms of the relative frequency of the occurrence of that event in the sample of chains. The difference between the two approaches lies in the mode of modification of the probabilities. In the first approach, the basic method, the modification is effected by a heuristic algorithm. Cross-impacts are rated on a nominal scale of −10 to +10. Modification of successive probabilities is

* Dr. Dalkey is Senior Mathematician at the RAND Corporation, Santa Monica, California.
This research is supported by the Advanced Research Projects Agency under Contract No. DAHC15 67 C 0141. Views or conclusions contained in this study should not be interpreted as representing the official opinion or policy of Rand or of ARPA.

[1] They actually have worked with four variations of the cross-impact technique. Two of these variations, the dynamic model and the scenario model, involve aspects of the interdependency problem (namely, strict time or order relationships) which are beyond the scope of the present paper.

Reprinted from *Technological Forecasting and Social Change*, 3 (1972) with permission of American Elsevier.

computed via a family of quadratic "adjustments," based on the cross-impact rating and the unmodified probability. The second approach, the likelihood ratio method, defines cross-impacts in terms of a factor by which the odds favoring the target event are to be multiplied, given the occurrence of the impacting event. The second approach is conceptually clearer than the first, and removes some of the arbitrariness associated with it.

Both approaches suffer from a lack of clarity concerning the purpose of the computation. The notion of "taking account of the interactions" is not adequate to answer questions such as "are the revised probabilities in fact more accurate estimates than the original ones?" In addition, as will be seen below, the Monte Carlo computation contains implicit assumptions concerning higher-order interactions that are not defined, and are surprisingly difficult to state precisely. (To say they are implicit is not to say they are not recognized by the developers of the method, only that the nature of the assumptions is not clearly stated.)

In this article an elementary model of probability cross-impacts is formulated that clarifies the notion of "taking account of interactions," and does this without requiring any assumptions concerning higher-order interactions. The model is based on fundamental postulates and theorems of probability. The model does not take into account all of the aspects of an interacting system that are pertinent—in particular, it neglects time as a parameter. More generally, it neglects nonprobabilistic order effects.

The approach is elementary in two respects. First, the type of probability system and the form of the cross-impacts (interdependence) assumed are of a very simple probabilistic form. Second, the notion of taking account of the interactions is elementary. It can be described as follows: If an individual or a group estimates a set of probabilities of events and this set contains interactive terms, then the set may be inconsistent. The purpose of computation, then, is to test the consistency of the set of estimates, and if the set is not consistent, to perform the smallest reasonable perturbation on the original estimates to create a set that is consistent. As soon as the consistent set is achieved—from this elementary point of view—the interactions have been "taken into account."

It might not hurt to amplify this point slightly. If we assume that the purpose of cross-impact analysis is to arrive at the best possible estimate of the separate probabilities of each of the events,[2] then regardless of how the original estimates are obtained, they should already include the interactions among the events. The basic assumption of cross-impact analysis is that the separate probabilities do take the interactions into account, but incompletely so that some modification is needed.

Another assumption is that the cross-impact estimates are more "solid" than the absolute probability estimates. There are several motivations behind this assumption. First, there is widespread, and probably generally sound, opinion that relative probabilities are clearer and easier to estimate than absolute probabilities. I do not know of any experimental data to support this, but it does appear that the more limited reference of a relative probability makes it psychologically easer to deal with. Second, there is an argument (which may or may not have logical justification) that narrowing the reference class in some way makes the probability more "correct." This is the basis for Reichenbach's recommendation [Ref. 2, p. 374] that in practical applications of probabilities (decisions) the narrowest reference class for which there is reliable

[2] There are a number of considerations which suggest that, for purposes of long-range planning, the absolute probabilities are of secondary interest, whereas the "scenario" probabilities, i.e., the probabilities of joint occurrence or nonoccurrence of many events, are more directly relevant. This topic will be discussed in the text in greater detail.

information should be used. Finally, and probably most important for cross-impact analysis, the notion of cross-impacts is new, and should receive greater emphasis.

None of the foregoing justifies the assumption that the cross-impacts are more solid than the absolute probabilities, but they do lend some heuristic weight to the computational structure. These considerations suggest that adjustments should be made in the absolute probabilities, not the cross-impacts. It will be shown below that this point of view cannot be maintained strictly. It is possible that inconsistencies appear in the cross-impacts as well as in the original estimates of absolute probabilities. However, this assumption can be maintained in a weaker sense if the cross-impacts can be adjusted without making use of the absolute probabilities, and then the absolute probabilities can be adjusted with fixed cross-impacts.

The results to be presented in this article, then, can be summed up by saying that given a set of estimates of absolute probabilities and cross-impacts in the form of relative probabilities, simple tests exist for determining the consistency of the cross-impact matrix, and for determining the consistency of the absolute probabilities given that the cross-impact matrix is consistent. If the set of estimates is consistent, then no further computation is required. If the set is not consistent, then a number of steps may be taken to adjust the set, ranging from simplified averaging techniques to reiteration of the estimates, given a display of the inconsistencies involved. The adjustment procedure used should depend on the nature of the inconsistencies and the opportunities for querying the estimators again.

The consistency condition derived below takes a particularly elegant form when the cross-impact matrix is expressed as a set of relative probabilities, that is, the probability of event e_i is p, given that event e_j occurs. With cross-impacts of this form, the consistency condition can be stated as follows: The n events define an n-dimensional probability space (strictly speaking an n-dimensional hypercube, since each probability can vary only between 0 and 1). If the cross-impacts are mutually consistent, they define a single line in this hypercube, which passes through the origin. A set of absolute probabilities consistent with the cross-impacts will then define a point lying on this line. As it turns out, the condition is relatively easy to test and allows a fairly simple description of methods of resolving inconsistencies if the estimates do not pass the test.

II. Consistent Systems of Probabilities

We will be concerned with an elementary system of probabilities:

A. A set of n events $e_1, e_2, \ldots, e_i, \ldots, e_n$.
B. A set of n absolute probabilities of these events, $P(e_1), P(e_2), \ldots, P(e_i), \ldots, P(e_n)$.
C. A set of $n^2 - n$ relative probabilities[3] of the form $P(e_1/e_2), \ldots, P(e_i/e_j), \ldots$ (read as "the probability of e_i given that e_j occurs").
D. A set of higher-order probabilities, illustrated by $P(e_i \cdot e_j \cdot e_k)$, where the period indicates joint occurrence (read as "the probability that e_i, e_j, and e_k all occur).

There is a large family of probabilities of type D. Since we will not deal formally with this set, they are not listed in detail.

[3] The notion of an absolute probability is sometimes misunderstood. For the purpose of this discussion, we simply assume a common universe of discourse for the events, and refer the probabilities to that. In this respect, the absolute probabilities of each event should reflect (in a completely buried form) all of the interactions between that event and all others. In particular, the absolute probability of an event is not interpreted in the Bayesian sense of an *a priori* probability.

The set of $n^2 - n$ relative probabilities of type C will be referred to as the cross-impact matrix.

The set of events $\{e_i\}$ could be a list of potential technological developments, a list of social or political events, a combination of these, or something entirely different, such as a set of symptoms of diseases among the total U.S. population.

It is easy to show that for a complete specification of a system, $2^n - 1$ probabilities must be given; these are independent except for a set of inequalities illustrated by $P(e_i \cdot e_j) \leqslant P(e_i)$. In general, the n absolute probabilities and the $n^2 - n$ relative probabilities are quite insufficient to completely specify the system.

Given a set of probabilities of the forms B and C, a simple question can be asked; namely, are they a consistent set? "Consistent" here means compatible with the usual calculus of probabilities. The question is meaningful because the set of probabilities in the forms B and C is redundant.

There are two kinds of redundancy. The first involves the probabilities of type C. All of the $n^2 - n$ relative probabilities in the cross-impact matrix can be replaced by the $n(n-1)/2$ joint probabilities of the form $P(e_i \cdot e_j)$. (The factor $\frac{1}{2}$ comes from the fact that joint occurrence is commutative.) In short, there are twice as many entries in the cross-impact matrix as are needed to specify all probabilities involving no more than two events.

The second type of redundancy involves the interrelationship of the absolute and relative probabilities. Even with a consistent set of relative probabilities, the absolute probabilities may not combine with the relative probabilities in accordance with the rules of the calculus of probabilities.

We will use three elementary postulates of the calculus of probabilities and one theorem. The three elementary postulates are:

P1. Normalization. $0 \leqslant p \leqslant 1$ for any probability p.
P2. Rule of the product. $P(e_i \cdot e_j) = P(e_i) \cdot P(e_j/e_i) = P(e_j) \cdot P(e_i/e_j)$.
P3. Rule of addition. $P(e_i \text{ or } e_j) = P(e_i) + P(e_j) - P(e_i \cdot e_j)$.

The theorem is one which I derived in my Ph.D. thesis, and is referred to in Reichenbach [Ref. 2, p. 112]. It will not be proved here, but simply stated.

THEOREM. *Rule of the triangle.*

$$P(e_i/e_j) \cdot P(e_k/e_i) \cdot P(e_j/e_k) = P(e_j/e_i) \cdot P(e_k/e_j) \cdot P(e_i/e_k). \tag{1}$$

The theorem states that for a set of three events, the product of the relative probabilities multiplying around the triangle in one direction is equal to the product of the relative probabilities multiplying in the other direction. (See Fig. 1.) The theorem is easily extended to four or more events, but the same effect can be achieved by treating the larger set in subsets of three.

Since we can assume that all the probabilities given in C and D are already between zero and one, the only role of P1 is to combine with P2 and P3 to give the weak inequality.

$$\begin{aligned} P(e_i) + P(e_j) - P(e_i) \cdot P(e_j/e_i) &\leqslant 1, \\ P(e_i) + P(e_j) - P(e_j) \cdot P(e_i/e_j) &\leqslant 1. \end{aligned} \tag{2}$$

The rule of the product has an immediate consequence.

$$P(e_j) = P(e_i) \cdot P(e_j/e_i)/P(e_i/e_j). \tag{3}$$

Eq. (3) can be interpreted as asserting that if $P(e_j/e_i)$ and $P(e_i/e_j)$ are fixed, then $P(e_i)$ and

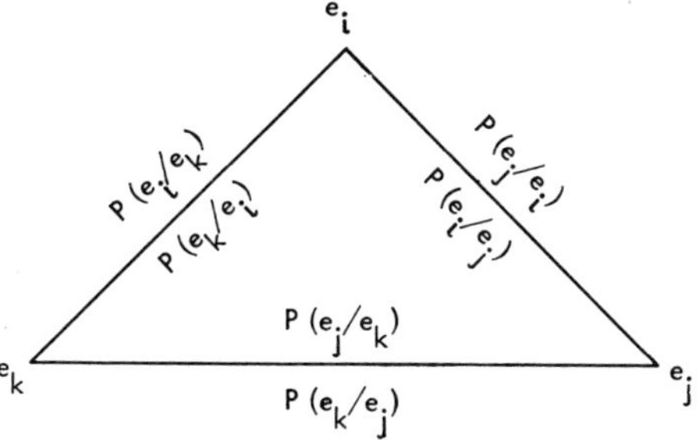

Fig. 1. Rule of the triangle.

$P(e_j)$ must lie on a straight line through the origin in the $P(e_i)$, $P(e_j)$ unit square, as illustrated in Fig. 2.

Similarly, for all other pairs of events, the rule of the product implies that these must lie on a straight line through the origin in their respective unit squares when the relative probabilities are given. Now, if we consider the space of all the absolute probabilities—the unit hypercube—the rule of the product in conjunction with the rule of the triangle assures that all of the absolute probabilities must lie on a single straight line within the hypercube. To illustrate this theorem, we first consider the unit cube defined by three events. Eq. (3) asserts that the probabilities must lie on the intersection of the planes defined by the straight lines in the respective unit squares.

Figure 3 illustrates the intersection of the two planes defined by the pairs $P(e_i)$, $P(e_k)$ and $P(e_j)$, $P(e_k)$. There is one additional plane defined by the pair $P(e_i)$, $P(e_j)$; and the intersection of it with the first two produce two additional lines. If the relative probabilities are consistent (by the rule of the triangle and the rule of the product) the three lines will coincide. Figure 3 portrays an inconsistent case.

To present the general consistency condition for cross-impact matrices, it is convenient first to establish a lemma concerning lines in n-dimensional Euclidean spaces. A line is defined by two points $\mathbf{X} = (x_1, x_2, \ldots, x_n)$ and $\mathbf{Y} = (y_1, y_2, \ldots, y_n)$. Any other point Z on the line is a linear combination of these two; i.e., $\mathbf{Z} = \alpha\mathbf{X} + (1-\alpha)\mathbf{Y}, -\infty \leqslant \alpha \leqslant \infty$. It is convenient to shift the origin to $\mathbf{Y}$, in which case $\mathbf{Z} - \mathbf{Y} = \alpha(\mathbf{X} - \mathbf{Y})$. Renaming $\mathbf{Z} - \mathbf{Y} = \mathbf{Z}'$ and $\mathbf{X} - \mathbf{Y} = \mathbf{X}'$, we have $\mathbf{Z}' = \alpha\mathbf{X}'$.

LEMMA 1. *A necessary and sufficient condition that a matrix* $\mathbf{S} = \{s_{ij}\}$, *where* $s_{ji} = 1/s_{ij}$, $s_{ii} = 1$,[4] *define a line* $\alpha\mathbf{X}$ *through the origin in n-dimensional Euclidean space, such that the slope* x_i/x_j *of the line projected on the two-dimensional subspace* (i,j) *is* s_{ji}, *is the triangle rule*

$$s_{ij} \cdot s_{jk} \cdot s_{ki} = 1. \qquad (a)$$

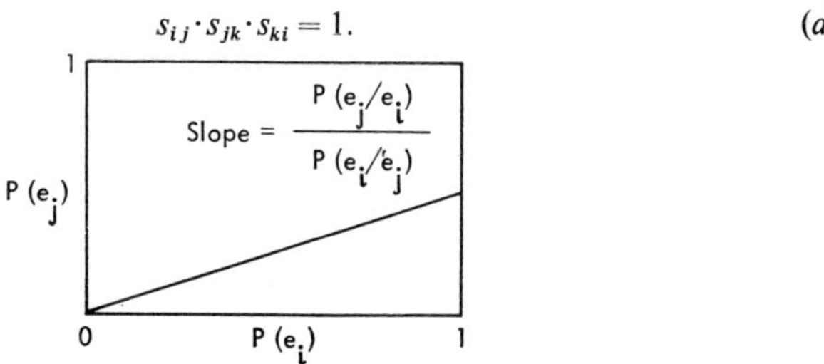

Fig. 2. Geometric representation of the rule of the product.

[4] To eliminate inessential special cases, s_{ij} is also assumed to be nonzero.

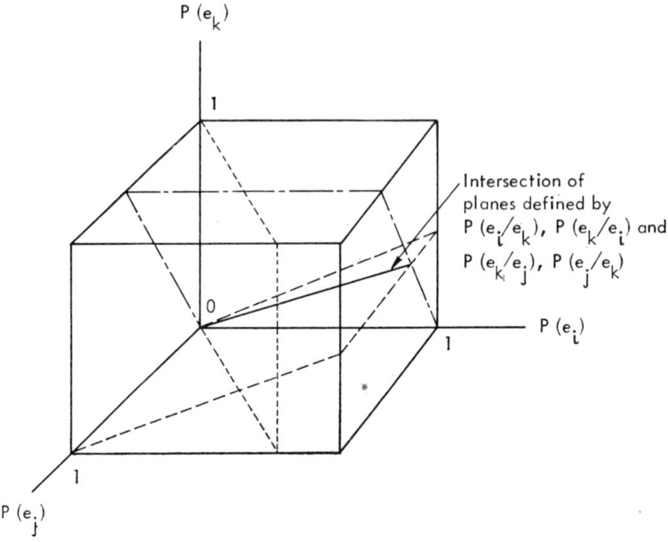

Fig. 3. Inconsistent relative probabilities for three events.

Proof. Necessity. Assume there is a line $\alpha\mathbf{X}$ with the hypothesized properties, then

$$\frac{x_j}{x_i}\cdot\frac{x_k}{x_j}\cdot\frac{x_i}{x_k}=1=s_{ij}\cdot s_{jk}\cdot s_{ki}.$$

Sufficiency. Consider a matrix $\mathbf{S}$ that fulfills the triangle rule. Define a point $\mathbf{X}$ by $x_i = s_{1j}$. The triangle rule of Eq. (*a*) implies that $s_{1j}\cdot s_{ji}\cdot s_{i1}=1$, where $s_{ji}=1/s_{1j}\cdot s_{i1}$, and since $1/s_{i1}=s_{1i}$,

$$s_{ji}=\frac{s_{1i}}{s_{1j}}=\frac{x_i}{x_j}. \tag{b}$$

Since for any other point $\mathbf{Y}$ on the line $\alpha\mathbf{X}$, $y_i/y_j=\beta x_i/\beta x_j = x_i/x_j$, Eq. (*b*) is completely general.—

Lemma 1 is applicable to a cross-impact matrix by defining a matrix $\mathbf{S}$, $s_{ij}=P(e_j/e_i)/P(e_i/e_j)$. The conditions $s_{ji}=1/s_{ij}$ and $s_{ii}=1$ follow immediately from the definition, and the triangle rule Eq. (*a*) follows from the rule of the triangle for relative probabilities.

The rule of addition can be invoked by determining the intersection of the lines defined by setting the inequalities to equalities in Eq. (2). Thus,

$$\left.\begin{aligned} P(e_i)+P(e_j)-P(e_i)\cdot P(e_j/e_i)&=1,\\ P(e_i)+P(e_j)-P(e_j)\cdot P(e_i/e_j)&=1. \end{aligned}\right\} \tag{4}$$

Solving these two for $P(e_i)$ and $P(e_j)$ gives

$$\left.\begin{aligned} P(e_i)&=\frac{P(e_i/e_j)}{P(e_i/e_j)+P(e_j/e_i)-P(e_i/e_j)\cdot P(e_j/e_i)},\\ P(e_j)&=\frac{P(e_j/e_i)}{P(e_i/e_j)+P(e_j/e_i)-P(e_i/e_j)\cdot P(e_j/e_i)}. \end{aligned}\right\} \tag{5}$$

It is simple to verify that the pair $P(e_i), P(e_j)$ lies on the line defined by Lemma 1.

The question still remains whether the limits imposed by different applications of Eq. (5) with different pairs result in the same limit. Thus, for example, we have

$$P(e_i) = \frac{P(e_i/e_k)}{P(e_i/e_k) + P(e_k/e_i) - P(e_i/e_k) \cdot P(e_k/e_i)}$$

In general, this is not the same limit as expressed by Eq. (5) above, thus the minimum of all the limits determined by all pairs must be selected. The minimum fixes a point **L** on the consistency line in the hypercube; any acceptable point is lower than or equal to that point.

This completes the set of consistency conditions. In sum, the consistency conditions define a line passing through the origin in the unit hypercube and a point on that line. To be consistent, the absolute probabilities must lie on the segment of that line between the origin and the given point.

The foregoing demonstrates the necessity, but not the sufficiency, of the conditions. A direct consequence of the rule of addition (P3) is that the probability of the disjunction of any subset of the absolute probabilities must be less than or equal to one. The disjunctive probabilities for subsets larger than two cannot be computed from the absolute probabilities and binary relative probabilities since the disjunctive probabilities of sets larger than two involve higher-order interactions.

III. Resolution of Inconsistencies

In previous sections we defined a set of conditions to be met for a set of absolute probabilities and a cross-impact matrix of relative probabilities to be consistent. The simplest application of the conditions for the matrix is to test all triplets of relative probabilities by the triangle rule. This test is tedious. There are $n(n-1)(n-2)/6$ triplets for n events. For 50 events there are 19,600 triplets.

However, it is necessary to check only a subset of the triplets, a convenient subset being the set of triangles having one event in common. The number of triangles in such a subset is $(n-1)(n-2)/2$. Thus, the number of independent triangles increases as the square of n, rather than as the cube which is the case for the total number of triangles. This happy situation is guaranteed by the following lemma.[5]

LEMMA 2. *If the rule of the triangle holds for all triangles that have one event in common, then it holds for all triangles.*

Proof. Consider any event e_l, and any triangle e_i, e_j, e_k, not including e_l. Assume that the rule of the triangle holds for all triangles containing e_l. Then, abbreviating $P(e_i/e_j)$ to $P(i/j)$ gives

$$P(l/i) \cdot P(i/j) \cdot P(j/l) = P(i/l) \cdot P(j/i) \cdot P(l/j),$$
$$P(k/l) \cdot P(j/k) \cdot P(l/j) = P(l/k) \cdot P(k/j) \cdot P(j/l),$$
$$P(i/l) \cdot P(k/i) \cdot P(l/k) = P(l/i) \cdot P(i/k) \cdot P(k/l).$$

Multiplying the three expressions on the left side of the equations and the three expressions on the right side, the terms containing l cancel and we arrive at

$$P(i/j) \cdot P(j/k) \cdot P(k/i) = P(j/i) \cdot P(k/j) \cdot P(i/k)$$

which was to be proved. —

[5] I would like to thank T. Brown and J. Spencer for helpful suggestions concerning this simplification of the consistency test.

For the absolute probabilities, assuming the matrix is consistent, the simplest test is first to calculate the limits imposed by the rule of addition. Assuming that test is passed, the individual probabilities can be tested by starting with $P(e_1)$ and computing the others by the rule $P(e_i) = P(e_1) \cdot P(e_i/e_1)$.

It appears likely that in most practical applications the sets of probabilities will not be consistent, and a question arises as to the method of proceeding. There is no "correct" method of resolving the inconsistencies. There are several directions that can be taken, depending on the interest of the study manager, on the availability of respondents for reestimation, and the like.

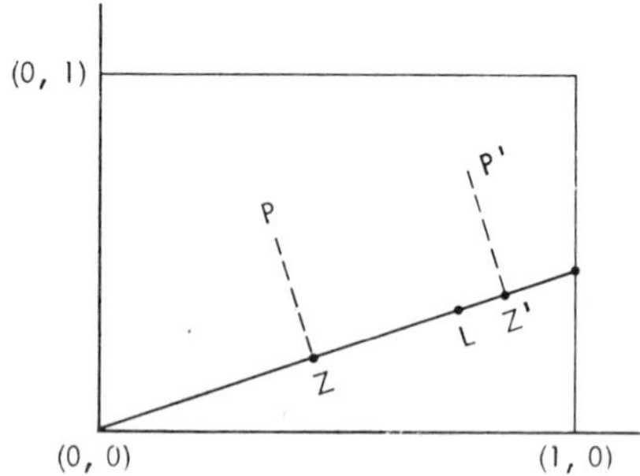

Fig. 4. Adjustment of absolute probabilities by computing the nearest point on the line defined by the cross-impact matrix.

Assuming no restrictions on reestimation, it appears desirable to present the information concerning inconsistencies to the respondents, and obtain reestimates from them. This poses the question of how the information is to be presented to be useful to the respondents.

One reasonable display can be obtained by computing the adjustments described below, and feeding back the original and adjusted estimates.

Assuming that the cross-impact matrix is consistent, but the set of absolute probabilities does not lie on the line, a natural and simple adjustment is to choose the point on the line that is nearest to the estimated point, as illustrated for two dimensions in Fig. 4. The computation of the nearest point is simple, given the line in parametric form. If **P** is the estimated point and $\alpha\mathbf{X}$ is the line, then the desired nearest point on the line **Z** is given by

$$\mathbf{Z} = \beta\mathbf{X}, \beta = \frac{\sum p_i x_i}{\sum x_i^2}. \tag{6}$$

If **Z** lies above the limit **L** imposed by the rule of addition (**Z**′ in Fig. 4), the most natural rule is to select **L** as the adjusted set of probabilities.

The situation is not quite as neat if the cross-impact matrix itself is inconsistent. The problem here is that inconsistencies result in a set of lines, which can become very large rapidly, and there is no obvious way of weighting these in the computation of a representative line. A relatively simple adjustment, and one that appears sufficient for the purpose—especially if the group is expected to reestimate—is the following

$$x_i = \frac{1}{n} \sum_j \frac{P(e_i/e_j)}{[\sum_k P(e_k/e_j)^2]^{1/2}}. \tag{7}$$

This computation arises from allowing each row in the matrix in turn to generate a potential line, and averaging these lines by summing their intersection with the unit sphere. Although Eq. (7) does not arise from any optimization rule, it does assure that in adjusting **X** the contributions of all other events to e_i will be "taken into account," and the contribution of e_i to all the other events enters in the normalization to the unit sphere.

A more satisfying variant of the procedure in Eq. (6) would be to make use of group weights on the probabilities. That is, each member of the group can be requested to evaluate how confident he is of his estimate (with some suitable rating scale) and a group measure of confidence—e.g., the average of the individual ratings—can be used to weight the absolute probability estimates.

If the weights w_i are normalized so that $0 \leqslant w_i \leqslant 1$, where 1 indicates certainty and 0 indicates sheer guess, then a reasonable weighted form of Eq. (6) would be

$$\beta = \frac{\sum p_i x_i w_i}{\sum x_i^2 w_i}. \tag{8}$$

A weighted adjustment for the relative probabilities is much more complex; a reasonable form has not yet been derived. One of the diffculties is that the consistency condition furnished by the rule of the triangle deals only with the ratios of the relative probabilities and not with the probabilities themselves. Ratios of estimated weights are somewhat awkward to use here.

IV. Scenarios

The cross-impact computation developed at the Institute for the Future not only produces a Monte Carlo estimate of revised absolute probabilities, but also gives a Monte Carlo estimate of the probabilities of joint occurrence of the total set of events. Each chain of events is a sample out of a large population of potential chains. The number of potential chains is $n!2^n$. The factor 2^n arises from the total number of joint occurrences (positive or negative) of n events, and the factor $n!$ from the manner in which the next event in the chain is selected, namely, by considering each remaining event equally likely to be next.

The computation involves a strong independence assumption, namely, that the change in the likelihood of occurrence of a given event is influenced only by the preceding event that has occurred. (In some forms of the computation the change is assessed in terms of occurrence or nonoccurrence of the preceding event.) In general probability systems, this assumption is not only not fulfilled; the assumption has as a consequence that the system is degenerate. A simple example may suffice to show this. If we make an assumption that all higher-order interactions are determined solely by the first-order interactions, then we would have

$$P(e_k/e_i \cdot e_j) = P(e_k/e_i) = P(e_k/e_j). \tag{9}$$

Applying the assumption to all interactions would imply that all the cross-impacts for a given target event are equal. The same result follows from other forms of the same assumption, for example,

$$P(e_k \cdot e_j/e_i) = P(e_k \cdot e_j), \tag{10}$$

$$\frac{P(e_i \cdot e_j \cdot e_k)}{P(e_i \cdot e_j \cdot \bar{e}_k)} = \frac{P(e_k)}{P(\bar{e}_k)}, \tag{11}$$

where the bar over an event indicates negation.

The Monte Carlo computation developed by Gordon *et al.* [1] for the likelihood method involves a slightly weaker assumption than Eq. (9). However, it is more complicated, and rather than try to deal with it analytically, a simple illustration will indicate the effect of the assumption. Consider the following cross-impact matrix:

	e_1	e_2	e_3
e_1	1	0.5	0.5
e_2	0.5	1	0.5
e_3	0.5	0.5	1

$$P(e_1) = P(e_2) = P(e_3) = 0.25.$$

Let

$$l_{ij} = \frac{1 - P(e_j)}{P(e_j)} \cdot \frac{P(e_j/e_i)}{1 - P(e_j/e_i)},$$

that is, l_{ij} is the factor by which the odds for e_j are multiplied to generate the odds for e_j, given that e_i has occurred. Then

$$l_{12} = \frac{0.75}{0.25} \cdot \frac{0.5}{0.5} = 3 = l_{ij} \text{ for all } i \neq j.$$

Each event has a positive impact on the others.

If the Monte Carlo computation is carried through, $P(e_i)$ is adjusted from 0.25 to 0.32. The computation suggests that in terms of the positive impacts the absolute probabilities should be higher. However, the original probabilities are fully consistent with the cross-impacts; the increase is, in this sense, unwarranted.[6]

For purposes of evaluating policies, the probabilities of the chains (where each chain can be considered one scenario) are of greater interest than the absolute probabilities of the separate events. We have indicated that to make the scenario computation useful it will be necessary either to obtain estimates of the higher-order interactions or find a more logically correct assumption concerning them. The first alternative is rather discouraging. The number of higher-order probabilities becomes somewhat astronomical with large sets of events. The second alternative—although more attractive in terms of the feasibility of obtaining estimates—is subject to the danger of artificiality.

Conclusion

This article presents an elementary approach to the estimation of complex probability systems. Above all, it contains no consideration of the structural properties of the subject matter involved. Generally speaking, it can be expected that a given application, whether it is technological events or health categories, will have certain underlying characteristics that will affect both the binary cross-impact probabilities and the higher-order interactions (e.g., with the health categories, it is very unlikely that any individual will simultaneously be afflicted with a large number of pathological conditions). Convenient methods of expressing these structural properties have not, to my

[6] Although I have not made the calculation, it appears likely that most of the adjustment in the cases presented in Ref. [1] are of this sort.

knowledge, been defined for the kinds of systems in which cross-impact analysis would appear to be applicable. My first impression is that such structural properties may furnish the required assumption concerning higher-order probabilities.

The resolution of inconsistencies in the cross-impact matrix expressed in Eq. (7) can probably be developed further and appears to be an interesting mathematical problem in its own right.

References

1 T. J. Gordon, R. Rochberg, and S. Enzer, *Research on Cross-Impact Techniques with Applications to Selected Problems in Economics, Political Science, and Technology Assessment*, The Institute for the Future, Report R-12 (August, 1970).

2 H. Reichenbach, *The Theory of Probability*, University of California Press, Berkeley and Los Angeles (1949).

V. C. An Alternative Approach to Cross Impact Analysis

MURRAY TUROFF

Abstract

This paper presents the theoretical justification for the use of a particular analytical relation for calculating inferences from answers to cross impact questions. The similarity of the results to other types of analogous applications (i.e., logic regression, logistic models, and the Fermi–Dirac distribution) is indicated.

An example of a cross impact analysis in an interactive computer mode is presented. Also discussed is the potential utilization of cross impact as: (1) A modeling tool for the analyst, (2) A consistency analysis tool for the decision maker, (3) A methodology for incorporating policy dependencies in large scale simulations, (4) A structured Delphi Conference for group analysis and discussion efforts and (5) A component of a lateral and adaptive management information system.

He took the wheel in a lashing roaring
hurricane
And by what compass did he steer the course
of the ship?
"My policy is to have no policy," he said in
the early months,
And three years later, "I have been controlled
by events."

"The People, Yes"
Carl Sandburg

Introduction

In Delphi[1] design, one of the major problems has been how to obtain meaningful, quantitative subjective estimates of the respondents' individual view of causal relationships among possible future events. Meaningful, in this context, is the ability to compare the quantitative estimates of one respondent with those of another respondent and correctly infer where they differ or agree about the amount of impact one event may have on another.

A number of design techniques, or question formats, have evolved as approaches to this problem. The particular formalism which has received comparatively wide usage, due to the ease of obtaining answers to a fairly involved problem, is the "cross impact" question format first proposed in a paper by Gordon and Hayward (see bibliography). However, the analytical treatments proposed as methods of either checking consistency or drawing inferences are essentially heuristic in nature and exhibit various difficulties. The Monte Carlo approach, which is in widest use, is particularly unsuited to obtaining a consistent set of estimates through individual modification. This is because the assumptions upon which the Gordon (Monte Carlo) approach is based imply inconsistency in the estimates provided. The analytical approach described in this paper was developed specifically for restructuring the cross

---

DR. TUROFF is associated with the Systems Evaluation Division of the National Resource Analysis Center, Office of Emergency Preparedness. Washington, D.C.

[1] See "The Design of a Policy Delphi" by Murray Turoff, in *Technological Forecasting and Social Change*, **2**, No. 2 (1970), for an explanation of the Delphi technique and a comprehensive bibliography.

Reprinted from *Technological Forecasting and Social Change*, 3 (1972) with permission of American Elsevier.

impact formalism in a manner suitable for use on an interactive computer terminal. This requires that the user be able to modify or iterate on his estimates until he feels the conclusions inferred from his estimates are consistent with his views.

The type of event one is usually considering in the cross impact formalism may not occur at all in the time interval under consideration. Furthermore, an event may be unique in that it can only happen once. Examples of the latter are:

The development of a *particular* new product
The occurrence of a *particular* scientific discovery
The passage of a *specific* piece of legislation
The outbreak of a *particular* war.

For this type of event there is usually no statistically significant history of occurrence which would allow the inference of a probability of occurrence. While there are sometimes historical trends for certain general items, such as the overall occurrence rate of scientific discoveries, specific scientific discoveries may fall outside this general trend. The probability of a cure for a particular type of cancer is one example. In this instance, one would make his estimate dependent upon a rather significant number of other events, involving such factors as the success of non-success of a large number of specific research projects that may provide information only on the nature of the disease and thereby influence in some manner the discovery of a cure. Also, of course, one would consider events related to the provision of funds necessary to start research projects in areas that should be, but currently are not, explored. This latter consideration may lead to the enumeration of additional events related to the economic and political environment determining the availability of funds.

It is quickly realized from the above example that the first step in the construction of a cross impact exercise is the problem of specifying the event set. At present, one workable and popular approach to this problem is to allow the individuals who will participate in the cross impact exercise to specify the set of events they feel are crucial to the problem under consideration. This process may be conducted in a face-to-face conference or committee approach, or in a Delphi exercise. The success of the exercises, in terms of specifying a good event set, is dependent upon the knowledge the group has about the problem, as is the value of the quantitative estimates that will be obtained. However, there are some situations where the formalism may be used as an educational tool on some groups to expose the complexities of the problem. This usually occurs when a non-expert group evaluates an event set generated by an expert group and the evaluation by the expert group is available for comparison.

Once an event set is specified, the first step is to ask a person what he estimates is the chance of that event occurring in the interval of time from now to some point in the future (i.e., ten years). An individual viewing an event dependent upon causal effects normally has a very discontinuous view of the event happening over time. If, for example, an event specified in the cross impact set is the expectation of receiving a raise in excess of a certain amount within the next year, then the individual concerned may feel that the event, from a time-dependent view, can only occur at certain sub-intervals of the year. The raise may normally only be possible at the completion of the mid-year and year-end reviews. To answer the question as stated, he must perform some averaging process taking into account the time dependence as well as causal effects from other potential occurrences.

Assume that one of the other events is the receipt of a letter from the head of the organization by his supervisor expressing a great deal of satisfaction with the results of the project this individual is currently undertaking. This can only occur after the project is completed and over that interval which is normal for review of the project. Let us hypothesize that the individual has a pessimistic estimate of this occurring. His view on these events and others in this particular event set represents his view or opinion of the world concerned with his getting a raise and is the first step in the cross impact procedure.

The next step in the cross-impact procedure is to perturb his view of the world (or to create a new world) by telling him to assume a certainty that one of the events will (or will not) occur and asking him to reconsider other events. In the example, assume that we tell him it is certain that the head of the organization will send a letter to his supervisor expressing satisfaction with the results of the project. This may cause him to re-evaluate upward his expectancy for getting the raise. More important for understanding the cross impact formalism, this may cause him to arrive at a completely new time dependency for the probability of getting a raise. In other words, a time interval during which he thought there was no probability of getting a raise because it occurred between the mid-term and year-end review could not become a very probable time interval for getting the raise because it occurs after the project is completed.

The important point to recognize now is that if we had extracted all the information contained in the time-dependent view of this event set, we could have used some of the standard relationships in probability theory to check the consistency of estimates at *each* point in time for *each* world view. This, as will become evident in the rest of the paper, would be an infeasible amount of information to ask an individual to provide for more than a few events. Rather, the cross-impact problem is to infer the causal relationships from some relationship among the different world views established by perturbing the participant's initial view with assumed certain knowledge as to the outcome of individual events. These different world views may represent, at least implicitly, different time-dependent distributions for the same event. Therefore, the probability estimates for the occurrence of each event, resulting from some subjective averaging procedure over the total time interval, do not conform to the definition of a unique probability space in the classical sense, to which the standard relationships between such concepts as prior and posterior probabilities may be applied. Rather, we are asking if there is some model or relationship, based upon causal effects, which can be used to relate a number of separate probability spaces.

We are faced with a situation analogous to some degree with the problem in quantum mechanics where, in order to measure the state of the system we must physically disturb it. In this case, in the process of setting up an instrument to measure the estimates of an individual's view of causal relationships we disturb those estimates. The concept of defining a measuring instrument is crucial, since in most cross impact exercises we wish to be able to compare estimates among different individuals. Unfortunately, unlike quantum mechanics, any analogy to a Planck's constant may differ from individual to individual and therefore we do not have available an analogous uncertainty relation.

The THEORY section of this paper attempts to describe and justify the analytical procedures for such a measurement device. It assumes that the reader has some familiarity with the literature on cross impact and the other analytical approaches which have been proposed for handling this problem. If this is not the case, the reader should perhaps read the EXAMPLE and APPLICATIONS section prior to reading other parts of this paper.

Theory

Structure of the Problem

Events to be utilized in a cross impact analysis are defined by two properties. One, they are expected to happen only once in the interval of time under consideration (i.e., nonrecurrent events) and two, they do not have to happen at all (i.e., transient events). If one holds to a classical "frequency" definition of probability than it is, of course, pointless to talk about the probability of a nonrecurrent event. We, therefore, assume an acceptance of the concept of a subjective probability estimate having meaning for nonrecurrent events. When dealing with recurrent events within the cross-impact framework, they should be restated as nonrecurrent events by either specifying an *exact* number of occurrences within the time interval or utilizing phrases such as "... will happen at least once." Any recurrent event may be restated as a set of nonrecurrent events.

If we are considering N nonrecurrent events in the cross impact exercise there are then 2^N distinct outcomes spanning the range from the state where none of the events have occurred to the state where all of them have occurred. If we are in a state where a particular set of K of the events have occurred, then there are at most $N - K$ remaining possible transitions to those states where $K + 1$ events have occurred. Since it is possible that no additional event will occur, the sum over these $N - K$ transition probabilities need not add to one. The amount by which the sum is less than one is just the probability that the system remains in that particular state until the end of the time interval. Once the system has moved out of a particular state, it will never return to it since each event is assumed to occur once and only once. The total number of possible transition paths and equivalent transitions probabilities (allowable paths) needed to specify this system of 2^N states is $N2^{N-1}$.[2] An example of all states and transitions for a three event set is diagrammed below, where a zero denotes nonoccurrence and a one denotes occurrence of the event. The events are, of course, distinguishable and it is also assumed two events do not occur simultaneously.

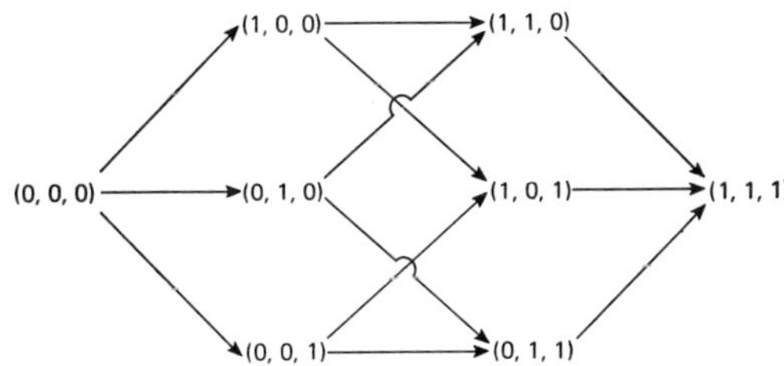

Transitions and states for three events (1, 2, 3).

One can see that as N gets larger than three it quickly becomes infeasible to ask an individual to supply estimates for all the transition probabilities. The cross impact formalism, as an alternative, has had widespread usage because: one, it limits itself to N^2 questions[3] for N events; and two, the type of question asked appears to parallel the intuitive reasoning by which many individuals view "causal" relationships among

[2] Assuming the system transition probabilities are independent of past history. The history or memory dependent case is discussed later.

[3] $N2^{N-1}$ is the number of questions one would have to ask to obtain quantitative estimates to completely specify the model.

events. However, it does pose a serious theoretical difficulty for extracting or inferring conclusions based upon the estimates supplied, since the answers supplied are both insufficient and different information from that required to completely specify the situation. This is easily seen by relating the answers to the cross impact questions in terms of the original transition probabilities. The first cross impact question which is asked for all N events ($i = 1$ to N) is:

(1) "What is the probability that an event, i, occurs before some specified future point in time?"

The answer to this can be related to the appropriate transition probability sum taken over all independent paths leading to all states in which event i occurs (i.e., one-half of the states). However, when the second cross impact question is asked for the remaining $(N-1)$ events relative to a j-th event:

(2) "What is your answer to question (1) if you assume that it is certain to all concerned that event j will occur before the specified point in time?"

we have in effect altered the original set of transition probabilities. This latter question is equivalent to imposing a set of constraints upon the transition probability estimates of the following form:

The sum over all the transition probabilities leaving a state in which the j-th event has not occurred must be equal to one.

The above must be the case since we cannot remain or terminate in a state for which the event j has not occurred. What we have done, at least subconsciously, to the estimator is to ask him, in the light of new constraints, to create a whole new set of prior transition probabilities. This creates an analytical problem in trying to relate the original transition probabilities estimated under the constraints. One may consider this as a problem in trying to relate different "world" views.

The so-called "conditional" probabilities derived from the second "cross impact" question are not the conditional probabilities defined in formal probability theory. Rather the answer to the second cross impact question might better be termed as a "causal" probability from which one would like to derive a "correlation coefficient" which provides a relative measure of the degree of causal impact one event has upon another. However, the term "conditional probability" has become so common in a lay sense that it is often easier to communicate and obtain estimates by referring to the answers to the second cross impact question as "conditional" probabilities.

The previous points may be illustrated by the following examples where the reasonable answers, to the cross impact questions, do not obey the mathematical requirements associated with standard conditionals or posterior probabilities. The first example is a "real" illustration and the second is an abstract urn representation of what is taking place. Consider the following two potential events:

Event 1

Congress passes a strict and severe law specifically restricting mercury pollution by 1975.

Assume the probability estimate of occurrence is e_1:

$$P(1) = e_1.$$

Event 2

At least 5,000 deaths are directly attributed to mercury pollution by 1975.

Assume the probability estimate of occurrence is e_2:

$$P(2) = e_2.$$

If it is certain that Congress will pass the above law by 1975, either Event 2 is not affected or its probability may decrease if the law is enacted soon enough to reduce levels of pollution before 1975. Therefore, the probability of Event 2 given that Event 1 is certain should be less than or equal to the original estimate,

$$P(2:1) = e_3 \leqslant e_2.$$

If it is certain that five-thousand people or more will die (Event 2) by 1975, then most rational estimators will increase their estimate of the probability for Congress passing the law,

$$P(1:2) = e_1 + \Delta \quad \text{where} \quad \Delta > 0 \text{ and } e_1 + \Delta \leqslant 1.$$

If $P(2:1)$ and $P(1:2)$ were standard conditionals (i.e., posterior probabilities), we would conclude that the probability of both occurring is

$$P(1,2) = P(1:2)P(2) = P(2:1)P(1).$$

However,

$$P(1:2)P(2) = e_1 e_2 + \Delta e_2,$$
$$P(2:1)P(1) = e_3 e_1 \leqslant e_2 e_1 < P(1:2)P(2).$$

Therefore, $P(1:2)$ and $P(2:1)$ are not the standard conditional probabilities.

A valid theoretical point may be made by arguing that the above problem would be eliminated by designing an event set consisting of mutually exclusive events as a basis vector from which a decision tree or table can be constructed and to which can be applied a Bayesian type analysis. However, in practice, economic, political, and sociological types of questions, often examined in the cross impact scheme, do not lend themselves to defining such a set, and, if they do, the number of events which have to be considered may grow too large for the purpose of obtaining estimates.

As the second example consider two urns, labeled urn one and urn two, in which are distributed a large number of black and white balls. An individual who is to estimate the chance of drawing a white ball from either urn has available two pieces of information:

(1) Two-thirds of the balls are white.

(2) Urn number two always contains at least one-quarter of the balls.

Given no other information, if the estimator were asked the probability of drawing a white ball from urn two (or urn one) his best estimate is two-thirds.

Now the estimator is supplied with a new piece of information:

(3) The probability of drawing a white ball from urn two is zero.

Then the estimator can infer from (2) and (3) that at least one-quarter of the balls, all of them black, are in urn two. From this and (1) the probability of drawing white ball from urn one lies between one (assuming all black balls are in urn two) and eight-ninths.[4] Assuming the distribution of probabilities between one and eight-ninths is uniform (no other information) the best estimate for the probability of drawing a white ball from urn one is half way between one and eight-ninths, or 17/18.

Suppose, on the other hand, instead of item three we supply the estimator with the following information:

(4) The probability of drawing a white ball from urn one is zero.

Now he knows that between none and all the black balls can be in urn one. This means the probability of drawing a white ball from urn two is between two-thirds and one. The midpoint estimate in this case is five-sixths.

These resulting four estimates are summarized in the following table:

Probability of drawing a white ball from	based upon (1) and (2) only	given (3) in addition to (1) and (2)	given (4) in addition to (1) and (2)
urn one	2/3	17/18	0
urn two	2/3	0	5/6

Note that, in this example, we have never drawn a ball from urn one or two; therefore, there is no posterior probability provided. The two probabilities calculated by assuming the information items (3) and (4) are new priors based upon assumed knowledge as to the state of the system.

The cross impact analysis problem in terms of the example is: Given the four estimates made in the table to what extent can the information items (1) and (2) be inferred analytically. In other words, will the relationships derived from the estimates provide a description of this system which behaves similarly or approximately like the system described by knowing explicitly items (1) and (2). The goal of the cross impact in this example would be to infer a model, from the four estimates provided, which will allow a prediction of the probability estimate of drawing a white ball from urn one if the estimator is given explicitly the probability of drawing a white ball from urn two, or vice versa. In essence, we wish to create an analytical model of his knowledge about the situation.

Another view of cross impact is to consider it as an attempt to obtain subjective estimates of correlation coefficients. Gordon's approach to this problem asks directly for these coefficients, while the approach in this paper is to ask for probabilities from which the correlation coefficients can be calculated. The transition from formal probabilities to subjective probabilities, or likelihood estimates, is not difficult to make. However, the formal theory of correlation coefficients in statistics does not specify a unique analytical definition of a correlation coefficient in the same sense that a unique measure of probability is defined. Therefore, the problem of defining subjective estimates of correlation coefficients to measure causal impact (whether direct, as in the Gordon approach, or indirect, as this approach) rests on a more intuitive foundation than does the concept of subjective probability.

The separate justifications presented on the following pages for a particular approach to the cross impact problem are all heuristic in nature. Since we are trying with N^2

[4] Assuming urn two contains only $\frac{1}{4}$ of all the balls, this leaves $\frac{1}{12} = (\frac{1}{3} - \frac{1}{4})$ of the black balls in urn one. Then $\frac{8}{9} = (\frac{2}{3})/(\frac{2}{3} + \frac{1}{12})$.

items of information to analyze a problem requiring $N2^{N-1}$ items of information for a complete solution, it would, therefore, seem that any approach to the analysis of the problem is an approximation. Also, there does not appear to be any explicit test which will judge one approach to be better than another. One significant measure of utility is the ease with which estimators can supply estimates and whether they feel the consequences inferred by the approach from their estimates adequately represent their view of the world.[5] The author feels that the method developed in this paper offers the estimator a greater opportunity to arrive at a consistent set of estimates and inferences than is available to him in the techniques currently reported in the literature on cross impact. The mathematical relationship developed is not new; it has been used in physics, statistics, operations research, and information theory in modeling situations where one is concerned with the probability of the outcome of random variables which can only take on zero or one values (see annotated bibliography) However, in many such cases one is dealing with a recurrent process where the model can be experimentally verified in at least some sense.

It should be noted that the state transition model of the interaction among events, which we have adopted to illustrate conceptually the meaning of the cross impact questions, provides only a lower bound ($N2^{N-1}$) on the number of parameters needed to completely specify the problem. It inherently assumes that once a given state (i.e., a specified set of events has occurred) is attained, the determination of the transition probabilities that leave that state (going to a state where one more event has occurred) is independent of the path used to move from the zero state (i.e., no events have occurred) to the state under examination (i.e., a system without memory[6]). If we had assumed the possibility of a completely different set of transition probabilities out of the given state, each set dependent upon the path that might have been used to arrive at the state, then the number of transition probabilities needed to completely specify the total problem would be:

$$\sum_{j=1}^{N} j! \binom{N}{j} \cong eN! \quad \text{while} \quad N \to \infty.$$

The real world, as modeled by a particular event set, is probably a mix of memory and non-memory dependent situations. Therefore the number of parameters one would theoretically require to specify all the information falls between the two limits. The following table contrasts the data demands of the cross impact analysis with the memory and no memory limits.

Number of Events	Cross Impact	No Memory System	Memory System
N	N^2	$N2^{N-1}$	$\approx eN!$
2	4	4	4
3	9	12	15
4	16	32	64
5	25	80	325
10	100	5120	$\approx$ 10 million

[5] This is not to say the estimator's view of the world may not be wrong, but that it may be overly presumptuous to expect the model to be able to correct the estimator's view. This is contrary to some cross impact approaches.

[6] Implies the system can be modeled as a Markov Chain.

Since most cross impact exercises deal with a range of 10 to 100 events, it is fairly obvious why no attempt is made to obtain estimates which would completely specify the problem.

The basic interpretation of cross impact conditionals as a new set of prior probabilities is not affected by the issue of whether or not one is dealing with a system that has a memory. This issue does arise, however, when one tries to describe or model the question of time dependence. This subject has not been adequately addressed in reported attempts to modify the cross impact formalism to allow variation of the time interval for the purpose of arriving at an explicit time dependent model.

In summary then, the cross impact approach in its most general context is an attempt to arrive at meaningful analyses of a system composed of transient, nonrecurrent events which may or may not be dependent upon history (i.e., memory). It is, however, fairly obvious that with event sets of the order of ten in size we have arrived at a point where it is desirable to find some sort of macro or statistical view of the problem as opposed to any attempt at enumerating all micro relationships such as the transition probabilities for all paths. This is analogous to the choice of trying to write dynamic equations for each particle in a gas or to utilize a set of relations governing the collective behavior of the gas.

The following five sections contain a number of alternative methods for arriving at the mathematical relationship used to model the cross impact problem. The explicit use of the resulting relationship for obtaining estimates from an individual is described in the EXAMPLE section of the paper. All the derivations provided are heuristic; however, the last two represent a fairly formal approach and provide some insight to the exact nature of the approximations being made.

Difference Equation

Given a set of events which may or may not occur over an interval of time, we assume that an individual asked to estimate the probability of the occurrence of each event will supply a "consistent" set of estimates. In other words, his estimate for the probability of the i-th event (out of a set of N events) includes a subjective assessment of the other events in terms of their probability of occurrence over the time frame and any "causal" relationships they may have upon one another. Under this assumption we may hypothesize that there exists a set of N equations expressing each of the probabilities (P_i for $i = 1$ to N) as a function of the other $N-1$ probabilities:

$$P_i = P_i(P_1, P_2, \ldots, P_k, \ldots, P_N) \quad \text{for} \quad k \neq i. \tag{1}$$

The above functional may also include other variables expressing the causal impact of potential events not specified in the specific set of N events.

If the individual making the estimate receives new information which would require a change in his estimates for any of the probabilities, then his changes should be consistent with the difference equation form of (1):

$$\delta P_i = \sum_{k \neq i} \frac{\partial P_i}{\partial P_k} \delta P_k + \frac{\partial P_i}{\partial \beta} \delta \beta, \tag{2}$$

where β is considered to be a collective measure of the impact of those events not included in the specified set. This will become clearer in later sections.

The boundary condition that must be satisfied by each of these equations is that if an event is certain to occur or certain to not occur then no change in the environment (as represented by the other events) can influence the outcome of the "certain" event, i.e., if

$$P_i = 1 \quad \text{or} \quad 0, \qquad \text{then} \quad \delta P_i = 0. \tag{3}$$

The simplest (in an algebraic sense) manner in which this can be satisfied is to assume

$$\frac{\partial P_i}{\partial X} = P_i(1 - P_i)\frac{\partial G_i}{\partial X}, \tag{4}$$

where X is any of the variables of differentiation on the right side of (2) and G is an arbitrary function. Therefore we may rewrite (2) as:

$$\delta P_i = P_i(1 - P_i)\left[\sum_{k \neq i} \frac{\partial G_i}{\partial P_k} \delta P_k + \frac{\partial G_i}{\partial \beta} \delta \beta\right]. \tag{5}$$

The next assumption is to consider the partial derivatives with respect to the P_k's as constants:

$$\frac{\partial G_i}{\partial P_k} = C_{ik}. \tag{6}$$

The whole string of assumptions to this point is based upon an appeal to simplicity.[7] We may now solve (5) as a differential equation to obtain

$$P_i = \frac{1}{1 + \exp\left(-\gamma_i - \sum_{k \neq i} C_{ik} P_k\right)}, \tag{7}$$

where the γ_i may be a function of unknown variables 'β' and also incorporates a constant of integration. One may easily verify that Eq. (7) is a solution to (5) by taking the total derivative.

This equation is recognizable as either the logistic equation which is often encountered in operations research or as a Fermi-Dirac distribution in physics. The implications of this will be discussed in later sections of this paper. The major difference in the assumptions leading to this result, as opposed to the Monte Carlo treatment of the cross impact problem developed by Gordon and others, is the crucial assumption that the hypothetical estimator of the occurrence probabilities is consistent in his estimates. In practice, an individual asked to estimate a significant number of related quantitative parameters is unlikely to be consistent on the first attempt. There must therefore be a feedback process for the individual in order to allow him to arrive at what can be viewed as a consistent set of values. In the Monte Carlo approach it is impossible for the individual to reasonably determine from the results of the calculations whether an inconsistent outcome (with his view) is merely a problem in his juggling of a large set of numbers or a basic inconsistency in his view of causal relationships. The primary advantage of Eq. (7) therefore is to provide an explicit functional relationship which presupposes consistency and thereby provides the estimator the

[7] There is no merit in attempting complex models for processes until the limits of validity for the simplest models are understood. Some of these limits will be discussed in a later section of the paper.

opportunity to arrive at consistency if he is provided with adequate feedback and opportunity to modify his estimates.

Underlying this view is the premise that the estimator would have a computer terminal available to exhibit the consequences of his estimates in terms of perturbations about the solution he initially provided. This then allows the estimator to determine if the resulting model adequately reflects his world view and to adjust his inputs accordingly. The lack of the ability for the individual estimator to first establish consistency for his own estimates is a major shortcoming in the current attempts to average in some manner the estimates of a group, as normally takes place in the cross impact Delphi exercises.

Likelihood Measure

Consider the following three measures which may be applied to the question of expressing the likelihood of the occurrence of a particular event (i.e., the i-th).

Probability: P_i,
Odds: $O_i = P_i/(1 - P_i)$, and
Occurrence ratio: $\phi_i = \phi(P_i) = \ln O_i = \ln [P_i/(1 - P_i)]$.

The boundary properties of these are summarized as follows:

	P_i	O_i	ϕ_i
Event certain to occur	1	∞	∞
Random occurrence (i.e., neutral point)	1/2	1	0
Event certain to not occur	0	0	$-\infty$

All the above measures are to be found in the literature of statistics. The "occurrence ratio" is commonly referred to as the "weight of evidence" when applied to two different, but mutually exclusive, events. It has the interesting property of being anti-symmetric about the neutral occurrence point. In other words, given two estimates of the occurrence: P_i and $P_i{}^*$ then if

$$P_i + P_i{}^* = 1 \quad \text{or} \quad (P_i - \tfrac{1}{2}) = -(P_i{}^* - \tfrac{1}{2}), \tag{8}$$

we have

$$\phi(P_i) = -\phi_i(P_i{}^*). \tag{9}$$

If we have a causal view with respect to the occurrence of an event, then we assume that the occurrence can be influenced by some policy, investment, or other type of "effort" directed at enhancing or inhibiting it. We would like, therefore, to establish some relationship between likelihood for the occurrence of the event and the effort invested in either promoting or preventing it (i.e., a negative or positive effort). We would also like this relationship to be such that if an equal amount of effort is devoted to both enhancing and preventing the occurrence of the event then the likelihood corresponds to a probability of one-half (i.e., random or neutral). In terms of the measures of likelihood commonly employed, the only one which might be assumed to be directly proportional to a measure of effort is the occurrence ratio. Therefore we assume:

$$\phi(P_i) \propto E_i{}^\tau. \tag{10}$$

Where E_i^T is the sum of all the effort invested in either bringing about the event (positive effort) or preventing it from occurring (negative effort).

Effort is the type of quantity which might be measured by the actual dollars invested in the goal. However, in many interesting cases we cannot model or measure the effort directly. We must, therefore, establish an empirical or indirect measure of effort. This can be done by assuming that the effort is measured by relating it to all other events which have a causal relationship to the i-th event:

$$E_i^T = \sum_{k \neq i} [D_{ik} P_k + F_{ik}(1 - P_k)]. \tag{11}$$

We may rewrite this, to correspond to our earlier notation, as

$$E_i^T = \gamma_i + \sum_{k \neq i}^{N} C_{ik} P_k, \tag{12}$$

where the γ_i may also include the events which have already been determined with respect to their occurrence of nonoccurrence as well as the events we are not specifying in the set of N events. We then have

$$\phi(P_i) = \ln [P_i/(1 - P_i)] = \gamma_i + \sum_{k \neq i}^{N} C_{ik} P_k. \tag{13}$$

This is the same result we arrived at earlier in Eq. (7). We note that while the contribution of the k-th event to the i-th event is additive in terms of the occurrence ratio, it is multiplicative with respect to the odds:

$$O_i \propto \prod_{k \neq i} O_{ik}, \tag{14}$$

where

$$O_{ik} = \exp(C_{ik} P_k).$$

Therefore any change in the probability of one of the events affecting event i changes the odds multiplicatively; i.e.,

$$O_{ij}(P_j + \delta P_j) = O_{ij}(P_j) O_{ij}(\delta P_j). \tag{15}$$

It should be observed that the conclusions expressed by Eqs. (14) and (15) could have been used as initial assumptions in deriving the cross impact relationship represented by Eq. (13). We also note a functional analogy between the odds in this problem and the partition function in quantum statistical mechanics.

Another aspect of relations (14) and (15) is that they satisfy a likelihood viewpoint of statistical inference in that the final odds may be written as the product of the initial odds times a "likelihood ratio."

Useful Relations

It is useful, at this point, to introduce some relationships involving the occurrence ratio which are needed to actually apply the results to obtaining estimates. If we assume an event (the j-th) becomes certain to occur, then we may define

$$R_{ij} = P_i \quad \text{for} \quad P_j = 1, \tag{16}$$

which is equivalently

$$\ln \frac{R_{ij}}{1-R_{ij}} = \gamma_i + \sum_{k \neq i, j} C_{ik} P_k + C_{ij}. \tag{17}$$

Subtracting Eq. (17) from Eq. (13) we have

$$C_{ij}(1-P_j) = \ln\left[R_{ij}(1-P_i)/P_i(1-R_{ij})\right] \tag{18}$$

or

$$C_{ij} = \frac{1}{1-P_j}\left[\phi(R_{ij}) - \phi(P_i)\right]. \tag{19}$$

Therefore, if we know P_i, P_j, and R_{ij} we may calculate C_{ij}. We note $C_{ij}=0$ if $P_i = R_{ij}$.
Similarly if we assume an event j becomes certain to not occur we may define

$$S_{ij} = P_i \quad \text{for} \quad P_j = 0. \tag{20}$$

Applying the same technique we obtain

$$C_{ij} = \frac{1}{P_j}\left[\phi(P_i) - \phi(S_{ij})\right]. \tag{21}$$

Therefore, if know P_i, P_j, and S_{ij} we may also calculate C_{ij}. Or by combining Eqs. (19) and (21) we have

$$C_{ij} = \phi(R_{ij}) - \phi(S_{ij}) \tag{22}$$

which may be used to calculate **S** or **R** given **C** and either **R** or **S** respectively.
If we have obtained values for all the C's then we can calculate γ_i by

$$\gamma_i = \phi(P_i) - \sum_{k \neq i}^{N} C_{ik} P_k. \tag{23}$$

This is in essence the normalization condition.

Eliminating C_{ij} from Eqs. (19) and (21) we have the following interesting relationship between **P**, **R**, and **S**:

$$\phi(P_i) = \phi(R_{ij}) P_j + \phi(S_{ij})(1-P_j). \tag{24}$$

This is plotted for some representative values on the following graphs. One may consider this last equation as a utility function relationship. In this instance we are not considering the utility of an event in terms of some winnings. Rather we are asking what is the utility of the j-th event to the occurrence of the i-th event. The occurrence ratio for the i-th event satisfies all the necessary properties of a utility function. One could have derived the cross impact relations by assuming the above utility relation and the condition that

$$\partial\phi(P_i)/\partial P_j = P_i(1-P_i)\, C_{ij} \tag{25}$$

in order to satisfy the boundary condition that the event j can have no utility for the event i when event i is already certain to occur or not occur. The C's, therefore, may be interpreted as *marginal utility factors* relating the utility of the j-th event to the i-th event. In a sense then, an alternative view of this cross impact approach is an assumption of a constant normalized marginal utility of one event for another.

Maximizing Information Added

Assuming we know the probability (P_i) that the i-th event will occur over some time frame, we wish to obtain two other probability estimates:

R_{ij} The probability of the i-th event, given the j-th event is certain to occur.

S_{ij} The probability of the i-th event, given the j-th event is certain to not occur.

The added information, over and above knowing P_i is defined as

$$I(i\,|\,j) = R_{ij}\ln(R_{ij}/P_i) + S_{ij}\ln(S_{ij}/P_i) + (1 - R_{ij})\ln\frac{1 - R_{ij}}{1 - P_i} + (1 - S_{ij})\ln\frac{1 - S_{ij}}{1 - P_i}. \tag{26}$$

It should be noted that the nonoccurrence of the event i is also considered significant information, hence the last two terms in the above equation. We also see

$$I(i\,|\,j) = 0 \quad \text{if} \quad S_{ij} = R_{ij} = P_i, \tag{27}$$

i.e., no added information.

We assume that if the values of R_{ij} and S_{ij} are correlated in any manner, then the correlation is such as to maximize the added information

$$\frac{\partial I(i\,|\,j)}{\partial R_{ij}}\,dR_{ij} + \frac{\partial I(i\,|\,j)}{\partial S_{ij}}\,dS_{ij} = 0, \tag{28}$$

which results in

$$dR_{ij}\ln[R_{ij}/(1 - R_{ij})] + dS_{ij}\ln[S_{ij}/(1 - S_{ij})] = (dR_{ij} + dS_{ij})\ln[P_i/(1 - P_i)]. \tag{29}$$

Recalling that the occurrence ratio is

$$\phi(x) = \ln[x/(1 - x)],$$

we may rewrite Eq. (29) as

$$\frac{dR_{ij}}{dS_{ij}}\phi(R_{ij}) + \phi(S_{ij}) = \left(\frac{dR_{ij}}{dS_{ij}} + 1\right)\phi(P_i) \tag{30}$$

or

$$\frac{dR_{ij}}{dS_{ij}} = \frac{\phi(P_i) - \phi(S_{ij})}{\phi(R_{ij}) - \phi(P_i)}. \tag{31}$$

The latter form indicates that $d\mathbf{R}/d\mathbf{S}$ must always be positive since if $P_i > \mathbf{R}$ then $\mathbf{S} > P_i$ or vice versa respectively.

The necessary assumption to obtain our earlier results is that

$$\frac{dR_{ij}}{dS_{ij}} = \frac{P_j}{1 - P_j} = \text{Odds}(j). \tag{32}$$

This behaves physically as one would desire, for if P_j is close to one, then a very large change in **R** is necessary to make a small change in **S**. Conversely if P_j is close to zero, a very large change in **S** is necessary to produce a small change in **R**. Also when $P_j = \frac{1}{2}$ the relative change in **R** and **S** is equal.

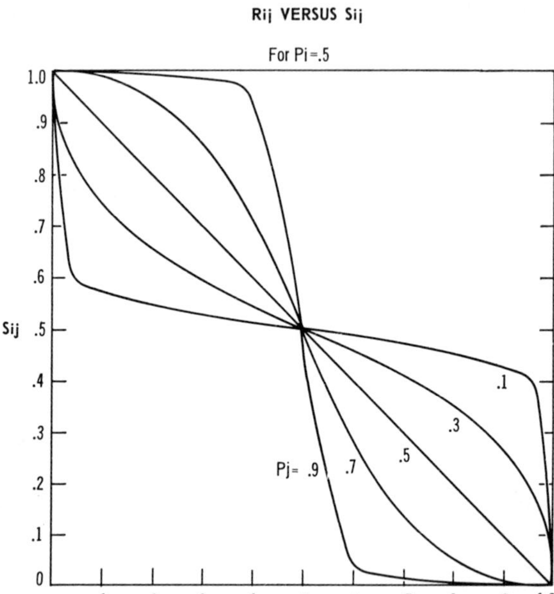
Rij VERSUS Sij
For Pi=.5
1.0
.9
.8
.7
.6
Sij .5
.4
.3
.2
.1
0
.1
.2
.3
.4
.5
.6
.7
.8
.9
1.0
Rij
Pj= .9
.7
.5
.3
.1

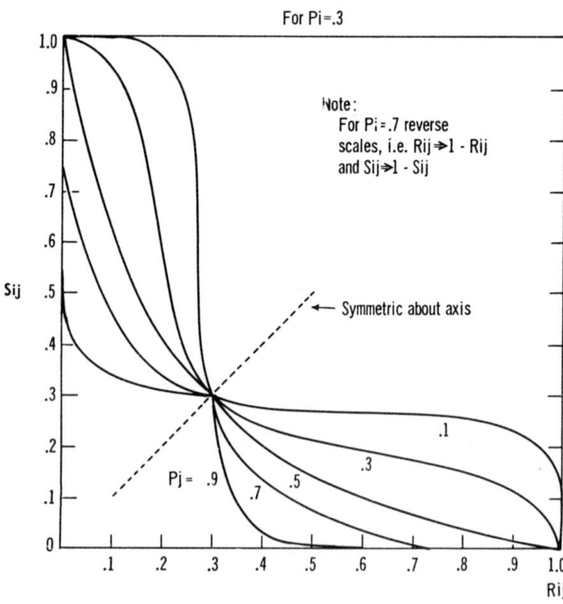
For Pi=.3
1.0
.9
.8
.7
.6
Sij .5
.4
.3
.2
.1
0
.1
.2
.3
.4
.5
.6
.7
.8
.9
1.0
Rij
Note:
For Pi=.7 reverse
scales, i.e. Rij⇛1 - Rij
and Sij⇛1 - Sij
Symmetric about axis
Pj= .9
.7
.5
.3
.1

This behavior is summarized in the following table where ϵ is a quantity close to zero and Eqs. (31) and (32) are linearized in ϵ.

If $P_j =$	ϵ	1/2	$1-\epsilon$
then $dR_{ij}/dS_{ij} =$	ϵ	1	$1/\epsilon$
and $P_i =$	$\frac{\mathbf{R}\epsilon + \mathbf{S}}{1+\epsilon} \cong S$	$\frac{\mathbf{R}+\mathbf{S}}{2}$	$\frac{\mathbf{R} + \epsilon\mathbf{S}}{1+\epsilon} \cong R$

So that as P_j approaches zero, P_i approaches S_{ij} and as P_j approaches 1, P_i approaches R_{ij}.

Substituting Eq. (32) into Eq. (30) or (31) we have

$$\phi(P_i) = \phi(R_{ij})P_j + \phi(S_{ij})(1 - P_j), \tag{33}$$

which is the earlier result, Eq. (24).

While this derivation is no less heuristic in nature than the previous sections, it does provide a fairly explicit statement of only two assumptions, i.e., Eqs. (28) and (32), necessary to obtain the cross impact relations developed in this paper.

A Mixed Statistical Mechanics and Information Theory Approach

Consider *all* events that may occur at some time in the future. We assume that each event may be described in such a manner that it is possible to evaluate at some future time the question of whether or not the event has occurred. This set of events in effect represents a state vector to define the "world" state of the system under observation. We may, in fact, explicitly define the state of this system as a binary message composed of one binary bit for each event, where the location (i.e., the i-th position in the message) corresponds to a particular event (i.e., the i-th event). A zero bit in the i-th position will indicate that the event has not occurred and a one bit will indicate that is has occurred. At the present time the message contains all zeros since we are referring to events that have not yet occurred.

We may further assume that there exists a set of prior probabilities (P_i) for the event set which indicates the likelihood of finding a one in each event position when we read the "message" at some specified future time. These probabilities are therefore an implicit function of the time interval which begins when we evaluate the values of the probabilities and ends when we plan to observe the content of the message.

As a result of the above conceptual model for the potential occurrence of events, we may write an expression for the information we know at the beginning of the time interval with respect to the content of our world message at the end of the time interval:

$$I = + \sum_i [P_i \ln P_i + (1 - P_i) \ln(1 - P_i)]. \tag{34}$$

The form of the above expression is based upon the fact that the event not occurring, as well as the observation that the event occurs, provides information. This expression has a minimum when all the P's are equal to one-half, corresponding to a completely random chance of occurrence of the events over the specified time interval—in other words, a complete lack of knowledge (i.e., a neutral position) about the likelihood of occurrence. The maximum occurs when all the P's are either zero or one which implies complete certainty as to the occurrence or nonoccurrence of the events.

The basic goal of the cross impact technique is to set up a measuring system whereby an individual's knowledge concerning a set of events can be quantified for the purpose of making a meaningful comparison among a set of individual estimates and collating the estimates into a group assessment. There are two aspects of this information or knowledge which are explicitly sought:

1. The prior probabilities of the events occurring given the world as the individual views it at the time.
2. The causal relationships, if any, whereby events may influence the occurrence or nonoccurrence of others.

In order to obtain a measure of this second phenomenon, we now take an approach analogous to a weakly interacting subsystems assumption in statistical mechanics.

For a set of N events, there are 2^N different outcomes in terms of the total message containing zero or one for each event. If the events are independent, the probability of receiving any particular message is

$$\prod_{l \in S} P_l \prod_{m \notin S} (1 - P_m), \tag{35}$$

where the index l ranges over those events which occur (subset S) in the message and m ranges over those events which do not occur (not in S). The sum of these probabilities over all the 2^N messages or possible world states is one.

Since our events are not necessarily independent and certain messages may be more or less likely than the quantity implied in Eq. (35), we introduce a set of statistical weights (W_k)[8] and define the probability of obtaining the k-th outcome of the 2^N as

$$\mathbb{P}(k) = W_k \prod_{l \in S} P_l \prod_{m \notin S} (1 - P_m). \tag{36}$$

If we could specify the actual physical interaction process between these events, then the W_k's would be obtainable from the analytical model of the process as is typically done in statistical mechanics. In our case they have to be viewed as quantities which can usually only be obtained by subjective estimates. It is still true, however, that the $\mathbb{P}(k)$'s must satisfy

$$\sum_{k=1}^{2^N} \mathbb{P}(k) = 1. \tag{37}$$

We now rewrite Eq. (37) utilizing (36) and a new set of 2^N constants (C's) as:

$$C^0 + \sum_{j=1}^{N} C_j^1 P_j + \sum_i \sum_{j>i} C_{ij}^2 P_i P_j + \sum_i \sum_{j>i} \sum_{k>j} C_{ijk}^3 P_i P_j P_k + \dots$$
$$\dots + C_{1,2,3\dots N} P_1 P_2 \dots P_N = 1. \tag{38}$$

Each of the C's in the above expression is uniquely defined as a linear combination of the W's in Eq. (36).

[8] Formally these weights may be viewed as made up of a complex expression of conditional probabilities.

We now view Eq. (38) as constraint upon maximizing Eq. (34) for the total information. Using the Lagrange approach we then have, for any particular event i (taking the differential with respect to P_i):

$$\ln\frac{P_i}{1-P_i}=\lambda\Bigg[C_i{}^1+\sum_{j\neq i}(C_{ij}^2+C_{ji}^2)P_j+\sum_{j\neq i}\sum_{k\neq i,j}(C_{ijk}^3+C_{jik}^3+C_{kji}^3)P_jP_k$$

$$+\text{ Higher Order in } P\text{'s}\Bigg]. \qquad (39)$$

Note that the right hand side of the above equation does not contain P_i.

It now becomes clear what sort of approximations are being made in the cross impact relation obtained earlier. In order to reduce Eq. (39) to the earlier result; e.g., Eq. (13), we do the following:

1. For any reasonable event set, it is infeasible to expect an individual to answer 2^N question in order to evaluate all the C's or W's. Therefore, we in effect ignore terms of P^2 or greater, hoping that the three- or four-way interactions are small.
2. The derivation is valid for the set of all potential events. Usually only a specific subset with a range of about 5 to 100 events is utilized. Therefore all events not specified in the application of the cross impact analysis are in effect lumped into the constants, since their prior probability of occurrence is assumed constant within the scope of the estimation process.

Under these approximations we may rewrite (39) as

$$\ln[P_i/(1-P_i)]=\gamma_i+\sum_{j\neq i}C_{ij}P_j, \qquad (40)$$

where the Lagrange multiplier λ has been incorporated in the constants and where both γ_i and C_{ij} are a function of the events not specified. In essence, the γ-coefficient may be viewed as a measure of the "temperature" of the environment created for the i-th event by the unspecified set of events.

The ratio

$$C_{ij}^*=\frac{C_{ij}}{\gamma_i} \qquad (41)$$

gives a good measure or indication of how sensitive the i-th event is to the j-th event as compared with the rest of the environment. Note that if γ_i is positive the unspecified events contributed to the occurrence of the i-th event and vice versa. Also if C_{ij} is positive, then the j-th event contributed the occurrence of the i-th event.

The effect of ignoring higher order interactions among specified events can be measured by asking a subjective question of the form:

> Given the most favorable (or unfavorable) set of circumstances for event i with respect to the occurrence or nonoccurrence the remaining specified events, what is your estimate of the resulting probability for the occurrence of the i-th event?

This allows one to calculate two other values for γ_i in addition to the one initially obtained. The range of γ_i defined by the difference of these two values in principle measures the inaccuracy in the approximation due to ignoring the higher order terms in the specified P's.

The three values of γ are defined by

$$\gamma_i = \phi(P_i) - \sum_{j \neq i} C_{ij} P_{j(\text{original})}, \tag{42}$$

where P_i is the original estimate

$$\gamma_i^1 = \phi(P_i^f) - \sum_{j \in C_{ij} > 0} C_{ij}, \tag{43}$$

where P_i^f is the favorable estimate

$$\gamma_i^2 = \phi(P_i^u) - \sum_{j \in C_{ij} < 0} C_{ij}, \tag{44}$$

where P_i^u is the unfavorable estimate.

For calculating γ_i^1 we have assumed $P_j = 0$ if $C_{ij} < 0$ and $P_j = 1$ if $C_{ij} > 0$. The converse is assumed for calculating γ_i^2. The explicit measure of the inaccuracy in ignoring the higher order terms would then be

$$\frac{\gamma_i^1 - \gamma_i^2}{\gamma_i}. \tag{45}$$

If one does choose to obtain values for γ_i^1 and γ_i^2 an interpolation procedure may be established to modify γ_i such that P_i will range between P_i^u and P_i^f as the other P's are allowed to vary in order to examine different potential outcomes for the set of events. Therefore, the effect of higher order "interactions" among the event set can at least be approximated.

This particular view of the cross impact leads one to the conclusion that two types of events should be specified in any cross impact exercise:

Dependent Events: Those whose occurrence are a function of other events in the set.

Independent Events: Those whose occurrence are largely unaffected by the other events in the set but may influence some subset of the other events.

These events may be obvious at the initial specification of the event set (e.g., the occurrence of a natural disaster) or they may be determined empirically when

$$\left|\frac{C_{ij}}{\gamma_i}\right| \ll 1, \text{ for all } j. \tag{46}$$

If the event is independent, then there is no need to ask for information on the impact of the other events on it. This has the benefit of reducing the estimation effort on the part of the respondents to the exercise.

While we can, therefore, obtain some idea of the significance of the unspecified events in terms of their impact on the specified set of events, there is no analytical guidance for resolving the fundamental question of what particular events should make up the specified set. This procedure is entirely dependent upon the group which will be supplying the estimates and the general problem area that is to be examined. However, the author does feel that the concept of *Dependent and Independent Events* should be introduced at the stage of actually formulating the event set.

Given an on-line computer system for collecting cross impact estimates, there is, in principle, no hindrance to extending the approach developed in this paper to allow estimators to express three way or higher order interactions when they think they are significant. Equation (39) may be used to specify the higher order cross impact factors. Also, the pairwise interactions can be evaluated and specific higher order questions can be generated about those pairwise interactions which appear to be crucial or dominant. However, extensions of this sort are feasible only with groups that will make regular use of such techniques and which have had some degree of practice with similar quantitative approaches.

Example

The following goes step by step through a cross impact exercise set up in an on-line user mode on a computer. The numeric quantities reflect the inputs of a young economist who felt that the behavior of the resulting model reflected his judgment. It took him three iterations (in terms of changes to the "conditional" probabilities) to arrive at this situation. For the sake of brevity the final inputs are presented as if they all occurred on the first iteration. The program also operated in a long or short explanation mode according to the users' option and did supply a verbal definition of probability ranges as well as an odds to probability conversion table. The first thing the user sees, if he wishes, is a list of the events. It is, however, not necessary to store the events themselves as they are referenced individually by a number throughout the exercise. All the user needs is a hard copy list which indicates the event number for each event statement. This is particularly useful where confidentiality of the events under consideration is of importance. The long form (i.e., full explanation) of the interaction is presented.

CONSIDER THE FOLLOWING EVENTS AND THE POSSIBILITY OF THEIR OCCURRENCE BETWEEN NOW AND THE YEAR 1980:

1. THE U.S. GETS IN A TRADE WAR WITH ONE OR MORE OF ITS MAJOR TRADING PARTNERS (JAPAN, CANADA, WESTERN EUROPEAN COUNTRIES).
2. COMPREHENSIVE TAX REVISION IS ENACTED WITH MOST PRESENT EXEMPTIONS AND EXCLUSIONS REMOVED, BUT WITH RATES LOWERED.
3. RIGOROUS ANTI-POLLUTION STANDARDS ARE ADOPTED AND STRICTLY ENFORCED FOR BOTH AIR AND WATER.
4. THE U.S. AVERAGES AT LEAST 4 PERCENT PER YEAR GROWTH RATE OF REAL GNP FOR THE TIME FRAME THROUGH 1980.
5. DEFENSE SPENDING DECLINES STEADILY AS A PERCENT OF THE FEDERAL GOVERNMENT'S ADMINISTRATIVE BUDGET.
6. THE U.S. EXPERIENCES AT LEAST ONE MAJOR RECESSION (GNP DECLINE IS GREATER THAN 5 PERCENT FOR A DURATION GREATER THAN 2 QUARTERS) DURING THE TEN-YEAR PERIOD.
7. A FEDERAL INCOME MAINTENANCE SYSTEM (E.G., NEGATIVE INCOME TAX) REPLACES ESSENTIALLY ALL CURRENT STATE AND LOCAL WELFARE PROGRAMS.
8. THE OIL IMPORT QUOTA SYSTEM IS PHASED OUT AND DOMESTIC OIL PRICES ALLOWED TO FALL TO THE WORLD PRICE.
9. THE U.S. AGRICULTURAL PRICE SUPPORT SYSTEM IS DISMANTLED.
10. A FEDERAL-STATE AND LOCAL REVENUE-SHARING PROGRAM IS ADOPTED WHICH ALLOCATES AT LEAST 5 PERCENT OF FEDERAL REVENUES TO STATE AND LOCAL GOVERNMENTS.

IN THIS EXERCISE WE WILL BE ASKING FOR YOUR SUBJECTIVE PROBABILITY ESTIMATES. FOR THOSE OF YOU WHO WISH TO THINK IN TERMS OF ODDS, THE FOLLOWING CONVERSION EQUATION AND EXAMPLES MAY BE OF USE:

ODDS = A:B EQUIVALENT TO PROBABILITY = A/(A + B)

EXAMPLES 1:99 = .01 1:9 = .1 1:4 = .2 3:7 = .3 1:2 = .33
2:3 = .4 1:1 = .5 3:2 = .6 2:1 = .66 7:3 = .7
4:1 = .8 9:1 = .9 99:1 = .99

A SEMANTIC EQUIVALENT TO THE NUMERIC PROBABILITIES MAY BE TAKEN AS:

VERY PROBABLE	> =.75
PROBABLE	>.5 BUT <.75
EITHER WAY	=.5
IMPROBABLE	<.5 BUT >.25
VERY IMPROBABLE	= <.25

TEAR OFF THE ABOVE LIST OF EVENTS FOR REFERENCE BY EVENT NUMBER THROUGHOUT THE REST OF THIS EXERCISE.

STEP 1: OVERALL PROBABILITIES

PLEASE SUPPLY YOUR BEST ESTIMATE FOR THE PROBABILITY THAT EACH OF THE EVENTS WILL OCCUR AT SOME TIME BETWEEN NOW AND 1980.

UNLESS YOU CHANGE THEM ALL, THE PROBABILITIES ARE INITIALLY SET TO .5 WHICH IS EQUIVALENT TO EXPRESSING A NO JUDGMENT FOR THE PARTICULAR EVENT WITH RESPECT TO THE ABOVE QUESTION.

ESTIMATES: 2, .3, 3, .6, 5, .4, 6, .3, 7, .6, 8, .2, 9, .1, 10, .6

SUMMARY STEP 1

EVENT	1	2	3	4	5	6	7	8	9	10
P =	.50	.30	.60	.50	.40	.30	.60	.20	.10	.60

IF SATISFIED HIT RETURN KEY, IF NOT TYPE SOMETHING FIRST.

STEP 2: CONDITIONAL PROBABILITIES

IN THIS STEP YOU ARE ASKED TO ASSUME FOR THE PURPOSE OF ANALYSES THAT YOU HAVE BEEN PROVIDED CERTAIN KNOWLEDGE AS TO WHETHER A PARTICULAR EVENT WILL OR WILL NOT OCCUR IN THE STATED TIME FRAME. BASED UPON THIS HYPOTHETICAL SITUATION, FOR EACH EVENT IN TURN, PLEASE INDICATE ANY RESULTING NEW ESTIMATE FOR THE PROBABILITY OF OCCURRENCE OF THE OTHER EVENTS.

UNLESS YOU CHANGE THEM, THESE CONDITIONAL PROBABILITIES ARE SET EQUAL TO THE OVERALL PROBABILITIES.

ASSUME EVENT 1 IS CERTAIN TO OCCUR. INDICATE YOUR ESTIMATES OF CHANGES IN THE PROBABILITIES OF OCCURRENCE FOR THE OTHER EVENTS.

ESTIMATES: 2, .25, 3, .55, 4, .4, 5, .3, 6, .4, 7, .55, 8, .1, 9, .05, 10, .55

At this point the computer calculates each C_{ij} from Eq. (19); or if the event had been assumed to not occur, Eq. (21) would have been used. If no change had been indicated, the corresponding C would be set to zero.

The computer informs the user about the occurrence or non-occurrence of an event according to how he specified the overall probabilities. If he specifies a probability of .5 or less, he is told to assume the event occurred; if more than .5, than he is told to assume it did not occur. This policy is arbitrary. In this example the user was told to assume occurrence for events 1, 2, 4, 5, 6, 8, and 9 and to assume non-occurrence for events 3, 7, and 10.

The user is allowed only two digit specification of a probability which must lie between (and including) .01 and .99. If he enters a zero or one, it is automatically changed to .01 or .99 respectively.

When the user has gone through all the events in the above manner and is satisfied with his inputs, then the γ_i's are calculated from Eq. (42).

The user is now presented a summary of his inputs and the converse "conditional" probability to the one supplied which is calculated from Eq. (22).

SUMMARY CONDITIONAL PROBABILITIES BASED UPON OCCURRENCE AND THEN NONOCCURRENCE, NC INDICATES NO CHANGE FROM OVERALL P

E	1	2	3	4	5	6	7	8	9	10
1	OV PR	.45	NC	.40	.45	.75	NC	.45	.45	NC
1	.50	.52	NC	.60	.53	.38	NC	.51	.51	NC
2	.25	OV PR	.28	.35	NC	.20	.38	.35	.35	.34
2	.36	.30	.33	.25	NC	.35	.20	.29	.29	.25
3	.55	.65	OV PR	.65	.70	.50	NC	.65	.65	.66
3	.65	.58	.60	.55	.53	.64	NC	.59	.59	.50
4	.40	.60	.51	OV PR	.55	.30	.53	.55	.55	.53
4	.60	.46	.49	.50	.47	.59	.45	.49	.49	.45
5	.30	.50	.39	.50	OV PR	.35	.47	NC	NC	.43
5	.51	.36	.42	.31	.40	.42	.30	NC	NC	.35
6	.40	.25	NC	.10	.25	OV PR	.27	.25	.25	.27
6	.22	.32	NC	.62	.34	.30	.35	.31	.31	.35
7	.55	.75	NC	.70	.75	.55	OV PR	NC	NC	.66
7	.65	.53	NC	.49	.49	.62	.60	NC	NC	.50
8	.10	.15	NC	.25	.25	.10	NC	OV PR	.30	.24
8	.36	.22	NC	.16	.17	.26	NC	.20	.19	.15
9	.05	NC	NC	.15	NC	.05	.15	.20	OV PR	.15
9	.19	NC	NC	.07	NC	.13	.05	.08	.10	.05
10	.55	.75	.59	.70	.75	.50	.66	NC	NC	OV PR
10	.65	.53	.62	.49	.49	.64	.50	NC	NC	.60

ASSUMING ALL THE OTHER EVENTS OCCUR OR DO NOT OCCUR SO AS TO ENHANCE THE GIVEN EVENT, THE MOST FAVORABLE PROBABILITY FOR EACH EVENT IS:

EVENT:	1	2	3	4	5	6	7	8	9	10
MFP =	.86	.73	.91	.89	.81	.85	.94	.78	.76	.95

ASSUMING ALL THE OTHER EVENTS OCCUR OR DO NOT OCCUR SO AS TO INHIBIT THE GIVEN EVENT THE LEAST FAVORABLE PROBABILITY IS

EVENT:	1	2	3	4	5	6	7	8	9	10
LFP =	.16	.06	.22	.11	.09	.02	.17	.01	.01	.13

FOLLOWING IS A TABLE OF THE RELATIVE CAUSAL WEIGHTS (CROSS IMPACT FACTORS) OF ONE EVENT (COLUMN) UPON ANOTHER (ROW) AND A MEASURE (GAMMA) OF THE EFFECT OF EVENTS NOT SPECIFIED APPEARS IN THE DIAGONAL ELEMENT. MINUS INDICATES AN INHIBITING EFFECT.

CROSS IMPACT TABLE

E	1	2	3	4	5	6	7	8	9	10
1	+.23G	−.29	0	−.81	−.33	1.57	0	−.25	−.22	0
2	−.50	−1.33G	−.23	.46	0	−.77	.90	.29	.25	.42
3	−.41	.31	−.30G	.43	.74	−.58	0	.27	.24	.68
4	−.81	.58	.07	−.05G	.33	−1.21	.33	.25	.22	.33
5	−.88	.58	−.14	.81	−1.02G	−.31	.74	0	0	.36
6	.88	−.36	0	−2.70	−.42	+.88G	−.38	−.31	−.28	−.38
7	−.41	.99	0	.88	1.16	−.29	−.91G	0	0	.68
8	−1.62	−.50	0	.58	.48	−1.16	0	−.97G	.60	.58
9	−1.49	0	0	.93	0	−1.07	1.25	1.01	−3.29G	1.25
10	−.41	.99	−.14	.88	1.16	−.58	.68	0	0	.74G

G INDICATES THE GAMMA FACTOR AS THE DIAGONAL ELEMENT.

The user may infer from the cross impact factors in the previous table the relative rank order with respect to the effect of one event upon another as interpreted from his judgments on the probabilities.

The next step is for the computer to present the user with a forecast of which events will occur. To do this it is assumed that the perception of the likelihood of the event occurring produces the causal effect, and not the actual time of occurrence. With this time independent view we can assume it is reasonable to apply a cascading perturbation approach to forecasting occurrence. This is done as follows:

(1) Examine the overall probabilities and determine which event or events is closest to zero or one.

(2) If the event is close to zero, assume it will not occur or if it is close to one assume it will occur (this is the smallest possible perturbation).

(3) Based on (2), calculate new probabilities for the remaining events.

(4) Begin step 1 again for those events which have not already been assumed to occur or not occur.

The above sequence is repeated until the outcome is established for all events unless the final probability is .5, in which case no outcome is forecast.

The following is what happens for the above example where each row is one cycle of the above cascade iteration procedure. The user can observe how the probabilities are affected. Note that the initial estimates on events three, seven, and ten are reversed.

FORECASTED CERTAINTY SEQUENCE, THE + INDICATES OCCURRENCE AND THE − NON-OCCURRENCE

E	1	2	3	4	5	6	7	8	9	10
0	.50	.30	.60	.50	.40	.30	.60	.20	.10	.60
1	.51	.29	.59	.49	.40	.31	.60	.19	−−−	.60
2	.52	.28	.58	.48	.40	.32	.60	−−−	−−−	.60
3	.55	−−−	.55	.43	.35	.36	.52	−−−	−−−	.52
4	.61	−−−	.45	.36	−−−	.46	.37	−−−	−−−	.37
5	.72	−−−	.36	−−−	−−−	.76	.26	−−−	−−−	.26
6	.86	−−−	.27	−−−	−−−	+++	.21	−−−	−−−	.18
7	+++	−−−	.24	−−−	−−−	+++	.18	−−−	−−−	.16
8	+++	−−−	.22	−−−	−−−	+++	.17	−−−	−−−	−−−
9	+++	−−−	.22	−−−	−−−	+++	−−−	−−−	−−−	−−−
10	+++	−−−	−−−	−−−	−−−	+++	−−−	−−−	−−−	−−−

YOU MAY REPEAT THE SEQUENTIAL ANALYSES WITH NEW INITIAL PROBABILITIES. YES (1), NO (2), CHOICE?

The user may now examine the sensitivities of this model by choosing to modify one or more of the overall probabilities and holding the rest and the cross impact factors constant. This would correspond to assuming a basic change in policies effecting the likelihood of a particular event. In this instant the user chose to increase the probability that defense spending decreased and then to separately view the effect of a major tax revision. The effects of these choices are summarized and compared to the original result above.

If the user is not satisfied with the behavior of the model he has built up, he may go back and make changes to the original overall probabilities and/or the conditionals until he has obtained satisfactory behavior.

If the activity were part of a Delphi or other group exercise, then once a user was satisfied with his estimates they would be collected in order to obtain a group response.

SUMMARY OF PERTURBED OUTCOMES
FOR CROSS IMPACT SET OF ECONOMIC EVENTS

THE + SIGN INDICATES WHICH EVENTS WERE FORECASTED TO OCCUR FOR THE GIVEN OVERALL PROBABILITIES	Overall Probability	Resulting Outcome	Change in Overall Probability	Resulting Outcome	Change in Overall Probability	Resulting Outcome
1. U.S. in trade war	.5	+				
2. Major tax revision	.3				.99	+
3. Rigorous anti-pollution standards	.6			+		+
4. GNP at 4% real growth	.5			+		+
5. Defense spending declines	.4		.8	+		+
6. "Recession" occurs	.3	+				
7. Income maintenance enacted	.6			+		+
8. Oil Import quotas eliminated	.2					+
9. Agricultural price support	.1					
10. Revenue sharing enacted	.6			+		+

The group response would be determined by a linear average of the cross impact factors and the gamma factors—not the probabilities. Then each individual would be able to see similar inferences as the above for the group view with the addition of a matrix which compared the number of individuals who estimated a positive, negative, or no impact relation between each event combination. In the group case one would also have to allow the estimator to indicate which cross impacts he has a no judgment position on. The computer would than supply for him, if he wishes, the average supplied by the rest of the group for that particular cross impact relationship.

Applications

The intriguing aspect of the cross impact formalism is its utility to a rather broad range of applications. The first application is as an aid to or tool for an individual in organizing and evaluating his views on a complex problem. The structure offers the individual more freedom in expressing the event set than the constraints of mutual exclusiveness imposed in decision tree and table type approaches. There also appears to be some compatibility between the pair-wise examination of causal relationships and the way many individuals think about causal effects. This is true to the extent that cross impact formalism may be utilized quite easily by individuals without any formal training in decision theory or probability. The author has, for example, gone through the creation and evaluation of a set of five events with a group of high school students within a one-hour period, using a computer terminal to perform the calculations. That particular exercise stimulated a great deal of class discussion as to under what economic conditions the students would plan to have children. The educational utility of the cross impact formalism, as well as other Delphi-oriented communication structures, has largely gone unnoticed.

The main problem encountered in utilizing the technique is that some individuals are so accustomed to the Bayes theorem that they will habitually apply it in responding to the cross impact questions.

Once some members in an organization have begun to employ the approach for their individual benefit then it becomes quite easy to introduce it as a communication form for expressing quite precisely to others in the group how they view the causal relationships involved in the problem under consideration. The benefit here is in allowing the group to quickly realize where disagreement exists in both the direction, as well as relative magnitude, of the impacts. This can eliminate a lot of superfluous discussion about areas of agreement.

Whether the evaluation of an event set is carried out in a committee, conference, Delphi, or some combination of these processes, it is mandatory that the group involved reach agreement and understanding on the specification and wording of the event set. In addition, the actual cross impact exercise may cause the group to desire modification of the event set.

In utilizing the technique for serious problems, there would appear to be benefits for groups of both decision makers and analysts. In addition, it may solve a problem that now exists in attempting to set up efficient communication structures between these two groups. The analyst attempting to build simulations or models of complex processes of interest to the decision maker very often encounters causal relationships dependent upon policy and decision options that defy any reasonable attempt at incorporation into the model, except in the form of prejudging the outcome of policy or decision options. At times these choices are so numerous that they are effectively buried and become hidden assumptions in the logic of complex simulations. The cross impact technique offers the analyst an opportunity to leave portions of the simulation logic arbitrary; thus, the users of the simulation may utilize a cross impact exercise to structure the logic of the simulation when they wish. While this application has not yet been demonstrated, it may turn out to be a major use of the cross impact technique. There is considerable advantage to be had from introducing a greater degree of flexibility in the application of the more comprehensive simulations being built to analyze various organizational, urban, and national problems.

As with many Delphi structures,[9] it is quite feasible to design an on-line conference version of the cross impact exercise which would eliminate delays in processing the group results and allow the conferees to modify their views at will. It would be necessary to tie this particular conference structure to a general discussion conference (such as

[9] See "Delphi Conferencing" by Murray Turoff, *Technological Forecasting and Social Change* **3**, No. 2, 1971. Also, "The Delphi Conference," in *The Futurist*, April 1971, provides a summary report.

the "Delphi Conferencing" system) in order that the group can first specify the event set and later discuss disagreement on causal effects.

If one considers the basic functions performed in the planning operation of organizations, whether they be corporate or governmental, there are two other types of conference structures that should be added to the general discussion format and the cross impact conference structure.

One is a resource allocation conference structure which allows a group to reach agreement on what is the most suitable allocation of the organizational resources to bring about the occurrence of the type of event which the organization controls or influences (i.e., controllable events). Various program options evaluated in terms of resources required and probability of accomplishment as a function of time and resource variability would evolve from this type of conference.

The other type of conference structure involves forecasting the environment in which the organization must function. This conference would be used to generate information on the uncontrollable events which specify the environment and or their likely occurrence over time.

The resource allocation conference may use various optimization techniques, such as linear programming, to aid the group members in their judgments. The environmental forecasting conference may use such tools as trend, correlation, or substitution analysis routines to aid the conference group.

The cross impact conference structure may now be viewed as a mechanism for relating uncontrollable events expressing potential environmental situations to controllable events expressing organizational options. The general discussion conference allows the group or groups involved to maintain consistency and resolve disagreements.

Initial design formats for all these conference structures already exist to some extent in the various paper and pencil Delphi's that have been conducted to date. It remains for some organization to piece these together within the context of a modern terminal oriented computer-communications system. Given such a system, represented in the accompanying diagram, an organization faced with a specific problem may first, and quickly, bring together the concerned group via the terminals and the general discussion conference format to arrive at specifications for the resource allocation and forecasting conferences. These two latter conferences may involve only subsets of the total group and may draw on added expertise as needed. Using the cross impact to correlate the results of the other two efforts, the variability of options versus potential environments can be examined. The sought-after result is a set of evaluated options suitable for providing an analysis basis for a decision.

One may envision simultaneous replication of this four-way conference structure focusing on different problems which may also relate to different levels of concern within the organization. A set of procedures could also be introduced for moving the results of one problem analysis to a higher level conference group or for sending requests to resolve particular uncertainties down to a conference group at a lower level.

The main advantage of such a system is the organization's ability to draw upon the talent needed for the problem on a timely and efficient basis, regardless of where it resides with respect to either geography or organizational structure. Also inherent in this type of system is the view that the individuals in an organization are the best vehicle for filtering the information appropriate to a particular problem out of established data management system and other constant-type organizational pro-

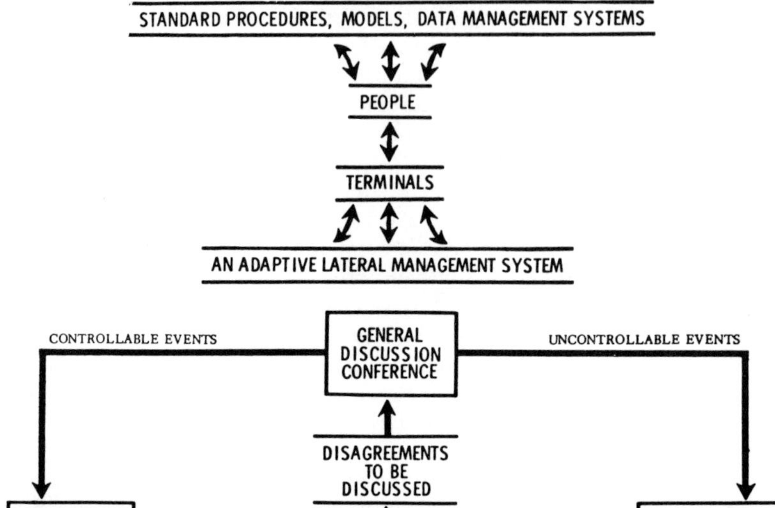

CONTROLLABLE EVENTS | GENERAL DISCUSSION CONFERENCE | UNCONTROLLABLE EVENTS

DISAGREEMENTS TO BE DISCUSSED

RESOURCE ALLOCATION CONFERENCE | CROSS IMPACT CONFERENCE | FORECASTING (ENVIRONMENTAL) CONFERENCE

CONTROLLABLE OPTIONS AND SCHEDULE | POTENTIAL ENVIRONMENTS AND LIKELIHOOD

EVALUATED OPTION CONSEQUENCES

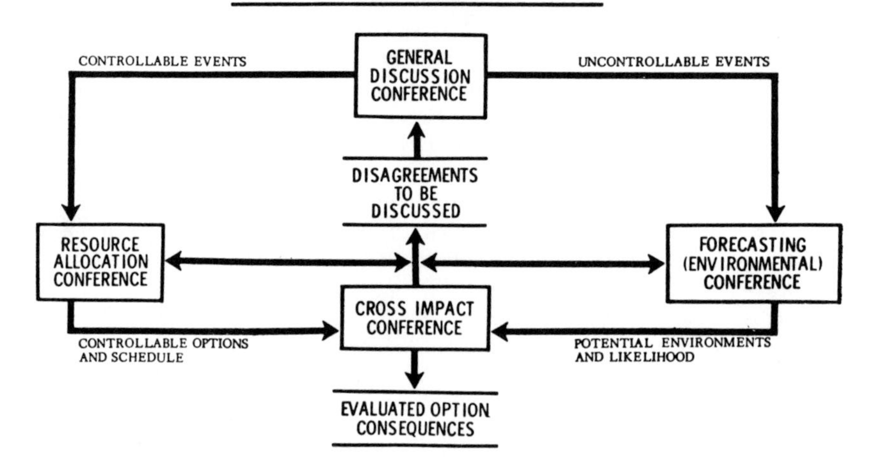

cedures. The mistake often made by designers of management information systems is the assumption that there is a standard algorithm which will continually transform the normal flow of organizational data into a form suitable for management purpose.[10] This is only true when the organization is faced with an unchanging environment, and very few organizations, unless they are deluding themselves, can claim that view in this day and age.

The author views a Management Information System as just this four-way conference structure existing in a design scheme which allows groups to easily shift from one format to another and which may be replicated either to improve lateral communication at various organizational levels or to tackle a multitude of problems. The concept is basically a lateral communication system and presupposes an organizational environment which supports or fosters lateral communication. It is also highly adaptive and able to respond to a changing environment which in turn may impose changing requirements on the organization. Many large organizations already have fairly extensive planning and forecasting efforts scattered through their various divisional or vertical organizational structures. It is less clear that these numerous segments of the organization can effectively relate to one another and the organization as a whole. Current methods of doing so, involving frequent travel and extended meetings, are often prohibitive on a time and effort basis. However, given the requirements facing

[10] This point is developed further in "Delphi and Its Potential Impact on Information Systems," by Murray Turoff, in the *Proceedings of the Fall Joint Computer Conference*, 1971.

organizations today, the growing availability of terminals, computer hardware and software to support conferencing, and the availability of digital communication networks providing reasonable communication costs, it can be expected that the system of the type described here will come into being over this next decade.

Acknowledgment

As one may suspect, this paper has evolved out of a number of earlier drafts. I would like to thank the following individuals for their aid in terms of comments and reviews (both pro and con): Dr. Ronald Bass, Dr. Norman Dalkey, Mr. Selwyn Enzer, Dr. Felix Ginsberg, Professor Jack Goldstein, Mrs. Nancy Goldstein, Professor Robert Piccirelli, Miss Christine Ralph, Professor Richard Rochberg, Dr. Evan Walker, Dr. John Warfield, and Mr. Dave Vance.

Annotated Bibliography

The initial paper on cross impact was published in 1968 by Gordon and Hayward. Other papers specifically on cross impact are those by Dalby, Enzer, Johnson, and Rochberg. Very closely related to the cross impact formalism is the cross support formalism in C. Ralph's work. This formalism in essence makes a clear distinction between dependent events, which are defined as goals, and independent events, which are related to resource allocation choices. Also related to the cross impact problem is the management matrix formalism referenced and discussed in the book by Farmer and Richman and in a paper by Richman. The measures of association concept in business problems (see Perry's paper for a review) is another variation of the same problem. The formal problem of defining measures of association or correlation coefficients are discussed in articles by Costner and Goodman.

The use of the logistic type equation in statistics for cases in which one is concerned with a binary outcome type process is reviewed quite well in the papers by Cox and Neter. The use of the logistic equation in economics (logic regression) is found in the works of Sprent and Theil. The Fermi-Dirac distribution in physics is discussed in Born and at a more advanced level in Tolman. The use of the logistic distribution in Technological Forecasting as a modeling tool is discussed in Ayres' book. The mathematical properties of the logistic equation and its usefulness to model population growth is reviewed in the book by Davis.

The point that all prior probabilities are conditional is brought out quite clearly in the book by Savage. Raiffa's book contains an excellent guide to the philosophical differences that surround the concept of subjective probability and inference. Tribus' book is one of the few works that treats the "weight of evidence" measure (e.g., defined in the paper as the occurrence ratio) in some detail and references earlier papers on this topic. The discussion of unsettled problems of probability theory in Nagel's book is also relevant.

Two papers by Ward Edwards appearing in one book edited by Kleinmuntz and one book edited by Edwards and Tversky review the psychological experiments to determine if humans make judgments on a Bayesian basis. Edwards asserts, on the basis of his work, that humans are conservative; i.e., always making more conservative estimates than would be implied by the use of Bayes theorem. A more recent experiment by Kahneman and Tversky appears to indicate that man "is not Bayesian at all." These authors propose a "representativeness" heuristic; wherein, "the likelihood that a particular 12-year old boy will become a scientist, for example, may be evaluated by the degree to which the role of a scientist is representative of our image of the boy." This view does not appear to be too far removed from the "causality" view adopted in approach of this paper to the cross impact problem. The average person deals almost everyday with at least a subconscious process for estimating non-recurrent and transient events. The Bayesian approach to modeling this subjective probability process does not appear to fit or explain the judgments made. However, the experiments to date do appear to confirm that some sort of universal or consistent model exists which humans of very different backgrounds and training are in fact using.

The paper by DeWitt, which deals with the philosophical problem of inferring reality from quantum mechanics, is an excellent review of what the author feels are analogous difficulties with justifying cross impact. The chapter in Bohm's book, conjecturing that the human mind may function with a quantum mechanical type thought process, may, to a limited degree, be viewed as circumstantial support for the propositions developed in this paper. If Bohm were correct, it should not be a complete surprise that a macro statistical quantum mechanical distribution (Fermi-Dirac) can be used to correlate measurements of subjective estimates by a group of humans. Walker's paper also explores the potential relationship of quantum mechanics to the nature of consciousness. Bohr also argues in his writings for the more universal application of quantum mechanics to the human thought process.

Also, Reichenbach in his book examines the relationships between quantum mechanics and the calculus of probabilities. In particular, Reichenbach interprets posterior probabilities as those resulting from measurements, and priors or "potential" as those arising from the physical theory. He further discusses the need for a three-value logic system to deal with "causal anomalies"—true, false and indeterminacy. The three-valued-logic proposition of Reichenbach leads to the speculation that if there is a rigorous foundation for the theory of cross impact, it may lie in the work taking place in "fuzzy" set theory (i.e., a class which admits of the possibility of partial membership in it is called a fuzzy set). In cross impact the set of all possible events may be considered as made up of two subsets, those that will occur and those that will not. Any particular event may have partial membership in either set. That which we have termed the probability of occurrence is referred to by the "fuzzy" set theory people as the membership function (i.e., see Zadeh).

Umpleby's work represents a first and interesting attempt to tie the cross impact formalism to the resource allocation problem in at least an automated game mode.

The problem of how to average probability estimates among a group is crucial to utilizing other cross impact systems. This is reviewed in Brown's paper and in Raiffa's book. The methodology proposed here at least explicitly avoids the question by averaging, for the group, correlation coefficients having a plus to minus infinity range.

In addition to the published literature there are at least three alternative methods under consideration or in use. These additional methods are proposed by Dalkey, or RAND, Enzer, of the Institute for the Future, and Kenneth Craver, of Monsanto. Extensive modifications to the original treatment by Gordon have recently been proposed by Folk, of the Educational Policy Research Center at Syracuse, and by Shimada, of Hitachi Ltd., Japan. Also the current work in the area of "Relevance Trees" represents attempts to tackle the same class of problems by unfolding the matrix structure into a tree structure. The concept of multi-dimensional scaling in psychology is also related to the cross impact problem (see J. D. Carroll's paper). †

† The Carroll-Wish paper is in the next chapter.

Ayres, Robert V., *Technological Forecasting*, McGraw-Hill (1969).

Bohm, David, *Quantum Theory*, Prentice Hall (1951); Analogies to Quantum Processes, Chapter 8.

Bohr, N., *Atomic Physics and Human Knowledge*, John Wiley & Sons (1958).

Born, Max, *Atomic Physics*, Chapter 8, Hafner Publishing Company (1935).

Brown, T. A., *Probabilistic Forecasts and Reproducing Scoring Systems*, RAND-RM-6299-ARPA (June, 1970).

Carroll, J. D., *Some Notes on Multidimensional Scaling*, in Proceedings of the IFORS meeting on Cost Effectiveness (May, 1971), Wash., D.C. (to be published by Wiley). †

Costner, H. L., *Criteria for Measures of Association, American Sociological Review*, **30**, No. 3 (1965).

Cox, D. R., *Some Procedures Connected with the Logistic Qualitative Response Curve*, Research Papers in Statistics, John Wiley (1966).

Dalby, J. F., *Practical Refinement to the Cross-Impact Matrix Technique of Technological Forecasting* (1969, Unpublished).

Dalkey, Norman C., *An Elementary Cross Impact Model*, RAND Report R-677-ARPA (May, 1971). ‡

Davis, Harold T., *Introduction to Nonlinear Differential and Integral Equations*, Dover Publications (1962).

DeWitt, Bryce S., "Quantum Mechanics and Reality," *Physics Today*, September 1970. See also April 1971 issue of *Physics Today*.

Edwards, W. and Tversky, A. (Ed.), *Decision Making*, Penguin Modern Psychology (1967).

Enzer, Selwyn, *Delphi and Cross-Impact Techniques: An Effective Combination for Systematic Future Analysis;* Institute for the Future WP-8 (June, 1970).

Enzer, Selwyn, *A Case Study Using Forecasting as a Decision-Making Aid*; Institute for the Future WP-2 (December, 1969). Also *Future*, **2**, No. 4 (1970).

Farmer, R. N. and Richman, B. M., *Comparative Management and Economic Progress* (1965).

Goodman, L. A. and Kruskal, W. H., "Measures of Association for Cross Classification I"; Am. Stat. J. (December, 1954); "Measures of Association for Cross Classification II," (March, 1959).

Gordon, T. and Haywood, H., "Initial Experiments with the Cross-Impact Matrix Method of Forecasting," *Futures*, **1**, No. 2 (1968).

Gordon, Theodore J., Rochberg, Richard, and Enzer, Selwyn, *Research on Cross-Impact Techniques with Applications to Selected Problems in Economics, Political Science and Technology Assessment*; Institute for the Future R-12 (August, 1970).

Johnson, Howard E., *Cross-Impact Matrices, An Exposition and a Computer Program for Solution*, Graduate School of Business, University of Texas WP 70-25 (January, 1970). Also *Futures* **II**, No. 2 (1970).

Kahneman, O. and Tversky, A., "Subjective Probability: A Judgment of Representativeness"; Oregon Research Institute: *Research Bulletin*, **II**, No. 2 (1971).

Kleinmuntz, B. (Ed.), *Formal Representation of Human Judgment*, Wiley (1968).

Nagel, E., *Principles of the Theory of Probability*, Foundations of the Unity of Science, **1**, No. 6, University of Chicago Press (1969).

Neter, John and Maynes, E. Scott, *Am. Stat. Assn.*, **65**, No. 330, Applications Section (June, 1970).

Perry, Michael, *Measures of Association in Business Research*; Research Report No. 9, Hebrew University of Jerusalem (September, 1969).

Raiffa, Howard, *Decision Analysis*, Addison Wesley (1968), Chapter 10.

Ralph, Christine, *An Illustrative Example of Decision Impact Analyses* (*DIANA*) *Applied to the Fishing Industry*, Synergistic Cybernetics Incorporated, Report (February 1969).

Ralph, Christine A., *Resource Allocation Logic for Planning Heuristically as Applied to the Electronics Industry*, International Research and Technology Corporation IRT P-22.

Reichenbach, Hans, *Philosophic Foundations of Quantum Mechanics*, University of California Press, 1965.

Richman, B. M., "A Rating Scale for Product Innovation," *Business Horizons* **V**, No. 2 (Summer, 1962).

Rochberg, Richard, "Information Theory, Cross-Impact Matrices and Pivotal Events," *Technol. Forecasting*, **2**, No. 1 (1970).

Rochberg, Richard, *Some Comments on the Problem of Self-Affecting Predictions*; RAND Paper P-3735 (December, 1967)

Rochberg, R., Gordon, T. J., and Helmer, O., *The Use of Cross-Impact Matrices for Forecasting and Planning*, Institute for the Future R-10 (April, 1970).

Rochberg, R., "Convergence and viability because of Random Numbers in Cross-Impact Analyses," *Futures*, **2**, 3 (Sept., 1970).

Savage, L. J., *The Foundations of Statistical Inference*, John Wiley Publishers (1962).

Shimada, Syozo, *A Note on the Cross-Impact Matrix Method*, Central Research Laboratory, Hitachi Ltd. Report No. HC-70-029 (March, 1971).

Sprent, Peter, *Models in Regression and Related Topics*, Methuen and Company Ltd. (1969).

Theil, Henri, *Economics and Information Theory*, Rand-McNally Publishers (1967).

† See also VI C in this book.

‡ See V B in this book.

Tolman, Richard, *The Principles of Statistical Mechanics*, Oxford University Press (1938), Chapter XII.

Tribus, Myron, *Rational Descriptions, Decisions and Designs*, Pergamon Press (1969).

Umpleby, S., *The Delphi Exploration. A Computer-Based System for Obtaining Subjective Judgments on Alternative Futures*, Report F-1, Computer-Based Education Research Laboratory, University of Illinois (August, 1969).

Zadeh, L. A., *Towards a Theory of Fuzzy Systems*, NASA Report CR-1432 (September, 1969).

Walker, Evan H., "The Nature of Consciousness," Math. Biosci. 7, No. 1/2 (February, 1970).

. D. A Primer for a New Cross-Impact Language—KSIM

(with Examples Shown from Transportation Policy)

JULIUS KANE

Editor's Note: *The understanding of the dynamics of complex systems for forecasting purposes has been stressed in the pages of this Journal. Forrester's article and the review of his* World Dynamics *in recent issues attest to its importance. Professor Kane's article shows that the behavior of nonlinear feedback systems can be demonstrated without mathematical sophistication. It is, therefore, of particular value for pedagogical purposes and for communication with individuals who do not possess advanced mathematical training but must, nevertheless, be concerned with these problems.*

Abstract

A new methodology/language has been developed which serves to make available the workings of cross-impact analysis available to a much larger audience in that no technical sophistication is required to become expressive in the new language. Unlike the procedures developed by Gordon, *et al* our methods stress the structural dynamics of the system, the geometry of the linkages rather than refining arithmetic estimates of future probabilities. However, while qualitatively and subjectively oriented, our procedures can be easily expanded to any degree of precision, providing the data and mechanisms are sufficiently well known. The key feature of our approach is that it allows one to work with data of any level—from subjective estimates to highly precise physical measurements—and the computer has the character of logical projections of basic hypothesis rather than dogmatic imperatives which is the nature of much of present social, economic, technological, and ecological modelling.

Introduction

Realistic problems involve a multiplicity of competing variables, presenting a complexity of behavior that usually dwarfs human capacity for comprehension. Consequently decisions are usually made in truncated spaces by sharply reducing the variables that will be considered. It has been the consistent endeavor of systems scientists to develop models which have the capacity of enlarging the scope of human comprehension.

Because of their mathematical nature, most simulation models suffer from a variety of problems. For one they tend to be excessively numerical, concentrating attention upon those variables which can be readily quantified, and tend to exclude variables which while important are basically subjective in nature. For example, traffic flow and traffic capacity are relatively easy to measure and are usually included within the scope of all transportation models. However such subjective notions such as status, comfort, convenience, and freedom of access are seldom included within the analysis except perhaps as marginal perturbations. This is particularly unfortunate since often

---

DR. KANE is Professor of Mathematical Ecology at the University of British Columbia, Vancouver, Canada.

Note: A reasonably complete bibliography of cross-impact analysis is given in Turoff's article in the first issue of *Technological Forecasting and Social Change*, Vol. 3 (1972) and will not be repeated here.

Reprinted from *Technological Forecasting and Social Change* **4** (1972) with permission of American Elsevier.

such subjective variables are the controlling parameters in the choice of transportation mode. Another shortcoming of most simulation models is that because of their highly technical nature they tend to inhibit policy makers from using them freely. Generally, simulation models are formulated and run by highly skilled "experts" with an elaborate and abstruse language. If anything, politicians are unlikely to read more than the abstract of the reports furnished by the "experts." As a result policy makers are denied the experience and intuition that comes with actual involvement with simulation models and thus tend to mistrust them. Thus a barrier is erected between those people who formulate and conceive simulation models and those who should ultimately use their output. It was the purpose of our research to try and design a simulation procedure —or better yet, a simulation *language* in which technically unsophisticated people could quickly become fluent in the logical expression of cross-impact concepts. In addition, the scope of this simulation language should be sufficiently powerful so that it would express the interaction of competing variables in a realistic and graphic fashion.

I. Aircraft Competition

The procedure we developed is simplicity itself. First, all the relevant variables are listed and given names. For example, in an aircraft interaction model, these might be:

SST: for the supersonic transport,
B747: for the Boeing jumbo jet and other large capacity aircraft of similar type,
JET: for conventional jets of the DC8/B707 variety,
STOL: for short take-off and landing planes,
VTOL: for vertical take-off aircraft, such as helicopters,
HSG: for high speed ground transportation, typically of the monorail variety, an
CM: for cost of money

It is immediately evident that these variables interact strongly with one another. In some cases they compete directly, for example, the *SST* and *B747* are in large part competitive. On the other hand, they also form alliances, for example, development of high speed ground transportation will probably be a big plus for the development of the SST because this would provide high speed rail links that could service a supersonic jetport located remote from urban regions.

To a first approximation the interactions can be simply summarized by a table of the following form (Table 1).

This matrix summarizes the interactions between the variables in the following fashion. At each location we enter the action of the column heading upon the row heading. A plus entry indicates that the action of variable *A* upon variable *B* is positive. In other words *A* encourages *B*'s growth and that such encouragement will be proportional to both the relative size of *A* and the magnitude of the interaction coefficient (not necessarily integer values). We have chosen most diagonal entries to be positive in accord with the idea that technology tends to foster its own growth. The sole exception is the self-interaction of *JET* which we have set as minus. This is to suggest that this variable has reached a stage of obsolescence in its evolution.

There is an extremely important pedagogical value in choosing the matrix entries as combinations of pluses and minuses rather than numerical entries. By not asking for numerical coefficients at the outset psychological barriers are greatly reduced, stimulating group participation and discussion. Furthermore *subjective* variables can very easily be introduced and there is no inhibition in making them play their proper role. Of course ultimately the pluses and minuses are translated into specific figures.

Table 1

Interaction Matrix of Transportation Modes[a]

Impact	SST	B 747	JET	STOL	VTOL	HSG	CM	INTERVENTION
SST	+	----	--	+	0	+	---	0
B 747	-	+	++	++	+	+	---	0
JET	--	----	-	-	0	-	0	0
STOL	++	+++	++	+	--	0	-	0
VTOL	+	+	+	--	+	-	-	0
HSG	+	+	0	--	--	+	--	++
CM	0	0	0	0	0	0	0	0

[a] Note that the *impact is column heading upon row entry*. Thus the impact of *JET* upon *B 747* = +2 while the impact of *B 747* upon *JET* is −4.

Note that the entries in the interaction matrix are not necessarily symmetric, the action of *A* upon *B* is not usually the same as *B* upon *A*. For example we note that *JET* upon *B747* = +2 while the impact of *B747* upon *JET* is −4. The strength and direction of the interaction can easily be adjusted by varying the appropriate interaction coefficient. In the matrix of interactions we have also added another column (temporarily blank) which is reserved for the action of the outside world. These are variables (say governmental intervention) which act upon the system but experience essentially no reaction in turn. This is a very convenient and important feature.

It is the nature of all variables encountered in human experience to be bounded. Invariably there is a minimum below which the variable cannot descend, and at the other extreme there is a maximum beyond which it cannot penetrate. With this in mind we can always scale the range of each of our variables between zero and one. For example, for an aircraft variable a value marginally above zero might indicate the raw conception of the vehicle, and at the other extreme, a value approximating unity would correspond to complete commercial success. With such a scaling in mind we could assign the following values for the present configuration:

SST = 0.2 because the vehicle has entered prototype phase but neither the Concorde or Tu-144 have been commercially introduced,

B747 = 0.35 because it has entered commercial service but not yet in appreciable quantity,

JET = 0.8 because this has entered commercial service in very significant numbers and is a proven yet not overwhelming financial success owing to the large numbers of competing aircraft,

STOL = 0.15 because while this has been tested in research models no viable commercial prototypes have yet appeared,

VTOL = 0.1 since *VTOL* shares all of the technical problems of *STOL* yet more so,

HSG = 0.1 because it is barely beyond the prototype phase and suffers from a multitude of operational problems.

CM = 0.9 inasmuch as the cost of money is hovering at historically high levels.

It will be noticed that both the entries and the interaction matrix and the initial values of the variables are somewhat arbitrary. There is considerable room for disagreement. For example, it would be easy to argue that high speed ground transportation should be assigned an initial value of 0.2 rather than 0.1. Likewise it could be argued that the action of the *SST* upon itself is negative rather than positive owing to the unfavorable publicity is has received. The ease of the model's formulation allows such contrary views to be expressed easily and in a self-consistent fashion. Often it will be found that the particular choice of a single interaction parameter is not terribly important. Our rationale is to give the policy maker free choice in designing his model without being burdened by mathematical and computational complexity. In other words, each policy-maker can change his conception of the structure of the system freely as his intuition into its behavior improves.

But once a particular interaction matrix and initial values have been chosen, then the future is set, continuously evolving from the initial configuration. The actual mathematics that achieves this goal is outlined in the appendix and will be discussed at length in a technical paper. Here, it suffices to say that each interaction is weighted proportionately to the strength of the interaction and also to the relative size of the variable producing the interaction. Also, and most important, growth and decay follow logistic-type growth variations rather than exponential ones, *automatically limiting reaction rates near threshold and saturation*. In a completely self-consistent way then the system will evolve from a knowledge of the binary interaction of its components.

MATHEMATICAL ASIDE (Can be omitted at first reading.)

The mathematical calculations are carried out on an iterative basis, avoiding the need for any explicit discussion of differential equations. To construct our model, we employ a simulation language (*KSIM*) with the following properties:

(1) System variables are bounded. It is now widely recognized that any variable of human significance cannot increase indefinitely. There must be distinct limits. In an appropriate set of units these can always be set to one and zero.

(2) A variable increases or decreases according to whether the net impact of the other variables is positive or negative.

(3) A variable's response to a given impact decreases to zero as that variable approaches its upper or lower bound. It is generally found that bounded growth and decay processes exhibit this sigmoidal character.

(4) All other things being equal, a variable will produce greater impact on the system as it grows larger.

(5) Complex interactions are described by a looped network of binary interactions.

With these conditions in mind consider the following mathematical structure. Since state variables are bounded above and below, they can be rescaled to the range zero to one. Thus for each variable we have

$$0 < x_i(t) < 1, \qquad \text{for all} \qquad i = 1, 2, \ldots, N \text{ and all } t \geqslant 0. \tag{1}$$

To preserve boundedness, $x_i(t+\Delta t)$ is calculated by the transformation

$$x_i(t+\Delta t) = x_i(t)^{p_i} \tag{2}$$

where the exponent $p_i(t)$ is given by

$$p_i(t) = \frac{1 + \dfrac{\Delta t}{2} \sum\limits_{j=1}^{N} (|\alpha_{ij}| - \alpha_{ij}) x_j}{1 + \dfrac{\Delta t}{2} \sum\limits_{j=1}^{N} (|\alpha_{ij}| + \alpha_{ij}) x_j}, \tag{3}$$

where α_{ij} are matrix elements giving the impact of x_j on x_i and Δt is the time period of one iteration.

Equation (3) guarantees that $p_i(t) > 0$ for all $i = 1, 2, \ldots, N$ and all $t \geqslant 0$. Thus the transformation (2) maps the open interval (0, 1) onto itself, preserving boundedness of the state variables (condition 1 above). Equation (3) can be made somewhat clearer if we write it in the following form:

$$p_i(t) = \frac{1 + \Delta t\,|\text{sum of negative impacts on } x_i|}{1 + \Delta t\,|\text{sum of positive impacts on } x_i|} \tag{4}$$

When the negative impacts are greater than the positive ones, $p_i > 1$ and x decreases; while if the negative impacts are less than the positive ones, $p_i < 1$ and x increases. Finally, when the negative and positive impacts are equal, $p_i = 1$ and x_i remains constant. Thus the second condition holds. To demonstrate conditions 3–5, let us first observe that for small Δt, Eqs. (2) and (3) describe the solution of the following differential equation:

$$\frac{dx_i}{dt} = -\sum_{j=1}^{N} a_{ij} x_i x_j \ln x_i \tag{5}$$

From Eq. (5) it is clear that as $x_i \to 0$ or 1, then $dx_i/dt \to 0$ (condition 3). Thus, the expression $x_i \ln(x_i)$ may be said to modulate the response of variable x_i to the impact it received from x_j. Considering x_j individually, we see that as it increases or decreases the magnitude of the impact of x_j upon any x_i increases or decreases (condition 4). Finally, it is seen that condition (5) holds since system behavior is modelled through the coefficients α_{ij}, each of which describes the binary interaction of x_j upon x_i. Although the previous discussion seems to imply that the impact coefficients are constants, this need not be so. In more advanced versions of *KSIM* any of these coefficients may be a function of the state variables and time.

To gain a greater intuitive understanding of this system of equations it is a worth while exercise to examine the one-variable system. The reader can easily check that in this simple case the system exhibits sigmoidal-type growth or decay corresponding to α positive or negative. Such growth and decay patterns are characteristic of many economic, technological, and biological processes.

DISCUSSION

Figure 1 shows the subsequent evolution of the variables from the assumed initial values. It will be noted that with the supposed interactions that *B747* and *STOL* have the brightest future whereas *HSG* and the *SST* are quickly driven to extinction. In other words, they just can't compete with alternate forms of transport.

The aircraft model suggests a number of very interesting intervention schemes. Figure 2 indicates the result of truly massive (and unrealistic) intervention of the federal

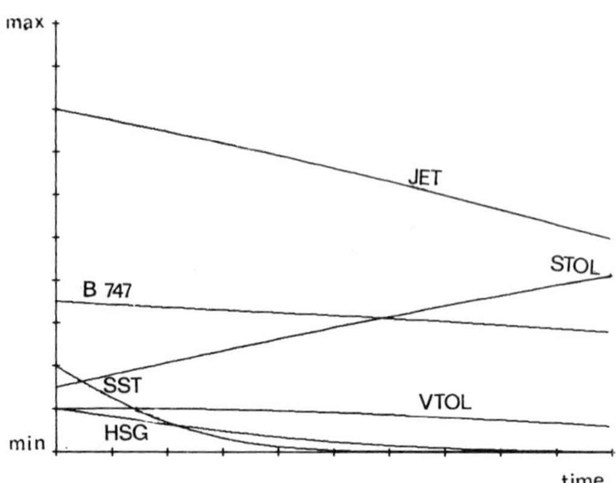

Fig. 1. Projected trends of interaction of transportation modes as given by Table 1.

government on *HSG* sufficient to reverse on the rest of the system except that the growth rate of jumbo jets, *B747*, are slightly enhanced, while *JET* and *VTOL* receive an adverse experience. (The dashed lines indicate their former positions in Fig. 1.)

We are considering much more refined versions of the present analysis including such important effects as environmental present analysis including such important variables as *EI*, environmental impact. These we are considering separately in a more sophisticated format, for example aircraft variables will be calibrated by the actual share of the market rather than subjective estimates. This work will appear separately together with a more elaborate version of the cross-impact language. In addition we are considering models which reveal social and economic consequences of various transportation policies. Owing to the large number of variables in such models it will be much easier to communicate the nature of our work by describing a much simpler model, one related to transportation planning in the greater Vancouver region. Namely, let us consider the possibility of diverting significant traffic away from the automobile

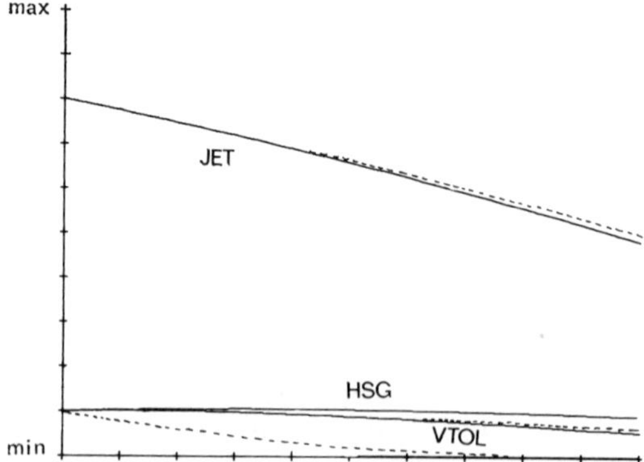

Fig. 2. Massive federal intervention on behalf of *HSG* produces slight declines in *JET* and *VTOL*. (Dashed lines indicate the old trends.) *B 747* prospects are slightly enhanced but too small to be indicated. There is no significant change in the other variables.

and towards rapid transit. As an explicit question may we ask, "Can massive funding of public transit achieve its goal of easy metropolitan access?"

II. A Metropolitan Transit Model

To answer this question we must consider not only economic factors but also subjective considerations, for example the freedom that comes with caruse (immediate availability, multiplicity of route choices, and easy diversion to alternate destinations). In public transit one is locked to timetables, rigid routing, and limited accessibility to destinations. Clearly *FREEDOM* should be a variable. We shall also include *USE* (for the relative fraction *auto* use), *C&C* for comfort and convenience, *COST*, and *SPEED*. In detail we define the variables in the following fashion:

1. *COST*[1]*:* This variable is scaled between zero and one and essentially measures the perceived cost of the automobile as a fraction of the total annual income of the individual. Consequently, an individual making $10,000 a year roughly figures (in his head) the automobile is costing him a thousand dollars per year—whether right or wrong—would have a value for this variable of 0.1.

2. *USE:* This is to be described as the use of the automobile as a transportation medium as compared to all competing means of transportation therefore a value of 0.95 of this variable indicates that the automobile is used 95% of the time as against all possible modes such as public transportation (buses, etc.) or walking.

3. *C&C:* Comfort and convenience and all other attributes of aesthetic desirability. We choose *C&C* to be zero when the use of the *automobile* is accompanied by acute discomfort. When C&C = 1 use of the automobile is with total satisfaction and luxury.

4. *FREEDOM:* This variable essentially measures the freedom of choice involved in travel. It includes such considerations as choice of alternate routes and restructuring of schedules to meet new situations as they occur. It will be obvious that the automobile —which is ready at almost a moment's notice—should have this particular variable set to values which are close to unity.

5. *SPEED:* This variable measures the *perceived* automobile speed between any two random points within the transportation net. Unity is described as the speed between two points under ideal circumstances—maximum speed and no traffic. During the day its value would be something like 0.9 for an automobile except during rush hours when it might drop as low as 0.7 or 0.6.

DISCUSSION OF MATRIX ENTRIES

In the transit model under discussion we have introduced five variables which interact with each other. In addition, there is always the outside world acting upon these variables for an additional five interactions. Accordingly there are five self-interactions, twenty binary interactions, and five external intrusions. Choosing these thirty parameters will define the system. A reasonable first approximation is given in Table 2.

[1] *Note.* We have run many versions with two cost variables, differentiating between direct operating costs (gas, oil, . . .) and implicit overhead costs (depreciation, insurance, . . .). The conclusions of such more refined approaches do not differ in any significant regard from the ones to be presented.

Table 2

An Interaction Matrix Describing the Present Pattern of Public/Private Transportation

Impact	COST	AUTO USE	C & C	FREEDOM	SPEED	OUTSIDE WORLD
COST	0	--	++	+	+	-
USE	++	+	++	++	++	+
C & C	+	--	0	0	0	+
FREE	+	---	0	0	0	++
SPEED	+	---	0	+	0	++

Let us write A : B for the action of A upon B and then we can describe our motivation for the above choices as follows.

AUTO USE:COST (– –). We argue that increased auto use diminishes *perceived* cost. Our rationale for this choice is that the costs of the automobile are largely implicit, most notably depreciation. For most people the major expense of having a car is consigned to the past, i.e. the date the car was purchased. Accordingly, the attitude of most is that a car standing idle in the garage represents an expense without any concomitant service. Thus people tend to use their car as a means of psychologically amortizing their high investment. This is in sharp contrast to the use of public transit where toll costs are highly explicit in nature. Whenever an individual uses the bus he is acutely reminded of its cost when he reaches into his pocket for a coin.

C&C:COST (++). Obviously the more luxurious a car, the more it will cost.

FREEDOM:COST (+). Clearly freedom exacts a price. For example increased reliability come with either better automobiles or better service and maintenance.

SPEED:COST (+). Faster and more manoeuverable cars generally cost more.

OUTSIDE WORLD:COST (–). Owing to technological advances and productivity gains, the *relative* cost of a car has been declining consistently for about forty years. The costs of car ownership have risen significantly less than wage rates or general inflation.

COST:USE (++). The more expensive a car the more anxious the individual will be to use it and to display his prized possession.

USE:USE (+). Use encourages more use and tends to be habit forming.

C&C:USE (++). The more comfortable and convenient the car (especially in contrast to mass transportation), the more its use will be encouraged.

FREEDOM:USE (++). The easier a car's access to arbitrary destinations the more its use will be encouraged.

SPEED:USE (++). The greater the relative speed of the car as compared to public transit the more it will be used.

OUTSIDE WORLD:USE (+). The outside world encourages the use of the automobile because the automobile is a much higher status mode of transport than public transit.

COST:C&C (+). The more a car costs in general, the more its comfort and convenience.

AUTO USE:C&C (– –). Clearly the more an automobile is used the less comfortable it will tend to be owing to mechanical deterioration as well as increased road congestion.

OUTSIDE WORLD:C&C (+). The outside world tends to encourage automotive comfort and convenience by virtue of freeway construction and other traffic engineering.

COST:FREEDOM (+). A brand new car will be more reliable than an old clunker. In general, we can expect that the more the cost the greater the freedom from service problems.

AUTO USE:FREEDOM (–). This is sharply negative because increasing auto use markedly increases traffic congestion which sharply diminishes freedom.

OUTSIDE WORLD:FREEDOM (++). The outside world will generally build freeways and provide alternate routing as traffic mounts, thus tending to restore freedom.

COST:SPEED (+). Jaguars can go faster than Volkswagens.

AUTO USE:SPEED (– – –). Auto use sharply inhibits speed because of increasing traffic congestion.

FREEDOM:SPEED (+). This is taken as a positive factor because the more choices an individual has in reaching his destination the higher his average speed can be. A car can choose an alternate route where a bus might be stalled in traffic.

OUTSIDE WORLD:SPEED (++). Again, because of the government's predilection to build freeways, it continuously encourages greater speed.

As initial values we choose

$$
\begin{aligned}
&\textit{AUTO USE} &&= 0.8\\
&\textit{C\&C} &&= 0.6\\
&\textit{FREEDOM} &&= 0.6\\
&\textit{SPEED} &&= 0.5\\
&\textit{COST} &&= 0.2
\end{aligned}
$$

These choices are *not* representative of the present situation and have deliberately chosen to handicap auto USE. It will be seen that the handicap is but a fleeting burden.

DISCUSSION OF COMPUTER OUTPUT

Once the initial values and the matrix of interactions have been agreed upon it is a simple procedure by our methods to project the future states of the system. Figure 3 illustrates the subsequent behavior that would emerge from our assumptions.

It will be seen that variable 2, *AUTO USE*, rises continuously until it reaches maximum. This is in spite of a slow but steady decline in comfort and convenience, freedom, and speed. If this is surprising then we must remember that as comfort, freedom, and speed decline, auto use begins to decline but this results in a decreased traffic congestion which ultimately encourages a net increase in *AUTO USE*. (An easy illustration of counter-intuitive behavior).

The behavior of the system cannot really be fully understood until a number of intervention strategies have been suggested and pursued to their conclusion. However any realistic manipulation will produce a barely perceptible effect. This stems from the

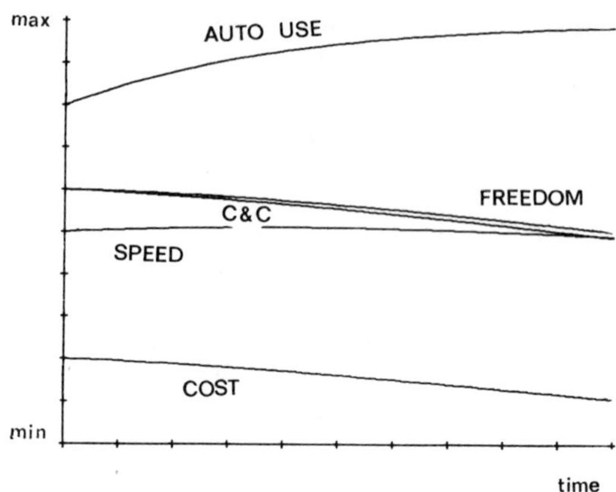

Fig. 3. Projected trends as set by the impact matrix given in Table 2.

intrinsic stable behavior of multiply-interacting systems. Even in our relatively naive model there are 21 nontrivial interactions and this complexity endows the system with a stubborn resistance to change. This of course is well known to ecologists (and some politicians) but the importance of this knowledge cannot be overstated. To illustrate the resistance to change we consider some extreme disruptions of the system.

TAX THE CAR TO DEATH?

In the next run we have made an extreme external perturbation, supposing that the outside world markedly increased the cost of the car instead of reducing it over time as before. We go to an extreme, making *OW:COST* = +9, rather than –1. As we can see from Fig. 4, *COST* instead of diminishing rises sharply to maximum. However, automobile *USE* still rises to saturation except that its rate of increase is no longer as sharp as it was in Fig. 3. It will also be noted that *C&C* and *FREEDOM* still decline. However *SPEED* now *rises*, slowly to be sure but significantly. This is largely a result of decreased congestion.

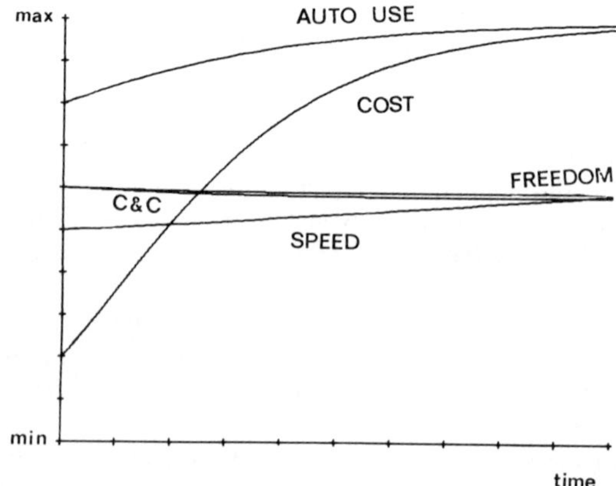

Fig. 4. By setting *OW:COST* = +9, rather than –1, we are hopefully "taxing the car to death." As can be seen, it manages to survive.

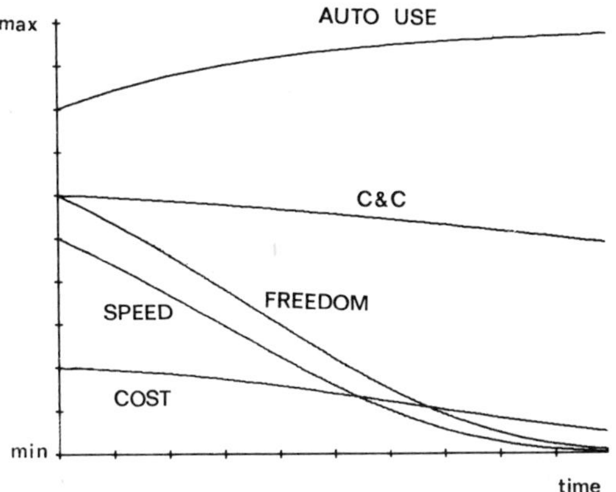

Fig. 5. Reversing freeway construction can be modelled by setting *OW:FREEDOM* = −2 and *OW: SPEED* = −2 instead of their former positive values. Even so, *AUTO USE* continues to rise.

HOW ABOUT HALTING FREEWAY CONSTRUCTION?

For our next run we have restored the original values in the interaction matrix and have made the following changes: The intervention of the outside world upon *FREEDOM* and *SPEED* instead of being doubly positive would be doubly negative. This would correspond to a strategy diverting funds from freeway construction to the construction of rapid transit. It will be seen (Fig. 5) that freedom and speed now decline rather sharply as well as comfort. However, despite these inhibitions auto USE, continues to rise to its saturation value.

DIVERTING FUNDS?

We now consider a different strategy, one of attempting to use negative feedback. In other words making the cost of the car tend to subsidize its opposite, public transit. This is achieved in the following way. Previously the impact of cost upon freedom and speed had been positive, in other words, man could buy more freedom and speed by spending more money. Now we reverse the situation and tax him proportionately in

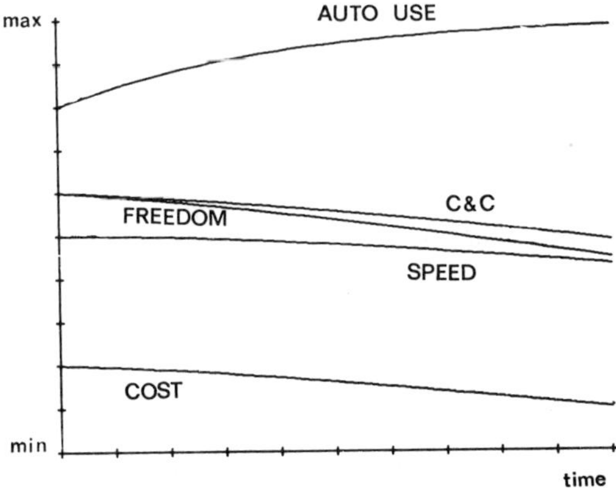

Fig. 6. Diversion of funds is introduced by setting *COST:SPEED* = −1 and *COST:FREEDOM* = −1.

accordance with *FREEDOM* and *SPEED*, diverting funds to improve rapid transit. Making the indicated changes results in Fig. 6.

As can be seen, *C&C* and *FREEDOM* are diminished but none the less *AUTO USE* continues to go up. The problem is that the negative feedback is linked through a variable which itself is rather small, namely cost. As long as the automobile occupies only ten or twenty percent of an individual's total budget increasing the cost of the automobile will be absorbed by other strategies than diminished *USE*. For example, the individual might be forced to drive a Chevrolet rather than a Buick, but he will doggedly continue to drive.

ANY HOPE?

Having considered a number of intervention models all of which implies the automobile's ultimate domination we may ask, can the model yield any other answer? The answer is yes, but only if the intrusions in the system are extremely strong and powerful, or a magnitude probably far in excess of that which the voting public would be willing to endure. What we are working against is the following linkage: In Table 2 the impacts *AUTO USE: FREEDOM* = –3 and *AUTO USE:SPEED* = –3 (congestion) serve to inhibit auto use. What seems to be a paradox, but is a typical example of the perverse behavior of looped systems, is that *almost anything we do to diminish AUTO USE makes AUTO USE more attractive because it decreases congestion.*

Conclusions

Certainly the naive models just considered are hardly conclusive. No doubt many readers will be infuriated, arguing for different choice of initial conditions or interactions. But this is just what we have sought, i.e., controversy and interaction. But also we have introduced a single and graphic model that any one can use, be they politician or citizen's action group. No mathematical or systems training is required. What is needed is the discipline to carefully consider *all* the interactions. Any one using the model must fill in all the matrix entries. Whatever his choice, we have found more often than not he will be hoist by his own petard. So often the system will either fail to respond or respond in an entirely perverse manner. What we want to communicate are two messages:

(1) *COMPLETENESS*. The need to consider *all* interactions. If we have n variables then there are n^2 interactions and these must be accounted for. Using purely qualitative considerations it is very easy to omit, forget, or underestimate significant interactions.

(2) The *STABILITY* that emerges from *COMPLEXITY*. The more interlocked variables the more resistant the system will be to arbitrary change. Thus, in a time of social upheaval, an elementary but reasonably comprehensive means of communicating, "Beware what you do, that you do not undo yourselves," is obviously needed, particularly if the model allows alternate strategies to be easily simulated.

A strategy that will work is to have the outside world strongly raise the cost of the automobile (*OW:COST* = +9) and at the same time to divert that cost to inhibiting auto use (*COST:USE* = –9). If this is done, the *COST* rises sharply as can be seen in Fig. 7 and *USE*, begins a slow but steady decline. Note however, that with this decline in *USE* that *C&C* and *SPEED* all rise.

In this presentation we have tried to communicate a conclusion that continuously emerges in working with simulation models: *The structure of the system* (the nature of

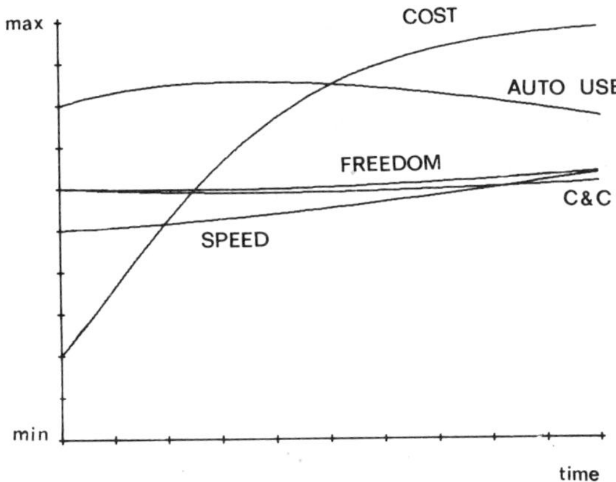

Fig. 7. A successful strategy: Half measures are ineffective. Truly massive cost increases are required, *OW:COST* = +9 and *COST:USE* = −9. Note even so that *AUTO USE* maintains its initial rise and then declines but slowly.

its interactions) *is far more important than the state of the system.* Any artificial increment of a variable is usually immediately dissipated unless concomitant structural changes are made. What is most important are the *linkages* of any variable to the other variables of the system. A major objective in devising our model is to communicate the overriding importance of *structure rather than state* to policy makers. As Jay Forrester has pointed out in a number of books, notably *Industrial Dynamics* and *Urban Dynamics* (MIT Press) interventions in complex systems often lead to results which are entirely at odds with the initial expectations. Any complex system defines an integrity of its own and strongly resists external changes, a fact well understood by ecologists. When complex systems change they seldom change continuously but rather flip suddenly into an entirely new configuration. These subjective conclusions can be made precise through the associated mathematics, and this will be the subject of a number of papers.

Extensions

In other papers (Kane, Vertinsky, and Thompson) we discuss the following elaborations of our language.

1. *CALIBRATION.* While the model easily accepts qualitative inputs, more precise methods of arriving at cross-impact parameters are needed. We are considering techniques based upon multiple correlation and the "Delphi method."

2. *HIGHER-ORDER SIMULATION MODELS.* For greater realism we have begun consideration of cascaded models of the variety described. This permits the interactions themselves to be functions of the state of the system. Clearly A need not always be B's friend. A's attitude towards B can be conditional on the relative status of their difference A − B or perhaps depend upon the state of a third individual C.

Perhaps much more important is the following refinement. Often we sense not a variable but rather *changes in a variable.* Our response to environment has much this character. Whether our locale is good or bad we quickly become acclimatized to it and then become sensitive only to gross changes. Such derivative interaction is very important and is the subject of other papers (Kane, Vertinsky, and Thompson).

Bibliography

J. Kane, I. Vertinsky, and Wm. Thompson, *Health Care Delivery*, Socio Economic Planning Sciences (in press).[†]

J. Kane, I. Vertinsky, and Wm. Thompson, "Environmental Simulation and Policy Formulation", *Proc. International Symposium on Modelling Techniques in Water Systems*, Vol. 1, A. K. Biswas, Ed., Ottawa (1972).

[†] **6** (1972) pp. 283–293

VI. Specialized Techniques

VI. A. Introduction

HAROLD A. LINSTONE and MURRAY TUROFF

Cross-impact analysis is only one of the interesting paths being pursued in the exploration of the frontiers of Delphi. In this chapter we consider other significant prospects uncovered in recent research.

The problem of "reduction" occurs again and again in Delphi as well as other analysis efforts. How similar are the goals, programs, objectives? Do these items overlap? Dalkey in his article illustrates the use of "cluster" analysis techniques to narrow the number of unique variables making up the "Quality of Life" as viewed by a Delphi panel. It is a very real problem in many Delphis to be able to reduce the scope of the items the respondent group must consider. One must carefully weigh the effort to be expected of a respondent group and establish a consistency with what actual time the panelists may have to devote to the effort. Techniques such as the one discussed by Dalkey can be of benefit to many applications.

Another technique to reduce the necessary range of considerations is that of Multidimensional Scaling. Although this technique is widely used in marketing research, it has yet to see application in a Delphi study. However, since we feel it has tremendous potential for application in Delphi design, we have asked Carroll and Wish to supply a review article on this subject. In essence, this is a method of systematically exposing underlying or hidden considerations (i.e., dimensions) about a set of multiple but related questions. In another sense this technique constitutes the development of a more sophisticated form of cluster analysis.

The third article in this chapter cannot strictly be subsumed under the label Delphi.[1] However, it introduces a very novel questionnaire concept of exploration of the future which may well have significant impact on the Delphi communication process. In one way it represents a rather pragmatic demonstration of some of the concepts put forth in Scheele's article. Adelson and Aroni use a set of images in the form of printed pictures rather than words. The latter are obviously inadequate in communicating holistic insights about complex systems. We fail dismally when we attempt to describe Picasso's "Guernica," Van Gogh's "Starry Night," Ravel's *Daphnis and Chloe*, or Olivier's *Othello* in words. Similarly we cannot depict a Soleri city of the future, Skinner's

[1] In particular, it does not have the iterative or feedback characteristic.

The Delphi Method: Techniques and Applications, Harold A. Linstone and Murray Turoff (eds.)
ISBN 0-201-04294-0; 0-201-04293-2

Walden II, Huxley's *Brave New World*, or Bell's post-industrial society satisfactorily by narratives. Holism is lost in a linear communication system. Pictures are not the ultimate answer; we may need life-size "live-in" models (à la Disneyland) to experience the true meaning of a future alternative. But pictorial images are a step in this direction and there is no justification for ignoring them in the development of Delphi as a communication concept.

In the final article John Sutherland constructs a formal process for scenario building which relies strongly on Delphi. Both normative (futures-creative) and exploratory scenarios are generated through a very disciplined use of Delphi. The author considers two means of forming panels: (*a*) clustering of individuals with similar backgrounds and strong internal consensus in each panel, implying wide differences among panels, and (*b*) stratification to obtain high diversity in each panel but small variance among the various collective panel views. In the first step of the process desired attributes are elicited; next a functional/structural relationship is established for them. A series of normative scenarios is thus derived with aggregate subjective indices of desirability and feasibility. In parallel an extrapolative scenario with indices of likelihood or probability of occurrence is created. The gap between any of the normative and the extrapolative scenarios is treated by a process of "qualitative subtraction," isolating differences and successively abstracting isomorphisms. The remaining generic or unreduced problems are then arranged in a causal order and allegorized into either hierarchical or reticulated networks. Such models then lead to action proposals or prescriptive policies which have a high *a posteriori* probability of facilitating a transformation from an extrapolative and less favorable to a normative and more desirable system state over some time interval. These proposals may be viewed as metahypotheses subject to correction by empirical trial. The process slowly changes the subjective probabilities of the Delphi-based normative scenario into objective probabilities.

The present and preceding chapters of the book present a set of papers that rest on reasonably sophisticated ideas and analytical considerations. If this may be interpreted as a sign of maturity for a professional endeavor, then Delphi has come of age. From a more serious viewpoint, many Delphi applications to date have not called for extremely sophisticated approaches to the analysis of responses. In some other cases, however, these techniques would have clearly been beneficial to the exercise.

VI. B. A Delphi Study of Factors Affecting the Quality of Life

NORMAN C. DALKEY

1. Earlier Related Studies

In a previous publication [1] several Delphi studies are reported using respondent groups to identify and estimate linear weights for those aspects of experience which they judged to be important in determining the quality of life or sense of well-being of an individual. The procedure was first to ask each respondent to list a certain number of aspects of experience which he thought were most important in determining the quality of life of an individual. This number varied from five to ten over the different exercises—depending mainly on the resources available to process the lists. In general, this open-ended round produced a long list of nominated factors—typically two hundred to three hundred. Inspection of the lists showed a fair amount of similarity among some of the items, but very few identical pairs. These long lists were compacted by the experimental team by aggregating items that appeared to have high similarity, producing an intermediate list of about fifty items. A matrix was formed and each respondent was asked to judge the similarity of each item to every other item. (In a 50 × 50 matrix, this required 1225 estimates, since the similarity relationship was presumed to be symmetrical.) The combined similarity matrices were then analyzed by the hierarchical clustering routine developed at Bell Telephone Laboratories by S. C. Johnson [2] and the hierarchy was truncated at lists of twelve or thirteen clusters.[1]

[1]In this procedure objects are clustered according to the (judged) similarity between them. The objects within a cluster are more similar to one another than to objects belonging to a different cluster. In addition, the procedure merges similar clusters into larger clusters in a step-wise fashion until all the objects are placed into a single cluster. This hierarchy of clusters can be used in its entirety, or, depending on the purposes of the experiment, truncated at some appropriate level to furnish a list of aggregated items.

Research reported herein was conducted under Contract Number F30602-72-C-0429 with the Advanced Research Projects Agency, Department of Defense.

The Delphi Method: Techniques and Applications, Harold A. Linstone and Murray Turoff (eds.)
ISBN 0-201-04294-0; 0-201-04293-2

The aggregated lists were then presented to the respondents for judgments of relative importance. In general, the exercises indicated that group relative importance ratings produce reasonable ratio scales, and that the reliability of such judgments across randomly selected groups is high.

2. Models of the Quality of Life

The general model of individual well-being underlying these studies is that quality of life is a function of the individual's location in a "quality space"; that is, the sense of well-being enjoyed by an individual depends upon the extent to which his experiences exemplify several basic characteristics. Inherent in this view is the existence of trade-offs among these characteristics; two dfferent individuals can enjoy about the same level of quality of life with highly different patterns of experiential rewards. One can be comfortable, receiving a great deal of love and affection, and living a routine existence; the other can be living an exciting life, with a high sense of achievement, and lonely, and each report about the same overall level of well-being.

What is not spelled out in this general view is how the contribution of each component is reflected in overall quality of life. There are two broad possibilities. One, by far the simpler of the two, is that overall quality of life is simply a weighted arithmetical sum of the individual's status on each component. In symbols

$$Q(i) = \sum_j q_{ij} w_j + C_i, \tag{1}$$

where $Q(i)$ is individual i's overall quality of life, q_{ij} is i's status on quality j, w_j is the relative importance of quality j, and C_i is a constant added to balance differences of scaling on Q and the q_{ij}. The essence of this model is that each quality makes a fixed relative contribution to overall Q, independently of where the individual stands on all the other qualities.

The second model assumes that the relationship is more complex, and the contribution of any one quality depends on where the individual stands with respect to all the others. That is, the relative importance relationships are not fixed, but are functions of the individual's location in the quality space. These two different hypotheses can be illustrated (using, for simplicity, only two of the thirteen qualities) as in Fig. 1 and Fig. 2. What is diagrammed in each case is the set of lines of equal quality of life. In both cases, quality of life (QOL) increases with a simultaneous increase in sense of achievement and in affluence. But in Fig. 2, the amount of increase from achievement depends on how affluent the individual is, whereas this is not the case for Fig. 1.

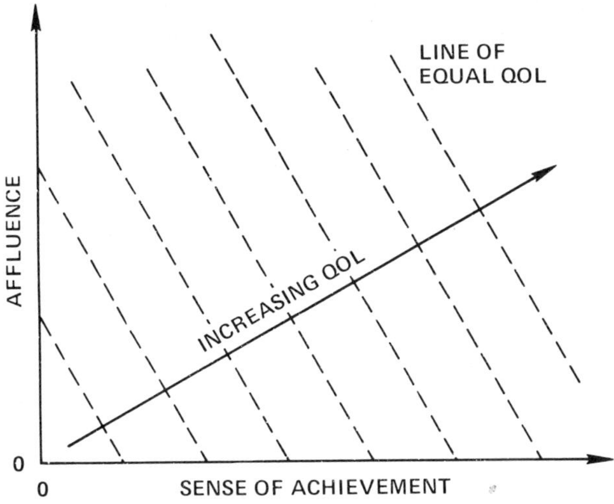

Fig. 1. Illustration of linear model of quality of life.

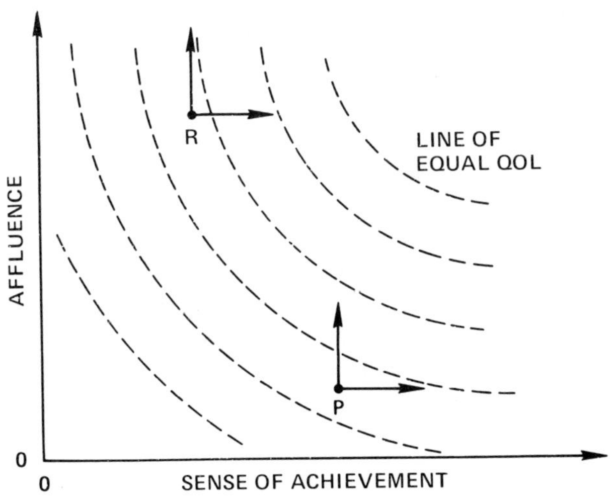

Fig. 2. Illustration of nonlinear model of quality of life.

Although existing data are not very helpful in deciding between these two types of models, it seems likely that a nonlinear model is, at least in principle, the more accurate representation. In Rokeach's survey study of American values [3], the low-income group ranked "comfort" at level 6, whereas the high-income group ranked it at level 15 (lower numbers indicating higher importance with a total range of 1-18); conversely, the low-income group ranked a sense of accomplishment at 12, and the high-income group ranked it at level 5. A possible interpretation of these data is expressed in Fig. 2 where the low-income individual P would gain more by a small increase in affluence (vertical arrow) than he would by a small increase in achievement (horizontal arrow), whereas the reverse is true for the high-income individual R.

Although the nonlinear model seems likely to be more correct, the linear model may be a fairly good approximation. From previous studies we can say that there is a large amount of uncertainty concerning the relative importance of qualities, and it appears reasonable to assume a fair amount of uncertainty in the individual's judgment of his own well-being. Under conditions of uncertainty on all parameters, a linear model often gives as good an approximation as a potentially more precise model where uncertainty in the data obscures the more complex interactions.[2]

3. Procedure

To investigate this possibility, and also to check the earlier quality of life studies with a nonstudent set of subjects, the entire exercise was replicated with a group of twenty-seven professional men involved in the Executive Engineering Program at U.C.L.A. These were mid-career engineers with a median age of thirty-seven, and upper-middle-class incomes. All had progressed at least through the bachelor's degree; more than half had obtained master's degrees.

The initial round of the experiment was identical to that of the earlier studies; each respondent was asked to list no more than seven aspects of experience that contribute in an important way to an individual's quality of life. The aggregation of the responses from this round was carried out in a way slightly different from that used with the students. After minimal editing to remove a few items that appeared to be out of context, a "raw" list of 227 items was presented to the subjects. These were reproduced on IBM cards, and each

[2]Since this section was written, Robin Dawes [4] has published the results of several studies showing that a much stronger statement can be made in this regard. This is taken up more fully in the final section of this paper.

subject had his own set of 227 cards. The subjects were requested to sort the set of cards into piles where each pile contained items describing roughly the same quality. The sorting activity can be thought of as a truncated similarity rating, where only ratings of 0 or 1 are employed. The resulting judgments were transformed into a group matrix and processed with Johnson's hierarchical clustering routine. The hierarchy was cut at a convenient level to furnish a list of twelve clusters.

4. Results

The resultant list of clusters is presented in Table 1 along with the medians and quartiles of the subsequent relative importance ratings of the group. The items listed are those most frequently mentioned. Table 2 gives the list of clusters identified in a previous study with forty upper-class and graduate students at U.C.L.A.

There is not a great difference between the two either in terms of clusters, or in terms of judged relative importance. Privacy and Sex do not appear as separate items in Table 1, and in Table 1 Achievement is separated from Respect and Prestige, which are combined in the student list. Other minor differences are apparent. There is probably not much to be gained by attempting to probe these differences. One evident difference is the number of ties for the top rating by the engineer group. There is no clear indication from the data why they should have been less discriminating among the high-rated clusters than the students.

In a subsequent round, the respondents were requested to rate their present status on each of the qualities. The ratings were expressed on a scale ranging from -100 to $+100$, where -100 meant that the present status of the respondent on that quality was negative with respect to its influence on his sense of well-being and "as bad as possible," and $+100$ meant that the influence of the quality was positive and "as good as possible." They also were requested to assess where they stood in an overall sense—that is, to assign a number between -100 and $+100$, where the extremes meant life in general is as bad as possible or as good as possible respectively. The respondents were also asked to rate themselves on the conventional survey rating scale, namely, whether they were *very happy*, *fairly happy*, or *not very happy*.

Figure 3 shows the distribution of the group on their global quality of life ratings, and the lower half of the figure indicates the location of the verbal happiness ratings. It is clear that the *fairly happy* category is not very discriminating, using the numerical ratings as a standard.

Table 1

Quality of Life Clusters and Relative Importance Ratings, Engineer Group

Quality	Relative Importance		
	Median	Lower Quartile	Upper Quartile
Self-confidence, self-esteem, self-knowledge, pride	80	70	100
Security, peace of mind	80	70	97
Sense of achievement, accomplishment, success	80	75	90
Variety, opportunity, freedom	80	60	90
Receiving and giving love and affection	80	50	90
Challenge, intellectual stimulation, growth	75	60	80
Comfort, congenial surroundings, good health	70	50	80
Understanding, helping and accepting others	60	30	75
Being needed by others, having friends	60	40	70
Leisure, humor, relaxation	53	50	75
Respect, social acceptance, prestige	40	20	70
Dominance, leadership, agression	30	10	50

The items are ordered first on the median—then on the upper quartile, and then on the lower quartile in case of ties.

Table 2

QOL Factors, Student Group

	Relative Importance (Median)
1. Self-respect, self-acceptance, self-satisfaction; self-confidence, egoism; security; stability, familiarity, sense of permanence; self-knowledge, self-awareness, growth	100
2. Love, caring, affection, communication, interpersonal understanding; friendship, companionship; honesty, sincerity, truthfulness; tolerance, acceptance of others; faith, religious awareness	96
3. Peace of mind, emotional stability, lack of conflict; fear, anxiety; suffering, pain; humiliation, belittlement; escape, fantasy	91
4. Challenge, stimulation; competition, competitiveness; ambition, opportunity, social mobility, luck; educational, intellectually stimulating	80
5. Achievement, accomplishment, job satisfaction; success; failure, defeat, losing; money, acquisitiveness, material greed; status, reputation, recognition, prestige	79
6. Sex, sexual satisfaction, sexual pleasure	78
7. Individuality; conformity; spontaneity, impulsive, uninhibited; freedom	76
8. Social acceptance, popularity; needed, feeling of being wanted; loneliness, impersonality; flattering, positive feedback, reinforcement	75
9. Involvement, participation; concern, altruism, consideration	72
10. Comfort, economic well-being, relaxation, leisure; good health	63
11. Novelty, change, newness, variety, surprise; boredom; humorous, amusing, witty	61
12. Dominance, superiority; dependence, impotence, helplessness; agression, violence, hostility; power, control, independence	58
13. Privacy	55

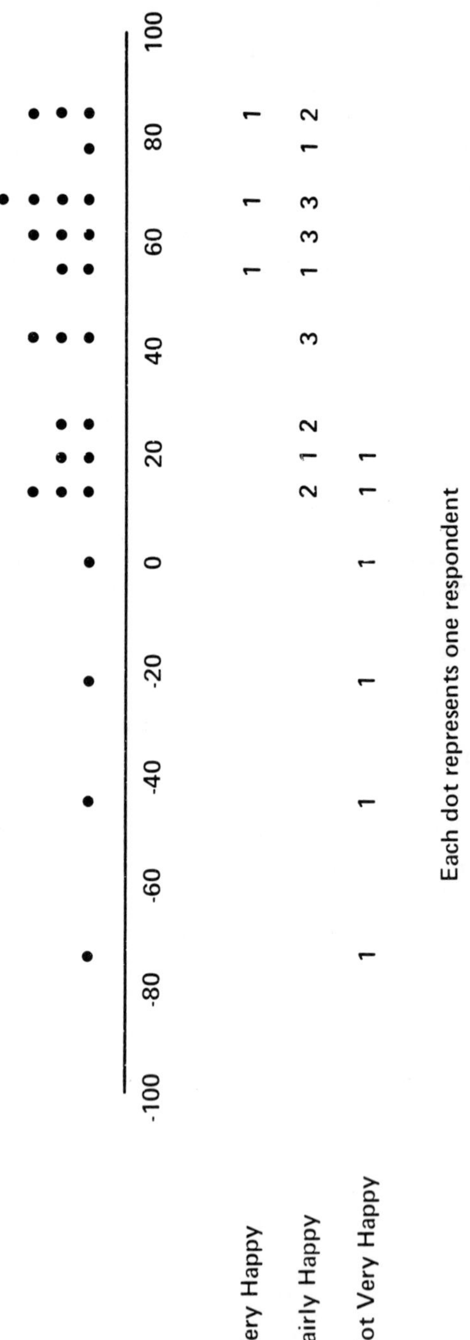

Fig. 3. Distribution of overall scores (present status judgments) and happiness ratings.

The engineer group is somewhat divergent from the general public in their verbal assessments. In particular, the proportion reporting *very happy* is definitely smaller than for the general public. [5] On the other hand, the numerical ratings appear to give a different picture—40 percent of the group rated themselves at 70 or higher. It would appear that a much more careful investigation is needed of the relationship between numerical and verbal ratings on the happiness question.

Given the individual self-ratings on qualities and on overall quality of life, several forms of the linear model can be tested. The general form of the test is to predict the overall ratings from the ratings on qualities. This can be done, using formula (1) where we can use for the w_j's either the individual's own relative importance ratings, or the group relative importance ratings. In addition, empirical weights can be computed by determining the best linear prediction of the overall ratings from the ratings on qualities.

When the overall ratings are computed using the individual's own relative importance numbers, the correlation between actual and computed scores is .57. When the same computation is made using group relative importance ratings, the correlation between actual and computed scores is .58. These correlations are not particularly impressive. They indicate roughly 33 percent (the square of the correlations) of the variation in the overall ratings has been accounted for by the predicted ratings. In addition, the group relative importance ratings do not give a better prediction than the individual relative importance ratings.

The results of the linear estimation computation are displayed in Table 3. The multiple correlation associated with this set of linear estimation weights is .84, hence the proportion of the variance accounted for is about 70 percent. This would generally be considered a favorable degree of accordance of the empirical linear model with the data. However, a number of considerations are raised by the computation. The weights bear little relation to the relative importance ratings of the group; the Spearman rank-order correlation between the two is −.16. The two items rated lowest and next to lowest by the group receive the third highest and highest estimation weights. In addition, two items that are given high weights by the computation, namely Respect and Variety, have negative coefficients—quite contrary to the perception of the group.

If the results of the computation were taken literally they would indicate that the group has a relatively clear perception of the basic ingredients of their sense of well-being, but that they have a very poor perception of the actual relative importance of the qualities. In particular, Dominance and Respect, which the group rates as least important, would be among the most important; and Security and Love and Affection, which the group ranks high, would be comparatively unimportant. Variety and Respect would make negative contributions to quality of life.

Table 3

Linear Estimation Weights for Qualities

Self-confidence	−0.29281
Security	−0.08595
Achievement	0.56667
Variety	−0.49070
Love and Affection	0.11297
Challenge	−0.06746
Comfort	0.51529
Understanding	0.12987
Needed	0.02671
Leisure	0.22044
Respect	−0.79423
Dominance	0.53622

5. Additional Analyses

There is no way to reject this interpretation on the basis of the data as it stands, and it certainly furnishes some intriguing hypotheses for further investigation. To cast some additional light on the stability of the qualities identified by the cluster technique, several factor analytic studies of the data were carried out. In the most extensive study, nine factors were identified. Only the first six factors were selected for rotation. (These six accounted for .807 of the variance; the next three increased the variance accounted for to .826.) The six factors and the loadings of the qualities on them is presented in Table 4. The first factor is a mixture of Self-confidence, Respect, and Understanding others—a generalized esteem factor. The second factor is almost entirely Dominance. The third might be considered a generalized success factor, involving Sense of achievement, Variety, and Comfort. The fourth is an interesting and rather puzzling combination of Love and Affection and Challenge. The fifth is primarily Leisure. The sixth is difficult to interpret. The two highest loadings—for Comfort and Being Needed—are negative.

All of the qualities that have high estimation weights in Table 3 have high loadings on no more than one factor. In addition, for all the factors except the rather mysterious factor six, there is a relatively sharp separation between the few qualities that load highly on the factors and all the others. In this respect, the factors are relatively "clean." This consideration lends some weight to the assumption that most of the aggregated qualities defined by the cluster analysis are fairly sharply defined.

Table 4

Factor Analytic Study

Variable	Factor 1	2	3	4	5	6
1 Security	0.25873	0.00800	0.30332	0.19134	0.33480	0.29012
2 Leisure	−0.02679	−0.03369	−0.03813	−0.09306	0.98470	−0.01437
3 Challenge	0.06144	0.20352	0.08204	0.72517	−0.05388	0.17499
4 Self-confidence	0.64276	0.03340	0.17468	0.26906	−0.02419	0.02358
5 Variety	−0.15179	0.10266	0.63550	0.09170	0.07230	−0.05646
6 Understanding	0.68341	0.09297	−0.13627	0.28061	0.18222	0.00691
7 Achievement	0.20224	−0.08240	0.94888	−0.05396	−0.07511	0.07892
8 Comfort	0.10102	0.00348	0.58909	0.11555	0.05377	−0.45189
9 Needed	0.12594	0.03455	0.20275	0.34421	0.28623	−0.42441
10 Love and Affection	0.07256	−0.04199	0.03769	0.87896	0.02770	−0.22396
11 Dominance	0.02457	1.03902	−0.04365	0.05515	−0.03478	0.00276
12 Respect	0.61814	0.33582	0.17243	−0.23114	−0.05515	−0.07596

The exercise is probably too small to draw any firm conclusions. It is possible that the counterintuitive results of the linear estimation analysis stem from nonlinear trade-offs. The variations within the group in self-ratings on the qualities are large (standard deviations range from 39 to 55) and thus the members of the group are scattered widely in the quality space, and nonlinearities in the trade-offs could make major differences in the contributions of the qualities to the overall score of different subjects. In short, the linear model may not be appropriate.

Another possibility is that despite the high multiple correlation obtained, an omitted variable is producing systematic effects. If the data on overall scores are analyzed by a 2 × 2 breakout on income and age, as in Table 5, a clear indication is obtained concerning the combined effects of these two. The numbers in the table are the median overall ratings of the members of the designated classes. The numbers in the small boxes refer to the number of cases. The younger, lower-income group tends to rate itself higher than the older, higher-income group. The relationship between age and income is so strong (the off-diagonal cells have few members) that it is not possible to distinguish the separate effects of the two variables. It is rather surprising that all the negative self-ratings occur in the high-income, older group, and all the ratings of *very happy* occur in the younger, lower-income group.

Table 5

Analysis of Overall Self-Ratings in Terms of Income and Age

		AGE	
		35 or less	Over 35
INCOME	20 K/year or less	70 [7]	50 [3]
	Over 20 K	50 [4]	30 [13]

The numbers in small boxes refer to the number of cases.

The relationship between age and income and overall ratings appears to be stronger than those found for the general public, and as far as income is concerned, counter to the relationship found in the general public, where increasing income is correlated with reports of higher happiness [5].[3]

Some of the anomalous features of the linear estimation weights in Table 5 may reflect the combined role of age and income. Thus, one would expect a fairly high correlation between age-income and prestige.

6. Discussion

The issue is left open in the exercise whether the individual, in making a judgment concerning his own quality of life, is estimating some directly perceived internal state (e.g., a "feeling tone") or whether he is, in effect, using his status on the various quality dimensions as a basis for estimating a nonperceived condition. In this respect, the QOL estimation task is formally similar to a number of perceptual or judgmental tasks—e.g., a psychiatrist making a diagnosis of mental illness on the basis of certain symptoms, or an individual estimating distance using various visual cues.

There is a fairly extensive literature on the type of model most appropriate for such multidimensional judgments, beginning with the work of Brunswick. For many perceptual judgments, objective measures exist by which the judg-

[3]However, the survey results may not be extrapolable to the income range involved here. Most of the survey data have a top category of $15,000 or over. A recent (March 1972) telephone survey conducted by the advertising firm Batten, Barton, Durstine, and Osborn indicated that individuals with high incomes rate themselves relatively less happy than those with middle income.

ments can be evaluated. For nonperceptual tasks—e.g., the psychiatrist's diagnosis—the situation is often similar to the QOL judgment in that there is no readily available objective measure.

Robin Dawes has recently reported on a series of studies of multidimensional judgments of both sorts—clinical estimation of degree of psychosis based on the set of test scores furnished by the Minnesota Multiphasic Personality Inventory, prediction of grade-point averages by faculty screening committees, and the like [4]. In these studies, weights for the putative parameters selected at random gave much more accurate results (measured by correlation with the judgments of a criterion team in the case of the clinical diagnosis and against subsequent actual grade-point averages in the case of the performance predictions) than the judgments of the individuals (or in the case of the clinical judgment, of the groups as well). Equal weights did even better. Dawes concludes that for these relatively poorly determined estimation tasks, human judgment may be relatively good in selecting the relevant dimensions, but relatively poor in determining the weights to be attached to those dimensions.

In this respect, Dawes' conclusions are quite compatible with the results reported earlier in the paper. Since writing the previous sections, the QOL exercise has been replicated with a group of thirty-two graduate students attending classes on Futures Studies at UCLA and UC, Irvine. Once again, the linear estimation of self-rated QOL gave a high R^2 (.75 in this case), but the estimation weights differed significantly from those obtained with the engineers.

There appear to be several issues raised by the present study as well as those of Dawes and others for Delphi methodology. One of the basic tenets of the Delphi approach to decision problems has been that for uncertain questions, where solid data or theories are lacking, informed judgment is the only resource available (leaving aside the option of postponing judgment until better information is available).

Dawes has suggested that for certain kinds of judgments, we may be better off with nominal assumptions than with human judgment. The evidence he presents warrants taking the suggestion seriously.

One possibility here—which I have to admit is highly tentative—is that an extension of what is often referred to as sensitivity analysis may be called for. For example, in the case of multidimensional estimates, if the correlation with some criterion is to be the figure of merit for the estimates, then the analysis may be highly insensitive to the precise form of the model or the weights attached to a linear estimation.

As a specific example, consider a two-dimensional problem where a quantity $Q = Q(x,y)$ is to be estimated. Suppose Q is a multiplicative function of x and y, i.e., $Q = xy$. How close an approximation is the linear function $L = x + y$? To make the computation simpler, assume that x and y have been normalized so that the region of interest is the unit square. Assume that x and y are uniformly distributed over the unit square. This is the least favorable assumption, since any linear correlation between x and y would increase the correlation between Q and L.

We have

$$\rho Q,L = \int_0^1\int \frac{(xy-\overline{xy})(x+y-\overline{x+y})}{\sigma_{xy}\sigma_x+y}\,dx\,dy.$$

Performing the integration and evaluating gives

$$\rho Q,L = \frac{\sqrt{6}}{\sqrt{7}} = .925.$$

As another illustrative example, let $Q = ax + by$ and $R = cx + dy$. Then

$$\rho Q,R = \frac{ac+bd}{(a^2+b^2)^{1/2}(c^2+d^2)^{1/2}},$$

where, if $a = 1$, $b = 2$, $c = 2$, $d = 1$,

$$P = .8.$$

Here, the error-free correlations are sufficiently high so that it would be questionable if the noisy kind of data met with in practice could distinguish the models.

These examples give some feeling why linear models are so "robust" when correlation with a criterion is the figure of merit. But more to the point, they suggest that ordinary human judgment would probably not do as well as nominal assumptions, if the underlying dimensions have been correctly identified. By a *nominal* assumption here is meant an assumption based on some quasi-logical rule such as symmetry or "equal ignorance." A good example is the assumption frequently met in statistical estimation that a priori probabilities are equidistributed over the set of hypotheses. Attempts to give a logical foundation to such an assumption (based on "insufficient reason" or equal ignorance arguments) have failed. And yet the assumption often works quite well in practice.

These comments whittle away the region for which human judgment is most appropriate. They do not, of course, directly affect the advantage of group judgment (Delphi) over individual judgment, for those cases where human judgment is the best available (or recognizable!) form of information.

Acknowledgments:

The author was aided in many ways in conducting the study by Daniel Rourke, now at Rockefeller University, New York. The author would also like to express his appreciation for the help of Professor Bonham Campbell of the Engineering Systems Department, UCLA and the members of the Executive Engineering Program for 1971–72. Methodological portions of the research were supported by the Advanced Research Projects Agency.

References

1. Dalkey, Rourke, Lewis, and Snyder, *Studies in the Quality of Life*, D. C. Heath and Co., Lexington, Mass., 1972.
2. S. C., Johnson, "Hierarchical Clustering Schemes," *Psychometrika* 32, No. 3 (1967), pp. 241-54. See also Carroll-Wish, Chapter VI, C, this volume, for a fuller discussion of cluster techniques and their role in Delphi studies.
3. M. Rokeach, "Values as Social Indicators of Poverty and Race Relations in America," *The Annals of the American Academy of Political and Social Science*, No. 388, (March 1970), pp. 97-111.
4. R. M. Dawes, "Objective Optimization under Multiple Subjective Functions," presented at the *Seminar on Multiple Criteria Decision Making*, Columbia, South Carolina, October 1972.
5. Paul A. David and Melvin W. Reder, eds., *Nations and Households in Economic Growth: Essays in Honor of Moses Abramovitz* (New York: Academic Press, Inc., 1974).

VI. C. Multidimensional Scaling: Models, Methods, and Relations to Delphi

J. DOUGLAS CARROLL and MYRON WISH

Multidimensional scaling (MDS) is a general term for a class of techniques that have been developed to deal with problems of measuring and predicting human judgment. These techniques all have in common the fact that they develop spatial representations of psychological stimuli or other complex objects about which people make judgments (e.g., of preference, similarity, relatedness, or the like). This means that the objects are represented as points in a multidimensional vector space. If this space is two- or three-dimensional it can be viewed all at once by various graphical display devices. If it is more than three-dimensional, it cannot, of course, be represented physically; but *subspaces* of two or three dimensions can be viewed, and can often yield very useful information. Furthermore, there are various plotting "tricks" (plotting higher dimensions by size or shading, for example) that make possible representation of higher dimensions (although it is not clear how many such dimensions a human user can comprehend all at once).

The crucial difference between MDS and such related techniques as factor analysis is that in MDS no preconception is necessary as to which factors might underlie, determine, or correspond to which dimensions. The only data actually required are the similarity judgments of one or more individuals, or some other measure of similarity or "proximity" of pairs of the entities under study. Simply put, and realizing there are many refinements of the technique, the basic similarity data are transformed (i.e., inverted in rank order) to measures of *dissimilarity* which may be viewed as (directly) monotonically related to Euclidean distances in some space of unknown dimensions. The greater the similarity of two objects, the closer they are constrained to be to each other in the multidimensional space or "map"; the more dissimilar, the farther apart. Then given for each object its distances to all other objects, we attempt to see how well the distances can be fitted (in a certain least-squares sense) by one, two, three, or more dimensions. The number of dimensions is increased until the addition of a new dimension makes very little improvement in the correlation (between the data and the distance) or other "goodness-of-fit" measure used. This usually occurs at considerably less than the N-1 dimensions that potentially could be needed to perfectly fit the distances among N objects. Since quantitative values for the coordinates of points in the space are obtained, one

The Delphi Method: Techniques and Applications, Harold A. Linstone and Murray Turoff (eds.)
ISBN 0-201-04294-0; 0-201-04293-2

may make plots of the results and observe if the dimensions (the X, Y, Z, etc., coordinates) can be interpreted as underlying psychological or judgmental factors, which then are presumed to account for the original subjective judgments. The contrast in trying to comprehend underlying factors based upon the raw data matrix as opposed to the plots obtained as a result of MDS is aptly demonstrated in the Morse Code example later in this paper.

The first order use of MDS in Delphi is to provide people with a graphical representation of their subjective judgments and see if they can ascribe meaning to the dimensions of the graphs. It is also possible to obtain a separate graphical representation for any subgroup or individual. Therefore the Delphi designer could exhibit for the respondents a graphical representation of how various subgroups view the problem—e.g., how much difference is there among the views of politicians, business executives, and social scientists. He could also easily compare results from different time periods, experimental conditions, etc.

A possible limitation of MDS in Delphi or other applications is that there is no guarantee that a meaningful set of dimensions will emerge. However, the results may be useful even if one cannot interpret the dimensions, since the graphical representation can greatly facilitate the comprehension of patterns in the data.

In MDS, as in other techniques which provide scientific insight, the results can be embarrassing or detrimental—that is, hidden psychological factors could be exposed which might disrupt or damage the group process if not handled with some skill. For example, in examining goals of an *organization* one might find out that certain individual goals like prestige constitute a hidden dimension playing a dominant role. The Delphi designer and the group must be of the frame of mind to face up to such possibilities.

Let us suppose that Delphi were being used to study nations. If one were using MDS as an auxiliary procedure, the Delphi respondents might be asked, first of all, to judge the similarity of pairs of nations. For example, each subject might be asked to judge the similarity of every pair of nations by associating a number between zero and ten with each pair (zero meaning "not at all similar" and ten meaning "virtually identical"). It is, of course, subjective similarity, not similarity in any objective sense, that is being measured. Furthermore, people will differ systematically in their judgments of relative similarity (as they will differ in preference and other judgments).

One of us (Wish) has actually undertaken studies in just this area. This particular study used the INDSCAL (for *IN*dividual *D*ifferences Multidimensional *SCAL*ing) method, developed by Carroll and Chang (1970). This method, which allows for different patterns of weights or perceptual importances of stimulus dimensions for different individuals or judges, will be discussed in detail at a later point. In the present case (in which the various nations were the "stimuli") the analysis revealed three important dimensions, shown in Figs. 1 and 2, which can be described as "Political Alignment and Ideology" (essentially "communist" *vs.* "noncommunist" with "neutrals" in

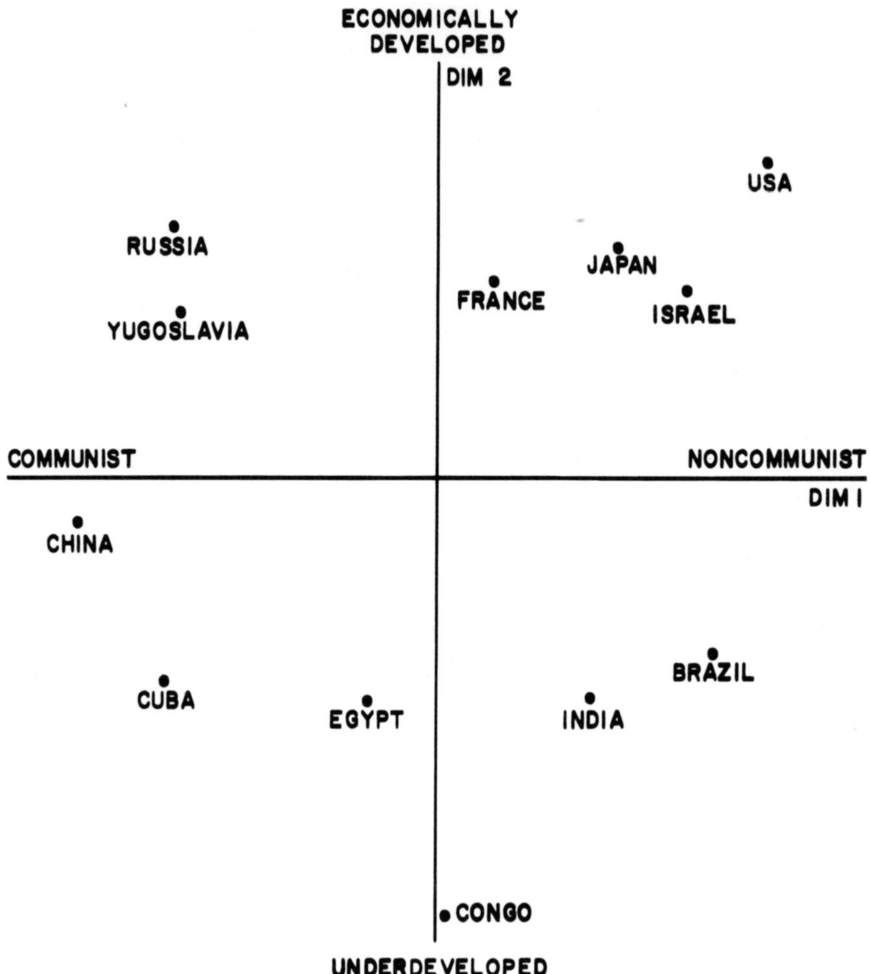

Fig. 1. The one-two plane of the group stimulus space for twelve nations (data dur to Wish). Dimensions one and two were interpreted by Wish as political alignment (communist-noncommunist) and economic development (economically developed-underdeveloped) respectively.

between), "Economic Development" contrasting highly industrialized or economically "developed" nations with "undeveloped" or "developing" nations (with "moderately developed" countries in between) and "Geography and Culture" (essentially "East" *vs.* "West").

The subjects in this study were also asked to indicate their position on the Vietnam war (these data were collected several years ago, it should be noted)

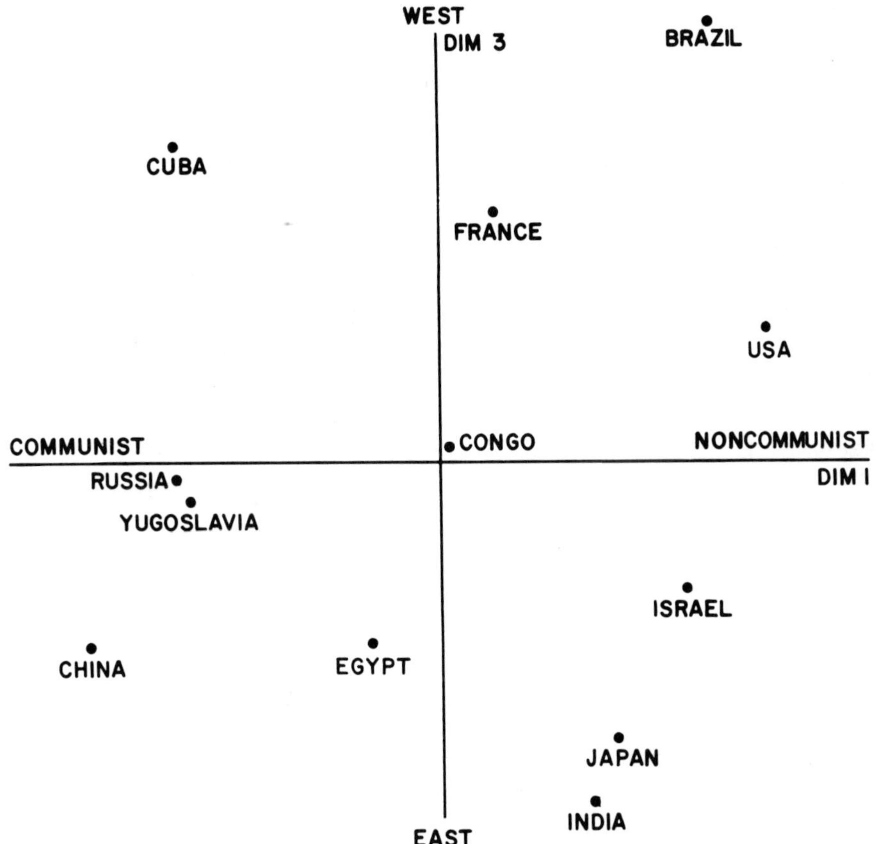

Fig. 2. The one-three plane of the group stimulus space for the Wish data on twelve nations. Wish interpreted dimension three as a geography dimension (east-west).

and based on this they were classified as "Doves," "Hawks," or "Moderates." As shown in Fig. 3, it was possible to relate political attitudes very systematically to the perceptual importance (as measured by the INDSCAL analysis) of the first two dimensions to these subjects. Specifically, "Hawks" apparently put much more emphasis on the "Political Alignment and Ideology" dimension than on the "Economic Development" dimension, while "Doves" reversed this pattern. A "Hawk" would see two nations that differed in ideology but were very close in economic development—e.g., Russia and the U. S.—as very dissimilar, while a "Dove" would see them as rather similar. In contrast, a "Hawk" would view Russia and Cuba (which are close ideologically but different in economic development) as very similar, while a "Dove" would

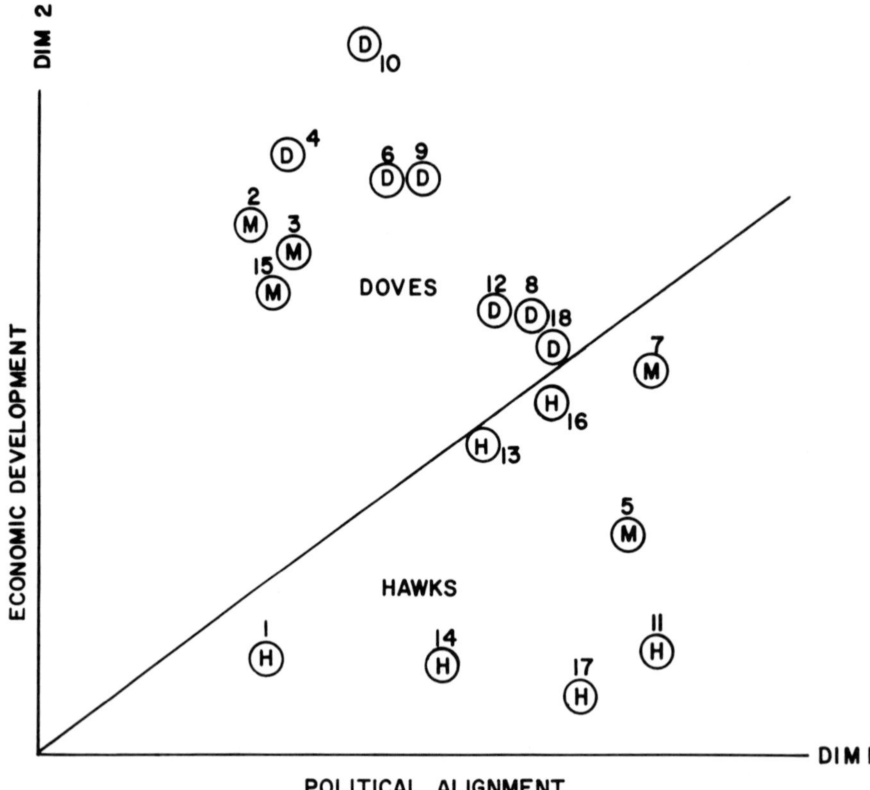

Fig. 3. The one-two plane of the subject space for the Wish nation data. *D*, *H* and *M* stand for "dove," "hawk," and "moderate" (as determined by subjects' self-report) vis-à-vis attitudes on Vietnam war. Forty-five-degree line divides "doves" from "hawks," with "moderates" on both sides.

perceive them as quite different. As can be seen in Fig. 4, the third dimension ("East-West") does not discriminate these political viewpoints.

This example serves to illustrate that MDS (particularly the INDSCAL method for individual differences scaling) can simultaneously yield information about the stimuli (nations, in this example) and about the individuals or subjects whose judgments constitute the basic data (similarities or dissimilarities of nations in this case). We have learned that the three most important dimensions underlying subjects' perceptions of nations, at least for graduate students, are ideology, economic development, and geography and culture (on an East-West dimension). [In later studies (Wish, 1972; Wish and Carroll, 1972), each of these single "dimensions" has shown itself to be multidimensional. To show this, however, required much more data and very high-

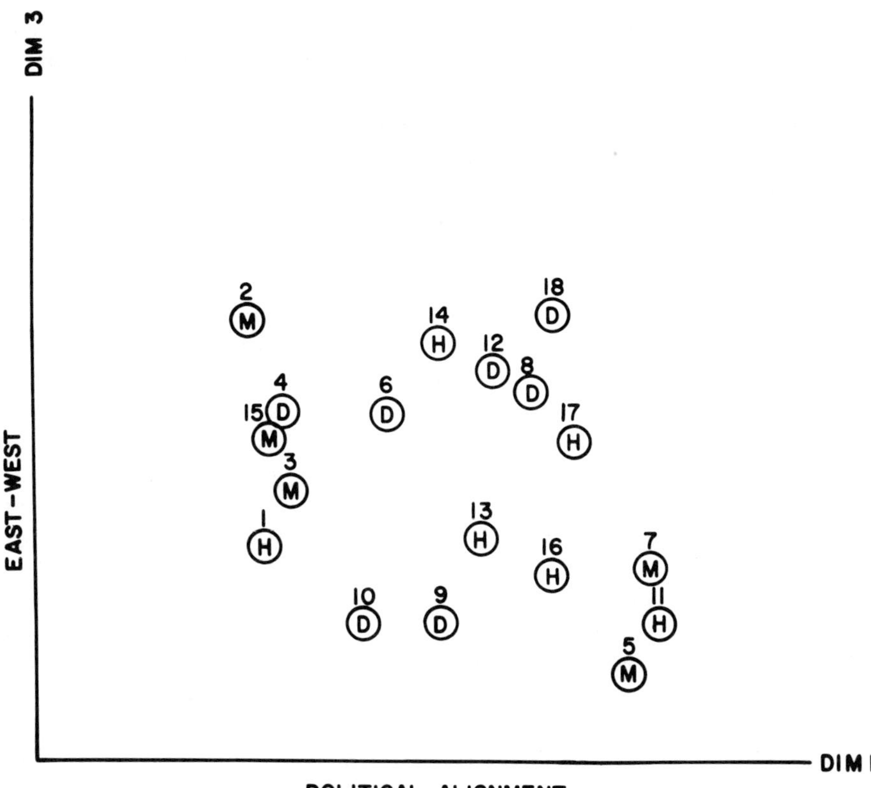

Fig. 4. The one-three plane of the subject space for Wish nation data. Coding of subjects is as in Fig. 3.

powered data analytic techniques which are beyond the scope of the present paper.] We also learn of a very interesting correlation between the importance or salience of these dimensions to individual subjects and these subjects' political attitudes (we cannot, of course, infer causality from this correlation, but we can certainly conclude that perception of nations and political attitudes about them are *not* independent). Finally, the graphical aspect of MDS lets the user (who may also be one of the subjects, especially in Delphi applications) see these relationships (and others) in a clear, concise, intuitively appealing, and highly informative manner. In a real-time computerized Delphi system, this should provide an excellent way of giving feedback about the judgments of a large number of respondents simultaneously (stressing the communalities in their judgments, but at the same time allowing each judge to see clearly the range of individual differences and, in particular, to see where *he* or *she* lies vis-à-vis the rest of the group). The respondent can also provide the interpretation of dimensions and the group can attempt to arrive at a collective view based upon discussions. In cases requiring special expertise available only to members of the respondent group, this may in fact be necessary.

Similarities or dissimilarities data are inherently *dyadic* data, in that they imply an ordering, or a scale, for the $n(n-1)/2$ pairs (dyads) of the n stimuli. There are many other kinds of dyadic data than similarities or dissimilarities, however. Usually these can be interpreted in some sense as proximities or antiproximities—that is, as related in some relatively simple way (e.g., monotonically) to *distance* in a multidimensional space.

To return to our example of nations, other such dyadic data that might be considered include friendliness; amount of trade or communication between; communality in ethnic, cultural, linguistic, or other background (these could broadly be interpreted as "proximity" measures, since a large value implies "closeness"). Such other dyadic measures as probability of war between, or amount of competition between (in trade, vying for third nations' favor, prestige in sports, technological feats, or the like), might be interpreted as antiproximity measures (for which a large value implies distance), although there is a very real sense in which some of the them might be viewed as "mixed" proximity and antiproximity measures (e.g., to be economic competitors may be taken to imply a closeness in size, degree of industrialization, and other economic factors, but distance or even opposition in national goals, ideology, and the like). Dyadic data generally are good "grist" for the MDS mill, particularly when interpretable as proximities or antiproximities. Even "mixed" proximity and antiproximity data can potentially be handled, however. Such data could be analyzed by INDSCAL, for example, if negative as well as positive weights are allowed for subjects (and if certain precautions, which will not be described here, are observed in preprocessing of the data).

A good example of a study using dyadic data which does concern (perceived) futures of nations has been provided by Wish (1971) who asked subjects to make two different dyadic judgments of friendship of pairs of nations. The first was of friendship "*now*" (i.e., at the time of the study) and the other of friendship "ten years from now." The ideology and economic development dimensions already mentioned also emerged as the principal dimensions underlying these "friendship" dyadic judgments. While ideology was the most important dimension for "present" friendship, economic development turned out to be far more important for "future" friendship. This indicated that (at least as perceived by those subjects) future alliances among nations were judged to be based much more on economic than on ideological considerations, although the opposite is now perceived to be the case. Not too much imagination is needed to see how other kinds of dyadic judgments could be used in this way to get a very general composite picture of the perceived future trajectories of nations (or of other entities of interest). The possible applications of MDS as an adjunct to Delphi are, indeed, virtually unlimited.

Another step in a Delphi application might be to ask the subjects (presumably "expert" judges of some sort) to make various judgments directly relevant to futures of nations. These might include things like expected level of industrialization or economic development in ten (or twenty-five or fifty) years, expected population, military development, probability of involvement in wars, or the like. These are all judgments that can be viewed as ordering the (single) nations on some attribute. As such, they are all formally analogous to preference judgments, analysis of which is discussed at a later point. Coombs (1962) coined the general term *dominance judgment* to cover the general class of judgments that order stimuli in terms of a single attribute (preference, wealth, size, etc.). Unlike data on similarities or other dyadic relationships between stimuli, multidimensional solutions are generally possible for dominance data only when more than one individual is involved, and when there are *individual differences* in their dominance judgments. Multidimensional methods for analyzing dominance data will be discussed at the end of this paper.

Having "whetted the appetite" (we hope) of our reader as to the potential of MDS used in conjunction with Delphi, let us now take a more detailed look (although an admittedly superficial and nontechnical one) at just what MDS is all about. We hope that the annotated bibliography at the end of this paper will satisfy the needs of readers desiring a deeper and more detailed view of the subject.

Ordinary or "Two-Way" Multidimensional Scaling

Multidimensional scaling of similarities, dissimilarities, or other dyadic data is usually based on some form of *distance* model [although not necessarily always; Ekman (1963) for example has proposed a scalar product model for similarities]. Generally speaking, the dyadic data (similarities, say) are assumed to relate in a simple way—usually via a linear or at least by a *monotonic* function—to distances in a (hypothesized, but initially unknown) multidimensional space. The multidimensional scaling method is usually called a *metric* one if this function is assumed to be linear, and *nonmetric* if it is assumed to be merely monotonic (or rank order preserving). The function, of course, will be assumed decreasing or increasing (whether linear or merely monotonic) depending on whether the data are proximities (e.g., similarities) or antiproximities (e.g., dissimilarities). These distances are usually, but not always, assumed to be *Euclidean* distances—the distances of ordinary physical space. (In the INDSCAL and some other individual differences models, "weighted" or other "generalized" Euclidean distances are assumed, but these can usually be represented as ordinary Euclidean distances in a "transformed" space for a particular individual.) The *process* of multidimensional scaling consists of deriving the multidimensional spatial representation *given* only the dyadic (e.g., similarities) data. It is in this sense the obverse of reading distances from a map. In MDS, we are given the distances (or numbers monotonically related to them) and are required to construct the map! As we shall see, this is quite feasible, especially if one has the aid of a high-speed digital computer.

To get a better feel for this approach, let us look at some data analyzed by multidimensional scaling methods. Fig. 5 shows a matrix of confusions between Morse Code signals, collected by Dr. E. Z. Rothkopf who is now at Bell Laboratories. Subjects who were learning the International Morse Code were

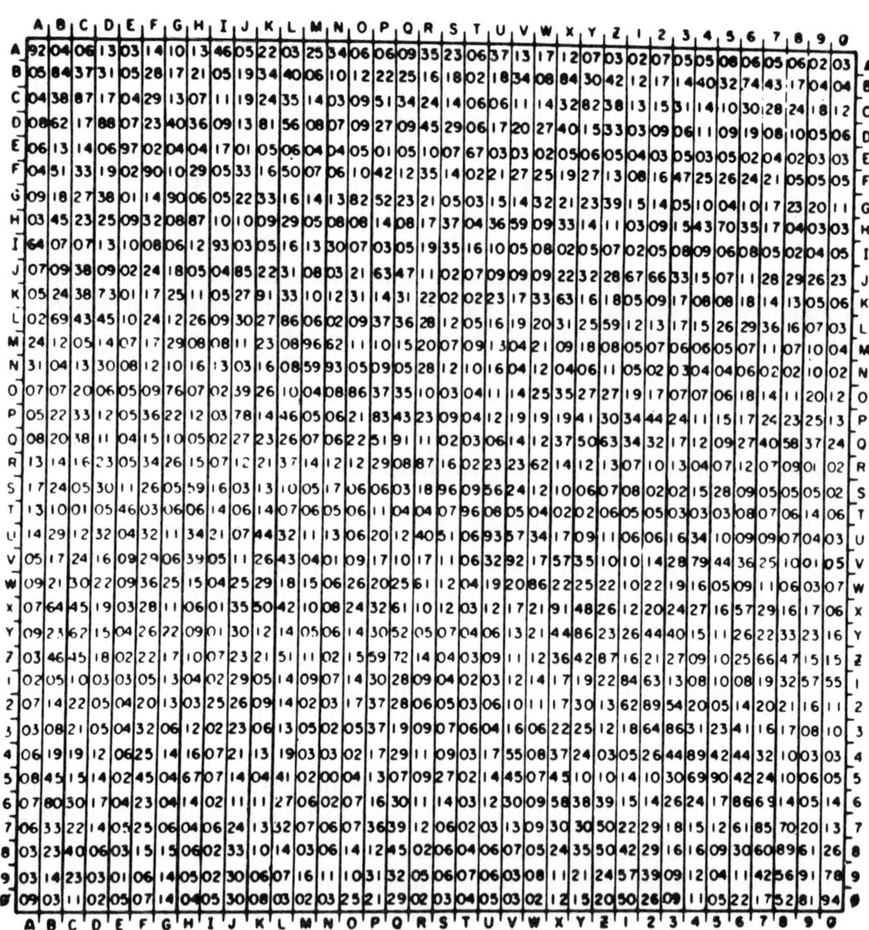

	A	B	C	D	E	F	G	H	I	J	K	L	M	N	O	P	Q	R
A	92	04	06	13	03	14	10	13	46	05	22	03	25	34	06	06	09	35
B	05	84	37	31	05	28	17	21	05	19	34	40	06	10	12	22	25	16
C	04	38	87	17	04	29	13	07	11	19	24	35	14	03	09	51	34	24
D	08	62	17	88	07	23	40	36	09	13	81	56	08	07	09	27	09	45
E	06	13	14	06	97	02	04	04	17	01	05	06	04	04	05	01	05	10
F	04	51	33	19	02	90	10	29	05	33	16	50	07	06	10	42	12	35
G	09	18	27	38	01	14	90	06	05	22	33	16	14	13	82	52	23	21
H	03	45	23	25	09	32	08	87	10	10	09	29	05	08	08	14	08	17
I	64	07	07	13	10	08	06	12	93	03	05	16	13	30	07	03	05	19
J	07	09	38	09	02	24	18	05	04	85	22	31	08	03	21	63	47	11
K	05	24	38	73	01	17	25	11	05	27	91	33	10	12	31	14	31	22
L	02	69	43	45	10	24	12	26	09	30	27	86	06	02	09	37	36	28
M	24	12	05	14	07	17	29	08	08	11	23	08	96	62	11	10	15	20
N	31	04	13	30	08	12	10	16	13	03	16	08	59	93	05	09	05	28
O	07	07	20	06	05	09	76	07	02	39	26	10	04	08	86	37	35	10
P	05	22	33	12	05	36	22	12	03	78	14	46	05	06	21	83	43	23
Q	08	20	38	11	04	15	10	05	02	27	23	26	07	06	22	51	91	11
R	13	14	16	23	05	34	26	15	07	12	21	37	14	12	12	29	08	87
S	17	24	05	30	11	26	05	59	16	03	13	10	05	17	06	06	03	18
T	13	10	01	05	46	03	06	06	14	06	14	07	06	05	06	11	04	04
U	14	29	12	32	04	32	11	34	21	07	44	32	11	13	06	20	12	40
V	05	17	24	16	09	29	06	39	05	11	26	43	04	01	09	17	10	17
W	09	21	30	22	09	36	25	15	04	25	29	18	15	06	26	20	25	61
X	07	64	45	19	03	28	11	06	01	35	50	42	10	08	24	32	61	10
Y	09	23	62	15	04	26	22	09	01	30	12	14	05	06	14	30	52	05
Z	03	46	45	18	02	22	17	10	07	23	21	51	11	02	15	59	72	14
1	02	05	10	03	03	05	13	04	02	29	05	14	09	07	14	30	28	09
2	07	14	22	05	04	20	13	03	25	26	09	14	02	03	17	37	28	06
3	03	08	21	05	04	32	06	12	02	23	06	13	05	02	05	37	19	09
4	06	19	19	12	06	25	14	16	07	21	13	19	03	03	02	17	29	11
5	08	45	15	14	02	45	04	67	07	14	04	41	02	00	04	13	07	09
6	07	80	30	17	04	23	04	14	02	11	11	27	06	02	07	16	30	11
7	06	33	22	14	05	25	06	04	06	24	13	32	07	06	07	36	39	12
8	03	23	40	06	03	15	15	06	02	33	10	14	03	06	14	12	45	02
9	03	14	23	03	01	06	14	05	02	30	06	07	16	11	10	31	32	05
0	09	03	11	02	05	07	14	04	05	30	08	03	02	03	25	21	29	02

	S	T	U	V	W	X	Y	Z	1	2	3	4	5	6	7	8	9	0
A	23	06	37	13	17	12	07	03	02	07	05	05	08	06	05	06	02	03
B	18	02	18	34	08	84	30	42	12	17	14	40	32	74	43	17	04	04
C	14	06	06	11	14	32	82	38	13	15	31	14	10	30	28	24	18	12
D	29	06	17	20	27	40	15	33	03	09	06	11	09	19	08	10	05	06
E	07	67	03	03	02	05	06	05	04	03	05	03	05	02	04	02	03	03
F	14	02	21	27	25	19	27	13	08	16	47	25	26	24	21	05	05	05
G	05	03	15	14	32	21	23	39	15	14	05	10	04	10	17	23	20	11
H	37	04	36	59	09	33	14	11	03	09	15	43	70	35	17	04	03	03
I	35	16	10	05	08	02	05	07	02	05	08	09	06	08	05	02	04	05
J	02	07	09	09	09	22	32	28	67	66	33	15	07	11	28	29	26	23
K	02	02	23	17	33	63	16	18	05	09	17	08	08	18	14	13	05	06
L	12	05	16	19	20	31	25	59	12	13	17	15	26	29	36	16	07	03
M	07	09	13	04	21	09	18	08	05	07	06	06	05	07	11	07	10	04
N	12	10	16	04	12	04	06	11	05	02	03	04	04	06	02	02	10	02
O	03	04	11	14	25	35	27	27	19	17	07	07	06	18	14	11	20	12
P	09	04	12	19	19	19	41	30	34	44	24	11	15	17	24	23	25	13
Q	02	03	06	14	12	37	50	63	34	32	17	12	09	27	40	58	37	24
R	16	02	23	23	62	14	12	13	07	10	13	04	07	12	07	09	01	02
S	96	09	56	24	12	10	06	07	08	02	02	15	28	09	05	05	05	02
T	07	96	08	05	04	02	02	06	05	05	03	03	03	08	07	06	14	06
U	51	06	93	57	34	17	09	11	06	06	16	34	10	09	09	07	04	03
V	11	06	32	92	17	57	35	10	10	14	28	79	44	36	25	10	01	05
W	12	04	19	20	86	22	25	22	10	22	19	16	05	09	11	06	03	07
X	12	03	12	17	21	91	48	26	12	20	24	27	16	57	29	16	17	06
Y	07	04	06	13	21	44	86	23	26	44	40	15	11	26	22	33	23	16
Z	04	03	09	11	12	36	42	87	16	21	27	09	10	25	66	47	15	15
1	04	02	03	12	14	17	19	22	84	63	13	08	10	08	19	32	57	55
2	05	03	06	10	11	17	30	13	62	89	54	20	05	14	20	21	16	11
3	07	06	04	16	06	22	25	12	18	64	86	31	23	41	16	17	08	10
4	09	03	17	55	08	37	24	03	05	26	44	89	42	44	32	10	03	03
5	27	02	14	45	07	45	10	10	14	10	30	69	90	42	24	10	06	05
6	14	03	12	30	09	58	38	39	15	14	26	24	17	86	69	14	05	14
7	06	02	03	13	09	30	30	50	22	29	18	15	12	61	85	70	20	13
8	06	04	06	07	05	24	35	50	42	29	16	16	09	30	60	89	61	26
9	06	07	06	03	08	11	21	24	57	39	09	12	04	11	42	56	91	78
0	03	04	05	03	02	12	15	20	50	26	09	11	05	22	17	52	81	94

Fig. 5. Matrix of confusabilities of thirty-six Morse Code signals (data from Rothkopf, 1957).

asked to listen to pairs of Morse Code signals and judge whether the two were "same" or "different." The elements in this matrix are simply the percentage of subjects in the sample who judged the two signals as "same." If the subjects were making errorless judgments, all the entries on the main diagonal would be 100 and off-diagonal elements would all be zero. We can see by the fact that there are a large number of nonzero off-diagonal elements in this matrix that, in fact, subjects were making quite considerable confusions among these stimuli. It is reasonable to assume, as Roger Shepard did in analyzing these data, that the more similar were two signals, the more likely it was that subjects would confuse them. Under this assumption, stimulus confusability, as measured by size of the entry for a pair of signals in this matrix, should be a direct measure of similarity of the pair.

Figure 6 shows the two-dimensional solutions Shepard attained on the basis of this matrix. The computer took the matrix in Fig. 5 as input, and produced the two-dimensional structure in Fig. 6 as output. The analysis was via Kruskal's (1964) nonmetric scaling procedure called MDSCAL, which stands simply for *M*ulti *D*imensional *SCAL*ing, while the *interpretation* of this figure, including the curvilinear coordinate system drawn there, is credited to Shepard (1963). The reasoning he used is as follows: it is easy to see by inspection that, as we move from the lower to the upper part of this figure, the aspect of the signals that is changing is number of components, ranging from those signals with one component at the very bottom of the figure to those with five components at the top of the figure. Thus, Shepard called this dimension (slightly curved in this figure) "number of components," As we move from left to right in the figure we can see that what is changing is the ratio of dots to dashes. That is, the figures on the left consist primarily of figures with all or mostly dots; those on the right, signals with all or mostly dashes. Shepard, therefore, labeled this dimension "dot-to-dash ratio." While subsequent analysis of these and related data (Wish, 1967, 1969) indicated additional dimensions relating to the precise sequence of dots and dashes, these two dimensions appear to account for a very large proportion of the variance in the "confusion" data.

It should be pointed out that, in addition to drawing the curvilinear coordinate system depicted in Fig. 6, Shepard had to orient the figure so that the two dimensions would correspond (more or less) with the horizontal and vertical directions. That is, to use the map analogy, it was necessary for him, as part of the interpretation process, to choose North-South and East-West directions on the map. The computer is able to construct the *configuration*, but not to orient the coordinate axes correctly. Although this may not be too much of a problem in two or perhaps three dimensions, it could be prohibitively problematical in four or more dimensions. This is, in fact, one of the principal obstacles to interpretation and use of MDS solutions, especially high-

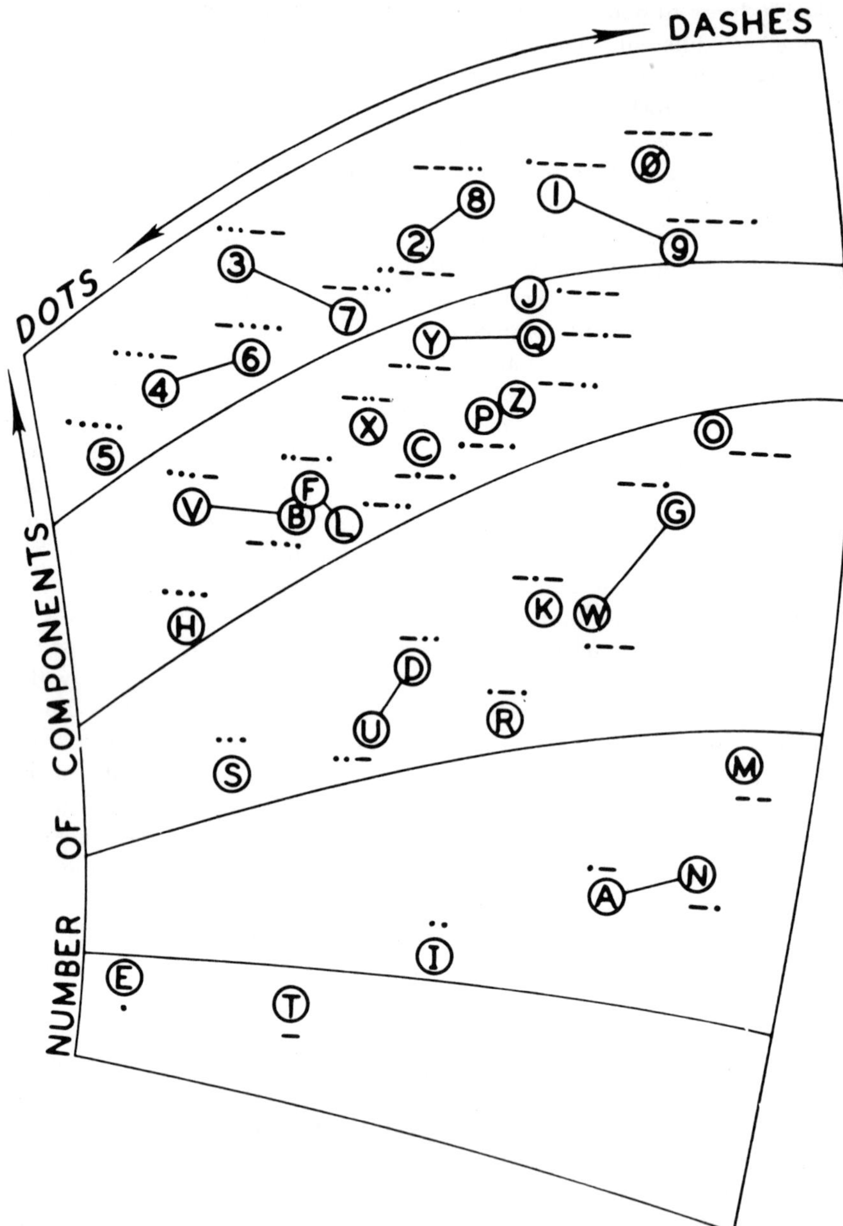

Fig. 6. Two-dimensional configuration resulting from multidimensional scaling analysis of Rothkopf's Morse Code confusabilities (analysis by Shepard, 1963).

dimensional ones. As will be pointed out later, however, the INDSCAL method offers a way out of this particular problem, as it is able to capitalize on individual differences among subjects or judges to orient the coordinate axes *uniquely*.

Figure 7 shows the monotone "distance function" that resulted from this analysis. On the abscissa are distances as computed from the two-dimensional solution shown in Fig. 6, while on the ordinate are the similarities data (in this case, the percentage of "same" responses to each pair) that served as input to the multidimensional scaling program. The monotone function is, in this case, roughly a negative exponential, which fits in very nicely with what we know about confusion data. There is good theoretical reason to assume a function of this general negative exponential type relating distance in a multidimensional

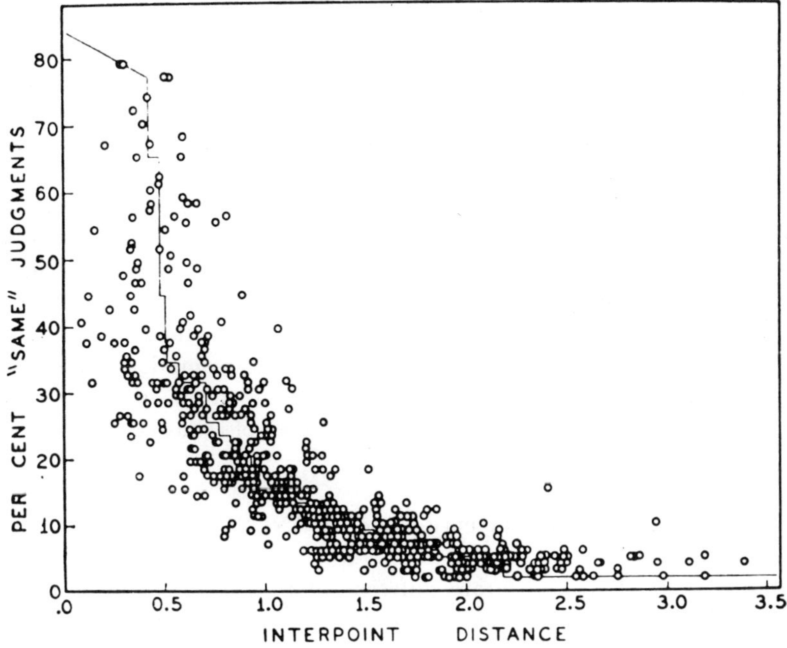

Fig. 7. Relation between mean percent "same" judgment and interstimulus distance in the derived spatial configuration for the Rothkopf data analyzed by Shepard.

space to stimulus confusability. The critical thing here is that the nonmetric scaling computer program *simultaneously* determined the form of this function and the structure of the stimuli in the two-dimensional multidimensional space, and without any knowledge about either the structure of Morse Code signals or about the psychology of stimulus confusion!

Looking back for a moment at the data matrix shown in Fig. 5, it can be seen that one can tell very little indeed about the stimuli or about confusability based on mere inspection of this matrix. However, the two-dimensional graphical configuration determined from scaling this matrix tells us a great deal. We know immediately that stimuli that are closer together in this two-dimensional space tend to be confused more often. Not only does this geometric picture summarize important aspects of the data, but it also tells us the psychological dimensions (or "cues") on which these confusions are based.

In this example we see illustrated dramatically two principal advantages of multidimensional scaling: a parsimonious and economical condensation of data, on the one hand, and insight into the psychological dimensions, factors, or "cues," underlying perception of the stimuli on the other. Nonmetric methods allow us to relax our assumptions about the relation between data and distances in the postulated underlying space, and at the same time vastly expand the variety and generality of data that can be considered for multidimensional scaling analysis. This "nonmetric breakthrough" has had a tremendous impact on the field theoretically, even though we are increasingly finding that for purposes of practical data analysis the distinction between metric and nonmetric methods is not as important as it once appeared. The nature of this "practical" insight is that, once one has correctly inferred the appropriate *dimensionality* for the analysis, the difference between metric and nonmetric solutions for the same data is not usually very great. To use statistical terminology, we can say that the metric methods are surprisingly "robust" against a wide variety of violations of the assumptions on which they are based. It required the development of the nonmetric methods, however, to provide the comparison that has made this degree of robustness apparent.

Individual Differences in Perception and the INDSCAL Model

Everyone knows that people are different. If they weren't, the world would certainly be very dull. Worse still, it would not be very functional! That people have different abilities, personalities, attitudes, and preferences is well accepted, and psychologists have been concerned for some time with their measurement. Measurement of individual differences in ability comes under the heading of intelligence, aptitude, or achievement testing. Differences in personality and attitudes are assessed by personality or attitude inventories. Quantitative methods for measuring individual differences in preferences have been developed only recently, and will be described somewhat at a later point.

Most people would agree that people differ in the aspects of human behavior we have so far discussed. One faculty, it would seem, in which people *do not* differ is that of perception. Surely, if we human beings have anything at all in common, it must be our perceptions, for the way we *see* the world forms the basis for all the rest of our behavior. If we perceive different worlds, can we in any sense be said to live and behave in the same world? Would not communication between different people be impossible and all social relations reduced to chaos? The answer, however, to which modern psychology is increasingly being led is that people do, indeed, differ in this all-important faculty, although there is necessarily a certain communality cutting across these differences in perception.

To produce an obvious example of individual differences in perception, consider the case of the color-blind individual, who must certainly perceive *colors* differently from most of the rest of us. A man who is red-green color-blind simply does not perceive a certain dimension of colors along which people with normal vision can clearly discriminate. He would judge two colors as very similar or identical that others would see as highly dissimilar. A man who is weak in red-green perception, but not totally red-green blind, would lie somewhere between the two extremes. To take another example, a tone-deaf individual perceives sounds differently from a musically inclined person. To take these examples to their logical extremes, a blind or deaf person most certainly perceives differently from a person with normal vision and hearing.

But these examples seem to concern obvious *deficiencies* in the peripheral equipment, the eyes or ears in these cases, that carry sensory signals to the brain. Surely, it will be objected, if no signal, or a diminished signal, gets through to the central nervous system, the individual must perceive in an impoverished way. Given the same signal from the peripheral receptors, the eyes, ears, nose, tongue, tactile, and kinesthetic receptors, two individuals *must* perceive the same way! Or must they?

There is mounting evidence that this last assertion is not true. Granted that it is difficult, and sometimes impossible, to "parcel out" perceptual effects between peripheral receptors and central nervous system processing of received signals (even taking into account complex interactions between the two such as central nervous system feedback to, and control of, incoming signals from receptor organs), there remains strong evidence for individual differences in perception that are purely a function of central activities.

In the field of comparative linguistics this fact has been embodied in the so-called "Whorfian hypothesis," formulated by the famous linguist Benjamin Lee Whorf. Stated in oversimplified terms, this hypothesis says that many of the differences between distinct languages, of the kind that often make it difficult to find a strictly one-to-one translation of important concepts and propositions from one language to another, are due to the fact that the languages are in fact talking about somewhat different *perceptual worlds*. This might well be a reflection of the fact that various perceptual dimensions have different relative saliences, or perceptual importances, for different linguistic communities or cultural groups. An example of this was reported by the anthropologist H. C. Conklin a few years ago. A Pacific tribal group called the Hanunóo, at first thought to be color-blind because they named colors quite differently from Westerners, proved on closer observation not to be color-blind at all, but to be emphasizing different dimensions in making color judgments. Their judgments rely much more heavily on brightness, for example, than do ours, so that a light green and a bright yellow are given the same name, but both are named differently from darker shades of the same hues. On the other hand, their color-naming behavior seems to reflect hardly at all what we usually think of as hue differences (that is, differences related to the physical spectrum). This is so even though there appears to be no physiological impairment of vision among members of this tribe.

We might suppose that such differences in naming of colors or other classes of stimuli, whether at the cultural or at the individual level, are correlated with differences in perception. Both kinds of differences could result in part from particular aspects of individual or cultural histories that have made some dimensions relatively more important for one person or group than for another. In the case of the Hanunóo, for example, their idiosyncrasies in color-naming seem to be related to the tribe's food-seeking behavior. Possibly some dimensions of color are more important than others for discriminating food from nonfood items, or differentiating among different kinds of foods in their island environment. Our interest, however, will center not so much on how perceptual differences may have developed, but on their general nature and how to measure such differences, assuming they exist.

Our approach to *measuring* individual differences in perception assumes that these differences are reflected in the way people make judgements of relative similarity of stimuli. It thus involves a model for individual differences, the INDSCAL model, that falls in the class of multidimensional scaling models for similarity (or other dyadic) judgments. As a matter of fact, the INDSCAL procedure was the one used to analyze the data on similarities between nations.

The INDSCAL model (Carroll and Chang, 1970) shares with other multidimensional scaling models the assumption that, for each individual, similarities judgments are (inversely) related to distance in the individual's perceptual space. Each individual is assumed to have his own "private" perceptual space, but this is not completely idiosyncratic. Rather, a certain communality of perceptual dimensions is assumed, but this communality is balanced by diversity in the pattern of relative saliences, or weights of these common dimensions. The salience of a perceptual dimension to an individual can be defined in terms of how much of a difference (perceptually) a difference (or change) on that dimension makes to the individual in question. For our hypothetical red-green color-blind individual a change all the way from red to green makes little or no perceptual difference, while a normal individual would easily detect a much smaller change, say from a brilliant red to a slightly less saturated reddish-pink. Perhaps a color-weak individual could not distinguish this difference, but could distinguish between a bright red and a very desaturated pink (that is, a greyish-pink).

The INDSCAL model handles this notion of differential salience of common dimensions by assuming that each individual's "private" space is derived from a common, or "group," space by differentially stretching (or shrinking) the dimensions of the group space in proportion to the subject's weights for the dimensions. The weights of the dimensions for each subject can be plotted in another "space," called the "subject space," in which the value plotted for a given individual on a particular dimension is just the stretching or shrinking factor, for that individual on that dimension (for technical reasons it is actually the square of that stretching factor that is plotted). A value of zero means that the dimension is shrunk literally to the vanishing point, which is to say that he just doesn't perceive the attribute corresponding to that dimension, or, in any case, "acts like" he doesn't perceive it.

After applying the stretching and shrinking transformations, as defined by the "subject space," each individual is assumed to "compute" psychological distances in this transformed space. His similarity (or dissimilarity) judgments are then assumed to be monotonic with the distances thus "computed." Of course, all of these statements about the INDSCAL model should be regarded merely as "as if" statements; that is, the individual acts "as if" he had gone through these steps. We do not literally believe, for example, that distances are computed and a monotone transformation applied to derive similarity values, only that there are psychological processes that have the same final effect "as if" these operations were performed.

Since we have been frequently invoking the example of color blindness in discussing individual differences in perception, it might be of interest to see how the INDSCAL method actually deals with data on color vision. Fortunately, data of the appropriate kind were collected by Dr. Carl E. Helm, now of the City University of New York, who asked subjects directly to judge psychological distances between ten colors. They did so by arranging triples of color chips into triangles so that the lengths of the three sides were proportional to the psychological distances between the colors on the chips at the vertices. By having subjects do this for every triple, Helm was able to construct, for each subject, a complete matrix of psychological distances. Helm's sample included four subjects who were deficient (to a greater or less extent) in red-green color vision. We applied the INDSCAL method to Helm's data, and obtained interesting results that serve very nicely to illustrate the INDSCAL model. The "group" stimulus space in Fig. 8 nicely reproduced the well-known "color circle" (as would be expected, since Helm used Munsell color chips selected to be constant in brightness, and very nearly constant in saturation) going from the violets and blues through the greens and yellows to the oranges and reds, with the "nonspectral" purples between the reds and the violets. Basically, the fact that colors of constant brightness and saturation are represented on a circle rather than on a straight line, as in the case of the physical spectrum, reflects

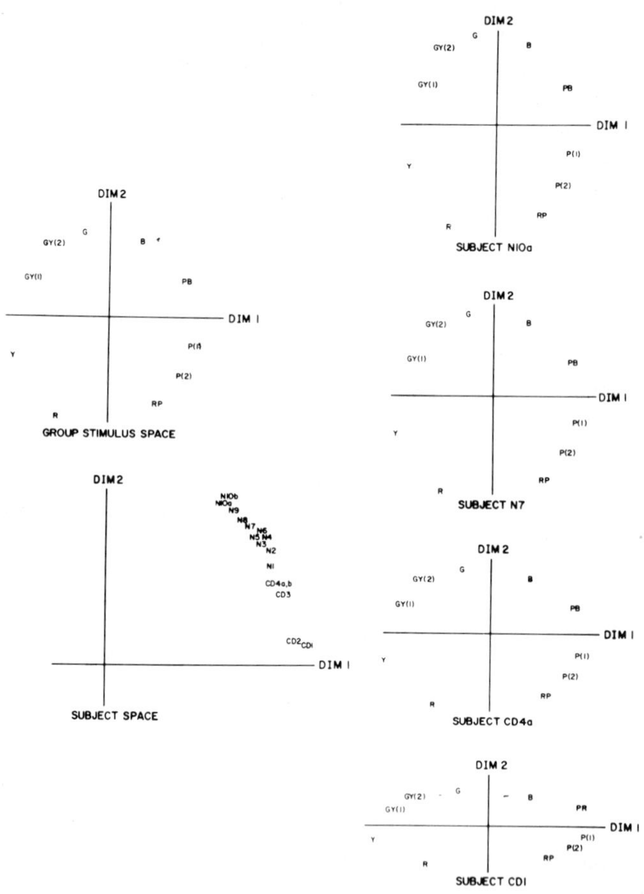

Fig. 8. The INDSCAL analysis of Helm's data on color perception produced the "group" stimulus space shown in *A*, which conforms quite well with the standard representation of the color circle, and the "subject space" shown in *B*. The coding of the stimuli is as follows: R=Red, Y=Yellow, $GY(1)$=Green Yellow, $GY(2)$=Green Yellow containing more green than $GY(1)$, G=Green, B=Blue, PB=Purple Blue, $P(1)$ =Purple, $P(2)$=a Purple containing more red than $P(1)$, RP=Red Purple.

the fact that violet and red, which are at opposite ends of the physical continuum, are more similar psychologically than red and green, two colors much closer on the physical spectrum, The circular representation also permits incorporation of the so-called "nonspectral" purples, which, psychologically, lie between violet and red. Of course, not all hues were actually represented in Helm's set of color chips, but it is fairly clear where the missing hues would fit into the picture. The precise locations of the two coordinate axes describing this two-dimensional space are of particular interest, since, as we have said, the orientation of axes is *not* arbitrary in INDSCAL. The extreme points of dimension 1 are a purplish blue at one end and yellow at the other. This dimension, then, could be called a blue-yellow dimension. Dimension 2 extends from red to a bluish green, and thus could be called a red-green dimension.

The "subject space" from the analysis of Helm's data shows that the color-deficient subjects (including one who appears twice because he repeated the experiment on two occasions) all have smaller weights for the red-green dimensions than do the normal subjects. Furthermore, the order of the weights is consistent with the relative degrees of deficiency, as reported by Helm, except for one small reversal (subject CD-3 was least deficient by Helm's measures, and CD-4 second least deficient, while the weight for CD-4 on dimension 2 is slightly higher than that of CD-3).

INDSCAL also allows construction of a "private" perceptual space for each individual. If we construct such a space for each of these subjects, we find that, while for normal subjects the "color circle" looks pretty much like a circle, for the color-deficient subjects it looks more like an ellipse whose major axis is the blue-yellow dimension and whose minor axis is the red-green dimension (the more deficient the subject the more elliptical—that is, the smaller the ratio of minor and major axes). Of course, there is considerable variation among the "normal" subjects themselves, with the more extreme (such as N-10) bordering on blue-yellow color deficiency (which is much rarer than red-green deficiency). For such an extreme individual the color circle is deformed into an ellipse whose major axis corresponds to the red-green dimension. The strength of the INDSCAL model and method as revealed in this analysis is that it accommodates all these perceptual variations among subjects in a very parsimonious way, and at the same time tells us a great deal about color vision, which, if not precisely new and unexpected, confirms in an elegant way what has been worked out over the years by vision theorists in much slower and more tedious ways.

Another example which illustrates particularly well the uniqueness property of INDSCAL dimensions (that is, the fact that one does not generally have to rotate axes to obtain an interpretable solution in terms of presumably fundamental perceptual dimensions) involves some data on perception of acoustical "tone" stimuli collected by Bricker and Pruzansky. Interest in the particular stimuli used in this experiment was generated in part by the fact that they were among possible candidates for "tone ringers," which will eventually replace the mechanical "bell ringers" of the present day in the all-electronic telephone of the future. This particular study is part of a larger project in which the perceptual dimensions of a wide range of acoustical stimuli are being studied and related to such other qualities as subjective preference, "attention-getting" properties, and the like. This project is bringing to bear a large number of multidimensional scaling and related techniques on the question of optimal selection of the sound which will ultimately become as familiar and omnipresent as today's telephone bell. It will help guide the decision on the "tone ringer" ultimately chosen, and also that of whether controls should be provided to allow individual telephone users to modify the sound of their ringers, and, if so, of what controls should be provided (i.e., what variables or dimensions should be modifiable). The stimuli in the present case were tones that varied in a systematic way in three well-defined physical dimensions. These were modulation frequency, for which there were four values (5, 10, 20, and 40 hertz), modulation percentage, with three values (3 percent, 10 percent, and 25 percent), and modulation waveform, which took on two values (sine wave *vs.* square wave). All combinations of values of these three physical dimensions were used, making a total of $4 \times 3 \times 2 = 24$ stimuli.

Since there were three physical dimensions, it should not be too surprising if there turned out to be three psychological dimensions corresponding more or less in a one-to-one fashion (or forming the "psychological concomitants") of the physical dimensions. It would be surprising if the "psychophysical mapping" from physical to psychological dimensions were perfectly linear, but on the other hand, it would be even more surprising if there were no correspondence at all between the two domains.

In fact an INDSCAL analysis did reveal three psychological dimensions underlying Bricker and Pruzansky's data, and these corresponded in a highly regular fashion to the physical dimensions (although the structure of the psychological space did reveal some small but systematic distortions reflecting interactions between these dimensions). The high degree of correspondence between the two kinds of dimensions was revealed by use of a mathematical technique for finding lines defining directions in the three-dimensional psychological space corresponding optimally to each of the physical dimensions. Although this technique placed no constraint at all on the locations of these lines, they turned out to correspond in an essentially one-to-one fashion to the coordinate axes of the "group" stimulus space from INDSCAL. This correspondence could be measured quantitatively by the fact that the angles between the "psychological" coordinate axes and the lines corresponding to the best physical concomitant of each psychological axis were all less than 15 degrees (the exact angles were 8.1, 5.7, and 14.5 degrees for modulation frequency, percentage, and waveform respectively), while the degree of correspondence of each physical dimension to its psychological concomitant was very high (even though the relations were somewhat nonlinear, the linear correlation coefficients were .956, .969, and .854 respectively, where a value of 1.0 represents a perfect linear relation). A more detailed discussion of this study, including graphical displays of the results, can be found in Carroll and Chang (1970).

This property of unique orientation of axes in the case of INDSCAL solutions is very important, for it means that the problem of rotation of axes that usually arises in interpreting multidimensional scaling solutions can often be avoided by reliance on individual differences to orient the "psychological axes" uniquely.

In these two examples, the perceptual dimensions of the stimuli were pretty well known beforehand, either through a long history of past study of the stimuli (in the case of the colors) or because the stimuli were actually generated by systematically manipulating some well-defined physical dimensions (in the case of the acoustical stimuli). It should be emphasized, however, that in neither case was this a priori knowledge about the dimensions "communicated to" the computer program that performed the INDSCAL analysis. The program "saw" only the raw data on similarity judgments of the various subjects, and constructed both the stimulus and subject spaces on the basis of this information alone. We will now discuss an example by Jones and Young (like the application involving nations) in which much less was known about the stimuli on an a priori basis.

The Jones and Young (1972) application should be of great interest to potential Delphi users, since it can be viewed as paving the way to a new approach to sociometry. Jones and Young were both at the Psychometric Laboratory of the University of North Carolina in 1969 and 1970, when this study was carried out. An interesting aspect of the study is that the "stimulus set" was the same as the "subjects"; namely, members of the Psychometric Laboratory (faculty, students, and staff). A subset of these members, acting as subjects or judges, were asked to judge the similarity of another subset of these members, and also to complete various rating scales relating to perceived interests, status, and the like, as well as to provide standard sociometric information. (The stimulus subset was actually contained in the subject subset, so that all stimulus persons were also judges, but not vice versa.) INDSCAL analyses were done separately for the similarity data collected in 1969 and 1970, as well as a composite analysis for the combined data from the two years (restricted to those stimulus persons included in the stimulus sample both years). The rating scale and similarities data were used as aids in interpreting the resultant dimensions.

The main results were exceptionally clear and consistent. Three dimensions emerged from the data for both years, and these agreed remarkably in the independent analyses of the data from the two years. The first dimension in both cases was, unambiguously, a status dimension (it correlated in the mid-to-high nineties with both rated "perceived status" and with academic rank). The second dimension was called "Political Persuasion"—it correlated highly with perceived position along a "left-right" or liberal-conservative political scale, and also with a lifestyle variable contrasting "unconventional" with "conventional" modes of dress, hair, behavior patterns, and the like. A third dimension had to do with "Professional Interests," and primarily contrasted those whose interests were statistical or methodological with those who were primarily concerned with specific content areas.

As with Wish's nation study, a very interesting pattern of subject weights emerged from this study. The subjects, as displayed in the subject (weight) space, fell into two clearly discernible clusters—one consisting of faculty members and the other of students. The faculty members tended to put much more weight on the "Status" dimension than on the "Political Persuasion" dimension, while the students put about equal weight on the two. The student and faculty groups did not differ in the weight for the "Professional Interests" dimension.

One notable aspect is that those who *had* greater status (i.e., were higher on the "Status" dimension in the stimulus space) put more weight on that dimension (i.e., were higher on the "Status" dimension in the *subject* space). Thus, in this sociometric application, where the stimulus and subject spaces may contain the "same" objects, there is reason to believe that the positions of these objects on at least this dimension (and quite possibly on other dimensions as well) will be correlated. The other aspect of this study that bears notice is the fact that "Status" did, indeed, come out as the first (and by far the strongest) dimension. While it may be a bit speculative to assume this will always be so, if it were this might suggest a very "neutral" way to measure status in a group—without ever asking for a direct judgment of status! That is, simply ask group members to judge similarity among themselves, and then take the first dimension from an INDSCAL analysis as a direct measure of status (we stress that this is only speculative at the moment, and would certainly not want to assume this would work in general without a great deal of additional study).

As was mentioned at the beginning of the paper, multidimensional methods can also be used to analyze preferences of other kinds of "dominance" data.

There are two different general kinds of multidimensional preference analysis, called by Carroll (1972) *internal* and *external*. In an internal analysis the positions of stimulus points and the directions of "subject" vectors are determined from the preference data alone. In an external analysis the stimulus points are given a priori (say from an MDS analysis of similarity or other dyadic data) and the vectors, ideal points, or other entities characterizing subjects are "mapped" into this predefined stimulus space. Such an "external" analysis called for mapping dominance judgments (vis-à-vis, say, probable futures of nations) into MDS spaces derived, say, from similarities data. For a complete discussion of multidimensional preference analysis, the reader is referred to the 1972 paper by Carroll (in the book edited by Shepard, Romney, and Nerlove [1972]).

A study by McDermott (1969) illustrates an internal analysis of preference data. In this study thirty-one subjects listened to speech transmitted over twenty-two simulated "telephone circuits" that had been subjected to various linear and nonlinear distortions. A preference ordering for the twenty-two circuits was determined for each subject on the basis of preference comparisons involving all possible pairs of circuits. The data were analyzed in terms of a "vector model" whose solution simultaneously provides information about the psychological factors underlying the stimuli and about the extent and nature of individual differences among the subjects. This model assumes that the stimuli (the circuits) can be represented as points in a multidimensional space, and that the subjects can be represented as vectors in the same space. A subject's preference ordering of the stimuli is assumed to correspond to the rank order of the projection of the stimulus points onto that subject's vector. The axes of a properly chosen coordinate system describing this multidimensional space can be thought of as corresponding to the psychological factors, or dimensions, along which the stimuli vary. The cosines of angles of a subject's vector with these coordinate axes measure the importance of that dimension for that subject's preference judgments. This vector model is illustrated in Fig. 9, in which the preference orderings of two hypothetical subjects for five hypothetical stimuli are shown.

Three dimensions, which were interpreted as "Degree of distortion in signal," "Degree of distortion in background," and "Loudness level" accounted very well for the preference judgments of all subjects. Although subjects preferred the circuits that had the least distortion, they differed with regard to the relative importance of signal and background distortion as well as in their preferred loudness level. The schematic diagram of the three-dimensional solution in Fig. 10, which shows only a subset of the twenty-two stimuli and only a few of the subject vectors, should make this interpretation clearer.

This "vector model" is only one of a large class of possible models for individual differences in preference, and is, in fact, the *simplest* model in a number of ways. Another model, often called the "unfolding model," postulates a different "ideal" stimulus for each subject, and assumes that preference decreases monotonically with increasing Euclidian distance from the "ideal." (The "vector model" is actually a special case of this, since it is equivalent to an "unfolding model" with "ideal" points infinitely distant from the origin.) Other models generalize the "unfolding model" by assuming , for example, different saliences for dimensions as well as different "ideal" points, or even that there is more than one "ideal," or highly preferred, region.

One thing that should be emphasized in summarizing analyses discussed in this paper is that the solutions were obtained by use of computer programs that were *in no explicit way* informed or instructed as to what the psychological (or physical) dimensions were. In the case of McDermott's analysis, for example, the computer was merely programmed to construct a three-dimensional solution involving stimulus points and subject vectors so as to account as well as possible, in a well-defined statistical sense, for the preference data from the thirty-one subjects (via the computer program, by Carroll and Chang, called MDPREF). As is true of all multidimensional scaling solutions, the computer produces the "picture" while the user furnishes the interpretation. This process of interpretation often involves what is termed "rotation of axes," which simply

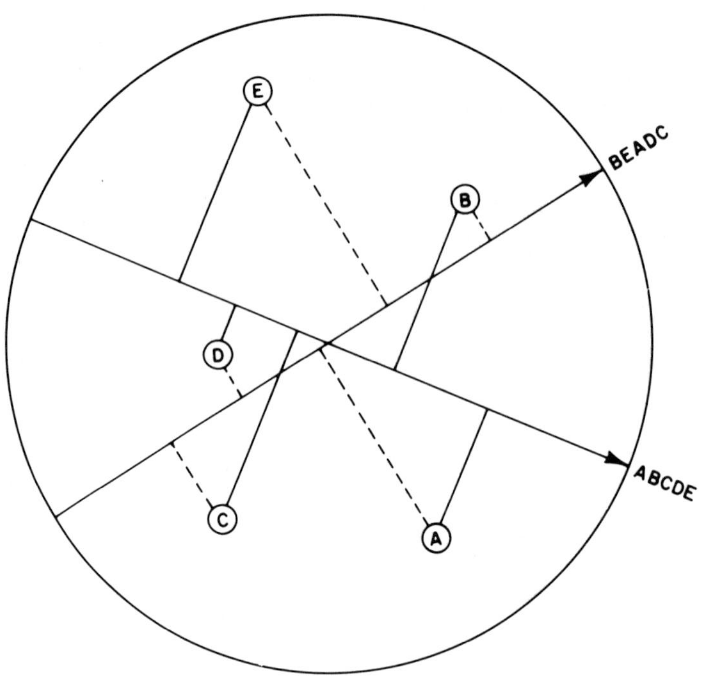

Fig. 9. Illustration of the vector model for preferences in a simple two-dimensional case. Five stimulus points are represented by the five letters *A* through *E*, and two subjects are represented by two vectors. Perpendicular projections of stimulus points onto a subject's vector are assumed to generate the preference order for that subject. Thus the first subject, for whom the lines of projection are represented as solid lines, has preference order ABCDE (that is, he prefers *A* to *B*, prefers *B* to *C*, and so on). The second subject, whose projection lines are dotted lines, has preference BEADC. Many more preference orders could be accommodated, even in this simple two-dimensional case, by allowing additional vectors pointing in still other directions.

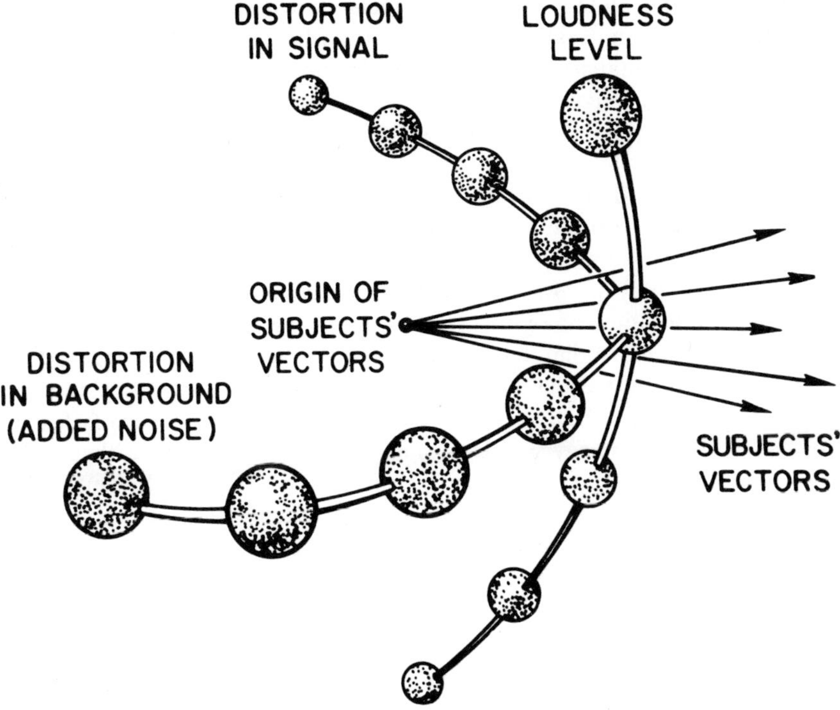

Fig. 10. A schematic representation of a subset of stimuli and subjects from the three-dimensional "vector model" analysis of McDermott's circuit quality data. The order of perpendicular projections of the stimulus points on a subject's vector reflects that subject's preference ordering.

means that, out of the infinite number of coordinate systems that *could be* used to describe the same multidimensional configuration, that particular one is sought that most nearly corresponds to interpretable or "meaningful" psychological dimensions. This process of rotation is analogous to orienting a map in terms of a north-south and east-west (rather than, say, a northeast-southwest and southeast-northwest) coordinate system. The problem of rotation to achieve maximum interpretability is one that often enters into, and sometimes complicates, applications of multidimensional scaling. As we saw earlier, Shepard had to rotate the two-dimensional Morse Code solution to achieve the interpretable solution he found. McDermott put in a great deal of time and effort to find the best orientation of axes for the three-dimensional circuit quality solution just discussed (in fact, this was by far the most difficult aspect of her analysis). It is, in fact, one of the principal virtues of INDSCAL that this rotational problem can usually be circumvented when this analytic technique is used. INDSCAL capitalizes on individual differences among subjects in similarity or other dyadic judgments to orient the coordinate axes uniquely. This property, we believe, makes it the method of choice whenever data of the right kind are available.

It would appear that both the Delphi approach and multidimensional scaling bear enormous promise for the future. While these two methodologies have much to offer separately, *in combination* they portend no less than a revolution in the study and rationalization of individual and group information processing and decision analysis. A great many of the fundamental topics examined in Delphi studies lend themselves to the MDS treatment. How similar or related are various alternative goals, objectives, values, plans, and programs? In essence, MDS can be used to determine if there exists for the Delphi group a morphology which clarifies the factors underlying their views on the subject at hand. The exposing of these factors, it is hoped, would allow such groups to come to a better collective understanding of their problem and thereby promote results of greater utility.

ANNOTATED BIBLIOGRAPHY[1]

I. Books

I. 1 Torgerson, W. S. *Theory and methods of scaling*. New York: Wiley, 1958 (particularly Chapter 11). This book is devoted to scaling in a very general sense; and most of it would be of primary interest to experimental or theoretical psychologists. Chapter 11, however, deals exclusively with the "classical" metric method of multidimensional scaling (based in large measure on papers published circa 1938 by Eckart, Young, Householder, and Richardson) with which Torgerson's name is associated.

I. 2 Coombs, C. H. *A theory of data*. New York: Wiley, 1964. This book is primarily of interest to theoretical psychologists and other social scientists. It deals with models more than methods, and many of the actual methods discussed have now been supplanted by more advanced computer-implemented procedures. It discusses the earliest form of nonmetric multidimensional scaling, which was invented by Coombs and his colleagues (in which not only the data, but also the solution is nonmetric). It also describes such highly useful models as the "unfolding" model of personal preference. The "Theory of Data" to which the title refers is a classification scheme for

[1]We thank Dr. Joseph B. Kruskal for his permission to borrow extensively from a previous jointly authored unpublished annotated bibliography on MDS for which he was principally responsible.

categorizing geometrically based measurement models of the kind assumed in MDS. This classification system may provide a useful conceptual tool for many readers. It is especially recommended to those interested in the theoretical and geometrical underpinnings of MDS models.

I.3 Green, P. E., and Carmone, F. J. *Multidimensional scaling and related techniques in marketing analysis*. Boston: Allyn and Bacon, 1970.

I.4 Green, P. E., and Rao, V. R. *Applied multidimensional scaling*. New York: Holt, Rinehart and Winston, 1972.

I.5 Green, P. E., and Wind, Y. (with contributions by J. D. Carroll). *Multi-attribute decisions in marketing*. New York: Dryden Press, 1973.

These three books, all having Paul E. Green of the Wharton School as first author, are fairly nontechnical, with emphasis on marketing applications. The first (Green and Carmone) may be the best book presently available to provide a general overview of the field to the nonspecialist (particularly one in marketing research or other fields of business or finance). It has a very long and helpful bibliography (although both the book and the bibliography may be somewhat out of date by now, owing to the rapid pace of developments in this field). The second (Green and Rao) is a kind of practitioner's guide to selection and comparison of algorithms. It uses the pedagogic device of applying virtually all the (then available) methods to a single core data set, so as to show directly the comparative virtues and weaknesses of the various methods. The third book (Green and Wind) focuses on "conjoint measurement" approaches, particularly to studies of "multiattribute decision making," but also uses a considerable amount of MDS methodology. It includes an appendix by Carroll that provides mathematical and technical descriptions of many of the most important MDS, conjoint measurement, clustering, and related multivariate analysis techniques (including descriptions of relevant computer programs).

I.6 Shepard, R. N., Romney, A. K., and Nerlove, S. *Multidimensional scaling: Theory and applications in the behavioral sciences*. Vol. I: *Theory*. New York: Seminar Press, 1972.

I.7 Romney, A. K., Shepard, R. N., and Nerlove, S. *Multidimensional scaling: Theory and applications in the behavioral sciences*. Vol. II: *Applications*. New York: Seminar Press, 1972.

This two-volume set contains fairly up-to-date contributions in both MDS theory and applications to many fields. Theoretical or methodological papers (in Vol. I) that should be particularly of interest include: Shepard's introduction and chapter on taxonomy of scaling models and methods; Lingoes' paper surveying the Guttman-Lingoes series of programs for nonmetric analysis; Young's paper on "polynomial conjoint analysis of proximities data"; Carroll's paper on individual differences, which discusses both the INDSCAL model for individual differences in similarities and other dyadic data, and a wide variety of models and methods for individual differences in preferences or other dominance judgments; Degerman's paper on hybrid models in which discrete clustering-like structure is combined with continuous spatial structure. Some of the applications chapters of major interest include: Wexler and Romney on kinship structures in anthropology; Rapoport and Fillenbaum on semantic structures; Rosenberg and Sedlak on perceived trait relationships; two different approaches (by Green and Carmone, on the one hand, and Stefflre, on the other) to marketing applications and the work of Wish, Deutsch and Biener on nation perception (which takes up where the "nation" study discussed here leaves off).

I.8 Carroll, J. D., Green, P. E., Kruskal, J. B., Shepard, R. N., and Wish, M. Book in preparation, based generally but not slavishly on Workshop on MDS held at University of Pennsylvania in June 1972. While book is not yet available, handout material for workshop can be obtained by writing to authors. Title and publisher not yet certain.

II. Basic Theory (Published Papers)

II. 1 Torgerson, W. S. Chapter 11 of *Theory and Methods of Scaling*. (See Books: I.1. for description.)

II. 2 Shepard, R. N. "The analysis of proximities: Multidimensional scaling with an unknown distance function. I." *Psychometrika*, 27, (1962), pp.125–39(*a*).

II.3 Shephard, R.N. "The analysis of proximities: Multidimensional scaling with an unknown distance function. II." *Psychometrika* **27**, (1962), pp. 219-46. (*b*).

These two articles by Shepard focus on the rank-order relationship between data and distances, and show it is often enough to determine the solution. They introduce what Guttman later called the "antirank-image principle" for estimating the relationship between distances and similarities.

II. 4 Kruskal J. B. "Multidimensional scaling by optimizing goodness of fit to a nonmetric hypothesis." *Psychometrika*, 29, (1964), pp. 1-27. (*a*)

II. 5 Kruskal, J. B. "Nonmetric multidimensional scaling: A numerical method." *Psychometrika* 29, (1964), pp. 115-29. (*b*)

These two articles focus on goodness-of-fit to explicit assumptions. They introduce computational methods from numerical analysis for systematically minimizing such a function. They introduce least-squares monotone regression for estimating the relationship between distances and similarities.

II. 6 Guttman, L. "A general nonmetric technique for finding the smallest coordinate space for a configuration of points." *Psychometrika*, 33, (1968), pp. 469-506. Introduced rank-image principle and goodness-of-fit functions based on it. Introduced new computational methods.

II. 7 Carroll, J. D. and Chang, J. J. "Analysis of individual differences in multidimensional scaling via an N-way generalization of 'Eckart-Young' decomposition." *Psychometrika*, 35, (1970), pp. 283-319. Introduced the generalization of multidimensional scaling called "individual differences scaling" (now called INDSCAL) which permits different data matrices to reflect different weights for the various axes. Introduced computational method which generalizes classical scaling, and is the basis for the INDSCAL method. Also includes discussions of analyses of the Bricker and Pruzansky "tone-ringer" data and of the Wish nation data described in the present paper.

III. Other Theoretical, Methodological, and Practical Contributions (Published Papers)

III.1 Shepard, R. N. "Analysis of Proximities as a technique for the study of information processing in man." *Human Factors*, **5**, (1963), pp. 33-48. Discusses among other applications of MDS, the analysis of Rothkopf's Morse Code data used as an illustration of two-way MDS in the present paper.

III.2 Tucker, L. R. and Messick, S. "An individual differences model for multidimensional scaling." *Psychometrika*, **28**, (1963), pp. 333-67. Given several matrices of dissimilarities, this approach uses factor analysis to cluster the matrices, and then analyzes an average matrix for each cluster. The relationship of this method to Carroll and Chang's individual differences scaling is discussed in their paper. (See Basic Theory II.7.)

III.3 McGee, V. W. "The multidimensional analysis of "elastic" distances." *British Journal of Mathematical and Statistical Psychology*, **19**, (1966), pp. 181-96. Introduces a kind of scaling similar to the Shepard-Kruskal approach, but with the squared differences multiplied by weights. Very similar results can be accomplished with MDSCAL, using the weights as permitted there.

III.4 Shepard, R. N. "Metric structures in ordinal data." *Journal of Mathematical Psychology*, **3**, (1966), pp. 287-315. Discusses general approach to deriving metric (i.e., numerical) information from nonmetric (i.e., merely ordinal) data. Nonmetric scaling and nonmetric factor analysis provide illustrations. Some early, but quite useful, Monte Carlo results are included.

III.5 Shepard, R. N., and Carroll, J. D. "Parametric representation of nonlinear data structures." In P. R. Krishnaiah (ed.), *Multivariate analysis*. New York: Academic Press, 1966, pp. 561–92. Introduces two new ideas. One, called "parametric mapping," relies on smoothness of mapping: the other is scaling based on a merely local monotonicity.

III.6 Krantz, D. H. "Rational distance functions for multidimensional scaling." *Journal of Mathematical Psychology* **4**, (1967), pp. 226-45.

III.7 Beals, R. W., Krantz, D. H., and Tversky, A. "The foundations of multidimensional scaling." *Psychological Review*, **75**, (1968), pp. 127-42.

III.8 Tversky, A. and Krantz, D. H. "The dimensional representation and the metric structure of similarity data." *Journal of Mathematical Psychology*, **1**, (1970), pp. 572-96.

These three articles are concerned with theoretical examination of the abstract mathematical foundations of multidimensional scaling.

III.9 Roskam, E. I. *Metric analysis of ordinal data in psychology.* Voorschoten, Holland: University of Leiden Press, 1968. A general overview and comparison of MDS methods and models up to 1968, with some original methodological contributions.

III.10 Kruskal, J. B. and Carroll, J. D. "Geometrical models and badness-of-fit functions." In P. R. Krishnaiah (ed.), *Multivariate Analysis. II.* New York: Academic Press, 1969, pp. 639-70. Properties and alternative choices for badness-of-fit, for scaling, parametric mapping, multidimensional unfolding, and tree structure. New theoretical results on stress function for scaling.

III.11 McDermott, B. J. "Multidimensional analyses of circuit quality judgments." *The Journal of the Acoustical Society of America*, 45, (1969), p. 774. Contains a full description of the circuit quality study described very superficially in this paper.

III.12 Carroll, J. D. "An overview of multidimensional scaling methods emphasizing recently developed models for handling individual differences." In C. W. King and D. Tigert (eds.), *Attitude Research Reaches New Heights.* Chicago: American Marketing Association, 1971, pp. 235-62. A nontechnical overview of MDS models and methods up to 1971, with emphasis on individual differences in perception and preference. Covers much of the same ground as the 1972 Carroll paper "Individual Differences and Multidimensional Scaling" in volume I of the Shepard, Romney, and Nerlove two-volume set, but in a much less technical way.

III.13 Jones, L. E. and Young, F. W. "Structure of a social environment: longitudinal individual differences scaling of an intact group." *Journal of Personality and Social Psychology*, 24, (1972), pp. 108-21. Describes in detail the "sociometric" study using INDSCAL discussed in the present paper.

III.14 Tucker, L. R. "Relations between multidimensional scaling and three-mode factor analysis." *Psychometrika* 37, (1972), pp. 3-27. Shows how three-mode factor analysis can be applied to analysis of individual differences in multidimensional scaling. The implied model is a generalization of the INDSCAL model, but lacks the dimensional uniqueness property of INDSCAL.

III.15 Carroll, J. D. and Wish, M. "Multidimensional perceptual models and measurement methods." In E. C. Carterette, and M. P. Friedman (eds.), *Handbook of Perception.* New York: Academic Press, 1974. A discussion of the geometrical models of MDS viewed as general theories of perception, and of a MDS methods as instruments for measurement of parameters of perceptual systems. Stresses the philosophical and mathematical foundations of MDS, relations to measurement theory, and to the basic theoretical underpinnings of psychology in general and perception in particular.

III.16 Indow, T. "Applications of multidimensional scaling in perception." In

E. C. Carterette and M. P. Friedman (eds.), *Handbook of Perception.* New York: Academic Press, 1973. Provides a number of applications of two-way MDS to data, principally on color perception and perception of "visual space."

III.17 Wish, M. and Carroll, J. D. "Applications of INDSCAL to studies of human perception and judgment." In E. C. Carterette and M. P. Friedman (eds.), *Handbook of Perception.* New York: Academic Press, 1974. Provides a wide variety of applications of the INDSCAL technique to areas of perception ranging from perception of color (including the analysis of the Helm data discussed in this paper), consonant phonemes, and "stress patterns" in English, through a very thorough study of nation perception.

III.18 Carroll, J. D. and Wish, M. Models and methods for three-way multidimensional scaling. In D. H. Krantz, R. C. Atkinson, R. D. Luce, and P. Suppes (eds.), Contemporary Developments in Mathematical Psychology. (Vol. II.: Measurement, Psychophysics and Neural Information Processing). San Francisco: W. H. Freeman, 1974, pp. 57-105.

Provides a general overview of INDSCAL, IDIOSCAL, Three-Mode Scaling, PARAFAC-2 and other three-way and higher way MDS models and methods, emphasizing the interrelationships among these methods, as well as their relationships to older methods for dealing with individual differences in similarities data, such as the Tucker-Messick "points of view" approach. Also discusses use of these methods for more general multi-way data analysis.

VI D. Differential Images of the Future

MARVIN ADELSON and SAMUEL ARONI

Introduction

A good deal of attention has been given recently to how the availability of alternative futures can affect decisionmaking in the present. Planners, decisionmakers, and designers all have a stake in the pattern the future will assume as they make their specific decisions today. Their actions will in part determine that future, and in part be chosen because of what that future is expected to be like.

Concern with the future has become a visible movement. One of the most popular best sellers in the past few years has been *Future Shock*, in which conjectures of future trends by many experts are systematized and interpreted, and some fears about society's ability to cope with the rate of future change are raised. In *The Greening of America*, another bestseller, a very different image of the future is generated. The literatures of technological forecasting, science fiction, and utopian planning provide many different kinds of such images.

One's action in the present is inevitably influenced by which images one carries with him of future prospects. One's choices, political behavior, priorities, and attitudes may well be different depending upon which set of images seems to him more convincing or acceptable. What and how much the farmer plants, how the businessman invests, what the student studies, how the general deploys his forces, what bills the legislator introduces, which experiment the researcher designs—all are determined in large part by expectations which *necessarily go beyond the facts*.

In an open society, and to some degree in any society, the character of the future is determined by innumerable decisions and actions interacting in rich and (in detail) indescribable ways. Such decisions are not made entirely, nor possibly even primarily, by experts. Nor are they random. They derive from the many ideas of future contingencies, possibilities, and certainties that individual people have generated, based on their past experience and present exposure to the world around them.

We can classify these ideas as images. Kenneth Boulding[1] has indicated the central role that images play in any thinking process. To the extent that the future is open to human control, information about people's images of the future is likely to be useful in dealing with both the present and the future.

[1]K. Boulding, "The Image," Univ. of Michigan Press, Ann Arbor, Mich. 1956.

The Delphi Method: Techniques and Applications, Harold A. Linstone and Murray Turoff (eds.)
ISBN 0-201-04294-0; 0-201-04293-2

Almost certainly, most people's future images are fragmentary, largely implicit, internally inconsistent, and mixed in terms of their time reference. Consequently, inferences about them based on "rational" grounds are dangerous, and should be approached cautiously.

Purpose

This pilot study was an attempt to discover if meaningful information about systematic similarities and differences in expectations of the future (next thirty-plus years) could be obtained from people's responses to a set of images presented in the form of printed pictures. If the answer was affirmative, we would then go on to try to relate those similarities and differences to choice behavior in the present and near-term future (zero to five years). Those choices might relate to investments, politics, lifestyles, dwelling, work, leisure, or purchasing patterns. Results of this kind could have significance for policy, planning, design, and investment decisions at many levels.

Background

The shape of the future is to a considerable degree subject to human control, but that control is not always well and wisely exercised. Attempts to deal more effectively with the future have tended to take the form of prediction, that is, anticipating what it will be like so as to enable the taking of appropriate and timely action. The more sophisticated attempts have acknowledged the importance of interactions among events as determiners of subsequent events. But too often, they have not involved the wills, wants, perceptions, and attitudes of "people" as either important or explicit variables in the prediction process.

Our basic premises are the very obvious ones, that while in some ways people resemble each other, in other ways they differ, and that there is *valuable information in their differences*, as well as in their similarities. The value of the difference information from a public-policy point of view is to enable steps that, to the extent possible, suit the differences. How do people differ *in the images they have of prospective futures*? They differ not only about *what* they expect to happen, but also about *when* they expect things to happen, how *important* different prospects are, and how *desirable* they are. These expectations influence their behavior, and hence also the expectations of others. Ignorance of these patterns of expectation, or images of the future held by others, is a major source of uncertainty for each of us in attempting to deal with the future. It is possible, in principle, to learn something about such expectations, and thereby improve by at least a little bit the basis for forecasting, and such related functions as policy-setting, decisionmaking, planning, designing, communicating, and interacting.

There is good reason to believe that almost all normal human choice behavior is based in part on anticipations or expectations of future events. Decision theory, management science, systems analysis, much learning theory, and related disciplines all involve use of "expected outcomes" as a basis for choice and action. Yet very little work has been done to elicit data on what those expectations are and how they relate to behavior, especially social behavior on a level broad enough to be meaningful in policy, planning, and design. Some surveys have been done (see below, this section). Generally, they have been couched in verbal terms, which is almost certainly not the best way to elicit some of the needed information. That approach is especially limiting with relatively nonverbal populations. Hence, the present study attempts, among other things, to demonstrate the feasibility of using a partly nonverbal technique. The use of pictures has the added promise of evoking more holistic and emotional associations.

As we remarked above, we view both consensus and dissensus on images of the future to be useful to understand. Consensus may increase the probability of—i.e., facilitate the process of reaching (or avoiding)—a particular future state of affairs, or increase the conviction that it will occur, but dissensus points up where issues are likely to arise, where incipient problems may lurk, where more information may be needed, or where the fact of diversity must be acknowledged and taken into account. In addition, it may stimulate synergistic thinking to resolve previously irresolvable differences in new creative ways. It is naturally interesting to relate diverse patterns of response on future images to independent variables describing individuals or groups.

The way to do this is simply to characterize individuals and groups of people in terms of their patterns of expectation, to the extent that those patterns are discernible. In other words, *to deal systematically with the future requires information on the statistics of future predictions*! But how can such statistics be acquired? Obviously, they cannot be obtained directly on every matter of interest that may arise. But it should be possible to gather "indicative" statistics that could be useful in predicting choice and other behavior. Such information should not only characterize whole populations, but should distinguish between identifiable subgroups, especially in a pluralistic system of decisionmaking. It is not hard to imagine how such differential statistics could be used to generate generally more satisfying conditions than present consensual approaches can. In fact, shared knowledge of such predictions may go far to alleviate intergroup misunderstandings, and serve to explain behavior which might otherwise appear incomprehensible to others.

There have been well-documented attempts to model mass reactions to future changes. These attempts are based in mathematical functions which extrapolate from empirical data and/or rational theories. Once the future is predicted in this manner, value judgments can be made by those in positions of authority, and action can be taken in the present to deal with it. Such a system could be greatly improved if a method could be developed which would allow the "people" to communicate their values, eschewing a false or forced consensus where a true one is not available and reflecting group differences in images of the future.

User preference studies have long been used to find out about values. These studies respond to a need for information to guide decisions, usually among alternatives of limited scope. Alternatives explored in such studies tend to be those which are practical today. The more distant future, with its more diverse set of alternatives, is deemphasized. Interest centers on micro-decisions. Images of the present, based more often than not on limited experience of the past, tend to be accepted as images of the future. The limitations of such images may exert a constraining effect on what possibilities the future can contain.

Two recent studies have successfully investigated ordinary people's expectations of the future. "Hopes and Fears of the American People" outlines people's images of future political and economic conditions across the United States. These opinions are divided into subgroups according to the respondents' race, age, education, optimism toward the future, and political party.

One of the most extensive studies of future expectations with emphasis on group differences was reported in 1968 by the Instituto de la Opinión Pública of Spain. Large samples from different parts of Spain, and from various foreign countries, were asked about their expectations concerning some 170 aspects of life and conditions in the year 2000. Results are reported in terms of sex, age, educational level, size of city of residence, region or country, marital status, occupation, economic sector, and income level. However, in both studies, only aggregate percentages are reported, and the patterns of individual response cannot be deduced. The pilot study reported here seeks response *patterns* via correlation analysis.

Hypotheses

People can coherently estimate the futurity, importance, and desirability of images of the future presented to them in picture form. In a substantial proportion of cases, they will either (a) respond in a way that relates with selected variables that they use to describe themselves, or else (b) show consensus on those images (as manifested in small variance in the distribution of responses).

Procedure

Self-administering questionnaires (see Appendix I, this paper) were distributed to students in the School of Architecture and Urban Planning at U.C.L.A. Each was asked to respond personally and also to give copies to parents (or others of the parents' age group who were willing to respond). Each questionnaire contained forty pictures representing various aspects of possible future living, learning, or working environments. They were selected to cover as wide a range of images as we could conveniently manage. We were able to collect responses on a group of fifty-four persons of both sexes, ranging in age from twenty to seventy-one years, and living in various parts of the country. We asked each of them to describe himself by answering a series of questions, including age, sex, marital status, occupation, population of the city from which he comes, region of the country from which he comes, personal and family income, amount of formal education, amount of time he spends thinking about the future compared to other people, the future "time window" of maximum interest to him, how he discounts the future, e.g., how much he would pay now for a guarantee of $1,000 in 1980, 1990, and 2000. Questions such as "Do you feel you are better off now than you were five (or ten) years ago," or "Do you feel you are better off now than you will be five (or ten) years from now," were also included.

We then asked each respondent to react to the images in the following five ways:

(1.) On the assumption that various environmental developments may grow or decline in importance over time, to estimate in what year the kind of thing represented by each picture would most likely reach its *peak or high point* in importance.

(2.) To estimate in which year, if any, it would have all but *disappeared*, or become insignificant.

(3.) To estimate its *importance* relative to "other things in its field" at the time it reached its high point.

(4.) To assess its *desirability* or undesirability.

(5.) To *caption* the image, so as to indicate concisely what it conveyed to him.

Some of the images derive from past lifestyles, built form, and human values, and some have no very firm precedent in the past. Since there is no guarantee that people interpret or perceive the images in the same way, we wished to get a feel for the range of interpretations they actually came up with when confronted with particular objective stimuli by asking for captions. Our intention was to do Delphi-like iterations, but we did not have the opportunity to do so. Clearly, the technique lends itself to Delphi procedures.

Results

Respondents report having difficulty in providing responses to many of the items, and they don't always have much confidence in the value of what they are doing. The pictures may, after all, be interpreted in several ways; the estimates we ask for tend to be subjective in the extreme; we do not always define terms, such as "important" or "desirable," leaving them for interpretation by each respondent. Under these circumstances, and assuming that individuals interpret given images rather differently, as indicated by their captions, one would expect little relationship to appear between response patterns and independent variables.

Correlation coefficients were computed between the various ratings each respondent gave for the futurity, importance, and desirability of the various images, and each of the independent variables (self-descriptions) used. In the case of the "sex" and "marital status" variables, point bi-serial correlation was used. For "occupation," an analysis of variance was used.

With all the shortcomings in the survey (and there are many), we found a goodly number of significant correlations and analyses of variance. A sample of the resulting table of relationships is shown in Appendix II of this paper. In that array, the column headings pertain to the pictures reproduced above them: "+" represents the "peak year" estimate; *i* is the "importance" estimate; *d* is the "desirability" estimate. A technical difficulty with the "phase-out year" estimate led to its omission from the results.

Squares represent negative correlations or inverse relationships, circles represent positive or direct relationships, both significant at the $\alpha \leqslant .01$ level. Solid half circles and half squares represent corresponding relationships at the $\alpha \leqslant .05$ level. Empty half circles and half squares are nearly significant ($.05 < \alpha < .055$). Obviously, in any large set of interrelations, on the average about 5 percent of them would appear significant at the 5 percent level, when in fact they would not be. At the $\alpha \leqslant .01$ level, about 1 percent of them would appear significant when in fact they wouldn't be, so the question arises, "Is the number obtained a larger number than those expectations?" The answer is, at the $\alpha \leqslant .01$ level, there are about three times as many significant relationships in this study as would be expected by chance. In the case of the $\alpha \leqslant .05$ level, there are some twice as many. Certain of the pictures lead to higher numbers of correlations than others, although variances were not dissimilar. Some correlations are up around .5 and .6 which, for a study of this kind, we consider very substantial.

Most of the correlation coefficients at the $\alpha \leqslant .05$ level ranged around .3 to .4, which accounts for about 10 percent to 15 percent of the variance. (These are not multiple *r*'s, but single *r*'s.) Although we have not examined them, multiple correlations should account for a good deal more variance than that. The matrix shows a larger number of significant relationships associated with the estimates of importance and desirability of the various future images than on the latency variable; that is, identifiable groupings of people seem to have more clear-cut differences in whether they like something, or think it is important, than on when they think it will happen. Considerable variation occurred in estimates of futurity; generally, people did not cluster their estimates closely around certain portions of the scale. A number of other trends could be identified:

(1.) Certain photographs produced high numbers of significant correlations. These were: 1, 4, 10, 19, 24, and 36.

These pictures may represent alternatives for the future dealing with relatively drastic and obvious changes with the past. They include two photographs of physical buildings: a trailer and a dome; three suggesting new social structure; and one of a person using a computer.

(2.) The means of the "importance" values were in the case of all but four pictures, above the midpoint of the scale ($i=5$), indicating that the pictures were not considered pointless.

(3.) The means of the "desirability" measures were in the case of twenty-six pictures above the midpoint of the scale ($d=0$).

(4.) Greater consensus occurred on images of the immediate future, as indicated by lower standard deviations associated with earlier dates.

(5.) Relatively *low* standard deviations on the "desirability" measure were associated with relatively *large* mean deviations from the neutral value of 0.

(6.) Relatively *low* standard deviations on the "importance" measure tended to be associated with *large* deviations from the midpoint of the scale (5).

(7.) On the occupation variable means were computed for each cell. People trained in the social sciences demonstrated consensus on images of very high organization (militarism and conventional office scenes) and relatively low organization (farm, potters).

Note: The *results* obtained do not depend upon any hypothesis about what people were actually responding to in each image; *interpretation* of the results *may* naturally be dependent upon such hypotheses.

Conclusions and Plans

Our initial expectations that response patterns are associated with independent descriptors appear to be confirmed. Some images produced stronger systematic differences between subgroups than others. In a few cases consensus on the future can be identified, especially in terms of importance and desirability. We conclude that this kind of survey tool offers promise for achieving the goals outlined earlier. Because of the shortcomings in the survey and the ill-defined sample, the next effort is using an improved survey instrument and a more carefully designed procedure. The new instrument is given as Appendix III of this paper. It covers a wider range of aspects of possible futures, and incorporates a number of other improvements. The new sample is an international one. We are now working on means to reduce the effects of cultural bias in selection of images.

Bibliography

Abt, Clark C. "Public Participation in Future Forecasting and Planning." Washington: World Future Society, July 1967. 12 pp. (mimeo.).

Adelson, Marvin. *The Technology of Forecasting and the Forecasting of Technology*. System Development Corporation, April 1968.

Banfield, Edward C. *The Unheavenly City. The Nature and the Future of Our Urban Crisis*. Boston: Little, Brown and Co., 1968.

Bernal, John Desmond. *The World, the Flesh and the Devil*. New York: E. P. Dutton and Co., 1929.

Boorstin, Daniel J. *The Image: A Guide to Pseudo-Events in America*. New York: Harper Row, 1961.

Boulding, Kenneth E. *The Image*. Ann Arbor: University of Michigan Press, 1956.

Cantril, Albert H., and Roll, Charles W., *Hopes and Fears of the American People*. New York: Universe, 1970.

Dalkey, Norman C. *The Delphi Method: An Experimental Study of Group Opinion*. Rand RM-5888-PR, June 1969.

David, Henry. "Assumptions about Man and Society and Historical Constructs in Futures Research." *Futures* **2**, No. 3 (September 1970), pp. 222-30.

Eldredge, H. Wentworth. "Education for Futurism in the United States: An On-Going Survey and Critical Analysis." *Technological Forecasting and Social Change* **2** (1970), pp. 133-48.

Enzer, Selwyn. *Delphi and Cross-Impact Techniques: An Effective Combination for Systematic Futures Analysis*. Institute for the Future, June 1970, W-8.

——. *Some Prospects for Residential Housing by 1985*. Institute for the Future, January 1971, R-13.

Gordon, T. J. *The Current Methods of Futures Research*. Institute for the Future, August 1971, P-11.

Hacke, James E., Jr. *The Feasibility of Anticipating Economic and Social Consequences of a Major Technological Innovation*. Stanford Research Institute, May 1967.

Hall, Edward T. *The Hidden Dimension*. Garden City, N. Y.: Doubleday, 1966.

——, *The Silent Language*. Greenwich, Conn.: Fawcett, 1959.

Helmer, Olaf. *Long-Range Forecasting: Roles and Methods*. Institute for the Future, May 1970. P-7.

——. *Social Technology*. New York: Basic Books, 1966.

Huntley, James Robert. "Gaps in the Future: The American Challenge and the European Challenge." *Futures* **2**, No. 1 (March 1970), pp. 5-14.

Kahn, Herman and Wiener, Anthony J. *The Year 2000: A Framework for Speculation on the Next Thirty-Three Years*. New York: Macmillan, 1967.

Kepes, Gyorgy (ed.). *Sign, Image, Symbol*. New York: Braziller, 1966.

Lindblom, Charles E. *The Intelligence of Democracy: Decision Making through Mutual Adjustment*. New York: Free Press, 1965.

Little, Dennis L. *Models and Simulation: Some Definitions*. Institute for the Future, March 1970, WP-6.

Lompe, Klaus. "Problems of Futures Research in the Social Sciences." *Futures* **1**, No. 1 (September 1968).

Martino, Joseph P. *Technological Forecasting for Decisionmaking*. New York: American Elsevier Publishing Co., 1972.

McHale, John. *The Future of the Future*. New York: Braziller, 1969.

McLuhan, Marshall. *Understanding Media*. New York: McGraw-Hill, 1964.

Polak, Fred L. *The Image of the Future: Enlightening the Past, Orienting the Present, Forecasting the Future*, Vol. 1: *The Promised Land Source of Living Culture*; Vol. 2: *Iconoclasm of the Images of the Future, Demolition of Culture*. New York: Oceana Publications, 1961.

——. *Prognostics: A Science in the Making Surveys the Future*. Amsterdam: Elsevier Publishing Co., 1971.

Rescher, Nicholas. "A Questionnaire Study of American Values by 2000 A. D." In *Values in the Future*, K. Baier and N. Rescher, (eds.), 1969, pp. 133-47.

Appendix I: Questionnaires for Images of the Future Study

ADDRESS______________________________________
ZIP CODE____________TELEPHONE______________________

AGE________SEX________ Marital Status____________________

Major Occupation or field of study____________________________

Income: Personal below 5 5-10 10-20 20-50 50+

Income: Family below 5 5-10 10-20 20-50 50+

Formal Education: number of years of school completed.________________

In your future planning have you seriously considered any other occupation or pastime?

Every person belongs to some major sub-groups; what are the most important ones you consider yourself a part of: (religion, race, political parties, nationalities, clubs, etc.)

Most people are interested in the future to some degree. Their main interest in the future usually extends beyond tomorrow, but not usually as far as hundreds of years from now. On the time scale below, please circle the period(s) in the future in which you are mainly interested:

Next Week	Next Month	Next Year	1973	1974–1979	1980–1989	1990–1999	2000

On the scale below indicate the amount of time you spend thinking about the future, compared with other people.
None a very small amount a moderate amount a large amount

In what year in the future do you feel you will know as much about the year 2000 as you now feel you know about the year 1950?

How much money would you be willing to pay right now for a guarantee of $1,000 in 1980? ______ For $1,000 in 1990? ______ For $1,000 in the year 2000? ______

Do you think you are better off now than you were 10 years ago? ______ 5 years ago? ______

Do you think you are better off now than you will be 5 years from now? ______ 10 years from now? ______

What is the population of the city in which you have lived for most of your life? ______
If it was a suburb of a major city what was the population of that city? ______ In what

region of the country have you spent most of your life (West, Midwest, North, South, etc.)? ______

People have been making a lot of predictions about the future recently. They have been talking about living styles, education, work and the role of technology in all of these. Most of us have some idea about what the future is going to be like. Taken together, these views will have an effect on shaping the future so they can be important. We want to find out something about your own views of what the future will be like.

We have assembled a group of pictures that represent various aspects of living, learning and working. Some of the things these pictures represent are growing in importance, becoming more common. Some are declining. Others will rise for awhile, then fall.

1. Tell us, for each picture, when you think what it represents will become most common, by writing a year in the (+) box. Think of the range of years between 1972 and 2000. If you think it will reach a highpoint after the year 2000, mark "2000±." If you feel it is or has been as high as it will ever go, mark "1972." If you feel it will never reach a highpoint write "never."
2. Similarly, tell us for each picture by what date you feel that kind of thing will be all but phased out. Do this by writing a year in the (−) box. Use the same range of years that was used for "highpoint" question.
3. When a development reaches its highpoint, it may be important or unimportant compared with other developments in its field. We would like to know your opinion of the importance of what each photograph represents when it reaches its highpoint. In the (i) box, write a number between 0 and 10: 0 would indicate little or no importance and 10 very high importance.
4. Also tell us how desirable you feel what each photograph represents is, whether you really like it or really hate it. Mark a number between −10 and +10 in box (d). If you really hate it, write −10. If you are neutral write (0), and if you really like it, write (+10). If you think it is reasonably desirable, you might use a number such as 3 or 4, etc.
5. On the page opposite the photographs, in the appropriate numbered box write a caption for each photograph. It should be as short as possible: preferably one word; two or three at the most. Try to pick a word that best sums up what the picture means to you. Please do not neglect to complete this part.

Try to answer conscientiously, but it shouldn't take much more than 45 minutes to complete the whole survey.

Images Used

The pictures used were divided into three groups representing a range of Living, Learning, and Working conditions, respectively. Briefly, the pictures may be described as follows:

Living

1. A mobile home on permanent supports.
2. A bedroom interior furnished with "stylish" Mediterranean pieces and bric-a-brac.
3. A young black man, holding a baby, sitting on a stoop of a high-density urban block.
4. A geodesic-like dome used as a private dwelling.
5. A Soleri "arcology" representing a massive double-pyramid city.
6. A barracks area exterior with a squad of soldiers in formation.
7. An original "futuristically-architected" house from Malibu, California.
8. A modern high-density "Scandinavian" apartment block, emphasizing stacking and packing.
9. A decidedly rustic scene with farm buildings, trees, and lake.
10. A group of twenty-seven lower-middle-class people of mixed ethnicity standing close together and smiling.
11. Moshe Safdie's "Habitat," randomly stacked dwellings from the Montreal Expo '67.
12. An aerial view of a California suburb showing curved streets and tract-area development.
13. A recent high-rise apartment building seen from below rising into an open sky.
14. A woman seated in the corner of a room with diamond-shaped grillwork intended to denote protection from intrusion.
15. A telephoto view of very uniform row housing.
16. A street of uniform four-story, medium-density dwellings.
17. A street of mixed older single-family and multiple-family units with fences, palms, and a prominently-displayed "Vehicles Prohibited" sign.
18. The rear portion of a private yacht with people at leisure.
19. A "hippie" couple with several others in a forest setting.
20. An aerial view of a high-rise cluster of a major city.

Learning

21. Rows of empty seats in a large modern lecture hall.
22. Intimate bubble-shaped interior of an inflatable structure with a handful of children relaxed but attentive.
23. An oriental print of Buddha in the lotus position with a child seated at his feet and looking up at his face.
24. A young man seated at a cathode ray tube computer console.

25. A group of soldiers, standing among trees on a campus, with gas masks and helmets on, firing rifles and a pistol.
26. A group of children in a school-yard with the high wire-fence prominently between them and the camera.
27. A wrist-television on a child's arm, with the caption "What about Learning" printed above it.
28. A couple of children seated on bicycles on a sidewalk watching a display through the open double side doors of a minibus equipped as a mobile educational unit.
29. A man's head with attached electrodes.
30. A typical college classroom showing rows of plastic chairs with armrests.

Working

31. A large, modern office interior with perhaps eight people working at desks.
32. A modern automated assembly line with a single person in attendance.
33. A helmeted, space-suited astronaut in a futuristic space vehicle (taken from "2001—A Space Odyssey").
34. A young couple working closely together making hand-thrown pottery on a wheel.
35. A business computer setting with two men and a woman working at various tasks.
36. A young couple dressed in a style reminiscent of India, standing outside a small (health food?) shop on the walls of which are painted two pictures of a girl in Indian garb.
37. Chimneys, electric power transmission lines and towers, a petroleum plant against a dark sky with lots of smoke and smog.
38. The very old, well-preserved section of a European city seen from within a very modern, very large office building.
39. A very large modern commercial office block with the sunlight glinting off the glass-and-steel facade.
40. A low brick-front commercial arcade entrance with potted plants, an overhead sign "La Ronda de las Estrellas," and several signs indicating the presence of offices and art galleries within.

Discussion of Results

Pictures used were divided into three groups: "Living" (20 pictures), "Learning" (10 pictures), and "Working" (10 pictures), covering three categories of human concerns.

We will discuss below the response patterns to six of the pictures, two from each category. Appendix II summarizes the statistical results, showing significant and near-significant correlations between responses and the independent variables that describe the respondents. The last column in Appendix II indicates the total number of significant relations for each independent variable with all forty images. Out of a total of 2040 relations, 54 (or 2.6 percent) are found to be significant at the $\alpha \leqslant .01$ level, and 219 (or 10.8 percent) are significant at the $\alpha \leqslant .05$ level, which far exceeds chance expectation.

Consider, first, picture #1, which shows a mobile home. We note that the "desirability" estimate correlates, $r = .39$ with age, and that the correlation is significant at the $\alpha = .01$ level. This means that older people in the study tended to rate the mobile home (and its implications?) to be more desirable than did younger people. So, if we go down the column of entries beneath picture #1, did people with higher *personal* income, although this relationship is significant at the $\alpha = .05$ level, but not at the $\alpha = .01$ level. However, people with higher *family* income rated this image as *less* desirable than those with lower family incomes. They also expected it to peak in importance sooner. To put it the other way, those with lower family incomes tended to rate the mobile home concept as more desirable and to expect it to grow in importance over a longer period of time ($\alpha \leqslant .05$). Those who were interested in the longer-term future tended to see the mobile home concept as less important than those whose interests peaked sooner (presumably this has something to do with age, although it did not show up as a significant correlation there). Those who were willing to pay more for a fixed sum of money in the future also ($\alpha \leqslant .01$) saw this concept as less important. Those who thought they would be less well off in five years than they are now rated the mobile home as more desirable than those who thought they would be better off ($\alpha \leqslant .05$). Those from larger cities also rated the concept as less desirable than those from smaller places ($\alpha \leqslant .01$). Married individuals also regarded the concept as less desirable, and those of different occupations rated it differently from each other ($\alpha \leqslant .01$). A just slightly less-than-significant relationship to sex of the respondent also appeared ($\alpha \leqslant .055$).

A second picture from the "Living" group, picture #19, shows a "hippie" couple with several others in a forest setting. Both its "importance" and "desirability" ratings correlated negatively with age, at the $\alpha = .01$ level. Thus older people in the study tended to rate this image as less important and desirable than did younger people. People with higher personal income, rated it less desirable than those with lower personal income, although only at the $\alpha = 0.014$ level. People who said they spend a large amount of time thinking about the future indicated that what this image represents will peak in importance sooner than did people relatively uninterested in the future ($\alpha \leqslant .05$). However, those who were willing to pay more for a fixed sum of money in the future also saw a later date for its peak importance ($\alpha \leqslant .01$). Those from larger cities rated the image as both more important ($\alpha \leqslant .01$) and more desirable ($\alpha \leqslant .05$) than those from smaller places. Married people thought that it will peak in importance sooner ($\alpha \leqslant 0.05$) but rated it more desirable, also at $\alpha \leqslant 0.05$. People having different occupations also showed different responses to its desirability ($\alpha \leqslant 0.05$).

Picture #24, from the "Learning" group, shows a young man seated at a cathode ray tube computer console. The correlation coefficients, with age, of the "importance" and "desirability" of this learning concept were $r = 0.33$ (at $\alpha \leqslant 0.05$) and $r = 0.52$ (at $\alpha \leqslant 0.01$) respectively. Thus older people tended to rate it more important and desirable than younger people in the study. People who spent a larger amount of time thinking about the future thought its importance and desirability (both at $\alpha \leqslant 0.05$) to be less than did those relatively uninterested in the future. Those who were willing to pay more for a fixed sum of money in the year 2000 also indicated greater importance attaching to this concept ($\alpha \leqslant 0.05$). They also thought it to be more desirable. Depending on the year for the future guarantee, the significance levels varied from $\alpha \leqslant 0.05$ for 1990 to $\alpha \leqslant 0.01$ for the year 2000. Those who thought they were better off now than they were ten years ago also rated this learning concept as more desirable ($\alpha \leqslant 0.05$). People from larger cities thought the concept to be less important and less desirable ($\alpha \leqslant 0.05$). Finally, married people also regarded it as less desirable ($\alpha \leqslant 0.05$).

A second picture from the "Learning" group, picture #27, shows a wrist-television on a child's arm, with the caption "What about Learning" printed above it. The correlation coefficient with age of the "desirability" of this concept was $r=0.31$ at $\alpha \leqslant 0.05$. Thus older people in the study tended to rate it more desirable than younger people. Those with higher education thought it to be less desirable than those less educated ($\alpha \leqslant 0.05$). Those who were interested in the longer-term future tended to see this concept peak in importance later ($\alpha \leqslant 0.05$) and be more desirable ($\alpha \leqslant 0.01$) than those whose future interests were of shorter-term. Those who thought they were better off now than they were ten years ago also rated the concept to reach its peak in importance sooner ($\alpha \leqslant 0.05$). Married individuals regarded this learning concept as less desirable ($\alpha \leqslant 0.05$). A just slightly less-than-significant relationship to sex of the respondent also appeared ($\alpha = 0.052$).

Picture #35, from the "Working" group, shows a business computer setting with two men and a woman working at various tasks. Again, older people in the study tended to rate this working concept as more desirable than younger people ($\alpha \leqslant 0.05$). People who spent a larger amount of time thinking about the future also thought its importance to be greater ($\alpha \leqslant 0.05$). Those who were willing to pay more for a fixed sum of money in the year 1980 also rated this concept as more important and more desirable (both at $\alpha \leqslant 0.05$). Those who thought they were better off now than they were ten years ago also rated it as more desirable ($\alpha \leqslant 0.05$). At the same time, those who thought they were better off now than they were *five* years ago thought that it will peak in importance later ($\alpha \leqslant 0.05$). On the other hand, those who thought they will be better off ten years from now thought that the peak importance will be reached sooner (less-than-significant $\alpha = 0.055$). Those of different occupations rated the image differently from each other ($\alpha \leqslant 0.05$). It should be emphasized that, unlike the other five images discussed in detail in this Appendix, none of the correlations of picture #35 reach the $\alpha = 0.01$ level.

A second picture from the "Working" group, picture #36, shows a young couple dressed in a style reminiscent of India, standing outside a small (health food?) shop on the walls of which are painted two pictures of a girl in Indian garb. Similarly to the "hippie" picture #19, both the "importance" and "desirability" correlated negatively with age, at the $\alpha = 0.01$ level. Thus older people in the study tended to rate this image to be less important and desirable than did younger people. So did people with higher education, although only at the $\alpha \leqslant 0.05$ level. To put it the other way, those with lower education tended to rate the image as both more important and desirable. People who spent a larger amount of time thinking about the future also thought its importance to be greater ($\alpha \leqslant 0.05$). Those who thought they were better off now than they were ten years ago rated this concept to be less important and desirable ($\alpha \leqslant 0.01$). The same applied to those who thought they were better off now than they were five years ago, but only with $\alpha \leqslant 0.05$. On the other hand, those who thought they will be better off ten years from now rated the image to be more important ($\alpha \leqslant 0.01$). Presumably, age was the main reason for the correlation with these last three independent variables. Those from larger cities rated the image as both more important ($\alpha \leqslant 0.01$) and more desirable ($\alpha \leqslant 0.05$) than those from smaller places. This is similar to the "hippie" picture #19. The importance of the image was related to the sex of the respondent ($\alpha \leqslant 0.05$), and married individuals expected it to peak in importance earlier than single ones ($\alpha \leqslant 0.05$).

Appendix II: Some Significant Correlations

EXPECTATIONS

LIVING LEARNING WORKING

	INDEPENDENT VARIABLE NUMBER	1			19			24			27			35			36			NUMBER OF SIGNIFICANT RELATIONS	
		+	i	d	+	i	d	+	i	d	+	i	d	+	i	d	+	i	d	1%	5%
age	1			0.39 0.008		-0.42 0.004	-0.61 0.001		0.33 0.025	0.52 0.001			0.31 0.036			0.35 0.017		-0.45 0.003	-0.44 0.004	9	21
personal income	2			0.37 0.046			-0.35 0.014													2	14
family income	3	0.33 0.023		0.31 0.033																1	13
education	4												-0.29 0.044					-0.37 0.013	-0.36 0.014	3	10
time window	5		.31 .029								0.29 0.047		0.43 0.005							2	7
time spent	6				0.30 0.031				-0.33 0.024	-0.29 0.040					0.31 0.029			0.36 0.016		0	11
1980 · present value of the future	7		.62 .001		0.46 0.009										0.43 0.017	0.34 0.047				3	11
1990	8		.66 .001		0.63 0.001					0.43 0.021										4	11
2000	9		.65 .001		0.65 0.001				0.41 0.042	0.57 0.006										4	11
10	10									[illegible]	0.34 0.023					0.32 0.028		-0.44 0.004	-0.42 0.006	5	14
5	11													0.37 0.012				-0.29 0.040	-0.32 0.029	2	10
+5	12			-0.30 0.031																6	12
+10	13													0.27 0.055				0.39 0.010		3	14
population of city	14			-0.48 0.001		0.37 0.009	0.35 0.014		0.34 0.010	0.31 0.032								0.41 0.007	0.32 0.030	4	17
sex	15			−0.22 0.051									0.24 0.062					0.26 0.037		0	12
marital status	16			0.46 0.001	0.28 0.021		0.26 0.026			−0.27 0.025			−0.29 0.025				−0.25 0.034			5	21
occupation	17			12.22 F RATIO			3.13 F RATIO									3.46 F RATIO				1	10
MEAN		1979.8	5.8	−1.5	1989.3	4.6	1.6	1997.0	8.2	5.5	1093.2	6.6	4.8	1997.0	9.0	4.7	1981.6	3.8	2.2	54	219
STANDARD DEVIATION		7.7	3.2	5.1	11.7	3.9	6.2	13.9	1.9	3.2	14.4	2.6	4.7	14.6	1.4	5.7	11.5	3.0	4.8		
No. of Significant Relations..α<1%		0	3	4	3	2	1	0	0	2	0	0	1	0	0	0	0	4	2		
No. of Significant Relations..α<5%		1	4	7	5	2	5	1	4	7	2	0	4	2	2	4	1	8	5	2040	

Appendix III: Expectations II: A Survey about the Future

We would like to know something about what you think will be happening in the world from now on. People's views of emerging conditions tend to affect their choices, and thereby can influence what the future will be like. Similarities and differences in assumptions, attitudes and estimates of the future across countries and cultures are therefore important to understand, although they tend to be difficult to discern. We have found the techniques used in this survey to be helpful.

We have assembled a group of pictures which seem to say something about the future. The images included are just a sample from a range of possible images representing future events. We would like *your estimates* as to when these events will become most common and how important and desirable they will be when they do.

You may find some of the pictures surprising. Others may seem to be obscure or difficult to interpret unequivocally. Don't be overly concerned. You can indicate what you are responding to by means of the caption you give to each picture. In each case, there is a good reason for inclusion.

Just a small amount of your time in filling out this survey will supply us with very valuable information on how different people view the future. You may not feel that these questions get at the heart of your views. Nevertheless, we can get some very important information from your responses, and we urge you to respond. We think the results might be interesting to you, too. We will send you information on the distribution of responses by others. Earlier research along this line has produced very interesting results, so we hope you will cooperate.

Thank you.

Marvin Adelson
Professor, School of Architecture and Urban Planning
U.C.L.A.

Samuel Aroni
Professor, School of Architecture and Urban Planning
U.C.L.A.

This study is supported by funds from the Committee on International and Comparative Studies, University of California at Los Angeles.

INSTRUCTIONS

Carefully look at each picture and decide what you think it is trying to show. Since images are often ambiguous we have asked you to write a caption for each picture so we can interpret results in terms of what you had in mind. Also, because we realize you may be very unsure of your responses to some of the images, we have included a confidence scale so you can rate how accurate you feel your answers are.

Here are the steps we would like you to go through for each picture:

1. Caption ________________________________
   Write a word or short title which sums up what you think the picture represents.

2. Date when development will peak

   A. Your Nation
   Mark the year between 1972-2000 when you think this development will become most common in your nation. Mark "+" if you think it will occur after the year 2000. Mark "-" if you think the peak has already passed.

   B. The World
   Mark the year when you think this development will become most common throughout the world as a whole.

3. Confidence Scale
   Rate how sure you feel about your answers to A and B. "1" means you are very unsure about your answers; "2" means you are fairly confident but not certain; "3" means you are very confident.

4. Importance
   Indicate by a check on this scale of 0-10 how important you feel the event will be compared with other developments in its field when it reaches its high point.

5. Desirability
   Rate how desirable, in your opinion, the development will be, using the scale of -5 to +5. (-5) would mean you hate it; (0) is neutral; and (+5) would mean you really like it.

Work rapidly. Don't take too much time on any picture. If you're not sure of a year or a rating, check your best guess and go on. If you change your mind later, you can come back to the picture and change your answer. Please answer all items.

EXPECTATIONS II
A SURVEY ABOUT THE FUTURE

FAMILY NAME ______________________ GIVEN NAME(S) ______________

ADDRESS ____

ZIP CODE __________ COUNTRY ______________ TELEPHONE __________

AGE __________ SEX _________ Marital Status ______________

(Optional:)
Race __________ Religion __________ Nationality of Ancestry __________

Major Occupation or field of study at present ______________________

Annual Income: Personal ______________ Family ______________

Formal Education: number of years of school completed. __________

In what city or town have you spent most of your life ______________,
City

______________________, ____________________.
State or Province Country

Population of that city. ____________

Most people are interested in the future to some degree. Their main interest in the future usually extends beyond tomorrow, but not usually as far as hundreds of years from now. On the time scale below, please circle the period(s) in the future in which you are mainly interested:

Next week   Next Month   Next Year   1974-1979   1980-1989   1990-1999   2000+

On the scale below, circle the amount of time you spend thinking about the future, compared with other people.

None   a very small amount   a moderate amount   a large amount

What proportion of your annual income would you be willing to pay right now for a guarantee of an amount equal to your present annual income in the year 2000? ______________

The scale below represents the condition of your life. The (0) represents a life where nothing is going right. The (10) represents a life where everything is going right. On this scale answer the three following questions. With a (•) indicate how you feel your life is going now. With a (X) indicate how you feel your life was going 5 years ago. With a (✓) indicate how you feel your life will be going 5 years from now.

0	1	2	3	4	5	6	7	8	9	10

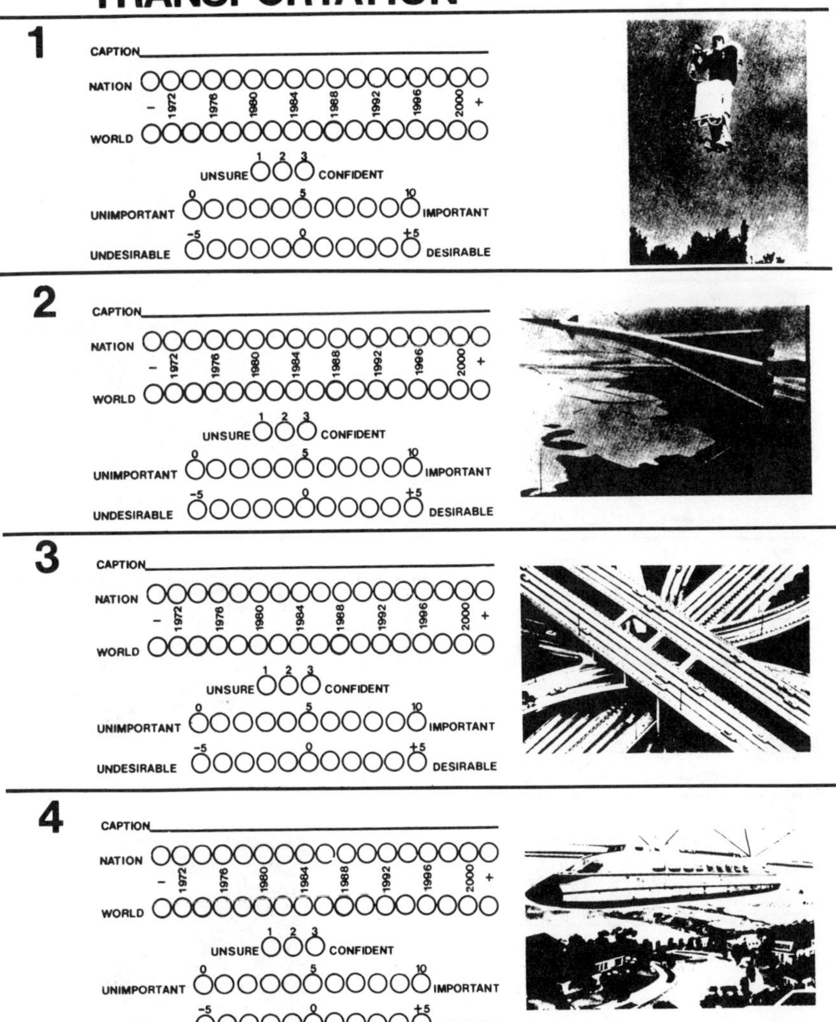
TRANSPORTATION
1
CAPTION
NATION
- 1972 1976 1980 1984 1988 1992 1996 2000 +
WORLD
UNSURE 1 2 3 CONFIDENT
UNIMPORTANT 0 5 10 IMPORTANT
UNDESIRABLE -5 0 +5 DESIRABLE
2
CAPTION
NATION
- 1972 1976 1980 1984 1988 1992 1996 2000 +
WORLD
UNSURE 1 2 3 CONFIDENT
UNIMPORTANT 0 5 10 IMPORTANT
UNDESIRABLE -5 0 +5 DESIRABLE
3
CAPTION
NATION
- 1972 1976 1980 1984 1988 1992 1996 2000 +
WORLD
UNSURE 1 2 3 CONFIDENT
UNIMPORTANT 0 5 10 IMPORTANT
UNDESIRABLE -5 0 +5 DESIRABLE
4
CAPTION
NATION
- 1972 1976 1980 1984 1988 1992 1996 2000 +
WORLD
UNSURE 1 2 3 CONFIDENT
UNIMPORTANT 0 5 10 IMPORTANT
UNDESIRABLE -5 0 +5 DESIRABLE

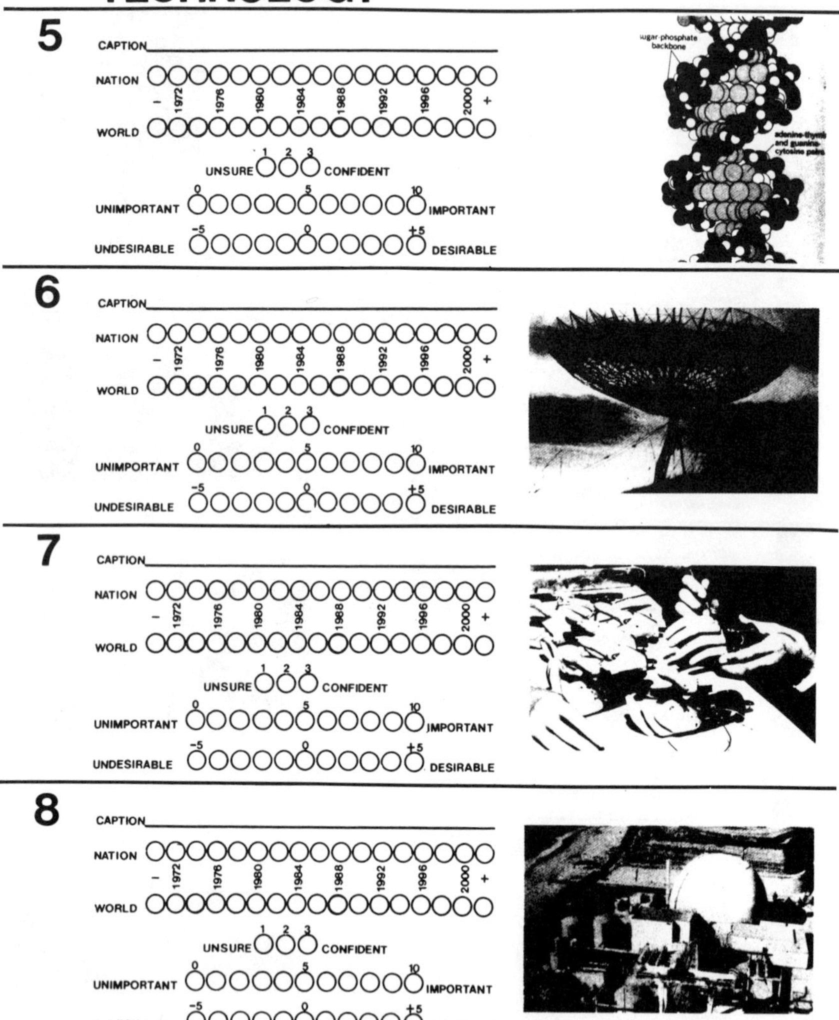
TECHNOLOGY
5
CAPTION
NATION
1972 1976 1980 1984 1988 1992 1996 2000
WORLD
UNSURE 1 2 3 CONFIDENT
UNIMPORTANT 0 5 10 IMPORTANT
UNDESIRABLE −5 0 +5 DESIRABLE
sugar-phosphate backbone
6
CAPTION
NATION
1972 1976 1980 1984 1988 1992 1996 2000
WORLD
UNSURE 1 2 3 CONFIDENT
UNIMPORTANT 0 5 10 IMPORTANT
UNDESIRABLE −5 0 +5 DESIRABLE
7
CAPTION
NATION
1972 1976 1980 1984 1988 1992 1996 2000
WORLD
UNSURE 1 2 3 CONFIDENT
UNIMPORTANT 0 5 10 IMPORTANT
UNDESIRABLE −5 0 +5 DESIRABLE
8
CAPTION
NATION
1972 1976 1980 1984 1988 1992 1996 2000
WORLD
UNSURE 1 2 3 CONFIDENT
UNIMPORTANT 0 5 10 IMPORTANT
UNDESIRABLE −5 0 +5 DESIRABLE

TECHNOLOGY

9

CAPTION__________

NATION ○○○○○○○○○○○○○○○○○○○○

– 1972 1976 1980 1984 1988 1992 1996 2000 +

WORLD ○○○○○○○○○○○○○○○○○○○○

UNSURE ○ 1 ○ 2 ○ 3 CONFIDENT

UNIMPORTANT ○ 0 ○○○○ ○ 5 ○○○○ ○ 10 IMPORTANT

UNDESIRABLE ○ −5 ○○○○ ○ 0 ○○○○ ○ +5 DESIRABLE

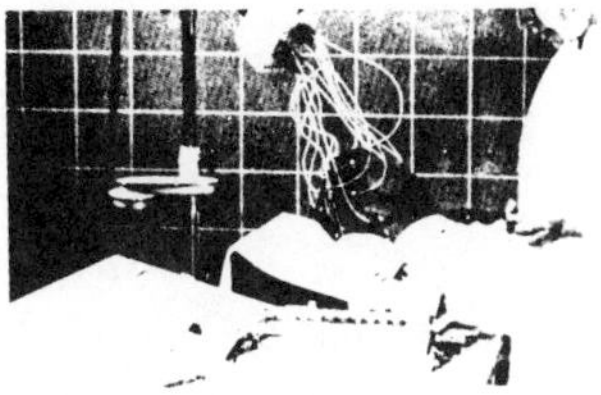

10

CAPTION__________

NATION ○○○○○○○○○○○○○○○○○○○○

– 1972 1976 1980 1984 1988 1992 1996 2000 +

WORLD ○○○○○○○○○○○○○○○○○○○○

UNSURE ○ 1 ○ 2 ○ 3 CONFIDENT

UNIMPORTANT ○ 0 ○○○○ ○ 5 ○○○○ ○ 10 IMPORTANT

UNDESIRABLE ○ −5 ○○○○ ○ 0 ○○○○ ○ +5 DESIRABLE

11

CAPTION__________

NATION ○○○○○○○○○○○○○○○○○○○○

– 1972 1976 1980 1984 1988 1992 1996 2000 +

WORLD ○○○○○○○○○○○○○○○○○○○○

UNSURE ○ 1 ○ 2 ○ 3 CONFIDENT

UNIMPORTANT ○ 0 ○○○○ ○ 5 ○○○○ ○ 10 IMPORTANT

UNDESIRABLE ○ −5 ○○○○ ○ 0 ○○○○ ○ +5 DESIRABLE

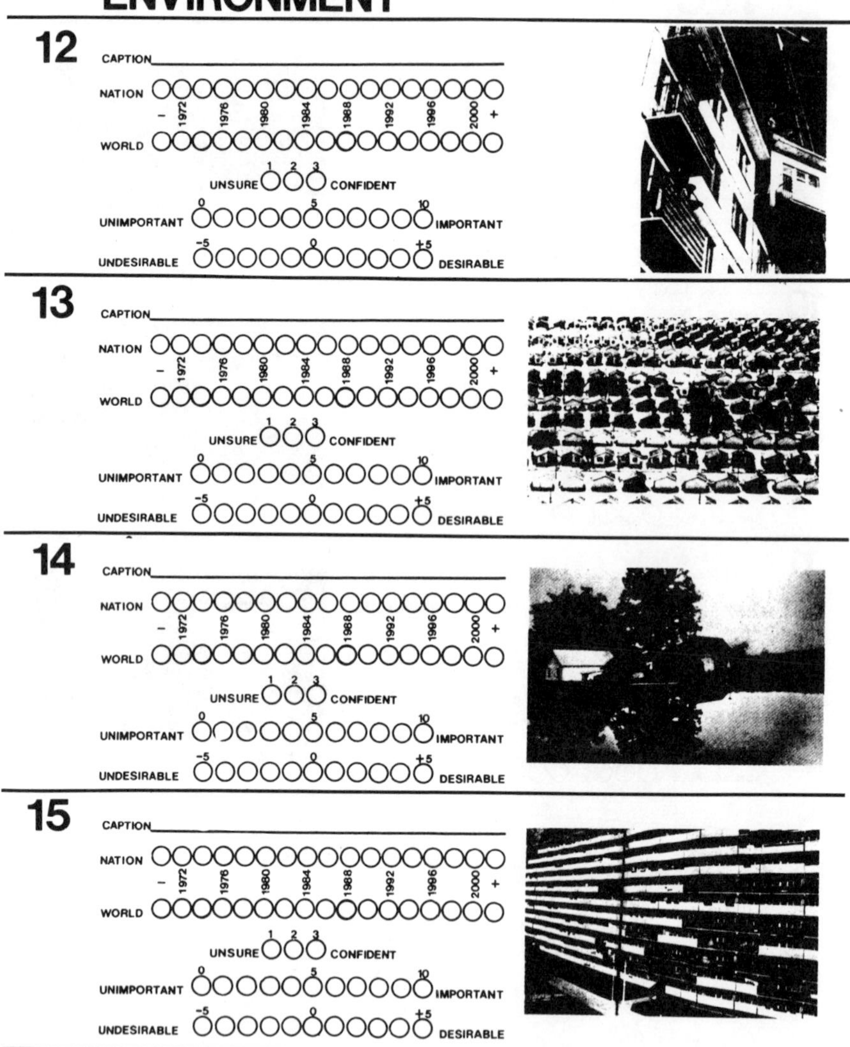

ENVIRONMENT

12

CAPTION ____________________

NATION
– 1972 1976 1980 1984 1988 1992 1996 2000 +
WORLD

UNSURE 1 2 3 CONFIDENT

UNIMPORTANT 0 5 10 IMPORTANT

UNDESIRABLE –5 0 +5 DESIRABLE

13

CAPTION ____________________

NATION
– 1972 1976 1980 1984 1988 1992 1996 2000 +
WORLD

UNSURE 1 2 3 CONFIDENT

UNIMPORTANT 0 5 10 IMPORTANT

UNDESIRABLE –5 0 +5 DESIRABLE

14

CAPTION ____________________

NATION
– 1972 1976 1980 1984 1988 1992 1996 2000 +
WORLD

UNSURE 1 2 3 CONFIDENT

UNIMPORTANT 0 5 10 IMPORTANT

UNDESIRABLE –5 0 +5 DESIRABLE

15

CAPTION ____________________

NATION
– 1972 1976 1980 1984 1988 1992 1996 2000 +
WORLD

UNSURE 1 2 3 CONFIDENT

UNIMPORTANT 0 5 10 IMPORTANT

UNDESIRABLE –5 0 +5 DESIRABLE

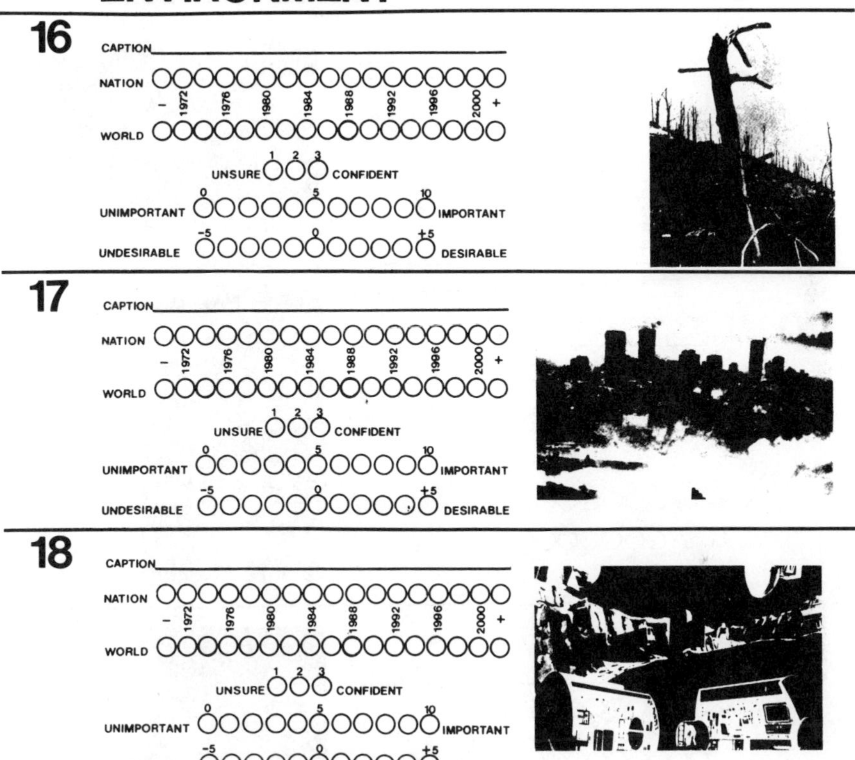

ENVIRONMENT

16

CAPTION ______

NATION

– 1972 1976 1980 1984 1988 1992 1996 2000 +

WORLD

1 2 3
UNSURE CONFIDENT

0 5 10
UNIMPORTANT IMPORTANT

-5 0 +5
UNDESIRABLE DESIRABLE

17

CAPTION ______

NATION

– 1972 1976 1980 1984 1988 1992 1996 2000 +

WORLD

1 2 3
UNSURE CONFIDENT

0 5 10
UNIMPORTANT IMPORTANT

-5 0 +5
UNDESIRABLE DESIRABLE

18

CAPTION ______

NATION

– 1972 1976 1980 1984 1988 1992 1996 2000 +

WORLD

1 2 3
UNSURE CONFIDENT

0 5 10
UNIMPORTANT IMPORTANT

-5 0 +5
UNDESIRABLE DESIRABLE

ENVIRONMENT

19

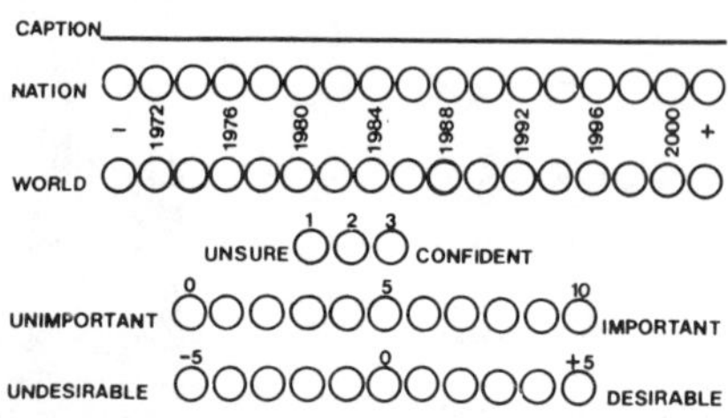

20

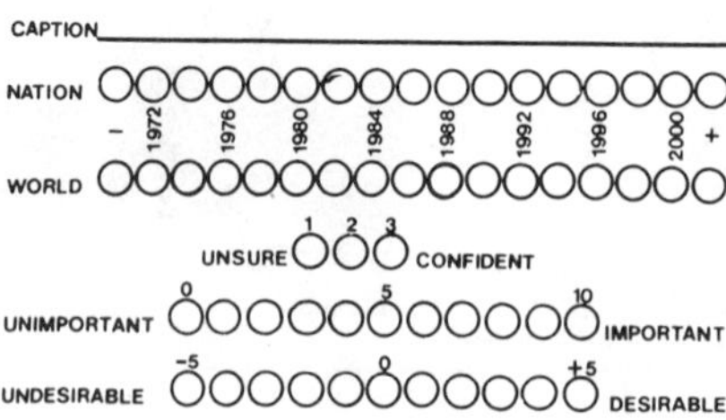

21

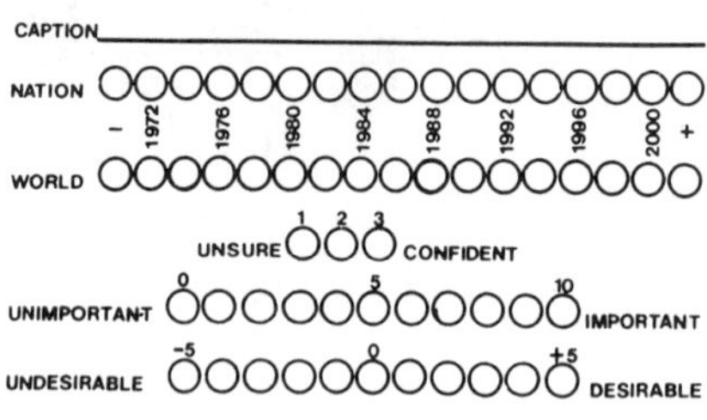

WORK

22

CAPTION______________________________

NATION ○○○○○○○○○○○○○○○○○

– 1972 1976 1980 1984 1988 1992 1996 2000 +

WORLD ○○○○○○○○○○○○○○○○○

UNSURE ○ ○ ○ CONFIDENT (1 2 3)

UNIMPORTANT ○○○○○○○○○○○ IMPORTANT (0 5 10)

UNDESIRABLE ○○○○○○○○○○○ DESIRABLE (-5 0 +5)

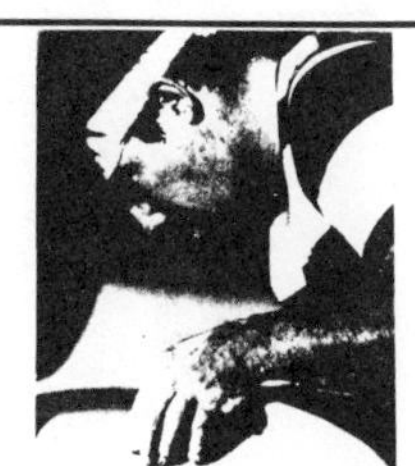

23

CAPTION______________________________

NATION ○○○○○○○○○○○○○○○○○

– 1972 1976 1980 1984 1988 1992 1996 2000 +

WORLD ○○○○○○○○○○○○○○○○○

UNSURE ○ ○ ○ CONFIDENT (1 2 3)

UNIMPORTANT ○○○○○○○○○○○ IMPORTANT (0 5 10)

UNDESIRABLE ○○○○○○○○○○○ DESIRABLE (-5 0 +5)

24

CAPTION______________________________

NATION ○○○○○○○○○○○○○○○○○

– 1972 1976 1980 1984 1988 1992 1996 2000 +

WORLD ○○○○○○○○○○○○○○○○○

UNSURE ○ ○ ○ CONFIDENT (1 2 3)

UNIMPORTANT ○○○○○○○○○○○ IMPORTANT (0 5 10)

UNDESIRABLE ○○○○○○○○○○○ DESIRABLE (-5 0 +5)

25

CAPTION______________________________

NATION ○○○○○○○○○○○○○○○○○

– 1972 1976 1980 1984 1988 1992 1996 2000 +

WORLD ○○○○○○○○○○○○○○○○○

UNSURE ○ ○ ○ CONFIDENT (1 2 3)

UNIMPORTANT ○○○○○○○○○○○ IMPORTANT (0 5 10)

UNDESIRABLE ○○○○○○○○○○○ DESIRABLE (-5 0 +5)

EDUCATION

26

CAPTION

NATION

– 1972 1976 1980 1984 1988 1992 1996 2000 +

WORLD

1 2 3
UNSURE CONFIDENT

0 5 10
UNIMPORTANT IMPORTANT

–5 0 +5
UNDESIRABLE DESIRABLE

27

CAPTION

NATION

– 1972 1976 1980 1984 1988 1992 1996 2000 +

WORLD

1 2 3
UNSURE CONFIDENT

0 5 10
UNIMPORTANT IMPORTANT

–5 0 +5
UNDESIRABLE DESIRABLE

28

CAPTION

NATION

– 1972 1976 1980 1984 1988 1992 1996 2000 +

WORLD

1 2 3
UNSURE CONFIDENT

0 5 10
UNIMPORTANT IMPORTANT

–5 0 +5
UNDESIRABLE DESIRABLE

EDUCATION

29

CAPTION________________________

NATION ○○○○○○○○○○○○○○○○○○○

– 1972 1976 1980 1984 1988 1992 1996 2000 +

WORLD ○○○○○○○○○○○○○○○○○○○

UNSURE ○ 1 ○ 2 ○ 3 CONFIDENT

UNIMPORTANT 0 ○○○○○ 5 ○○○○○ 10 IMPORTANT

UNDESIRABLE −5 ○○○○○ 0 ○○○○○ +5 DESIRABLE

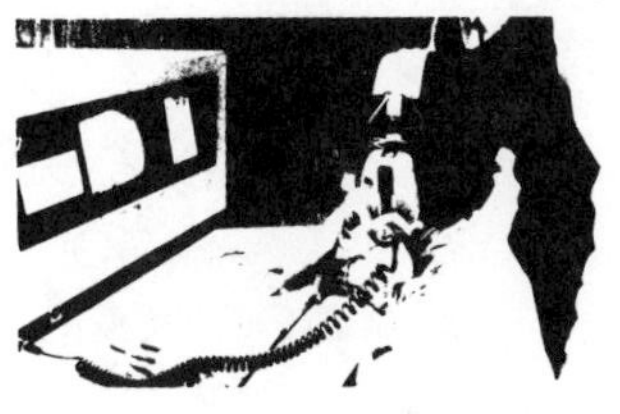

30

CAPTION________________________

NATION ○○○○○○○○○○○○○○○○○○○

– 1972 1976 1980 1984 1988 1992 1996 2000 +

WORLD ○○○○○○○○○○○○○○○○○○○

UNSURE ○ 1 ○ 2 ○ 3 CONFIDENT

UNIMPORTANT 0 ○○○○○ 5 ○○○○○ 10 IMPORTANT

UNDESIRABLE −5 ○○○○○ 0 ○○○○○ +5 DESIRABLE

31

CAPTION________________________

NATION ○○○○○○○○○○○○○○○○○○○

– 1972 1976 1980 1984 1988 1992 1996 2000 +

WORLD ○○○○○○○○○○○○○○○○○○○

UNSURE ○ 1 ○ 2 ○ 3 CONFIDENT

UNIMPORTANT 0 ○○○○○ 5 ○○○○○ 10 IMPORTANT

UNDESIRABLE −5 ○○○○○ 0 ○○○○○ +5 DESIRABLE

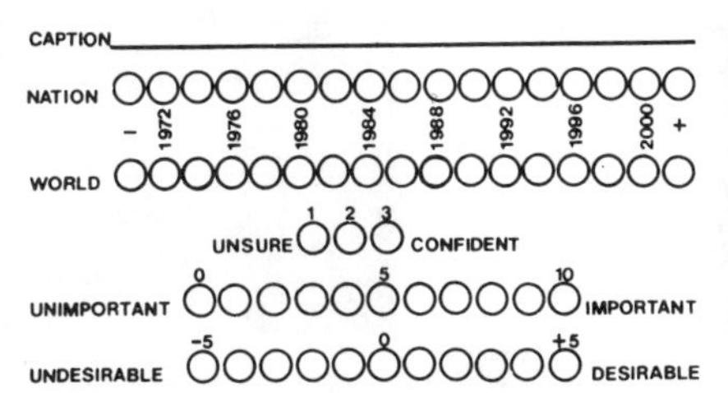

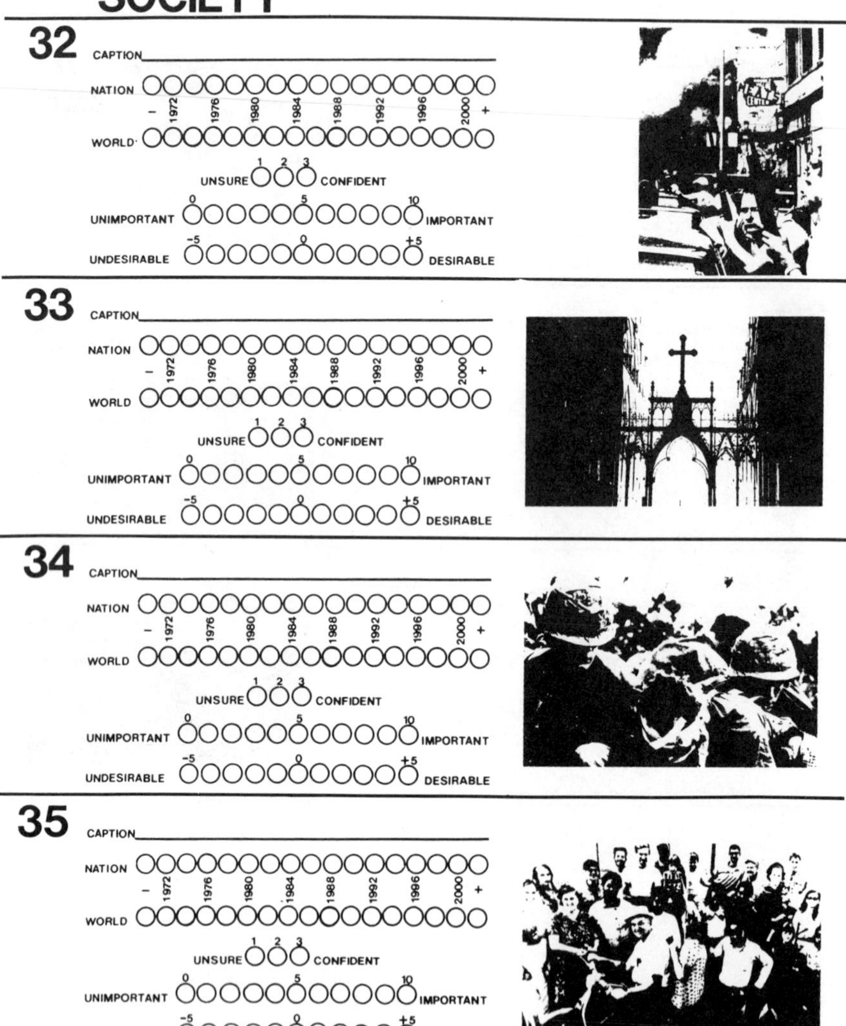

SOCIETY

32 CAPTION____________________

NATION − 1972 1976 1980 1984 1988 1992 1996 2000 + WORLD

UNSURE 1 2 3 CONFIDENT

UNIMPORTANT 0 5 10 IMPORTANT

UNDESIRABLE −5 0 +5 DESIRABLE

33 CAPTION____________________

NATION − 1972 1976 1980 1984 1988 1992 1996 2000 + WORLD

UNSURE 1 2 3 CONFIDENT

UNIMPORTANT 0 5 10 IMPORTANT

UNDESIRABLE −5 0 +5 DESIRABLE

34 CAPTION____________________

NATION − 1972 1976 1980 1984 1988 1992 1996 2000 + WORLD

UNSURE 1 2 3 CONFIDENT

UNIMPORTANT 0 5 10 IMPORTANT

UNDESIRABLE −5 0 +5 DESIRABLE

35 CAPTION____________________

NATION − 1972 1976 1980 1984 1988 1992 1996 2000 + WORLD

UNSURE 1 2 3 CONFIDENT

UNIMPORTANT 0 5 10 IMPORTANT

UNDESIRABLE −5 0 +5 DESIRABLE

SOCIETY

36

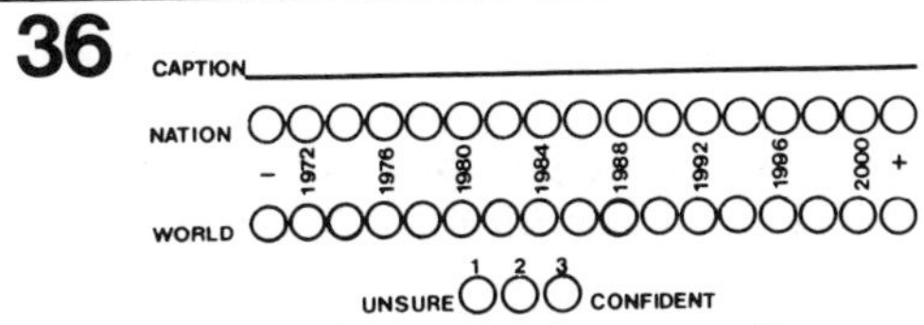

CAPTION ____________________

NATION
– 1972 1976 1980 1984 1988 1992 1996 2000 +
WORLD

UNSURE 1 2 3 CONFIDENT

UNIMPORTANT 0 5 10 IMPORTANT

UNDESIRABLE -5 0 +5 DESIRABLE

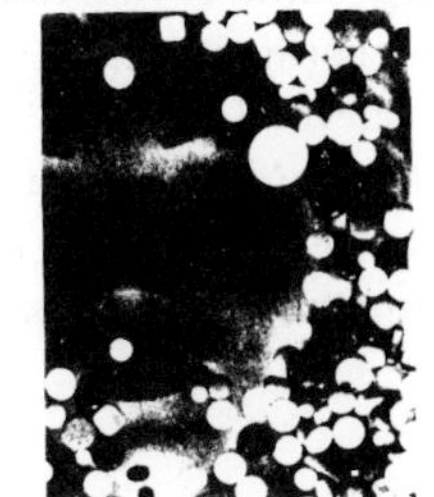

37

CAPTION ____________________

NATION
– 1972 1976 1980 1984 1988 1992 1996 2000 +
WORLD

UNSURE 1 2 3 CONFIDENT

UNIMPORTANT 0 5 10 IMPORTANT

UNDESIRABLE -5 0 +5 DESIRABLE

38

CAPTION ____________________

NATION
– 1972 1976 1980 1984 1988 1992 1996 2000 +
WORLD

UNSURE 1 2 3 CONFIDENT

UNIMPORTANT 0 5 10 IMPORTANT

UNDESIRABLE -5 0 +5 DESIRABLE

39

CAPTION ____________________

NATION
– 1972 1976 1980 1984 1988 1992 1996 2000 +
WORLD

UNSURE 1 2 3 CONFIDENT

UNIMPORTANT 0 5 10 IMPORTANT

UNDESIRABLE -5 0 +5 DESIRABLE

Acknowledgments

The authors acknowledge with pride and appreciation the work of Joseph Valerio (now on the faculty of the School of Architecture, University of Wisconsin at Milwaukee) for his work in designing and producing the original "Expectations" survey form and preparing the results for analysis; and of Betsy Morris (at Quinton Budlong) in assembling and producing the "Expectations II" survey form.

The authors also acknowledge the financial support provided, in connection with "Expectations II," by the U.C.L.A. Committee on International and Comparative Studies.

VI. E. Architecting the Future: A Delphi-Based Paradigm for Normative System-Building

JOHN W. SUTHERLAND

I. Prelude

The purpose of this brief paper is to present some essentially tentative steps toward the development of a system-based "technology" for policymaking. This technology would take the form of a procedural paradigm which, it is hoped, would assist in accomplishing the following:

(a) Removing policymaking from the realm of rhetoric to which it has been historically condemned—by introducing instruments to lead to a consensus of opinion and value on "objectified" issues (whereas existing policies often reflect only casual, suboptimal compromises among competitive a prioristic or "irrational" positions).

(b) Removing, to the extent possible, the ad hoc character of many existing policy-setting processes, by offering a system-based alternative to the "management by crisis" administrative modality under which most social, political, and economic systems are currently administered (which so often causes policies to become temporal panaceas, producing surprises and embarrassment more frequently than structural cures).

(c) Substituting the "hit or miss" character of many existing policy-setting processes with a disciplined "learning" attribute—one in which all policies are viewed as hypotheses which subsequent experience and experiment must validate, invalidate, or modify. That is, the analytical procedures in establishing policies are now complemented by a monitoring-feedback system which constantly evaluates their impact relative to expectations, and which constantly audits the validity and rectitude of the premises from which the policies were derived. Thus, policymaking becomes an exercise in action-research.

From the standpoint of the system theorist, the process of rationalizing policy decisions may best be approached within the context of a somewhat immature but rapidly emerging discipline—*normative system design*. For policies are generally broad, usually long-range, formulation-based constraints on action. In effect, then, they represent a restraint on the action-space available to individuals or systems, or in some way serve to limit the repertoire of desired behavior. In operation, policies serve as a broad strategic envelope within which decisions are made, and, in effect, provide certain a priori premises for the normal decisionmaker. Thus, for example, a state or nation may have an

The Delphi Method: Techniques and Applications, Harold A. Linstone and Murray Turoff (eds.)
ISBN 0-201-04294-0; 0-201-04293-2

environmental policy that demands conservation; the existence of this policy then constrains the decision-space available to individual businessmen, government agents, etc., or anyone else dealing with any aspect of the environment. It thus serves to limit the prerogatives available to systems which might have an impact on the environment. Without going into great detail, most proper policies would be of this sort, and all have a very special property as a class of phenomena: all policies reflect assertions about desirable system properties and, one hopes, serve the cause of transition from less favorable to more favorable system states. In short, policies are *normative* in nature, and always involve a preference given to one state alternative at the expense of all others which might have been elected as most desirable. Thus, *normative systems* may be thought of as ends (systems states) which policies are designed to obtain, and thus reflect "ordered" sets of "utopian" attributes pertinent to some future point in time. To this extent, policy-setting becomes the rational activity associated with the isolation of those "actions" which should prove most effective and efficient in causing a transition from less favorable to more favorable system states through some bounded interval in time.

The rationality involved in systematic normative future-building suggests that we employ some "scientific" method which, while offering procedural discipline, does not a priori constrain the substance of the future. This, basically, is the problem with schemes predicated on crude extrapolation, historical projection, or analogy-building—where we run the risk of restricting the *results* of our analysis by the *instruments* of our analysis. This is not acceptable. What would be acceptable, however, is a method which is Janus-faced, with one face turned toward imagination, evaluation, and axiological inputs, while the other face scans the empirical domain for relevant facts, experiences or objective "data" [1]. Our speculative flights might thus set the initial and tentative investigatory trajectories, whereas the objective information which emerges would serve the cause of validation, invalidation, or modification. Thus, the method appropriate for normative system-building would rest somewhere between the abject intuition of the science fiction writer, and that empiricist-positivist platform which dominates modern science and which knows only facts and figures. It is in the interstice between imagination and observation—between concept and percept—that the "science" of policy-setting, future-building, and normative system-building seems to rest. Specifically, the methodology might well take the form of a procedural paradigm like that shown in Fig. 1.

II. The Scenario-Building Logic

Our methodology must, in the first instance, generate a *syncretic scenario* which will, as nearly as possible, accommodate a diversity of needs or desires. This scenario should serve to emphasis similarities in axiological or affective predicates, not merely explicate differences. This means that it should be constructed

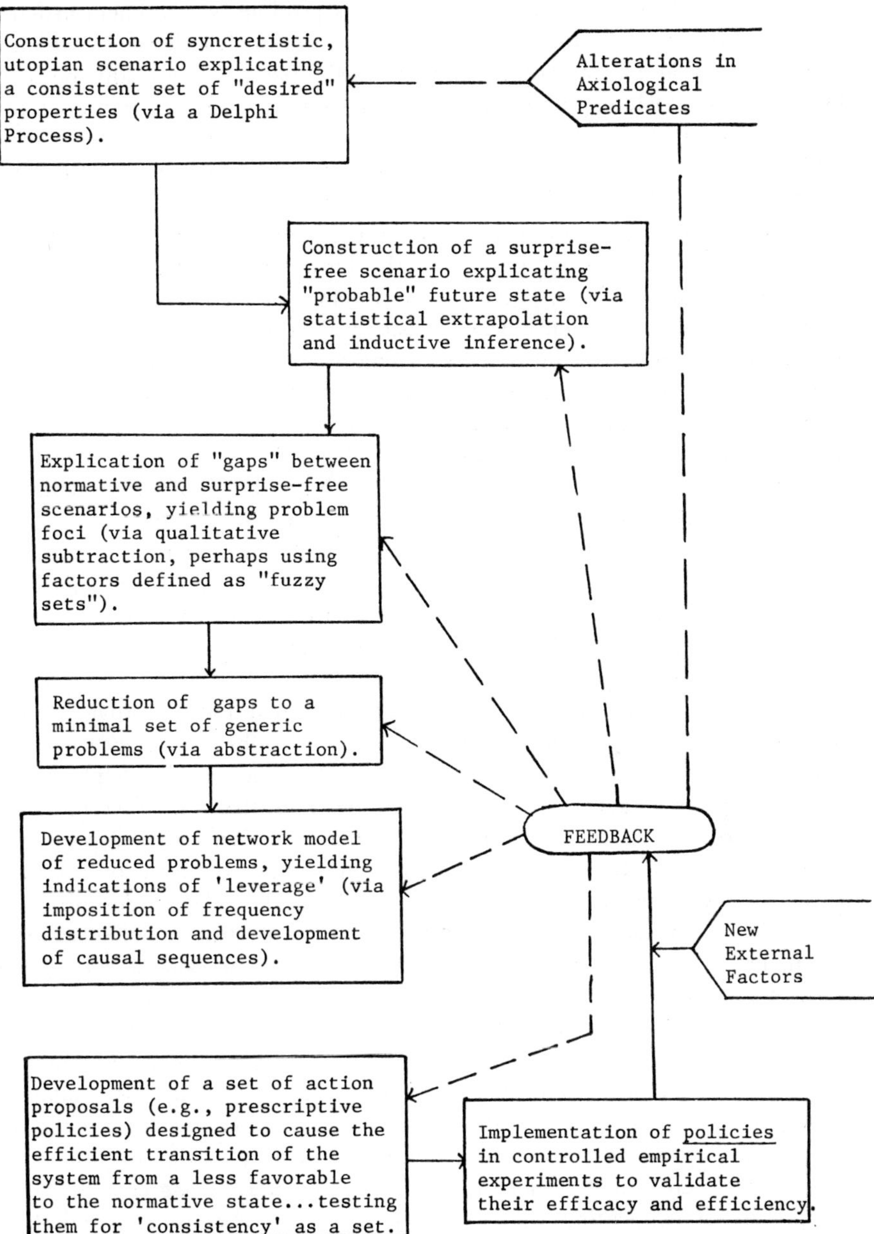

Fig. 1. Normative system-building paradigm.

at a level of abstraction which, as nearly as possible, works backward from superficial positions to fundamental societal attributes. The device by which we may do this would be a simple variation on the basic theme of this book, the Delphi process.[1] Here, rather than searching for opinions or for probabilistic expectations about events, we would try to structure a set of properties which could be integrated into a normative future—properties based on the criterion of desirability rather than likelihood, for example.

Given this set of utopian or normative future properties, we might set about convening a set of expectations about the future independent of any action which we might take. This might be the development of a "business as usual" scenario, or the type which Harold Linstone has referred to as *surprise-free* [2]. The normative and the surprise-free scenario would differ in an important respect: the former would be predicated largely on hypothetico-deductive instruments (e.g., defactualized or idiographic "information"), whereas the surprise-free scenario might legitimately be based on largely extrapolative or projective instruments such as those used to generate the "limits of growth," etc. In sum, the normative scenario would be a product of intellect and discernment, whereas the latter would be largely a mechanical exercise in statistical analysis coupled with limited uses of ampliative inductive inference (given that ampliative inductive inference simply extends the properties inherent in some empirical or experiential data base, introducing changes which are matters of *degree* rather than of quality).

The papers preceding this, and some which will follow, provide ample elucidation of the procedures and promise of the Delphi process, so I need not repeat them here. I simply want to note that while the normal Delphi process will value any divergence of opinion which emerges, that which we must seek in constructing these scenarios is a *consensus*; hence the syncretistic aspect of the normative scenario. In any exercise of any significance we must expect to begin with a considerable divergence of opinion about the relative desirability of alternatives. In most cases, these divergences will arise from axiological biases which, for many participants, may be empirically inaccessible. When these exist, there will generally be several uniquely different scenarios which emerge when participants are asked about a normative future through the Delphi process. The points of divergence among the various scenarios should enable us to locate the major axes of recrimination and, in the process, make the axiological predicates of the scenarios visible and manipulable. This is the first step in gaining a consensus where divergences are a matter of values and idiosyncratic perceptions rather than matters of objective disagreement. For in the process of explicating what are otherwise tacit assumptions underlying desired events, there is the opportunity for generating syncretic models at a next level of abstraction, that is, one step removed from the level at which the

[1]For a discussion of some precedents for the use of Delphi processes in scenario-building exercises, see Chapter III, of this volume.

axiological arguments hold sway. This would be a process similar to that described by Ackoff, and extended by Mitroff and Turoff, concerned with "measuring the effect of background assumptions" [3].

Where divergences are a matter of objective argument—such that the engines of differentiation are explicable in universalistic rather than idiosyncratic terms—then a syncretic solution can usually be generated by exchange and modification of causal expectations rather than by the indirect exchange of assumptions associated with cases of axiological argumentation. Here, again, the Delphi process could be used to generate these adverse scenarios, with these subsequently becoming the target for consensus operations.

Thus, in sum, a consensus as to the normative scenario will generally involve several interations of the Delphi process, with the rate at which a consensus is achieved (or the rate at which we are converging on a syncretic scenario) being measured by the rate at which the raw variance in responses declines. Thus it becomes important to introduce questions into the Delphi process which are susceptible to some sort of ordered response (e.g., asking questions which permit the respondent to answer in terms of a probability of likelihood or degree of desirability). Thus, for example, we may pose questions where the response is to be made as follows:

100 75 25 Ø (*Probability / Desirability Index*)

Each respondent, pertinent to each question, is asked to place a check mark at the point on the continuum which best describes his expectations about probability or desirability of the event (or item) in question. The aggregate of all such responses, for each iteration of the Delphi process, will thus represent a frequency distribution with proper statistical properties, proper in the sense that it will yield a formal measure of variance. The questions should be phrased in such a way that affective or objective origins of disputation may be isolated (e.g., asking "leading" questions or questions with a strong but subtle valuational bias). When essentially the same issues are raised, perhaps, however, in different form, on subsequent iterations of the Delphi process, a measure of the decrease (one hopes) in variance can be obtained, which should enable us to evaluate roughly the effectiveness of the strategy or tactics by which we are trying to establish a consensus. Thus, through the course of a scenario-building exercise, it is the variance of responses with which we shall be concerned, for this is the best guide we have as to the means by which a consensus might be obtained. I shall not specify any of these means here, except to note that behavioral scientists concerned with consensus and conflict elimination have methods which are at our disposal, and the comparative effectiveness of these methods can be audited by the above procedure. Thus, the consensus-seeking process might be viewed as an action-research experiment in its own right, shifting instruments in response to empirically derived variance estimates.

It should be mentioned here that the response-by-continuum (e.g., asking for responses in the form illustrated above) lends us another capability which is extremely important, and which I have introduced in greater detail elsewhere [4]. Specifically, the frequency distributions which emerge from the successive iterations of Delphi processes aimed at securing indices of desirability pertinent to future properties may be thought of as subjective (a priori) probabilities, and may be treated as such. In short, in operational terms, expectations about the likelihood of an event occurring, and estimations of the desirability of events, become intelligible in exactly the same terms. In fact, we may usually expect to generate fairly accurate surrogates for desirability of events by asking about their probability of occurrence, this in the sense that respondents are often likely to assign the highest likelihood estimates to events which they themselves value.[2] There is some advantage in this surrogation approach especially where social conventions might act to impede true responses (e.g., a respondent might really prefer that his future minimize contact with members of minority groups, but would be reluctant to be tied to such a statement; asked, however, about the likelihood of homogeneous communities, we may get a reasonable basis for inference about desirability). Naturally, some local testing would have to be done to determine the extent to which expectations about probability of occurrence and estimations of desirability are truly correlated, with the reliability of inferences being adjusted accordingly. As a general rule, however, there should be some opportunity during the course of the Delphi process to gain indices of both likelihood and desirability.

The net result of this process will be a scenario which, at each point, may be thought of as involving a selection from among all alternatives which might have been included. In this sense, a scenario emerges as a proper model, with each of its components being assigned some index of either desirability (in the case of the normative scenario) or probability of occurrence (in the case of the extrapolative or surprise-free scenario). In graphic terms, we may think of each component of the scenarios as being represented by an event-probability distribution, as shown in Fig. 2. In the left-hand illustration, the events on which the probability distribution is imposed may represent anything we wish pertinent to some scenario under construction (that is, the e_i's may be variables, relationships, functions, complex events, parameter values, etc., or any other component). Of the two curves drawn there, the a priori (presumed to be that which exists after the first Delphi iteration and prior to the first attempt at consensus) is less favorable in terms of variance than the a posteriori (which is presumed to exist after some n iterations of the Delphi and consensus-seeking process). The transformation from the a priori to the a posteriori curve as a result of our polling and consensus-seeking operations results in the favorable

[2]For a discussion of this phenomenon, see J. R. Salancik's "Assimilation of Aggregated Inputs into Delphi Forecasts: A Regression Analysis," *Technological Forecasting and Social Change* 5, No. 3 (1973), pp. 243-47.

learning curve at the right. We may now interpret these constructs in the following two ways:

(1.) *For the Normative Scenario*: Suppose that the event set [E] is an array of mutually exclusive attributes pertinent to some aspect of the future, such that the two curves should be read as products of respondents' estimations of the desirability of these alternatives. In the a priori state there is a considerably greater divergence than in the a posteriori, with the numerically calculated variance acting as an imputed measure of the degree of consensus obtained. The "learning curve" on the right-hand diagram thus reflects the rate at which this consensus has been gained, such that the curve in that diagram may now be thought to represent the variance in estimations of desirability among those contributing to the scenario-building exercise.

(2.) *For the Extrapolative Scenario*: Here the a priori and a posteriori curves represent expectations about the likelihood of occurrence of the events constituting set [E], such that the a posteriori curve assigns respectively higher probabilities of occurrence to a sharply decreased set of alternatives. Thus the "learning curve" in the right-hand diagram may be read as written: measuring the probability of expected error associated with the aggregation of predictions. In short, we have converged gradually on a consensus of opinion about what will occur, whereas in the normative exercise we converged on a set of opinions about what is desirable.

In both of the above cases, we have used a statistical surrogation process to discipline the inputs from Delphi participants, and to gain an empirical appreciation of the *efficiency* of our model-building and consensus-seeking process. The normative scenario will, then, constitute a complex model which at all points has been assigned specific indices of consensus as to the desirability

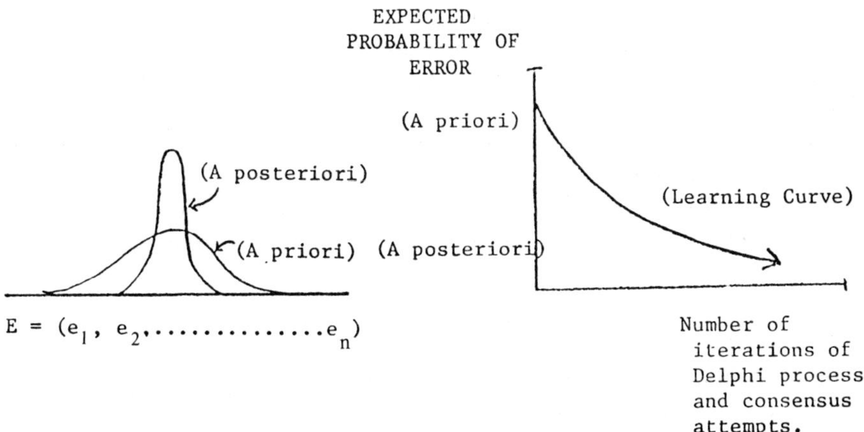

Fig. 2. Event-probability distribution and associated learning curve.

of the components entered; for the extrapolative or surprise-free scenario, we have a model which, at all points, contains those components which have been assigned the highest probabilities of occurrence by the participants. The net result should be a normative scenario with the highest logical probability of accurately encompassing the desires of participants, and an extrapolative scenario which, relative to all others we might have built, has the highest probability of accurately reflecting the future *in the absence of any therapeutic or directional actions we might take*. In other words, this latter scenario represents what we think will happen if we do not interfere, or if we abrogate our opportunity to create the future.

There are two other aspects of the scenario-building process which we should note here. The first is that the normative and extrapolative scenarios should represent logically consistent models. Thus, as Joseph Martino suggests [5]:

> A scenario is more than just a set of forecasts about some future time. It is a picture of an internally consistent situation, which, in turn, is the plausible outcome of a sequence of events. Naturally there is no rigorous test for plausibility.... The scenario thus occupies a position somewhere between a collection of forecasts whose interrelationships have not been examined, and a mathematical model whose internal consistency is rigorously demonstrable.

While there are no hard and fast rules for establishing consistency, it is nevertheless useful to have at hand some sort of instrument for examining the nature of the interaction between system components, an exercise which is naturally aided by having an interdisciplinary flavor to the scenario-building team and concentrating some attention on the interfaces which emerge among components of the emerging models.

Second, the utility of this normative scenario-building process stems from a simple realization: if we were able to start from scratch, very few systems which we have developed would be designed the way they have evolved. Therefore, it sometimes pays simply to ignore what exists and turn instead to a disciplined exploration of what should exist. Little has been gained from the large number of incredibly costly studies which aimed at "finding out what's wrong with system *x*." The response which the normative system-builder might make to such an assignment is this: "Compared to what?" Without that normative reference, studies of existing systems lack reference and therefore can accomplish little (yet empirical science's domination of research technology has lent such studies a patina of legitimacy despite their long record of sterility).

III. An Amplification of the Scenario-Building Technology

In terms of an amplification of the procedural logic we have developed, we first asked that a Delphi process be inaugurated which will gradually converge on a set of attributes deemed, by consensus, to be most desirable at some future point for the system in question. That is, we are searching for a set of state-variables,

$X=\{x_1.......x_n\}$, which as a set exhaust the desires of the model-building participants. Now, associated with this set will be an event-probability distribution[3] derived from a frequency distribution (histogram) reflecting the number of mentions each factor received, should we have provided a facility for participants to give a numerical priority expressing the importance of the particular factors. At any rate, through the first iteration, we will emerge with an event-probability distribution where, instead of probabilities, per se, we are using *indices of desirability*. This first distribution will normally be a high-variance one, which successive iterations of the Delphi process must attempt to reduce to a low-variance one. The degree to which the variance declines is an estimate, at all points, of the efficiency of the Delphi process in its role as a syncretic instrument, as Fig. 2 indicated.

At any rate, the output from this first Delphi process should be some set of n attributes which, as a set, contains both a manageable number of elements and, for that n, maximizes the aggregate desirability. In short, the n-factors selected are those which have been assigned the individually highest indices of desirability (as computed either by number of mentions or by some weighted measurement including a ranking variable). This set of x_i's (attributes) now becomes the universe of properties available to us for the second of our analytical exercises—scenario-building. The construction of a proper scenario becomes intelligible in two steps: (a) A selection, from among the universe of x_i's, of some consistent subset of attributes (consistency meaning that all are mutually compatible or "synergistic") and (b) The imposition of an "order" of some sort on the attributes selected, which in effect means that we establish some functional/structural relationship among them [6]. When these two tasks have been completed, we will emerge with a proper normative system model—where the attributes included represent state-variables, and the "ordering" represents the necessary interaction conditions among them.

There may be several scenarios, each representing specific interests or ambitions or desires, and these may differ eiter in terms of the specific subset of attributes elected for inclusion, or in terms of the functional relationship imposed on the elected subset. Thus we might emerge with something like this:

ALTERNATIVE SCENARIOS:

$$A=f_a[X_a]$$
$$B=f_b[X_b] \qquad \textit{Where: } X_a, X_b \text{ and } X_m \subset X \text{ (the universe)}$$
$$\vdots \qquad \text{and where } f_a \neq f_b \neq f_m.$$
$$M=f_m[X_m];$$

[3]We will continue to use this term, even though we are here concerned about desirability rather than probability—and we will continue to use continuous rather than discrete formulations for these distributions, even though in most cases a discrete form would be more appropriate. The logic, in either case, remains the same.

Alternative scenarios are thus those where either the subset of attributes considered is different than that considered in other scenarios, or where the relationship (the f_i) among those state attributes is unique. In either case, scenarios represent normative system models defined at the state-variable and relational levels, and hence represent qualitatively unique system states. Associated with each of these scenarios will thus be some index of desirability, which will be computed by the number of responses which favor a particular formulation (or, again, by some scheme which incorporates a ranking function).

This aggregate desirability index is really a product of indices of desirability (and feasibility) generated at lower levels—pertinent to the individual structural or functional components which are to be entered into the emerging scenarios. For example, some components (e.g., normative attributes) will receive a high consensus, given the other components of the scenario under construction, while others will receive barely enough "votes" to get them included. The same will be true of the functional relationships which are imposed on the elected attributes, causing the generation of a "system" model, per se.

Thus, for *each element* in the alternative scenarios, there will be an event-probability distribution which reflects the desirability of that attribute and ideally one also reflecting its "feasibility." These individual component indices may be formulated as follows:

(*a*) DESIRABILITY $[x_i|x_i^c]$, for $\forall x_i \in X_a (\& X_b \ldots X_m)$;

(*b*) FEASIBILITY $[x_i|x_i^c]$, for $\forall x_i \in X_a \ (\& X_b \cdots X_m)$.

The first formulation simply asks about the desirability of some attribute x_i in company with all other attributes constituting the set X_a (this being a measure of the conditional desirability of the factor). The feasibility formulation then asks about the "consistency" of having x_i appear in concert with the other attributes constituting the set X_a (e.g., the set of attributes *qua* state-variables for the A-scenario). The same task would be performed by each of the clusters, ultimating in scenarios where each of the x_i's has been assigned some index of desirability and feasibility. Essentially the same logic would hold for the derivation of the functional relationship among the x_i's pertinent to each scenario. In this sense, each pair of interacting attributes (*qua* state-variables) will have to have an assertion made about the nature of the influence or interface conditions. That is, we have to set the conditions for the following:

$f_{i,j}[x_i|x_j]$, for all interacting pairs of attributes or state variables in $A, B \ldots M$, where all $f_{i,j} \in f_a$.

Thus, when f_a (the functional relationship imposed on the set X_a, constituting the elements of A) is finally evolved, there will be possibly some divergence of opinion about the validity of each of the individual $f_{i,j}$'s of which f_a is comprised. And, if we are to be strictly formal in our approach to the problem of establishing a relationship function, we would again have to consider the relationships among successively more encompassing sets of variables, eventually emerging at the point where relationships are considered simultaneously among all variables (factors) connected even indirectly. Only in rare occasions, however, will we resort to this depth of analysis (except when the system at hand is an effectively "mechanical" one, permitting these relationships to be set empirically through controlled manipulation of successively larger subsystems, etc.).

In general, however, a *clustering* effect will emerge in most scenario-building operations. This will tend to find individuals with like interests, ambitions, backgrounds, or axiological predicates forming subgroups which have homogeneous concepts of what the future should look like. As a result, there is a natural tendency to emerge with scenarios which have a high degree of consensus as to the desirability or feasibility of their components (and which, as a result, are constructs exhibiting little internal variance of opinion). To a limited extent, we can counteract this clustering effect by *stratifying* panels (where we put individuals with substantially different backgrounds, interests, skills, or axiological positions together into scenario-building teams). Nevertheless, the alternative scenarios will usually result from the fact that there will not be unanimity among panel members as a whole, and that there will usually be divergences of opinion about the rectitude of model components *within* strata or clusters.[4] But the clustering and stratification schemes differ in important respects, as follows:

	WITHIN SCENARIO VARIANCE	BETWEEN SCENARIO VARIANCE
Clustering:	LOW	HIGH
Stratification:	HIGH	LOW

That is, under a clustering scheme, we tend to put similar panel members together into a model-building team (e.g., all economists or all sociologists or all environmentalists); under the stratification scheme, we tend to form teams of opposites. The net result is that the models built by clustered teams will vary greatly in aggregate, but will have high degrees of consensus (e.g., low variance) internally ... for the premises under which clustered teams operate vary greatly *between* the teams, but little if at all *within* the teams. Just the opposite is true of the stratification effect. There we will have a high degree of

[4]Nevertheless, we assume that when it comes to "voting" for one or another of the alternative normative scenarios, there will be unanimity among cluster members in preferring their own to any other.

divergence within teams, and a low degree of divergence among teams, with the net result that the alternative scenarios will vary little between each other, but have high degrees of internal variance.

As a general rule, clusters will tend to form naturally, whereas strata have to be deliberately invoked ... something which may prove to be *dysfunctional* if recrimination and acrimony occurs within groups.

For the sake of "efficiency" in the normative scenario-building process, it is a usually better strategy to permit the formation of these spontaneous clusters explicitly, and then concentrate on the elimination of the clear-cut differences between constructs as wholes at some later point. For the models which are built by clustered teams will be internally consistent and carry a high degree of consensus as to desirability, even though their conclusions may be hotly disputed by members of other clusters. Thus we have reasonably disciplined, "finished" products on which objective argument may be directed, whereas groups containing acrimonious components will have severe difficulty in actually *completing* anything substantial. To make this point somewhat clearer, consider Fig. 3. What we tend to find, then, is that the aggregate probability distribution functions for alternative normative scenarios (viewed as heuristics) tend to be products of the *means* of the event-probability distributions reflecting the degree of consensus among members of a cluster as to the desirability (or feasibility) of their construct.

Thus, for clusters, we have a situation where a larger number of alternative scenarios are proposed (such that the aggregate distribution is greater for the "flatter" clustered case), but where the consensus as to the individual desirability of these alternatives is strong (which means that each of the alternatives in the clustered case will have a narrow, peaked, internal variance). Just the opposite situation prevails with respect to the stratified distribution: there we have fewer "events"[5] assigned significant probabilities—with less variance in the aggregate distribution due to their tendency to be redundant with one another due to the stratification process—but each of the alternatives carries with it a much higher internal variance, and consequently shows less consensus) than do the alternatives in the clustered case. That neither of these situations is really satisfactory may be seen by calculating the overall variance associated with the constructs (e.g., computing variance as the product of the internal and external variance) [7]. The net result is that both cases would carry about the same total variance, with the lower internal variance associated with the clustered case (indicated by the fact that $[b-a]<[d-c]$) being offset by the lower aggregate index of desirability (or feasibility) associated with any single alternative—as is indicated by the "flatter" aggregate distribution.

What we eventually want to emerge with is, of course, an aggregate event-probability distribution like that shown in Fig. 4. We could achieve this

[5]Alternative scenarios.

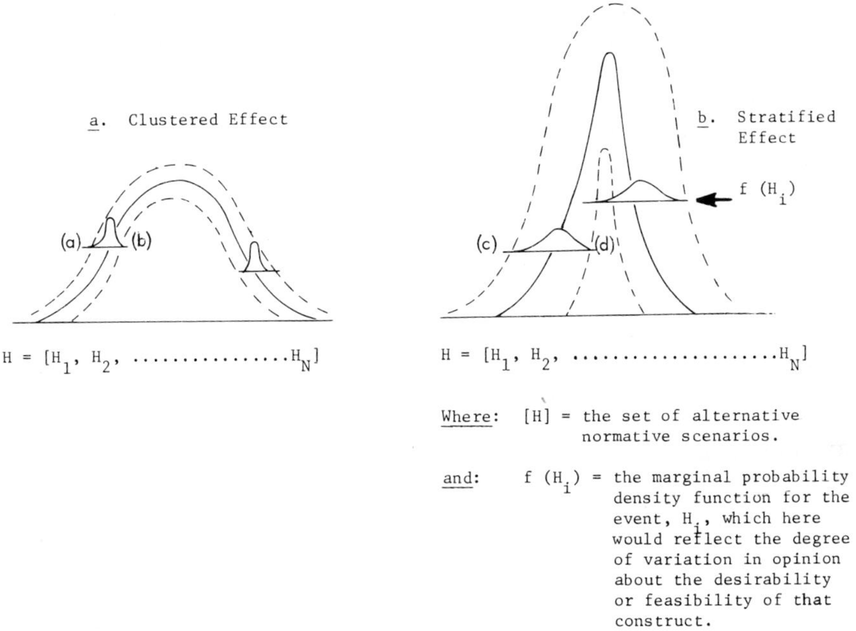

Fig. 3. (*a,b*). Results of clustered as opposed to stratified model-building structures.

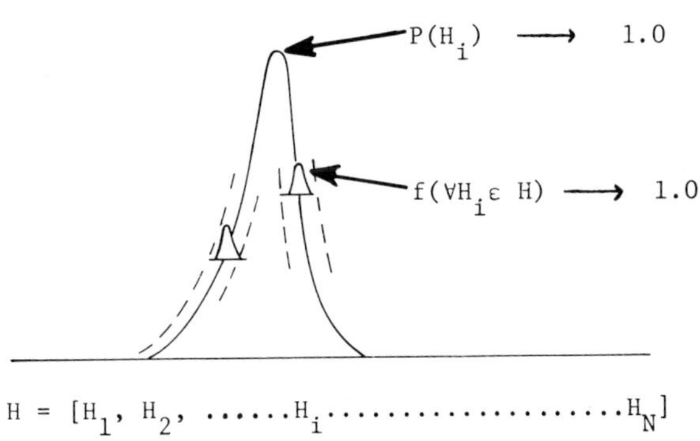

Fig. 4. The idealized event probability distribution.

distribution in either of two ways, relative to the distributions of Fig. 3: (*a*) We could inaugurate Delphi processes which will reduce the *within-scenario* variance associated with each of the few alternatives remaining in the stratified distribution (by forcing a consensus within the scenario-building groups). This, in effect, means forcing the marginal desirability/feasibility density functions for all remaining H_i's to converge toward 1.0 (which reduces the expected dysfunctionality/infeasibility associated with each alternative, and hence results in more "peaked" superimpositional probability functions). (*b*) We could take the low "internal variance" alternatives of the clustered distribution and inaugurate Delphi processes which cause the aggregate desirability density function to "peak," thereby eliminating some of the less expectedly desirable/feasible alternatives and correspondingly increasing the relative desirability/feasibility of those remaining. The new result, in either case, is the same: we have reduced a high-variance situation to a lower-variance situation, with the end result that a single, most expectedly desirable and feasible normative system model is gradually converged on. Thus, in the above figure, both the aggregate and marginal density functions reflect an improvement over the cases illustrated in Fig. 3 *a* and *b*, respectively.

In many cases, and we shall not belabor the point here, the ultimate construct (e.g., the H_i which is finally agreed to be the most desirable and feasible system state) might emerge as some sort of compromisive "hybrid," where agreement is obtained at the expense of either precision or actionability. The result of *democratic* model-building processes is often a construct which, while it has very little probability of proving absolutely wrong as a representation of some phenomenon, also has almost no probability of being "optimal." And where such constructs are to become premises for decision or policy action, the fact of a subjective convergence or consensus will have little bearing on the actual rectitude of the model, and hence may be cases where science disserves its larger community. A very important distinction, then, would be between models which obtain a convergence through compromise, and those which are truly *syncretic* or rationally (and sapiently) discerning. These strictly compromise hybrids, like so many essentially political formulations, will generally trade-off ultimate rectitude against short-run procedural expedience. And there is really no defense against this except for the fundamental integrity of the team leaders, so we can offer no "technology" for escaping these degenerative Delphi processes—just a caution to be alert for their presence. Thus it is important not to force an artificial or fabricated convergence where there is no natural basis for syncretistic accommodations, it is better to let the normative model-building process proceed along two or more parallel trajectories. The advantage of using the paradigm presented here will then be simply the fact that all parties have employed basically the same methodology, and this, in itself, may eventually provide some accommodative basis which might otherwise elude us.

At any rate, essentially the same logic would hold for the construction of the extrapolative scenario—that system model which simply projects existing system properties into the statistical future. We would largely just replace indices of desirability/feasibility with indices of likelihood (e.g., probabilities of occurrence), but would still have to go through a consensus-seeking process of the type imposed on the normative scenarios. Here, however, use of "stratified" as opposed to clustered panels would probably be indicated, as there will be greater reliance on experience and factual knowledge (objective "data"), and "differences" could, it is hoped, be resolved "rationally." That is, where divergence is more likely to be a matter of fact rather than a matter of value, accuracy of forecasts should increase to the extent that extrapolative teams are drawn from broadly different experiential and professional sectors.

IV. Constructing the Problem Network

The completion of the normative and extrapolative scenarios now allows us to concentrate on the differences between what we want to occur and what we expect will occur in the absence of action on our part. These differences thus represent "gaps" which must gradually be closed if the properties of the normative scenario are to be realized. It is important, therefore, that the normative and extrapolative scenarios be built on exactly the same dimensions, such that they represent essentially different states for the same essential system. In short, expectations and estimations of desirability should be generated only in pairs with respect to a set of system components which remain constant through both exercises.

Thus, for each aspect of the future, there will be a set of ordered pairs of properties, with the first element in the pair representing the desired property, and the second representing the property expected to obtain given inaction. I won't go into detail here, but it should be obvious to systems scientists that defining properties in terms of ordered pairs lends us some valuable mathematical capabilities and allows us to treat the two scenarios as formal state alternatives for a formally defined system [8]. Where the elements of each pair are essentially similar, the "gap" will be negligible and this particular system aspect may thus be removed from immediate attention. However, before we finalize our action proposals, we must again review these cases held in abeyance, for new conditions might be introduced which will prevent the extrapolative value from converging on the desired. The logic here is simple but critical: in interfering with those aspects of the system deemed to be dysfunctional, we might very well introduce changes which prevent desired events taking place through the extrapolative mechanism. In short, when we change any aspect of the system, we must reassess the impact these changes might have on aspects of the system with which we did not wish to interfere.

At any rate, the generation of the "gaps" which do exist might be viewed as a problem in *qualitative subtraction*, where we are basically interested in isolating that set of properties which are *not shared* by the two scenarios (relative to each of the dimensions of the scenarios or to each system component being considered). As far as formal procedures for this exercise, there is an increasing literature on what is known as "fuzzy" set theory, which is the attempt to deal mathematically with qualitative variables [9]. The serious student of scenario-building will find some guidance here, though recourse to such discipline will hardly be necessary in most cases. The result of this subtractive exercise should, then, be a set of ordered differences which now may be thought to represent problems, per se, toward which the remainder of our work will be directed.

But for a scenario of any significant size, much less for those directed at large-scale systems such as regional socioeconomic systems or urban areas, etc., the number of problems *cum* gaps will be enormous. True, we have eliminated for the moment those aspects of the system where desired and extrapolated ends were the same (e.g., where the trend is adequately favorable or where a favorable system aspect is expected to remain secularly stable), but we must now perform a further reduction.

We want, in effect, to arrive at an array of problems which exhausts the "gaps" isolated above, but which contains the fewest possible elements or entries. We do this by looking for ways to phrase problems such that all those which we arrive at will be mutually exclusive and pertain to the largest possible number of specific instances (e.g., gaps). In a sense, then, the problem definitions which are sought here will be arrived at by a process of *successive abstraction*. Under this process, we are constructing generic problem referents which perform much the same function as theories or laws: they introduce efficiency into our explanatory process by allowing us to account for the largest number of specific (unreduced) phenomena with the fewest possible unique formulations. In operational terms, this successive abstraction process would be conducted by searching the several "gaps" for elements of *isomorphy* or attributal similarities. The search for, and exploitation of, instances of isomorphy is one of the best described procedures of systems science, so there are ample references to which the reader may turn for specific guidance [10]. The general strategy, however, relies mainly on the intellectual tenacity and sensitivity of the participants, and on the ability to absolve the problems *cum* "gaps" of their contextual or superficial properties so that instances of isomorphism may be uncovered. In short, the method here would be to take the "gaps" previously isolated and gradually strip away as many *modifiers* as possible. The result should be the fewest number of mutually exclusive generic problem definitions which, in substance, exhaust as fully as possible the "gaps" used to generate them.

The next step is the analysis of these generic problem referents, and the imposition of some sort of *causal order*. This may be done in two steps. The first move would be to calculate the number of specific "gaps" (e.g., unreduced problems) to which each of the generic problems refers, thus giving us a rough measure of the degree of abstraction associated with each element of the reduced problem array. When this is done for all generic problems, we may develop a simple frequency distribution which will often have a valuable property for us. For when the various generic problems are arrayed as the event-set in the frequency distribution, the amplitude of the resultant curves imposed on each element in that set (each generic problem) gives us a measure of the relative *system leverage* it exerts. That is, the frequency distribution will point out those problems which, when operated on or treated, are likely to have the greatest overall therapeutic effect on the system as a whole. In short, the measure of abstraction associated with each of the generic problem referents is also a measure of its influence on the *state* of the system, with those generic problems encompassing the largest number of specific "gaps" being those which carry the greatest expected leverage. Hence, in the histogram below, Problem [p_2] carries imputedly greater leverage than any of the others, being the generic referent for the largest number of unreduced problems (or having appeared as a factor in the largest number of specific "gaps").

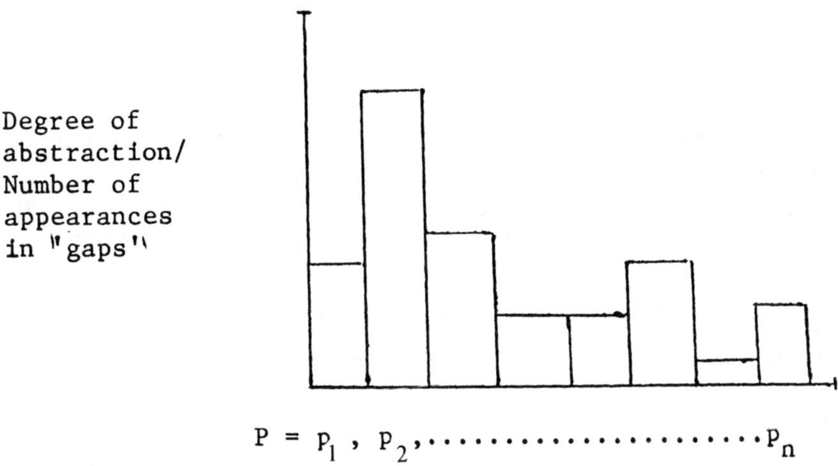

Fig. 5. Sample histogram.

The second move is to search for some sort of *causal order* among the various generic problems we have defined. All must be directly or indirectly related, else the system with which we are working is not a *proper* system. The task for us now is to allegorize the nature of these relationships, and these will generally fall into one of two broad categories: hierarchical or reticulated networks. If the ordering of influences among problems is of the former type, then we will be able to develop a causal *tree* (e.g., a partially ordered set) where problems may be arrayed on levels such that problems on a lower level are deemed to be "causes" of problems on the next level, and so forth. The net result of such a structure is a rather neatly linear relationship among the various problems, with micro-problems gradually giving way to intermediate level problems which finally merge into a macro-problem. In order to be a true hierarchy, we might impose a constraint that prohibits higher-order problems from affecting lower-order, or perhaps a constraint which acts against reflexivity among the various problems (as reflexivity would abrogate the linearity and uni-directional causality we like to see in proper hierarchies). At any rate, there are many sources of information about the properties of hierarchies, properties which must be met by the generic problems if the hierarchical modality is to be employed to lend them an ordering [12].

The reticulated case represents a more complex set of relationships, one where the various problems are equipotential (i.e., all may affect each other). A reticulated network thus does not possess the neatly algorithmic properties of proper hierarchical structures, and so becomes very much more difficult to model. For, here, problems may be related reflexively, recursively, and with omni-directional potential trajectories of influence. Under this structure, one must conceive of our problems being related in a network fashion where each of the various problems represents a *node*, from which many different paths of interaction may radiate. As the reader might rightly expect, then, problems related in reticular form generally arise from systems which are more inherently complex than are systems which enable us to impose a hierarchical ordering on the generic problems. That this is so may easily be verified by comparing some of the formulations of modern network theory against their hierarchical counterparts [13].

With the completion of these causal orderings, the results from the frequency distribution should be reexamined. The problems which received the greatest number of appearances in the histogram of the first step should be those with the greatest number of connections in the reticular model (or, where the hierarchical structure is imposed, should be at the lowest levels of causality, i.e., influencing the greatest number of other problems). In this way we have a check on our derivation of problems from "gaps" and an additional verification of the consistency of our reduction logic.

V. Generation of Action Proposals (Prescriptive Policies)

The models just generated will serve as an input to the next phase of the policy-setting/future-building process—the generation of an array of action proposals which, as a set, have the highest a posteriori probability of causing the transformation of a less favorable to a "normative" system state over some interval. These action proposals will generally take the form of prescriptive policies (strategic instruments) which are expected to achieve the effect of narrowing the gap between existent (or extrapolative) system properties and those that are deemed desirable, for each of the several problems comprising the reduced problem network just developed. There is no easy algorithm for the generation of strategic decisions—this depends on the wit and sensitivity and assiduity of the systems analysts and policymakers—but we can say something about the *order* in which action proposals should be generated and the analytical setting in which they should be evaluated.

Obviously, those problems shown to exert the greatest leverage (e.g., those with the highest number of appearances in the histogram, those representing most "highly connected" nodes in the reticular formulation, or those at the lowest causal levels in the hierarchical structure) should be the first attacked. Thus, with respect to the frequency distribution just set out, problem $[p_2]$ would be the first for which an action proposal (solution strategy) would be developed. The causal ordering model—in either the reticular or hierarchical form, whichever was deemed most appropriate—would then be used as the basis for a *simulation* of the effects of the solution strategy on the system as a whole. The results of this first iteration of the simulation model would then be to isolate any redefinitions in the other problems associated with the simulated implementation of the solution to the first problem. In short, we want to know what effects our action with respect to problem $[p_2]$ would have on the other problems, so that, for each of them, a solution strategy could be devised in light of the solution to the prior problem. We would then have to iterate the simulation until we were able to incorporate the entire array of problems to be solved, together with the parameters of their solution strategies. This is simply to be certain that, in solving one problem, we make every feasible effort to see that we do not create others or exacerbate the system situation in some way (in short, that we act to minimize the probability of occurrence of unexpected consequences or adverse second-order effects).

In the beginning, with that problem deemed to exert the greatest leverage on the system, we may emerge with an aggregate solution strategy (pertinent to the simultaneous solution of all problems) which promises to be most effective and efficient in making the desired system transition. In some case, however, particularly where a hierarchical problem structure was uncovered, we may think about introducing solutions *sequentially*—implementing them in a sequence which, at each step, isolates that problem which at that point in time promises to exert the greatest therapeutic leverage when solved. In either case, iterations of the simulation model should give us a *logical* pointer to the necessary conditions for each solution step, and the indication as to which particular problem will emerge as the best target for attack at that iteration. Thus, the causal orderings on the problems give us the basis to make a most efficient transition, such that expenditures of developmental resources will produce an expectedly "optimal" effect, both in aggregate and at each point in the solution process.

VI. The "Learning" Aspect

It should now be mentioned, again, that all the work we have done thus far results in constructs which are just hypothetical in nature, and which have generally only "subjective" indices of accuracy assigned them. Nothing, at this point, is fixed. The "hypothetical" character of these components simply reflects the system dictate that, in the face of a complex environment, linear procedures (those that anticipate a direct drive toward some invariant objective) are to be avoided with assiduity. Rather, our analytical behavior must be heuristic in nature, denying the impulse prematurely to stop analyzing and start acting, and denying the impulse to use what are essentially unvalidated constructs as action premises. So the policies *qua* action proposals which we now must generate themselves become hypotheses, and serve basically as "manageable" units of analysis for the empirical learning loops which will lend some "objective" discipline to what until now have been largely speculative exercises. It is here, then, that we make the switch from the hypothetico-deductive modality to the empirical modality.

The empirical "learning loops" which we now inaugurate will have basically two foci. First, there is the matter of validating—in the immediate short-run—the rectitude of the hypothetico-deductive components constituting the scenarios, and their problem and policy derivatives. Second, there is the long-range aspect to the learning context—that which finds even those constructs validated in the short-run being constantly monitored and evaluated in the long-run. The net result is a totally heuristic process, one where as little as possible is taken for granted, including not only the analytical constructs themselves but the basic "values" and axiological premises that existed at the time of the initial analysis. There is a distinct advantage here, for in the world of the future, approached heuristically rather than algorithmically: "... one may solve one's problems not only by getting what one wants, but by wanting what one gets."[6] In a world where we know only how to seek goals, and have not the capability to adjust them constantly, the actual future will hold only frustration and suboptimality. Thus, while the forecaster or traditional predictor of futures may bemoan the fact that there is no determinism out there, the "rational" system will recognize that where there is no determinism, there is also no lack of opportunity.

[6]W. R. Reitman, "Heuristic Decision Procedures, Open Constraints and the Structure of Ill-Defined Problems," in *Human Judgements and Optimality*, Shelly and Bryan (eds.), Wiley, New York, 1964.

At any rate, we begin the empirical learning loops by recognizing that our normative scenario is simply a "metahypothesis," a product of many different hypothetico-deductive elements linked together in a consistent system of concepts (where the consistency here may be considered to mean that the laws of deductive inference were followed more or less closely in relating the various components). In a similar way, the extrapolative (surprise-free) scenario is also a metahypothesis, though it is not strictly a hypothetico-deductive one because at least some of its substance is owing to the projection of empirical data and experiential bases. Now, the problem network models were evolved directly from these two scenarios through the processes of qualitative subtraction and successive abstraction (described earlier). Thus, in the strictest sense, the problem network model may be thought of as a *surrogate* for the normative scenario, for it simply has caused a reordering and redefinition of elements which were not found in the union between the normative and extrapolative constructs (e.g., those properties associated with the normative scenario which were not present in the extrapolative) [14]. Thus, our manipulation of the problem network model in the face of simulated solutions is roughly the equivalent of operating on the normative scenario itself, though the number of factors associated with the former are vastly greater than those comprising the reduced network model. In this way, very large scale systems may usually be reduced to more manageable entities without too much loss in rectitude—providing that the normative scenario can ultimately be *resynthesized* from the problem network models to which it was reduced. This can be done, for our reduction procedure was a fairly algorithmic one, bearing not only replication but retroconstruction. Hence, the reduction process we went through not only allowed us to isolate those aspects of a potentially very large system which demand treatment if some normative system state is to be realized, but has enabled us to develop a considerably simplified surrogate on which we can perform analytical operations such as simulation, thus allowing us to economize greatly on the *costs* of analysis.

The added attraction of this reduction process is that we have a unit of analysis (e.g., the network model) which is highly amenable as a referent for empirical experiments aimed at validating (in surrogate form) the hypothetico-deductive components of the scenario and our expectations about therapeutic actions we might take. In this sense, the strategies we evolve in the form of action proposals now become hypotheses, per se, and the process of their implementation now becomes intelligible in terms of a reasonably well-controlled *experiment*. The results of these empirical trials will then be fed back to the network model, and if our expectations were in error, appropriate modifications must be made in that model. And because the reduced network model can be resynthesized into the normative scenario, we are able to allow our experiments on solution strategies to have a direct bearing on our normative construct way up the line, allowing modifications necessary in the network model to resonate back toward their origin in the much larger, less empirically tractable construct. The result of the hypothesis-experimentation-feedback process is what we have sought all along: the gradual transformation of initially hypothetical constructs into empirically validated ones, which in effect means that gradual transformation of the subjective probabilities associated with the normative scenario into objective probabilities.

It is this latter property which lends the paradigm outlined here its status as an *action research* platform, one which enables us to *learn*, in a disciplined way, while still having a hopefully positive effect in the world at large.

VII. Summary

The process we have gone through should be that which assures that the policies which we finally implement are those which have the highest a posteriori (i.e., objective) probability of proving both effective and efficient in causing desired system transformations. The logic of the process, is summarized in Fig. 6.

In short, we have gone through this rather demanding exercise to make sure that the actions we initiate are those which are the very "best" we can arrive at, and not simply expedient products of casual or strictly opportunistic policy-setting procedures. And, finally, because the normative system design paradigm is essentially an exercise in *action-research*, our most logically probable policy prescriptions will be tested within an empirical envelope, using structured (and, one hopes, reasonably well-controlled) experiments designed to lend some empirical validation to the essentially hypothetico-deductive constructs from which the action proposals (or prescriptive policies) were derived. The results of these empirical experiments will then be fed back to the hypothetical constructs, either modifying, invalidating, or reinforcing them. In any case, a Bayesian transformation system will allow us to make a priori indices of desirability, feasibility, or likelihood responsive to a posteriori, objectively predicated "data," thus allowing us gradually to metamorphosize our deductive constructs into positivistic ones (at least to the extent that the nature of the system and its operational context permits). And it need hardly be mentioned that even those empirically validated aspects of the normative, extrapolative, or prescriptive constructs must simply remain hypotheses against which emerging realities are constantly compared—especially those "realities" relevant to the operational effects of the prescriptive policies we implement and whose effects may be far spread and resonate very widely throughout a range much broader than that occupied by the system of primary interest itself.

In summary, this normative system-building process—and this system-based approach to policy-setting—is a very demanding and somewhat "idealistic" construct in its own right—much like the scenarios it is intended to help generate. Thus, it may represent more what could or should be done than what is immediately feasible or immediately likely. And at least it represents a start toward a methodology useful for those who believe that we control the future and are responsible for it, a position which, thankfully, seems to be taken more and more often (especially as regards environmental issues).

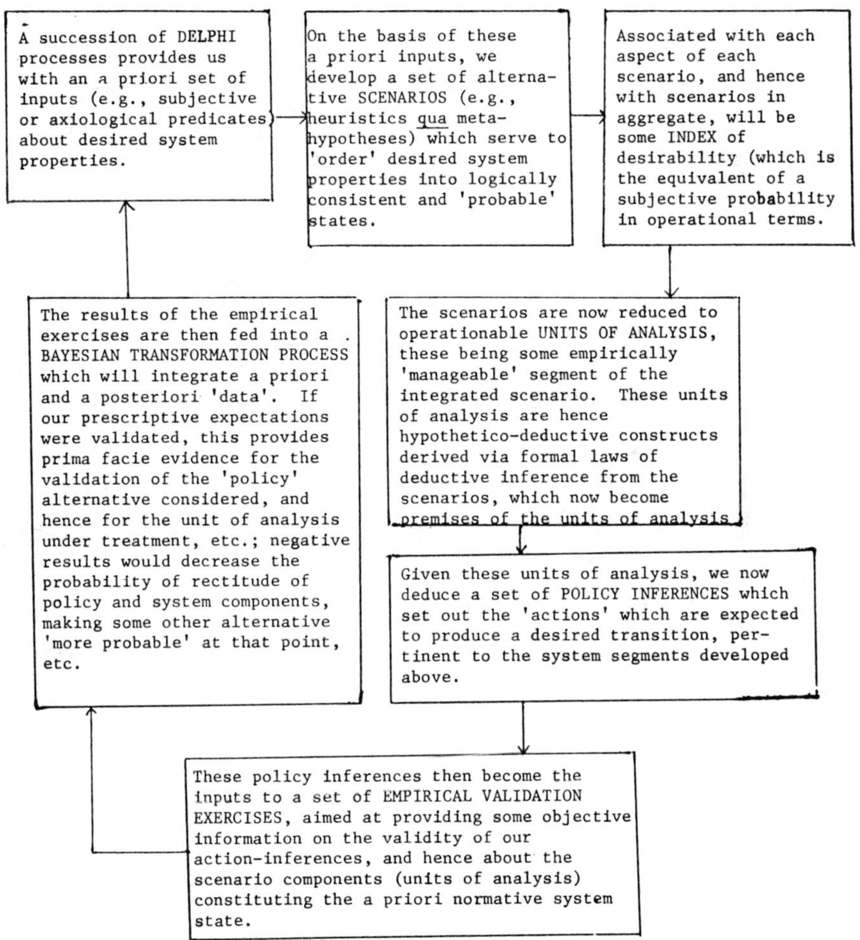

Fig. 6. The normative system architecture process.

Notes and References

1. For some indication of the importance of Janus-faced models in the social and behavioral and political sciences, see Chapter 2 of my *A General Systems Philosophy for the Social and Behavioral Sciences* (New York: George Braziller, 1973).
2. Harold Linstone, "Four American Futures: Reflections on the Role of Planning," *Technological Forecasting and Social Change* **4**, No. 1 (1972).
3. Cf., Mitroff and Turoff's "On Measuring Conceptual Errors in Large-Scale Social Experiments: The Future as Decision," *Technological Forecasting and Social Change*, Vol. **6** (1974) pp. 389–402.
4. "Beyond Systems Engineering: The General System Theory Potential For Social Science System Analysis," *General System Yearbook* **XVIII** (1973).
5. Joseph P. Martino, *Technological Forecasting for Decision Making* (New York: Elsevier, 1972), p. 267.
6. For some illustration of the techniques available for generating "consistency" in scenarios, see Russell F. Rhyne's *Projecting Whole-Body Future Patterns—The Field Anomaly Relaxation (*FAR) Method (Stanford Research Institute Project 6747, Research Memorandum #6747-10; February, 1971). Also *Technological Forecasting and Social Change*, Vol. **6** (1974), p. 133.
7. See especially Jamison's excellent paper, "Bayesian Information Usage," in *Information and Inference*, Hintikka and Suppes (eds.), (Dordrecht, Holland: D. Reidel Publishing Co., 1970), pp. 28-57.
8. In essence, formal systems are treated as giving rise to ordered sets of inputs and outputs. In this respect see L. A. Zadeh's "The Concepts of System, Aggregate, and State in System Theory," *System Theory*, Zadeh and Polak (eds.), (New York: McGraw-Hill, 1969).
9. L. A. Zadeh, "Outline of a New Approach to the Analysis of Complex Systems and Decision Processes," *IEEE Transactions on Systems, Man and Cybernetics* **SMC-3**, No. 1, (January 1973).
10. A good conceptual introduction to this area of systems is given in Anatol Rapoport's "The Search for Simplicity," *The Relevance of General Theory*, Ervin Laszlo (ed.), (New York: George Braziller, 1972).
11. This concept of system leverage was developed from Albert O. Hirschman's discussion of "development leverage" in his *The Strategy of Economic Development* (New Haven: Yale University Press, 1958).
12. For example, see *Hierarchical Structures*, Whyte, Wilson and Wilson (eds.), (New York: Elsevier, 1969), or *Hierarchy Theory*, Howard H. Pattee (ed.), (New York: George Braziller, 1973).
13. In general, the type of network which would evolve from the reduction process we have been describing could be portrayed in terms of graphs, the theory of which is essential to the construction of switched or reticular models.
14. For this reason, it is important to try to construct normative scenarios in positive terms, such that they can be translated eventually into structural-functional properties. In short, negative scenario elements (e.g., elimination of prejudice; less unemployment) are to be avoided in the context of the present paradigm.

VII. Computers and the Future of Delphi

VII. A. Introduction

HAROLD A. LINSTONE and MURRAY TUROFF

If *homo ludens* is the man of the future, one would expect the computer to be his favorite instrument. We find, however, that in recent years it has become quite popular to deride the computer for its dehumanizing effect on various aspects of society: the automated billing and dunning procedures, the inability to correct computerized information, the feeling of being quantified, etc. While most of the applications the average citizen has encountered justifiably provide this impression of impersonalism and rigidity, there is some hope that this is a transition phase in the utilization of computers. In part, this is due to the lack of a "Model-T" product or service for "everyman." There is, however, considerable merit in the proposition that the use of the computer as an aid to human communication processes will correct this situation. The current generation of computers, associated hardware, software, and particularly terminals, now begins to provide on an economic basis the capabilities for considerable augmentation of human communications. When this technical capability is coupled to the knowledge being gained in the area of Delphi design, all sorts of opportunities seem to present themselves. Underlying this view (or bias) is the assumption that Delphi is fundamentally the art of designing communication structures for human groups involved in attaining some objective.

The work taking place today in this area and involving computers often appears under the titles of computerized conferencing or teleconferencing. The latter is a more general term including such things as TV conferencing.

In its simplest form, computerized conferencing is a system in which a group of people who wish to communicate about a topic may go to computer terminals at their respective locations and engage in a discussion by typing and reading, as opposed to speaking and listening. The computer keeps track of the discussion comments and the statistics of each contributor's involvement in the discussion. In effect, one may view this process as a written version of a conference telephone call. However, the use of the computer provides a number of advantages in the communication process, compared to the use of telephones, teletype messages, letters, or face-to-face meetings.

In the use of telephones and face-to-face meetings, the flow of communication is controlled by the group as whole. In principle, only one person may speak at any time. With the computer in the communication loop, each participant is free to choose when he wants to talk (type) or listen (read) and how fast or slowly he wants to engage in the process. Therefore, the process would be classified by psychologists as a self-activating form of communication. Also, since all the individuals are operating asynchronously, more information can be exchanged within the group in a given length of time, as opposed to the

The Delphi Method: Techniques and Applications, Harold A. Linstone and Murray Turoff (eds.)
ISBN 0-201-04294-0; 0-201-04293-2

verbal process where everyone must listen at the rate one person speaks. Furthermore, because the computer stores the discussion, the participants do not have to be involved concurrently.

The discussion may take place over hours, days, weeks, or be continuous. Therefore, an individual can choose a time of convenience to him to go to the terminal, review the new material, and make his comments.

When compared to letters or teletype messages, the first item to note is the common discussion file available to the group as a whole. Having this file in a computer allows each individual to restructure or develop subsections of the discussion that are of interest to him. Normally the computer supplies each participant with whatever he has not yet seen anytime he gets on. In addition, the participant may choose to ask for certain sets of messages which contain key words or for the messages of certain specific individuals in the group. The computer also allows users to write specialized messages which may be conditional in character: (1) private messages to only one individual or to a subgroup of the conference; (2) messages which do not enter the discussion until a specified date and time in the future; (3) messages which do not enter unless someone else writes a message that contains a certain key word; (4) messages which enter as anonymous messages, etc.

The possible variations are open-ended once one incorporates directly into the communication process the flexibility provided by computerized logical processing.

The next dimension the computer can add to the communication process is that of special comments, which allow the participants to vote as a group. For example, a comment classed as a proposal would allow the group to vote on scales of desirability and feasibility. The computer would automatically keep track of the votes and present the distribution back to the group. Based upon discussion, the individuals can shift votes and reach a consensus or better understanding of the differences in views.

The computer also allows the incorporation of numeric data formats and the ability to couple the conference to various modeling, simulation, or gaming routines that might aid the discussion in progress.

In practice, one should view computerized conferencing as the ability to build an appropriate structure for a human communication process concerning a specific subject (problem). One can consider different conference structures for different applications—project management, technology assessment, coordinating of committees, community participation, parliamentary meetings, debates, multi-language translation, etc.

If the individuals enter such a discussion with fake names then we have de facto a "Delphi discussion." The computer allows us to go from this complete Delphi mode to various mixed modes such as that in which the conferee is able to decide whether an individual comment is signed with the person's real name. It is the view of the editors that the question of anonymity or its degree is less crucial to the definition of Delphi than the concept of designing the human communication structure to be used. The hundreds of meaningful paper-and-pencil Delphis that have been done represent a storehouse of knowledge on the design of human communication structures for implementation on modern computer communication systems. The area of computerized conferencing, in effect, provides an important option as an addition to the few available mechanisms people have to conduct communications: telephone, face-to-face, letter, teletype, video. In particular, the growing cost of travel has increased the concern for examining in greater depth the questions surrounding transportation/communication substitutability.

Studies by the Computer Science Department of the New Jersey Institute of Technology show that this type of service, computerized conferencing, can ultimately be brought to the user at a computer cost of one to two dollars an hour, by utilizing dedicated mini-computers as the conferencing vehicle. This compares with 15 dollars an hour for the large general-purpose time-sharing system. Also, the communication requirements are significantly less than those demanded by video or picture-phone-type systems. In addition, many applications demand the hard copy proceedings available through the use of the computer.

In general, computerized conferencing appears to be a more attractive alternative than other forms of communication when any of the following conditions are met: (1) the group is spread out geographically; (2) a written record is desirable; (3) the individuals are busy and frequent meetings are difficult; (4) topics are complex and require reflection and contemplation from the conferees; (5) insufficient travel opportunity is available; (6) A large group is involved (15 to 50); (7) disagreements exist which require anonymity to promote the discussion (e.g., Delphi discussions) or free exchange of ideas.

As a result of the foregoing, one can call to mind an almost endless list of specific potential applications:

- Project management efforts among geographically dispersed groups
- Communications for the deaf
- Educational seminars for home-bound handicapped
- Preparing agendas for weekly or monthly committee meetings
- Salesmen in a company (spread out across the country) conducting a continuous sales conference on techniques and competitor information
- Joint seminar classes or discussions involving courses at different educational institutions
- Researchers in a given area staying in contact and presenting and reviewing current findings
- Coordination conferences among community groups
- Technology Assessment working groups spread out in different departments
- Sensitivity sessions and staff development training conferences
- Delphi forecasting and exploratory conferencing

The list could go on. All the foregoing examples would require different types of communication structure and different options available to the participants. One can, for example, create a conference that would follow *Roberts' Rules of Order*; another might be structured along debating lines with judge and jury, as well as the debaters.

In essence, computerized conferencing, being an alternative form of communication, can be applied to almost any area about which human beings desire to communicate.

We can forecast with some confidence that at some point in the 1980s an individual at home can in the evening decide to phone via his home computer terminal a list of ongoing conferences on specific topics. By joining one of these which deals with a subject of interest to him he suddenly has a method of easily finding other people in the society of similar interests. In the long run this could have a dramatic effect on society itself.

The collection of articles gathered in this section represents only a small sample of many steps underway to improve human communications with the aid of computers. By design they are chosen to represent diverse directions and even differing underlying philosophies. We hope, via this mechanism, to expose for the reader the richness of this area and whet his appetite to investigate it in more depth. These articles deal not only with the utilization of computers as vehicles for improving the Delphi technique, but either explicitly or implicitly they impact on the more general issue of improving the human communication process.

The first article in this chapter, by Charlton Price, is an excellent review of some of the significant work in this area over the last five years. Mr. Price has, in the role of classical reviewer, delineated the strengths and weaknesses underlying these past efforts and raised for examination many of the questions this new area of communications stimulates. He also contributes a number of concepts, both quantitative and qualitative, on measuring the effectiveness or efficiency of communications among human beings.

For any future evaluations of computerized conferencing systems it is vital to consider the whole gamut of group communications using electronic media. In fact, a great many classical group communication experiments should be replicated in the new media and the results compared to the earlier experiments using traditional media. The paper by Johansen, Miller, and Vallee considers this larger context of research. It introduces a descriptive theory leading to a clarification of the social effects and reviews the literature of group communication in the light of several major research projects in the field, both in the U.S. and abroad. The philosophy and history of each project is summarized and its relationship to the current trends in media development and assessment is described. Laboratory experiments, field trials, and survey research are three directions that social scientists have pursued and the overall relevance of each approach to specific problems is analyzed.

A new taxonomy of mediated group communication situations is proposed on the basis of these trends and an effort is made to relate it to ongoing work, pointing to the increasingly important need for systematic appraisal of existing communications media.

The third article, by Sheridan, introduces the concept of the "portable" Delphi and the use of Delphi as a vehicle for improving the local community meeting. While there is a great deal of writing on the need and possibility of using technology to enhance the democratic process, our choice of Sheridan's work is representative of the fact that he and his associates have carried out a great deal of careful experimentation in this area and are continuing to do so. The whole concept of "participatory democracy" and the decentralization of political power implies a need for effective *two-way* communication which cannot be met by current mass-communication media. This area is likely to see greatly increasing use of computer and communication technology over the next decade.

The fourth article by Johansen and Schuyler, represents a scenario for the use of real-time Delphi in a university of the future. It is, however, based upon real experiments carried out by them at Northwestern University. Implicitly the views of these two researchers and the work they have been doing are an attack on the concept of Computer-Assisted Instruction (CAI) as commonly held by most people in the CAI field. That is to say, neither CAI nor education should be viewed as a process of "programmed instruction" by the computer (or the teacher). Rather, education must be primarily viewed as a communication process.

It is not the objective of this book to take up issues surrounding the philosophy of the education process but merely to point out that a Delphi can serve an educational function for its participants and potentially, therefore, can be developed into a standard educational tool.

While many educators have utilized Delphi for planning and assessment purposes, very few other than Johansen and Schuyler have grasped the concept of Delphi as an educational tool. We foresee a potential here not only at the universities but also in high schools and elementary schools.

To this point, the view of the impact of the computer has been somewhat rosy. However, lest the reader become too comfortable and complacent, the final article in this chapter is guaranteed to raise a few doubts and some concerns. It is a scenario intended to convey an impression of what might happen if we do not attempt to improve human communications. It represents a society in the future which extrapolates to the logical extreme some of the dehumanizing trends brought about by some current organizational and technological developments.

There is, of course, more to the future of Delphi than its automation. Let us consider three disciplinary advances and one sociological change which may have a striking impact on Delphi.

For the former, we may consider major achievements in (*a*) holistic communications, (*b*) fuzzy set theory, and (*c*) psychological measurement research.

(*a*) If Delphi is primarily a communication system among human beings, then we must admit that it misses the vital nonverbal, nonliterate components of interpersonal communications entirely. The work of Adelson *et al.* (Chapter VI, D) suggests one step beyond the familiar literate stage. But we must move much further in communicating images. Holistic gestalt or pattern recognition and transmission as well as induction of altered states of consciousness can vastly expand the ability to communicate. What will be the impact of either concept on Delphi?

(*b*) An area of mathematical research which is potentially of great significance to Delphi is the theory of fuzzy sets. It is an attempt, largely stimulated by L. A. Zadeh,[1] to deal quantitatively with the imprecise variables commonly used in social systems (e.g., big, happy, important, likely). In essence, the theory develops algorithms which enable us to operate more systematically in communicating complexity.

(*c*) Psychological methods are undergoing an enormous revolution, made possible, in part, by the computer. Techniques such as multidimensional scaling offer hope of understanding the process of human judgment and utilizing such insight effectively.

Finally we turn to a sociological change which could have an enormous impact on our subject. There has been growing discussion of the possibility that Western society is in a period of transition from a uniformity-seeking type to one which emphasizes heterogeneity. Maruyama[2] has provided one concise description of what underlies these two concepts of society. They are briefly summarized by the following lists of characteristics contrasting the traditional uniformity type to the emerging heterogeneity type:

Traditional	Emerging
uniformistic	heterogenistic
unidirectional	mutualistic
hierarchical	interactionist
quantitative	qualitative
classificational	relational
competitive	symbiotic
atomistic	contextual
object-based	process-based
self-perpetuating	self-transcending

[1]See, for example, L. A. Zadeh, "Outline of a New Approach to the Analysis of Complex Systems and Decision Processes," Electronics Research Lab., Memo, ERL-M342, July 24, 1972, College of Engineering, University of California, Berkeley.

[2]M. Maruyama, "Commentaries on the 'Quality of Life' Concept," unpublished.

Delphi is exceptionally well adaptable to this emerging logic:

(*a*) It is interactionist rather than hierarchical. Anonymity hides the hierarchical status of the participants.

(*b*) The feedback can be positive as well as negative, thus amplifying differences as much as dampening them out. There is no a priori reason that convergence must and should result during the Delphi process. Differences can be crystallized and heterogeneity sharpened. Jones (Chapter III, B, 4) conducts his Delphi in a way to maintain the distinctiveness of the four participating groups and highlight differences in their attitudes on the same subject, while Schofer (Chapter IV, C) suggests means to measure the degree of polarization.

(*c*) The process is mutualistic rather than unidirectional. Ideas can originate with any participant and need not flow from a single source.

(*d*) The method is symbiotic rather than competitive. There is no grading of an individual's responses or comparison with others on some value scale. The feedback explains and clarifies, thus facilitating symbiosis.

(*e*) Good Delphis tend to seek relational and contextual representations of a problem and avoid preclassification or rigid atomistic considerations or structures.

(*f*) Qualitative input are normal in Delphis which involve personal judgment and subjective views.

In applying this societal shift to quality of life, a topic recently subjected to Delphi studies by Dalkey (Chapter VI, B) and others, Maruyama writes:

> The definition of the quality of life must come from specific cultures, specific communities, and specific individuals, i.e., from grass roots up. There still persists among the planners the erroneous notion that "experts" must do the planning. Many of them, when talking about "community participation" still assume that the "experts" do the initial planning, to which the community reacts. There was a time when it was fashionable to think that Ph.D.'s in anthropology were experts on Eskimos. This type of thinking is obsolete. The real experts on Eskimos are Eskimos themselves. I have run a project in which San Quentin inmates functioned as researchers, not just data collectors but also as conceptualizers, methodology-developers, focus-selectors, hypothesis-makers, research-designers, and data analysts. Their average formal education level was sixth grade. Yet their products were superior to those produced by most of the criminologists and sociologists.... We can use the same method in creating criteria for quality of life in specific cultures and specific communities.[3]

Mitroff and Blankenship have expressed a similar view in their guidelines for holistic experimentation:

> The subjects (general populace) of any potential holistic experiment must be included within the class of experimenters; the professional experimenters must become part of the system on which they are experimenting—in effect the experimenters must become the subjects of their own experiments.

---

[3]Maruyama, *op. cit.*

Corollary: The reactions of the subjects to the experiment and to the experimenters (and vice versa) are part of the experiment and as such must be swept into its design (i.e., conceptualization).

They add that "it is becoming increasingly clear that as much as the structural research situation contributes a great deal to our precise understanding of the responses and properties of human subjects, by its very nature, structured research also tends to inhibit and distort many of our subject's responses."[4]

The key is to be found in the role played by the staff which prepares the Delphi, gathers the participants, analyzes the feedback, and synthesizes the output. The preceding observations point to the importance of having the "subjects" themselves take a major part in the staff activity. The Delphi process has the potential of becoming a basic tool of society, representing a new level of ability to use information from all segments of the society. This view reflects the Singerian inquiry philosophy, i.e., a need to sweep psychological aspects into most societal problem analyses. If we intend to take full advantage of Delphi for such mutualistic symbiosis we must resist one major impediment: subversion of the process by traditional modes of thought and staff procedures.

[4]I. I. Mitroff and L. V. Blankenship, "On the Methodology of the Holistic Experiment" *Technological Forecasting and Social Change*, **4**, pp. 339-354 (1973).

VII. B. Conferencing via Computer: Cost Effective Communication for the Era of Forced Choice

CHARLTON R. PRICE

Introduction

To keep our high-technology, fast-changing, urbanized society going, new ways are needed to share information, solve problems, anticipate the future decide, and manage. "Business as usual" in these respects seems likely to become even less acceptable than at present in an era of scarce human and physical resources and inflationary cost increases. We are entering an Era of Forced Choice.

For some time the available ways of pooling information, planning, and determining what had to be done next have been ill-fitted to the problems and the pace of change. Much of this increasingly unmanageable change was triggered by earlier decisions based on inadequate information, too limited assumptions, and too little anticipation of consequences. Thus the energy "crisis" turns out to be the forerunner of a probably chronic condition: resource shortages and/or unacceptable costs requiring fundamental changes in many of the ways we work and live. These ways have been built on the assumption that fuel for heating, cooling, and moving people would continue to be plentiful and cheap. Now the shortfall in petroleum seems only one symptom and consequence of a much more fundamental shortcoming: the lack of ways for preventing possible alternatives from becoming forced choices.

Reexamining trade-offs between transportation and communication is one kind of reassessment which is now unavoidable. For examples, nowadays more is heard of substituting telecommunications for work-related travel: the kinds of things that have to be communicated about require improved ways of linking people, and costs of moving people from place to place are becoming more and more burdensome. Choice of communications over transport, it would seem, will be more and more necessary. Innovations that can make communication a preferable alternative to transportation are especially needed. Some such technologies, and the new ways of behaving they encourage, are just now emerging.

One of these is conferencing via computer. As a way of linking people and organizations for many purposes, computer conferencing promises more effective use of energy and resources for many group tasks, especially those previously assumed to require that participants be in each other's presence at the

The Delphi Method: Techniques and Applications, Harold A. Linstone and Murray Turoff (eds.)
ISBN 0-201-04294-0; 0-201-04293-2

same time (and therefore with a need to travel to the same location). In the Era of Forced Choice, such assumptions will be increasingly questioned. Innovations that offer a better way, not just a second-best alternative, will be especially significant. Conferencing via computer shows all the signs of being one such innovation.

This brief discussion is to describe how conferencing via computer works, how it might be applied more widely, and why it is a significant innovation for an era in which new and better ways of problem-solving and managing are being forced upon us.

What Is Computer Conferencing?

Conferencing via computer is a readily available, but as yet little publicized and not widely tested, medium for communications and problem-solving. It consists of linkages between individuals who "connect" by each conferee having available a remote terminal (a keyboard with letters, numbers, and symbols linked to the control computer) plus a cathode ray tube (CRT) display device and/or a printer. The central computer is programmed to sort, store, and transmit each conferee's messages. The individuals linked in this way may interact at the same time, or, more typically, at their convenience, with the computer holding all messages until accessed.

Linkage via computer provides significant advantages for conducting many kinds of information exchange and problem-solving, as compared with other media (face-to-face meetings, mail, conference telephone, closed-circuit television). It also calls for new ways of behaving, and creates some problems for those unaccustomed to the fast and complex information flow that computer conferencing encourages. Among these new conditions are the relative novelty of communicating via a keyboard and mastering a set of conventions for operating such systems. These can be surmounted with practice. More subtle, and of greater significance, is the degree to which conferencing via computer will offer more cost/effective, more flexible, and informationally richer alternatives for managing and problem-solving in groups and organizations—if those who try it can permit themselves to abandon some more traditional communications habits.

The following discussion of several versions of computer conferencing emphasizes very recent developments that have not yet been reported in the small body of professional literature on this subject. Most of the information came from personal and telephone interviews with some of the principal innovators and practitioners of computer conferencing, plus a review of existing literature both on computer conferencing and on relevant discussions of social technology and organizational behavior. (The literature on computer conferencing per se is not large, and new developments are considerably ahead of publications on the subject.)

Because developments in this field are occurring very rapidly, with a number of different groups exploring the technique with various goals or emphases, there is substantial disagreement among the various practitioners on what should be considered "true" computer conferencing. One view is that conferencing via computer is a new primary communications technology, like the teletype or the telephone, offering the possibility of many new kinds of communicative behavior, as well as virtually infinite potential for supplanting or supplementing other media [35, 39]. Another view is that computer conferencing capability is but one example of a large family of tools for "dialogue support" to facilitate connectedness between "knowledge workers" [9-12].

Computer conferencing is seemingly only a small part of the field of computer networking and time-sharing, even though all time-shared systems in principle could offer the person-to-person interactive feature (direct communication between those accessing the network). But computer time-sharing developments have emphasized person-machine interfaces: that is, most attention has been paid to how the operator interacts with the computer, making inputs and getting outputs from the computer's models and data bases [42]. Relatively little attention has been devoted to how the computer could be used for *throughput*, establishing links between individuals, between an individual and a group, or between groups. It becomes evident that different ways of structuring these communications are appropriate to various kinds of intellectual or managerial tasks. Use of the techniques of conferencing via computer forces explicit consideration of how the interaction should be managed and in what ways the participants should interact. These implications of computer conferencing as a technique or communications medium require separate consideration from classical time-sharing applications of computers (the transfer or melding of *data*, without the aspect of mutually influential interaction between *persons*).

The use of the computer for conferencing makes the computer a true communications medium, supplementing its use as a medium for storage and computation. Because of the features of storage, retrieval, and data processing, not available through other media, computer conferencing is a significant advance over other communications media. According to criteria for a true communications medium stated by Gordon Thompson [35], computer conferencing:

- offers an easier and more flexible way to access and exchange human experience
- increases (virtually to infinity) the size of the common "information space" that can be shared by communicants (and provides a wider range of strategies for communicants to interrupt and augment each other's contributions)
- raises the probability of discovering and developing latent consensus. (The enriched information base and heightened interconnectedness increases the

chances that each conferee can receive unexpected and/or interesting messages.)

These are the principal versions of computer conferencing now in use or being developed:

● Civilian (nondefense, nonaerospace) agencies of the federal government, mostly through the now disbanded Office of Emergency Preparedness in the Executive Office of the President, have used links via computer to carry on information exchange and coordination activities in a number of situations requiring rapid information flow and collaboration [38, 39]. In this case, computer programs for conferencing developed from a need for better communications for crisis-management. The simplest version, PARTY LINE, operates with participants involved simultaneously, the equivalent in electronic data processing technology of a conference telephone call. PARTY LINE has the added advantage that a full record of the inputs by each participant is preserved. An elaborated version of this computer program, EMISARI, makes it possible for specialized forms of information to be included, such as tables, bulletin-board items, and special computations. EMISARI also includes participation by a conference monitor who directs attention to key issues, guides discussion, and "brokers" between different interests of participants. This feature takes account of various roles participants may play, as in any conferencing situation [28]. Some will be *stakeholders*, with an interest in or responsibility for decision-making. Others may be *experts*, with relevant knowledge that needs to be accessed or revised. Still others, including the monitor, may be *process specialists*, whose concern is to enrich the total information base or level of collaboration. All such roles are played in effective face-to-face conferencing; EMISARI and other computer conferencing require that these roles be played more explicitly, leading to more effective results from the total exchange. Participants can interact at times of their own convenience over hours, days or weeks, and messages can be retrieved, edited or rearranged by the conference monitor or other participants. In short, the EMISARI monitor —borrowing from Delphi studies techniques (see below)—interfaces between the system and the users, and wires them up as needed. The computer becomes in effect a huge blackboard capable of handling different types of messages. Development of these systems is principally the work of Dr. Murray Turoff, now in the Computer Sciences Department at the New Jersey Institute of Technology, where he is continuing design, development, and testing of computer conferencing programs, as well as evaluating effectiveness of their use for a variety of purposes.

● At the University of Illinois (Champaign-Urbana), Stuart Umpleby and his co-workers have devised the DISCUSS conferencing program. DISCUSS operates on the PLATO system, originally a device for computer-assisted instruction developed by Dr. Donald Bitzer. DISCUSS augments PLATO by adding conferencing features. This new development is being continued with support from the National Science Foundation. The PLATO system uses a plasma tube to display messages, which can be in the form of graphics as well as alphanumeric symbols, and can access text files as well as other stored data [45].

● The Institute for the Future (IFF) in Menlo Park, California, is the center for a long-range program to develop computer-conferencing techniques. IFF also has received a grant from the National Science Foundation to study the effects of conferencing via computer on the attitudes and behavior of users of the system. This work evolved from a search for improved ways of conducting Delphi studies: the systematic interrogation of experts in forecasting, and pooling of their judgments. Drs. Robert Johansen, Richard Miller, Hubert Lipinski, and Jacques Vallee are tied to the network linking researchers who participate in the work of the federal Advanced Research Projects Agency (ARPANET). Through this computer network the IFF program is able to tap into data files elsewhere by the use of codes punched into the remote terminals, and to combine these data sources and computational capabilities for problem-solving as well as information exchange [23]

● At Northwestern University, the Computer Aids to Teaching Project in the School of Education is building a computerized learning exchange linking students, teachers, and researchers with a program called ORACLE. The basic premise of this system came from seeing education not as programmed instruction, but as a process of communication. This combines the conferencing feature with computer-assisted instruction (a program called HYPERTUTOR), access to information files, and computational capability. The total system consists of some thirty terminals on the Evanston and Chicago campuses, about 15 of which are used for CAI and conferencing. Dr. James Schuyler reports that software has just been inserted in the system that will give reliable and comprehensive cost data on system performance. The Northwestern group also participates in conferencing the PLATO system centered at the University of Illinois [33].

● A commercial application of conferencing is the MAILBOX program developed by Scientific Time Sharing Corporation (STSC), Bethesda [2,6]. STSC started in 1969 and in early 1974 had about eighty employees. These individuals are at about two dozen locations; the sales offices are in twenty other cities; about half the people are at the Bethesda headquarters. MAILBOX links all of these individuals at whatever location and the several hundred customers of the firm. Common data that can be accessed by any number of users is stored on IBM 370/158 computers in Bethesda. MAILBOX is used for technical discussions as well as exchange of information with customers. Customers may also, by subscribing to the time-sharing service, use MAILBOX for communicating with each other and with STSC, under controlled-access conditions permitting any degree of confidentiality desired. STSC also has a parallel program—NEWS—designed for information exchange only; in effect, a computerized news service capable of being "customized" in a variety of ways.

● Bell Northern Research, a part of the Bell Telephone Company in Canada, is another center for research and product/service development in computer conferencing. Gordon Thompson of Bell Northern is exploring ways to turn the conferencing feature into a generalized information utility. Also under development at Bell Canada is a linkage of participants in a telephone conference call to a multitrack tape recorder, so that asynchronous conferencing could be done by telephone. As with printout or CRT, callers will be able to send and receive messages encoded to specific individuals or to the conference at large, by using touch-tone codes [36]. Use of voice links will circumvent the main present barrier to widespread computer conferencing—the inability or disinclination to type in a remote terminal (or use a secretary as an intermediary) [33].

● The Augmented Knowledge Workshop at Stanford Research Institute [10], headed by Douglas Englebart, has been in existence for more than a decade and is an important parallel development to computer conferencing systems. Inventions by the SRI Workshop potentially could augment computer conferencing by enriching the information environment through CRT display of data files, linkage of conferees via TV, and text editing using a computer storage and three-dimensional displays of data. The SRI Workshop is mainly devoted to increasing the richness of face-to-face interactions between conferees. Now these techniques could be grafted to programs that permit interaction between conferees at a distance from each other. The SRI group is building from this a "computer utility for knowledge workers," primarily in business and industrial organizations, who as subscribers can make use of the special facilities developed in the Augmented Knowledge Workshop. Some twenty organizations have already subscribed to this service [12].

Costs and Benefits, Financial and Otherwise

According to Turoff [40], there are five situations in which conducting a conference via computer is better than the alternatives:

● when the individuals needing to contribute knowledge to the examination of a complex problem have no history of adequate communication, and the communication process must be structured to ensure understanding;

● when the problem is so broad that more individuals are needed than can meaningfully interact in a face-to-face exchange;

● when disagreements among individuals are so severe that the communication process must be refereed;

- when time is scarce and/or geographic distances are large, inhibiting frequent group meetings; and/or
- when a supplemental group communication process is needed.

The most significant feature of all of these versions of computer conferencing is that they have developed in response to communications challenges for which conventional media were either inadequate or did not occur until the computer conferencing mode became available. In no sense do the systems so far developed represent unproven or experimental uses of available hardware or software. They are well within the state of the art of computer equipment and programming capabilities. But flexible and fluent use of computer conferencing is an art, as are other communication forms. The limitations of the method are determined more by the skills of users than by the capacity of the technology. The problem, as Englebart has said, is how to apply and customize the service to particular uses, and for users to develop facility in using such systems so that the conferencing mode melds well with other kinds of communications and facilitates work, rather than representing merely an exotic use of computers to be pursued for its own sake [12].

A frequent comment regarding new media is that each new medium offers problems without really compensating advantages. Estimates of the value of a medium of communication (both in terms of effectiveness and in terms of cost savings or other benefits) must be in terms referring to how the system is to be used and what the insertion of the medium into a situation does to other variables in that situation [7]. Thus for computer conferencing it is not enough to compute the financial costs of the technology itself in comparison with other technologies. Attention must also be given to what the use of the medium costs and what it gains—in terms of user time and productivity, effectiveness of communication, etc.—in comparison with other modes.

Some financial data compiled by Turoff are a good starting point. He estimates [44] that the installation costs of a remote terminal are now in the neighborhood of $2,000 using a cathode ray tube (CRT) display, or about $3,000 with the capability of producing hard copy. These costs are declining rapidly, and by 1980 can be expected to be equivalent to the present cost of a large color television set. Thus, basic equipment for computer conferencing capability is well within the reach now for even fairly small organizations using time-sharing computer services, and is still easier for those medium and large organizations that have their own computers. (There are approximately 300, 000 to 500,000 terminals now in use, and about forty manufacturers of terminals.)

The use of computer time (for processing of messages in the computer), another cost element, is minimal even for fairly continuous conferencing. Scientific Time Sharing Corporation, which owns its own computer, estimates that the cost per message in central-computer-processing-plus-telephone-circuit time is somewhere between twenty-five cents and fifty cents, which is far cheaper than long-distance telephone and is competitive with mail when typing time and postage are considered. An added saving comes from the fact that the same message can be communicated to a large number of people at no significant extra cost per recipient [6].

There are these additional data available on costs:

> Published commercial rates for timesharing costs of a Univac 1108 gives an upper limit of 19 cents per second for central processing time, $5 for an hour of terminal time, and $1.20 a month for 10,000 characters in storage. [One conference] conducted over 13 weeks used approximately an hour of central processing time, 100 hours of terminal time, and less than 100,000 characters in storage. If we assume an hour of thinking time for every hour on the terminal, 25 days of effort were donated, which is a minimum of eight consulting trips, not counting travel and per diem costs, telephone, and clerical processing costs in the conventional mode [48].

Other analyses by Turoff show that, when the value of the time of participants is taken into account, the savings of computer conferencing over face-to-face meetings become dramatic. Cost comparisons should be made in terms of communication effectiveness for various media per dollar invested in communications. For example, effectiveness can be defined as the ratio of (1) amount of information exchanged face-to-face to (2) amount of information shared using the computer as a medium. Turoff assumed a cost of computer time of seven dollars per hour per participant. If the number of participants is more than a dozen, and the value of the time of each participant is even as low as five dollars per hour, his analysis showed that conferencing via computer was no more costly than: (1) a meeting requiring travel, (2) a meeting with all participants at the same location, or even (3) a conference telephone call. Of course, computer conferencing becomes far more cost/effective when any of the following increase: value of individual participant's time, conference duration, or number of participants.

Thus, cost of hardware and computer time aside, computer conferencing in some circumstances can result in considerable savings of staff time as well as greater use of information and ideas potentially available. Some of the ways in which this works are illustrated from this account of an experience of the computer conferencing group at the Institute for the Future:

> We had come back to the office from a large-scale test of our conferencing system. The test had involved twelve terminals and there were many difficulties. To get all the points considered and to begin to solve the problems that had been identified, the four of us who had been at the test first went to remote terminals and had a teleconference with each other. We just went to the terminals and poured out all our ideas and observations. There was no waiting for "air time" (as would be the case in a face-to-face meeting). After about an hour of that we had ten or twelve pages of printout. Then we proceeded to sort out the ideas. After an hour or two we had an ordered array of all the topics to be discussed and all the

decisions that had to be made. Then we needed a blackboard; the teleconferencing broke down. But normally these steps would have taken a full day. Instead, it took a total of about two hours, but we did need both media.

...We had discussed what needed to be done while on the way back from the test, and in that preliminary discussion we thought we had much the same information and reactions. But in putting all our thoughts on the terminal we found that each one of us had forgotten certain aspects, and some points were clearer than others. Looking at the printout we had a much fuller view of what had gone on in the test [23].

Origins of Computer Conferencing

In a 1968 article [25] computer conferencing was proposed as a means of speeding up the rate of interaction and the data-processing operations involved in Delphi studies. The Delphi technique is a method that has been developed to pool expert judgments while attenuating or removing some biasing effects such as the influence of the reputation of one panelist on others. The meaning of Delphi has now been broadened to include all forms of structured communications processes involving information exchange and the pooling of judgments [27].

The monitor of a Delphi exercise presents a series of assertions (predictions of future occurrences, normative statements regarding proposed actions, statements of intention, etc.) to the panel, which is selected for its special knowledge of the subject. The panel also may include invididuals with expertise outside the field under consideration but with special capabilities for analysis or judgment between alternatives. Panelists vote on these alternatives, return the information to the monitor, and the data are combined and arrayed, then resubmitted to the panelists individually. At this point each panelist knows the judgment of the group and can compare the distribution of group judgments with his own choices. New data (e.g., additional assertions from the monitor) may also be introduced at this point. Panelists may then modify their judgments or not, but in any case they proceed from a larger information base than in the first round. The process may continue through several rounds in the same manner.

The difficulty with the Delphi process conducted in face-to-face conferences or by mail is that the turn-around (sending and receiving mailings plus processing of the data and preparing each round for the panelists) can be very time-consuming, resulting in a loss of initiative and focus. Adding computer capability makes it possible to cut turn-around time to a minimum, thus permitting the introduction of many more iterations and more material per iteration.

It became clear that communication via computer could be useful for many other kinds of geographically dispersed groups. All computer networks in principle could have interactive features akin to conferencing. That is, the operators of timesharing systems could use the system to transmit operator-to-operator messages as well as to interchange data or run routines on remote computers [20]. But the emphasis in conferencing was originally for real-time exchange of information, proposals, and instructions in situations requiring fast response time and the coordination of the efforts of many different kinds of actions at locations widely separated geographically. Gordon Thompson observes [36] that computer conferencing works best when the problem is clear but information exchange is complex and pressured (as in coping with a geographically dispersed crisis). Such was the case in 1970 when the Office of Emergency Preparedness needed data on steel-industry capacity and production to do a rapid national assessment of the industry's performance and capabilities [40]. One outcome of this experience was the realization that the system should be automated, so that information and technical assistance could be exchanged rapidly between a large number of participants. The PARTY LINE program evolved into EMISARI.

This augmented capability was in place at the Office of Emergency Preparedness a year later when the ninety-day wage-price freeze was announced. It was then necessary to have information rapidly from every section of the country on enforcement activities, problems and questions encountered in the field, rapid dissemination of decisions by the Cost of Living Council, and coordination of the Office of Emergency Preparedness headquarters efforts with those of the OEP regional offices, and with field representatives of the Internal Revenue Service, who, along with representatives detailed from other agencies, had been charged with field enforcement. Conferencing using EMISARI went on throughout the three-month period, starting with about thirty participants (i.e., terminals; each participating terminal might serve a number of individuals at a particular location) and rising at the end to more than seventy [42].

In the OEP response to the wage-price freeze, EMISARI became in effect an electronic blackboard of unlimited size, on which any of the participants could insert data, raise queries, or request revisions. These interactions could occur at times convenient to each of the participants. When coming on-line, a participant could receive all messages held in the computer's storage and also see what had occurred throughout the information array since the last participation. This information was available on printed-out hard copy, or a CRT display, or both.

On the "blackboard" were identifications of all the participants, their names and responsibilities. Associated with each of these participants were quantitative data and other messages. Those who had programmed the system were also participating. Thus the programmers could be queried directly by those in the field who did not understand an instruction or thought that something might be incorrect about information presented. Text files could also be accessed by participants. The text files contained "bulletin board" announcements, abstracts of documents, policy decisions or directives (e.g., from the Cost of Living Council), and press releases. The total system provided a real-time monitoring and communications device with fast turn-around to track and direct the wage-price freeze on a national basis. Turoff sums up the experience [43] as follows:

> We took all the people who were gathering data regionally, those who were trying to respond to information requests in each region, those preparing staff reports from these data, the middle management level in the agencies, and the data grubbers. They were all in contact and everyone could know what everyone else was doing. The system allowed them to pass information back and forth, and also allowed monitoring of the process and the passing on of instructions from high up, and inputs from such groups as the Cost of Living Council, such as the type of information the White House might want in any particular week.

Since 1971, EMISARI and its derivatives have been used for further work on economic controls, and for rapid assessment of fuel stocks and production capabilities early in the energy crisis [43]. In August of 1973, the software for conferencing aspects of EMISARI, created by Language Systems Development Corporation of Bethesda, became available as a computerized conferencing system at nominal cost through the National Technical Information Service, Department of Commerce.

At the Institute for the Future, computer conferencing began to be used as early as 1969 for the conducting of Delphi studies, of the type outlined above. IFF started with a strong bent toward classical Delphi (pooling of estimates and forecasts by experts) but moved toward less-structured versions of the technique as the use of computer conferencing permitted greater freedom. IFF also began recently a long-range research program on the behavior of users: what kinds of people and what kinds of problems prove most amenable to the computer conferencing technique, and what problems are associated with the use of the medium.

Little analyzed so far is the impact that computer conferencing can have on the style and structure of organizations that use it. For example, at Scientific Time Sharing Corporation, using MAILBOX has caused the STSC organization to be shaped by the interaction patterns that have grown up as a result of the use of the system [2]. Instead of a pyramidal form of the conventional organizational hierarchy, with specialized groups reporting and through individual line managers, the organization has taken the form of constantly changing clusters of groups or teams; each may be formed quickly for a particular project or problem, and disband just as quickly when the issue at hand has been dealt with. There are also relatively more permanent groups, such as those concerned with continuing development of MAILBOX itself. The experience of working with MAILBOX is described as follows by Lawrence

Breed, a vice president of STSC in Palo Alto, California, who has been largely responsible for developing the system:

> It has a flavor which is quite unlike any other form of communication, for example face-to-face exchange. For instance, you have whatever time is needed after receiving a message to get your own thoughts together and come back with a fairly incisive and coherent reply. You don't have the effect of thinking up snappy comebacks ten seconds too late. On the other hand, the messages go back and forth so rapidly, perhaps several times a day between any two particular participants—that it is completely unlike first-class mail, which is so slow that you really lose the interactive characteristic. It is unlike the telephone in a couple of important ways, too. I can communicate with someone at a time of my choosing, not wait for him to answer the phone or get filtered through his secretary. Nor is he interrupted by the message in what he is doing. ...The way that this mode of communication affects the company is that it has a very strong democratizing effect. It would be almost impossible to enforce communication through the channels defined by an organizational chart. My boss may send a message to me with a copy to the people who report to me, and like as not one of them will have replied to it before I even see my "mail." This can make some people pretty uncomfortable. ...This kind of communication is not a substitute for face-to-face communication, as we've been discovering. We've got to get together occasionally to have round-table discussions. ...But the benefits are pretty substantial. Most time-sharing services that cover a wide geographic area have something similar. I think ours is one of the easier to use, and most flexible. We transmit 175 to 225 messages a day—that is the serial listing, because each of the messages might go to a large number of people [6].

Dr. Philip Abrams, director of research at the Bethesda office, says that it would be impossible to operate STSC without MAILBOX [2]. Users report that participating in computer conferencing is a powerful democratizer. It is very hard to keep information from flowing freely and to reduce the ability of persons to participate, once they have mastered the basic skills needed to operate the keyboard of a remote terminal.

Computer conferencing was used in the spring of 1970 to conduct a Delphi on the design of computer conferencing and to evaluate potential applications [39]. The data processed were a total of ninety-nine assertions about various features of the technique and its potential applications. The panel could give judgments on the importance, degree of confidence, validity, desirability, and feasibility of each statement. The printout following analysis of these data showed the distribution of group judgments and the degree of dissent from majority views on each statement. Some of the findings were:

- A strong majority of the panel felt that the system could be used to convene one-day conferences, given the current availability of terminals and printers.
- Because length of messages in the version examined is necessarily limited, the panel did not feel that computer conferencing could replace work by

committees, and that it is not suitable for joint editing of draft texts or reports.

● The panel liked the idea of incorporating a "wait" feature which would trigger the terminal when a new item or message has been entered from another terminal. This would enable use of the system for an intensive, scenario-type of simulation lasting perhaps a day, with the monitor feeding in events and getting reactions.

In reviewing the experience, Turoff commented on the expansion of effectiveness coming from joint consideration of the system by a variety of analysis and system designers: "Many of the suggestions made for improvement of the system appeared, once stated, to be so obviously beneficial that this part of the discussion was not always rewarding to the designer's ego. Behind this observation is one of the difficulties in getting decision-makers to risk Delphi exercises." These remarks relate to the observations from STSC regarding the democratizing effect of wide information sharing, and the threat this seems to pose to some people who attempt to exercise influence by restricting information flow [17, 18].

Some Possible Futures for Computer Conferencing

The widespread extension of time-sharing capabilities and computer networks [8, 26] should further accelerate the practice of conferencing via computer. Turoff explores the implications of these probable developments as follows:

> ...the forthcoming wide introduction of digital data networks will probably provide computerized conferencing systems with another order of magnitude edge in costs when compared to conventional verbal telephone conference calls. Thus, economic pressure may force the various (technical) problems (of computer conferencing) to be resolved favorably, so that computerized conferencing becomes a major application on such networks. The anticipated future reduction of costs, by the late seventies, for computer terminals with CRT, perhaps placing them in the same purchase range as home color TV receivers, suggests a picture of future society in which a major substitution of communications for transportation can take place [40].

It should be clear from the preceding discussion that conferencing via computer is still in an embryonic state. A number of systems, each with distinctive emphases, are in various stages of development. Some data on costs compared with benefits already available are intriguing enough to merit further and more definitive analysis. Such analysis should, for example, be sufficiently cogent and detailed to enable an organization to make considered choices in planning or installing teleconferencing capability (computer, telephone, closed-circuit television, or some combination of these). The reasons for making such considered choices are increasing daily. A recent study by Richard Harkness, on the future of telecommunications, gives some of the reasons why use of such alternatives (conferencing via computer and others) may be expected to accelerate:

> ...what appears to be emerging is a competition between communications and transportation facilities for servicing a large number of contacts that now require travel but which might possibly be made electronically....it appears that communications services can be extended and improved more readily than can transport services, since communication facilities have low visibility and are largely controlled by private firms whereas transport facilities are highly visible, controversial, and dependent on a public-political decision-making process for implementation. Therefore, it is possible that communications will tend to improve faster than transport and a relative shift from real travel to tele-travel should result [22].

The broad implication of this and similar views is that through telecommunications there may be more *considered* choice—in research and problem-solving, in organizational activity generally, and in the form and functioning of communities—so that *forced* choice will be less necessary. As teleconferencing is more widely considered and tried, it can emerge as not only a second-best alternative to "being there" but as a positive advantage in many kinds of information-sharing, problem-solving, and managerial processes. Some of these possibilities are sketched below. These are both extrapolations from what is already happening, and prophecies that might turn out to be self-fulfilling.

1. Research and Problem-Solving: Many research problems today are not only multidisciplinary but "multi-epistemological" in character. For topics such as collaborative physical design of buildings and other spaces, assessing the potential impact of new technologies, or deciding among various priorities for effort in a research problem, not just "apples and oranges" but "pears, pineapples, and pine trees" as well must somehow be reconciled. Different experts, stakeholders, and process people in such projects will have different mind-sets, different ways of seeing the world, and different bases of expertise to draw upon in coming up with the required solutions. Good solutions, increasingly, cannot be imposed by fiat but must result from heuristically, iteratively combining information and ways of thinking about information.

For this, highly interactive systems that can involve many participants over extended periods are needed. The system must be capable of highlighting and explicating differences in point of view, so that "who is thinking in what way about the problem" is considered as well as "what is the problem." [18]. As Christopher Alexander has noted, there needs to be a continual interplay of figure and ground, form and context; this becomes essential in a society where problems are "nested" and workable solutions require *induction* from iteration of many factors in various combinations [1].

For this process to occur, technology such as that available in computer-based conferencing is required, so that the record of interactions—providing enough detail to show thoughtways as well as data—can be examined by facilitators of the conferencing process and other individual participants.

After all, there is no such thing as "the facts." A fact is a statement about an empirical phenomenon which implies a conceptual scheme or frame of reference. For problems that force contrast and consideration of different perspectives, people need to learn to make distinctions between various possible versions of "the facts," distinctions that are not inherent in the facts themselves but which derive from the purpose at hand. Is urban transportation "primarily" a source of pollution or a process of interaction to make communities operate? Is "the energy crisis" a threat to "our way of life" or an incentive for considered choice? These are instances of broad problems in which the frame of reference makes an enormous difference in the kind and cogency of the solutions reached. There are a host of detailed research and problem-solving tasks in which there needs to be a way of making explicit how the definition of the problem implies solutions. Linkage of workers on the problem with a continuous record of their interactions is essential to such situations.

Computer conferencing opens up the possibility for researchers and policymakers of an electronic newsletter-like service, one which would be much more interactive and information-rich than the printed version. The newsletter "subscribers" would constitute a permanent conference of variable membership; each conferee could receive messages sent to all members, additional news or greater detail in an augmented service, or answers to queries—electronic letters-to-the-editor. Such a system would be particularly useful for integrated complementary research programs at a variety of locations, or tying together special-interest subgroups within professional and technical societies.

Computer linkages could be used to speed and enrich the presently rather pedestrian process of research contracting, from the request-for-proposal stage through completion of the work and even including dissemination of the results [29]. For example, an agency requesting qualifications of potential contractors to carry out a given assignment might query a number of them via computer conferencing on a quick assessment of the proposed task and a rough estimate of costs. Proposers offering the five most promising methods of approach and preliminary estimates of cost might be funded minimally to prepare more detailed technical proposals and more careful cost estimates. A review panel or panels might also be convened via computer, and even interact through the organization requesting proposals as the proposals come in.This would be done in real time so that the whole procedure would be much speeded over present practice. A side effect could well be the emergence of new, collaborative interest groups. Potential contractors and reviewers, by having interacted on the original proposal, might well discover other shared interests and additional bases for information exchange. And it would still be possible to protect proprietary data by encoding messages for various degrees of confidentiality (as in EMISARI and MAILBOX).

2. Organizational Behavior and Management: The capability of conferencing via computer, along with other kinds of telecommunications and information processing, can undergird a new definition of management—techniques by which complex systems are designed, developed, operated, and evaluated. For about ten years students of organizational behavior have been asserting that organizations will have to become more "temporary," "adaptive," "flexible," "team-based," and "responsive" or they will be inadequate tools for dealing with new demands and opportunities in a turbulent environment characterized by rapid change [5,30,35]. More sophisticated and sensitive ways of sharing information, making decisions, and coordinating different kinds of specialized undertakings are required to make such organization forms actual rather than, as at present, largely programmatic.

There is considerable indication in the literature on innovation and technology transfer that small organizations are more innovative than larger, and that the barriers to the diffusion of innovation within large organizations are considerable [3, 4, 31, 32]. Within large organizations, it is the individual or the small working group that is the source of innovation more often than not [19, 37]. This would suggest that for the management of innovation, the stimulation of creativity, and the diffusion of innovations achieved, it would appear practical to augment the capabilities of these small organizations or organizational units by adding to their working equipment the kinds of computer conferencing resources discussed here. Similar opportunities would be appropriate to provide for technical associations that have a need for information exchange and problem-solving on a regional or national basis, such as meteorologists-plus-air-traffic-controllers, newsgathering-processing-dissemination organizations, those responsible for building-code interpretation and enforcement, the probably emerging "community" of weather modifiers, members of widely separated project teams in engineering and construction organizations, and a host of similar groups who need to exchange current lore or benefit from each other's innovations.

In underdeveloped countries, the cost of linking villages by computer is well within the means of present technology and would represent a cost per village equivalent to providing a small truck to each community [44]. The network, once installed, could be used for bid-and-barter exchanges, information inputs to regional or national policymaking [13, 14, 15, 24], queries of government officials (which would make it necessary for government to be more responsive), and even personal messages or short letters. In some instances this might mean that a developing area or country could bypass telephone technology and leap directly into the more flexible, higher capacity types of communications technologies now available [13, 24]—much as in many such countries there has been a leap from oxcart to aircraft in transport development. The main constraints would appear to be psychological, social, and cultural. As in the case of the erosion of bureaucracy in so-called developed countries, the social and institutional structure of more traditional, less industrialized societies might be seriously strained by the introduction of such highly interactive communications systems on a large scale. Yet the possibility that such innovations can be installed forces a consideration of social and psychological costs and benefits, and at least some of those concerned with regional or national development in the less industrialized world will want to make such assessments.

3. Considered Choice in Communication and Transport: Ten years ago Melvin Webber outlined the concept of "urban realms," a view of life and work in an urban society in which urbanity would be defined less by geographic concentration and more by the intensity of interpersonal linkages.

> ...an urban realm...is neither urban settlement nor territory, but heterogeneous groups of people communicating with each other through space....the special extent of each realm is ambiguous, shifting instantaneously as participants in the realm's many interest communities make new contacts, trade with different customers, socialize with different friends, or read different publications....the population composition of the realm is never stable from one instant to the next [47].

Energy shortages and other difficulties in operating urban communities because of continued emphasis on concentration should revive this idea and give it new significance. Recently, for example, Webber has urged that the thinking about urban transportation be put in a broader frame [46]. It is necessary to take seriously the dictum in every Geography 101 course that transport shapes, as well as is shaped by, the city. Both communications and transport need to be assessed and considered choices made; the options for transport should not be limited to servicing foreseeable "demand" with all else held constant.

If such a view were really to take hold, it would undermine much current "comprehensive" urban planning, which assumes current patterns of land use, work-home travel, and existing transport and communication facilities. As Harkness notes, costs and benefits (even in economic terms alone) may lead to a use of telecommunications as a tool for reshaping urban form-office locations, urban travel demands, transportation investments, and a host of related variables [21].

The same theme surfaces in recent discussions of population distribution and efforts to see how "urban blight" might be relieved by revitalization of smaller cities and towns. William Ewald, for example, extends the view of urbanism-as-interaction and shows how this could be accelerated by public policies and programs that would increase electronic linkage as an alternative to transport:

> [There has been] scant attention to the enormous potential impact of telecommunications on lifestyles and settlement patterns...a force as potent in its potential effects as the automobile and the expressway... It is in changing the meaning of time and space that the technology of transportation and communication can have its most profound effect on population dispersion. Government policy could "interpret" this technology with important effect. Every time the

> federal government writes a regulation, it is, in a sense, creating artificial gravity. Especially with the capabilities of information retrieval, computation, education, servicing instruction and conferencing via telecommunications, does propinquity lose some of its reason for importance.... Settlements exist primarily as a reflection of ...efforts to increase opportunities for interaction. It then follows that both individual locational behaviors and overall spatial structures are mirrors of communication. With the changing patterns of communication that are imminent, we can expect that the individuals' locations and that overall spatial structures will also change—possibly in very dramatic ways.... For it is interaction, not place, that is the essence of the city and of city life [16].

This last section of the discussion has emphasized the importance of several new forms of telecommunication. Conferencing via computer is only a small and (so far) relatively little-used example of such new techniques. But computer conferencing is a "leading-edge" or "bellwether" technology within telecommunications as a whole, because it is especially well suited to the kinds of information exchange and problem-solving that will have to go on in order to bring about the needed new ways of operating organizations and communities.

The probably rapid decline in the cost of remote computer terminals, alluded to earlier, is perhaps the key development to accelerate wider use. If the cost of a remote terminal approaches that of a color television set, a logical expansion to expect would be the installation of such terminals in many homes and most businesses. Wide access to computer communications via this medium could have a major impact on postal services. Communication via computer might well be the medium of choice for the vast majority of first-class business and personal mail.

Another implication: this wide availability of terminals would make it possible for the first time for individuals and groups to connect with each other or discover each other's existence on the basis of shared interests, rather than by job titles, organizational purposes, or personal introductions of mutual acquaintances. Wide consequences of such a possibility can be readily imagined not only for work habits but also for new leisure and entertainment pursuits. In work life it would clearly accelerate the decay of organizational loyalties; people's main primary affiliations would be more likely to be with those with shared interests, wherever they might be and whatever organizational affiliations they might have.

It would seem that computer conferencing has the potential of becoming a communication/problem-solving medium that offers a believable way to directly challenge the current rapid drift toward an impasse caused by too much information, too little time to process it, and too little capability within human beings alone to interrelate and evaluate information even if processed [34]. There really is no alternative to more considered choice of methods for information exchange and problem-solving: better ways of sorting and routing information, easier access to information, and augmented capacity to act. There need not be an Era of Forced Choice. But avoiding this condition will require energetic effort to make wider and more creative uses of such technologies as conferencing via computer.

References

1. Christopher Alexander, *Notes on the Synthesis of Form*, Harvard University Press, 1964.
2. Philip Abrams, Scientific Time Sharing Corp., Bethesda, Md. Personal communication (1973).
3. Homer Barnett, *Innovation: The Basis of Cultural Change*. McGraw-Hill, 1963.
4. Samuel N. Bar-Zakay, "Technology Transfer Model," *Industrial Research and Development News* 6:3 (1972). United Nations Industrial Development Organization, Vienna.
5. Warren G. Bennis, *Changing Organizations*, McGraw-Hill, 1966.
6. Lawrence A. Breed, Scientific Time Sharing Corp., Palo Alto, California. Personal communication (1973).
7. Rudy Bretz, *A Taxonomy of Communication Media: A Rand Corporation Report*. Educational Technology Publications, 1971.
8. Lawrence H. Day, "The Future of Computer and Communications Services," *AFIPS Conference Proceedings* **43** (1973), pp. 723-34.
9. D. C. Englebart, "Intellectual Implications of Multi-Access Computer Networks," *Interdisciplinary Conference on Multi-Access Computer Networks*, Austin, Texas. April 1970.
10. ——, "The Augmented Knowledge Workshop," *1973 Joint Computer Conference, AFIPS Conference Proceedings,* **42**.
11. ——, "Coordinated Information Services for a Discipline-or-Mission-Oriented Community," *Second Annual Computer Communications Conference*, San Jose, California. January 1973.
12. ——, Augmentation Research Center, Stanford Research Institute. Personal communication (1973).
13. Amitai Etzioni *et al.*, *Preliminary Findings of the Electronic Town Hall Project (MINERVA)*. Report to the National Science Foundation, 1972.
14. William R. Ewald, Jr., *ACCESS to Regional Policymaking*. A report to the National Science Foundation. July 27, 1973.
15. ——, *GRAPHICS for Regional Policymaking, A Preliminary Study*. A report to the National Science Foundation. August 17, 1973.
16. ——, *Hinterlands and America's Future*. A paper prepared for Resources for the Future, Inc., 1973.
17. Erving Goffman, *The Presentation of Self in Everyday Life*. Viking, 1962.
18. ——, *Interaction Ritual*, Doubleday Anchor Books, 1972.
19. William H. Gruber and Donald G. Marquis (eds.), *Factors in the Transfer of Technology*. M.I.T. Press, 1969.
20. Thomas W. Hall, "Implementation of an Interactive Conferencing System," *AFIPS Conference Proceedings*, **38** (1971), Spring Joint Computer Conference. Pp. 217-29.
21. Richard Harkness, "Communication Innovations, Urban Form and Travel Demand: Some Hypotheses and a Bibliography," *Transportation* **2** (1973), pp. 153-93.
22. ——, *Telecommunications Substitutes for Travel: A Preliminary Assessment of their Potential for Reducing Urban Transportation Costs*, Ph. D. dissertation University of Washington, 1973. Catalog No. 73-27, 662, University Microfilms, Ann Arbor, Michigan.
23. Robert Johansen, Institute for the Future, Menlo Park, Calif. Personal communication (1973).
24. Norman Johansen and Edward Ward, "Citizen Information Systems: Using Technology to Extend the Dialogue Between Citizens and Their Government," *Management Science* 19:4 (December 1972) Part 2, pp. 21-34.
25. J. C. R. Licklider, R. W. Taylor, and E. Herbert, "The Computer as a Communication Device," *International Science and Technology* **76** (April 1968), pp. 21-23.
26. Jack J. Peterson and Sandra A. Veit, *Survey of Computer Networks*, MITRE Corporation, September 1971.
27. D. Sam Scheele, Reality Construction as a Product of Delphi Interaction," Chapter II C.

28. D. Sam Scheele, Vincent de Santi, and Eduard Glaaser, *GENIE: Government Executives' Normative Information Expediter*. Report prepared for the Office of Economic Opportunity and the Governor's Office, State of Wisconsin. Singer Research Corporation, 1971.
29. D. Sam Scheele, Social Engineering Technology, Los Angeles. Personal communication.
30. Warren Schmidt (ed.), *Organizational Frontiers and Human Values*. Wadsworth Publishing Co., 1970.
31. Donald A. Schon, *The Displacement of Concepts*, London: Tavistock Publications, 1963.
32. ——, *Technology and Change, the New Heraclitus*. Norton, 1966.
33. James Schuyler, Computer Aids to Teaching Project, Northwestern University. Personal communication (1973).
34. *Synergy/Access*. No. 1, August 1973, Wes Thomas (ed.). Twenty-First Century Media, Inc., 606 Fifth Avenue, East Northport, N. Y. 11731, p. 5.
35. Gordon Thompson, "Moloch or Aquarius," *THE*, No. 4. (Bell Northern Research of Canada), February 1970.
36. ——, Bell Northern Research, Ltd., Ottawa, Ontario, Canada. Personal communication (1973).
37. Victor A. Thompson, *Bureaucracy and Innovation*. University of Alabama Press, 1969.
38. Murray Turoff, "The Design for a Policy Delphi," *Technological Forecasting and Social Change* **2**:2 (1970), pp. 149-72.
39. ——, "Delphi and Its Potential Impact on Information Systems," *AFIPS Conference Proceedings*, Vol **39** (971) Fall Joint Computer Conference. pp. 317-26.
40. ——, "Conferencing via Computer," *Information Networks Conference*, NEREM, 1973. Pp. 194-97.
41. ——, "Delphi Conferencing: Computer-Based Conferencing with Anonymity," *Technological Forecasting and Social Change* **3** (1973), pp. 159-204.
42. ——, "Human Communication via Data Networks," *Ekistics* 35:211 (June 1973), pp. 337-41.
43. ——, Department of Computer Sciences, New Jersey Institute of Technology, Newark, N. J. Personal communication (1973).
44. ——, *Potential Applications of Computerized Conferencing in Developing Countries,* Rome Special Conference on Future Research, 1972. (appears in *Ekistics*, Vol. 38, Number 225, August 1974).
45. Stuart Umpleby, "Structuring Information for a Computer-Based Communications Medium," *AFIPS Conference Proceedings* **39** (1971) Fall Joint Computer Conference, pp. 337-50.
46. Melvin Webber, *Societal Contexts of Transportation and Communication*. Working Paper No. 220, Institute of Urban Studies, Berkeley. November 1973.
47. ——, "The Urban Place and the Nonplace Urban Realm," in *Explorations into Urban Structure*. University of Pennsylvania Press, 1963, pp. 79ff.
48. R. H. Wilcox and R. A. Kupperman, "EMISARI: An On-Line Management System in a Dynamic Environment," in Winkler (ref. 49).
49. Stanley Winker (ed.), *Computer Communications: Impacts and Implications*. Proceedings of the First International Conference on Computer Communication, Washington, D. C., IEEE, 1972.

VII. C. Group Communication through Electronic Media* Fundamental Choices and Social Effects

ROBERT JOHANSEN, RICHARD H. MILLER, and JACQUES VALLEE

How will a given medium of communication affect the way in which groups of people communicate? What are the most promising near-future directions for research considering this question?

Our own incentive for exploring these issues began with a more specific concern about the probable social effects (and utility) of communication through a computerized conferencing system called FORUM, which is now under development at the Institute for the Future. The starting point for our inquiry was to consider computerized conferencing as a medium of communication, just as the telephone and face-to-face conversations may be considered media of communication. Not surprisingly, the criteria for evaluation of a medium of communication typically involve (either consciously or unconsciously) comparison with other media. Since the medium most familiar to the majority of us is face-to-face communication, there is an inherent tendency for this to become the standard of judgment.

One needs to exhibit great care when doing this, since computerized conferencing and other telecommunications media are not necessarily surrogates for face-to-face communications. It seems more likely that each medium will have its own inherent characteristics which should not be expected to mimic face-to-face patterns. On the other hand, comparison with face-to-face communication is often crucial in order to understand a new medium. While most of the work in this area to date has been applied to conferencing media such as TV and voice systems, some of it has direct bearing on any future work in the computerized conferencing area. For instance, Anna Casey-Stahmer and Dean Havron developed a mathematical ratio to aid in assessing teleconferencing systems [1]. In this study, each system under assessment involved groups of people gathered at stations and communicating with groups at other stations.

---

*Reprinted by arrangement with original publisher from *Educational Technology* Magazine, August, 1974, Vol. 14, No. 8. Copyright © 1974 Educational Technology Publications, Englewood Cliffs, New Jersey.

This is a paper from the FORUM project at the Institute for the Future. FORUM is being developed and evaluated by a research team composed of Roy Amara, Hubert Lipinski, Ann McCown, Vicki Wilmeth, Thad Wilson, and the authors of this paper. This research is supported by the National Science Foundation under Grant GJ-35 326X.

An analysis was made of the amount of electronically mediated communication between stations and the communication within the face-to-face groups at each site. The point was to look at the ratio of between-station communication over within-station communication. This ratio has offered interesting data in this case, but needs to be used with great care in order to avoid the assumption that face-to-face communication is the ideal medium.

In turning to the literature of group communication, however, one does not readily discover general principles or procedures which are easily adopted as "standard." Instead we find a very scattered literature—often parochial and littered with jargon—which is impressive in its lack of coordination. Individual researchers (and often "schools" of thought) provide rich and provocative information within strikingly narrow frames of reference. Also, the social dynamics which have been explored in these research efforts are concentrated almost exclusively on face-to-face communication. As evidence, one finds only six entries dealing with media other than face-to-face among the 2,699-entry, generally acclaimed bibliography on small-group research by McGrath and Altman [2]. Beyond the literature of face-to-face group process research, very little has been done which attempts to apply derived principles of face-to-face group communication to other media.

In 1963 Alex Bavelas offered this summary appraisal of the research in face-to-face communication as it relates to research in electronically mediated group communication:

> In consequence, the findings are, in most cases, only remotely related to teleconferencing. The significant contribution of this work lies instead in the methods and techniques of quantitative study that have been developed, and in general hypotheses about social process in terms of which specific propositions relating to teleconferencing may be formulated [3].

Bavelas went on to say: "It appears that published information bearing directly on teleconferencing is practically nonexistent" [4]. Thus it is clear that most of the directly relevant research has been done within the last ten years—with the added comment that Bavelas' observation has not changed radically since the time that he made it.

Certainly the literature of group process is broad and provocative, and the potential for transfers into communication research seems real—though obviously complicated by multiple factors. Alex Reid, while recognizing this fact, offers an optimistic view of near-future possibilities: "There seems every opportunity for a fruitful transfer of both theory and experimental method from social psychology to telecommunications engineering, a transfer that will be particularly valuable as the telecommunications system moves away from simple one-to-one voice communication toward more sophisticated visual and multi-person systems" [5].

One of the first research efforts which considered teleconferencing directly was done under the auspices of the Institute for Defense Analysis (IDA), beginning in the early 1960s. The focus of this work was on the possible use of such communications media as telephone, teletypewriter, and/or television in international relations. Of special interest was the potential for using teleconferencing in crisis-negotiation situations. This series of studies, which has only recently been released to the general public, can be considered as a kind of methodological forerunner of the work which is described in this article.

The theoretical work done by the IDA is still instructive for research design involving group communication. Figure 1 shows the key elements identified in these studies. Their approach involved simulated crises in laboratory situations and field trials using different combinations of media. Since another of the purposes of the IDA studies was to "assemble and review information relevant to teleconferences and teleconference research and to draw implications therefrom for the long-range teleconference research program" [6], this seems an excellent starting point in surveying the current situation.

The IDA studies offer findings which are uniquely geared to international crisis situations, but which can also be generalized to some degree. For instance, the speed of communication offered by telemedia was thought to be an advantage, but was later found to have negative effects in those negotiation situations where participants needed time to think before responding [7]. Also, media which encourage rigid behavior patterns (e.g., the teletypewriter, where

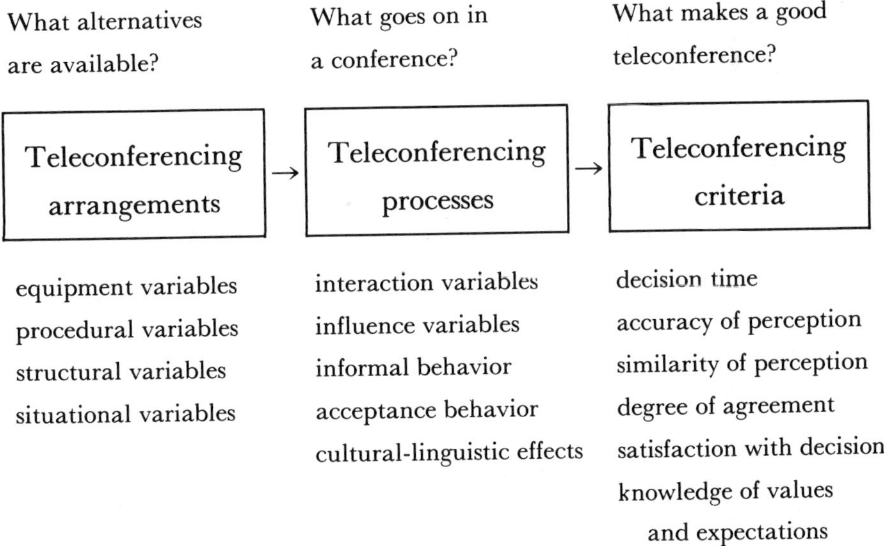

Fig. 1. Typical predictive system for the experimental study of teleconferencing (from IDA teleconferencing studies [9]).

all communication is via print) were found to increase the need for parallel channels to provide for informal and symbolic exchanges. "It appears, then, that historical biases, formal and stilted language, and a disposition to defend one's position to great lengths are characteristics that are potential drawbacks to effective communications by written message [8]. However, the IDA simulation of a teletype message forwarding system does not allow the extrapolation of these conclusions to the computerized conferencing systems which permit more than two people to interact simultaneously.

Current Research Approaches

The social research which is currently being done with personal communications media can be divided into three, sometimes overlapping, categories: laboratory experiments; field trials; survey research.

Each of the approaches has certain benefits and weaknesses, some of which we shall try to point out. Also, such an arbitrary division seems necessary as a first step in defining possibilities for comparing results and interpreting the research findings which will become increasingly common over the next few years.

Laboratory Experiments: The most classic of these research approaches arises out of the traditions of experimental psychology. The goal here is the control and manipulation of certain key elements (independent variables), while monitoring the resultant effect on other elements (dependent variables). Because of the problems in monitoring the many variables surrounding a social situation, laboratories are used to establish a controllable environment. From this point, attempts are made to design the laboratory in such a way that it replicates (or at least approximates) the "real world."

In the case of communications research, the problems of control have been magnified. In even the most "simple" instances of interpersonal communication, multiple complexities are always present. A researcher must attempt to isolate the effects of a communications medium from the interrelated effects of such things as group dynamics, personal attitudes, and topical content of the communication. In a situation such as this there is the constant danger of simplifying the "real world" to meet the limitations of the laboratory.

Bell Laboratories has produced much work in communications and information theory [10] and this work has continued, using variations in experimental methodology. The work at Bell Labs is frequently tied to the development of new communications technology. There is, however, a renewed interest in the exploration of basic communications processes—apart from the application of a specific technology. For instance, ongoing work at Bell Labs is now concentrating on the behavioral dimensions of two-person, face-to-face communication, with an eventual goal being the development of a procedure for comparing and evaluating different media of communication. This work is strengthened by interesting applications of statistical techniques (particularly multidimensional scaling) to the unique characteristics of interpersonal communication through electronic media [11].

In 1970 the Communications Studies Group (CSG) was founded in London, England, with direct support from the Civil Service Department and the Post Office [12]. CSG has now become a major center of telecommunications research, using a style begun with a base in laboratory experiments and mathematical modeling. CSG has also begun exploration of attitudes toward various communications media—a dimension of communications research which has been largely neglected. The experiments done by CSG are, of course, considered within the context of actual problems and planning in the English government.

Several general conclusions from CSG's Final Report (September 1973, Vol. 1) provide an overview of some current results:

- Two criteria characterize tasks whose outcome is likely to be affected by medium of communication: the task must necessitate interaction and must be such that personal relationships are relevant to the outcome. Thus communication involving negotiation or interpersonal relations between the participants forms areas of sensitivity. Information exchange and problem solving were two important purposes for which the outcome, in two-person tasks, was found to be insensitive to variation in the medium of communication.

- Attitudes toward media are dependent on the tasks for which they are being used.

- A substantial number of business meetings which now occur face-to-face could be conducted effectively by some kind of group telemedia (usually not merely the telephone) [13].

As can be seen, the conclusions are quite general at this point, but CSG also has a growing amount of data on particular experimental situations.

Alphonse Chapanis at the Johns Hopkins University has been doing laboratory research "aimed at discovering principles of human communication that may be useful in the design of conversational computers of the future" [14]. Though his facilities are quite limited at this point, Professor Chapanis has added much toward identification of dependent variables for evaluating communication patterns in laboratory settings. To date, his experiments have been exclusively centered on two-person communication and experimental tasks which have a defined solution. His plans, however, are to move into group experiments with more open-ended experimental tasks.

Chapanis has done a series of laboratory experiments comparing audio, handwritten, teletypewritten, and face-to-face communication. The tasks were carefully selected to be credible "real-world" situations, but the two communicators were always identified as "seeker" and "source." Thus the experiments actually use information-seeking and information-giving tasks [15]. In this test environment, the results showed that the oral media (audio and face-to-face) were clearly much faster for solving the test problems than were handwriting and typewriting. Much more information could be passed back and forth in the oral modes. (Also, a general finding has been that level of typing skill per se has little effect on the generally slower communication time.) These conclusions suggest that nonverbal communication media offer a more restricted environment than do the more common oral media. Replication studies which broaden the group and task components of the Chapanis work seem crucial for the near future, however, if these results are to be accepted on a more general level.

It should be noted that laboratory experiments involving communication process have typically concentrated on two-person communication, with clearly defined tasks. (The limitations of this approach are discussed later in this article.) Thus, time to solution of the task is often a major criterion. Also, the inherent problems of simulating the "real world" in a laboratory are especially intense when trying to facilitate a "natural" communication process in an artificial environment. There is rarely any continuity in this environment, meaning there is no prior communication or follow-up to the actual communication situation being evaluated. These factors raise validity questions about the experimental approach, though the approach certainly has its appealing aspects (e.g., higher degree of control and ability to isolate key factors).

Field Trials: In order to clarify the distinction between laboratory and field experiments, it seems most appropriate to touch briefly upon the theoretical characteristics of a quasi experiment:

> There are many natural social settings in which the research person can introduce something like experimental design into his scheduling of data collection procedures (e.g., the *when* and *to whom* of measurement), even though he lacks the full control over the scheduling of experimental stimuli (the *when* and *to whom* of exposure and the ability to randomize exposures) which make a true experiment possible. Collectively, such situations can be regarded as quasi-experimental designs [16]."

For our purposes here, field experiments are defined as explorations of actual "real-world" situations with a minimum of experimental manipulation. In this sense they are quasi-experiments, though considerations such as randomized sampling are usually not involved. Thus, in general, some of the techniques of the laboratory are applied under less controlled circumstances.

Such a field experiment in electronically mediated group communications was performed at Carleton University (Ottawa, Canada) under the auspices of the Department of Communications, Canada. Jay Weston and Christian

Kristen were the principal investigators in this exploratory attempt at developing "appropriate methodologies and measures for evaluating the behavioural effects and effectiveness of broadly defined teleconferencing systems" [17]. The Weston-Kristen effort involved direct comparison of three communications media (face-to-face, mediated video plus graphic video, and mediated audio plus graphic video), which were used as a basic part of the pedagogy in a human communications course. Thus, the experiment was actually done as part of the students' normal academic program, as they participated in the group-conferencing sessions. The data were then gathered from self-reporting questionnaires, analysis of verbatim transcripts, and analysis of split-screen videotapes of the sessions. Of course, the techniques for performing these analyses are in some cases rather undeveloped and exploratory in themselves (e.g., content analysis of transcripts and videotapes). However, our research has revealed very few comparable efforts at analysis of group communication through alternate media. Thus this effort should become an important prototype.

Our own work at the Institute for the Future has been moving toward a field experiments model in the analysis of the social effects of computerized conferencing [18]. These experiments take a somewhat different tack, since they begin with the task of developmental work on this new medium of communication—while still striving for established criteria which can be used to compare computer teleconferencing with other media. With a developing medium, research problems are further complicated, since the results will almost certainly vary as the characteristics of the medium evolve. (In fact, the results of tests will actually influence this evolution.) Also, one of our supporting grants comes from the National Science Foundation to explore the potential for using computerized conferencing to improve interaction among experts. Not surprisingly, it is difficult to explore "expert interaction" in laboratories, since one of the major characteristics of "expertness" is the availability of personal resources (files, library, etc.)—even assuming that we could convince experts to come into a laboratory setting. Thus the experimental problems surrounding the development of new media add greatly to the already pressing problems of group research using electronic media.

The research design for the above project began with the gathering of folklore about the new medium, since we really didn't even know what questions to ask without biasing reactions. Then the approach gradually became more systematic, working toward the field experiments model. Thus our procedure moved from gathering anecdotal information to field experiments with defined user groups and systematic collection of results using multiple measures of effectiveness—primarily user reaction and attitudes, content analysis where appropriate, and analysis of communications patterns [19]. (See Fig. 2.)

Another example of the field experiment approach is that applied by the

MEASURES OBTAINED BY:

MEASURES APPLIED TO:	AUTOMATION	HUMAN/COMPUTER COMBINATION	HUMAN ANALYSIS
COMPUTER	CPU* usage/conference Connect time/CPU usage Connect time/conference Disk access Cost in CPU, core, disk/ conferee System response time		User evaluation of system operation (e.g., interviews, attitude scales, etc.)
INDIVIDUAL	Inputs/conferee Errors/conferee Commands/conferee Editing commands/conferee User response times Length of comments # of private messages sent and received # of anonymous messages Amount of connect time/ conferee Sequence of errors Words per entry		Self-evaluation of FORUM sessions User evaluation of system operation Bales' indices from "Interaction Process Analysis" Direct access to resources Indirect access to resources Generalized status index Index of control over situation Directiveness of control Familiarity with computers Typing ability Age Expectations of FORUM Attitudes toward group Attitude toward task
GROUP	Inputs/conference Errors/conference # of each type of error Command requests/conference Editing commands/conference Time between each input Length of comments # of private messages # of anonymous messages Amount of connect time/ conference Sequence of errors	Communication patterns Analysis of content Attention profile Psycholinguistic measures Semantic differential Adjective checklist Free association Type-token ratio Rate of verbal output Tense analysis	Categorization of prior and subsequent comments Index of difficulty of communication (Bales) Index of expressive/malintegrative behavior (Bales) Index of total differentiation (Bales) Contingency analysis "Survey of Organizations" approach (CRUSK) Communication patterns Affective language Attention profile Achievement of end results References to previous inputs Psycholinguistic measures Word frequency measures Topic classification Overall subjective evaluation

Fig. 2. Categorization of measurement techniques for analysis of computer conferences (Institute for the Future).

British Columbia Telephone Company in their initial tests of the Confravision system for video conferencing. This system, given the same name as that used by the British Post Office, provides live television links between two conference rooms located in Vancouver and Victoria. Under the direction of Anders Skoe, a process for initial field experiments was developed and implemented before the system was to be made generally available. These field experiments involved selected types of users who were representative of those expected to use the system. The evaluation procedure employed the following techniques: preexperiment interview; live observation during the test with report by independent consultant; posttrial questionnaire (based on the "Teleconference User Opinion Questionnaire" developed by Communication Studies Group); questionnaire/interview some weeks after test session [20].

In this way the British Columbia Telephone Company group hoped to begin assessing the behavioral impact of the medium on the people who would be using it. Also, this behavioral goal was linked to initial technical tests of the system and longer-range socioeconomic forecasts.

An important characteristic of field trials is the ability to run tests over a longer period of time. Thus the test itself becomes more credible, since it is conducted with a sense of continuity and integration with everyday experiences. One example of such a field trial is that being conducted by the New Rural Society Project in Stamford, Connecticut. In this trial, two banks located in Stamford and New Haven are connected via an audio conferencing system, with a studio at each location. The time period for the trial is six months, and a battery of questionnaires was developed to assess both expectations of the system and reactions at various points in time.

The field trial approach, then, attempts to apply experimental research procedures to the degree which they can be effectively used in the "real world." Though the controls are limited, the goals are to gather a maximum amount of systematic data in actual group communication situations. These field trials can vary from quasi-experiments, which have a higher degree of control over variables, to the more open-ended kinds of field trials.

Survey Research: The basic tools of survey research are the old reliables (?): questionnaires and interviews. In communications research of this sort, however, unique problems are added to the routine dilemmas of the survey researcher. For instance, subjects might be asked to evaluate their needs for media which they have not yet experienced. Still, the techniques of survey research remain the most basic of the social sciences, and can be used creatively to gather information on both reactions to existing media and speculations about future needs.

A good example of survey techniques in telecommunications research is found in the study by Dean Havron and Mike Averill, which had as its goal "to develop a plan and instrument for a survey of needs of Canadian Government managers for teleconference facilities and equipment" [21]. This goal was pursued by designing a questionnaire which was administered to potential teleconference users, asking them a series of questions related to their present conferencing style and the possibilities for employing teleconferencing. The specific strategy of the questionnaire was to ask the respondents about a meeting they attended recently which required travel on the part of some group members. From this base, respondents were then asked to project what (if any) teleconferencing facilities could be used to conduct a future meeting of the same sort.

A similar project was directed by James Kollen of Bell Canada [22]. Kollen focused on existing travel patterns between Montreal, Quebec City, Ottawa, and Toronto. A questionnaire was administered to businessmen traveling between these cities, with the goal of determining why respondents felt they needed to travel (rather than use alternate communications media) to achieve their objectives. In this case, the overall purpose of the project was to explore the possibilities for substituting electronic communication for travel in interurban exchanges.

Dean Havron, with Anna E. Casey-Stahmer, was also involved in a project which used survey techniques to assess existing telecommunications systems [23]. In an effort to establish a general approach to research in teleconferencing, they developed a grid to classify teleconferencing system dimensions and characteristics through information gathered in interviews with actual users. Also involved were a series of interviews with users of four existing teleconferencing systems.

The techniques of survey research are certainly relevant to the social evaluation of communications media, and are frequently employed—even in more controlled situations such as those mentioned earlier in this article. Figure 3 summarizes the various research approaches described earlier.

Isolating Key Elements in Mediated Group Communication

The basic research approaches outlined here adopt different methodologies for approaching common research problems. In efforts to assess any communications medium, however, some comparison with other media (usually with face-to-face) is typically assumed and is certainly of central importance. Yet in order to make such comparisons, a duplicate series of techniques is less important than commonality in the research philosophy which prefaces the choice of methodology. In fact, any of the general approaches outlined in this article could be justified as valid tactics, and methods which cross these suggested categories may also be appropriate. The problem in comparison comes in the adoption of a general taxonomy which can be employed across media in various group communications situations.

RESEARCH GROUPS

DOMINANT RESEARCH STYLE	Inst. for Defense Analyses Studies	Communications Studies Group	Bell Labs (Murray Hill)	Chapanis	Human Sciences Research	Bell Canada	British Columbia Telephone	Weston/Kristen	New Rural Society	A T & T	Institute for the Future
CONTROLLED LABORATORY EXPERIMENTS	■	■	■	■							
QUASI-EXPERIMENTS (less control of variables)			■					■			■
DIRECTED FIELD TRIALS	■								■		■
OPEN - ENDED FIELD TRIALS		■				■	■		■	■	■
SURVEY RESEARCH (not arranged by researchers)		■			■	■				■	
MODELS		■									

NOTE: The classifications shown here represent our perception of the dominant research styles of various groups which have been active at some time.

No doubt there will be some overlap in the categories.

Fig. 3. Summary of approaches.

The taxonomy which we are suggesting would simply frame the most fundamental questions which must be asked in order to assess a particular communication situation involving any group. These questions must apply across media (including face-to-face) and they must be flexible enough to encourage development of a broad range of research techniques.

The existing taxonomies of group process, however, tend to be oriented toward dyadic communication (only two persons), and extrapolation from dyadic patterns to group patterns of communication seems to be questionable. The principles simply cannot be assumed to be transferable, though they certainly should not be ignored.

Our own examination of the literature reveals that perhaps the most useful taxonomy was developed in the context of the recently declassified teleconferencing studies by the Institute for Defense Analyses [24] which were mentioned earlier. These studies examined the possibilities for using various kinds of teleconferencing systems in international crisis communications of the sort typical of NATO. Thus their efforts focused on the careful evaluation and comparison of various teleconferencing systems in relation to the peculiar characteristics of crisis negotiations (e.g., translation problems, time constraints, etc.). The key variables were divided into independent variables (e.g., teleconference arrangements, dimensions of crises, and social determinants), intervening variables (e.g., interaction process), and criterion variables (e.g., group satisfaction and group outcome) [25].

Our own attempt to construct a taxonomy of this sort is similar to the IDA effort, but attempts to be more precise and does not assume the crisis orientation. Also, our taxonomy of elements does not attempt to incorporate the dynamic aspects of the communication. Through this does, of course, need to be analyzed, the taxonomy merely attempts to isolate the elements in a communication situation *before* the interpersonal personal process begins.

The taxonomy (shown in Fig. 4) is arranged to suggest a varied weighting among five key factors. None of the factors will be completely discrete. For instance, if members of a given group have a very high need to communicate, they are more likely to make appropriate efforts to gain access to the chosen medium—even if it is difficult to use or unfamiliar to them. Conversely, however, familiarity with a particulat media is likely to be a very important factor in choice of that media, unless some other factor becomes more important.

The problem in constructing a taxonomy of this sort is making it flexible enough to include all options, while still keeping its utility in terms of decisionmaking. The taxonomy should provide basic advice about such things as choice of media when a series of options are available. For instance, Rudy Bretz has constructed a taxonomy of communications media which divides them into eight classes according to the coding process which is being employed

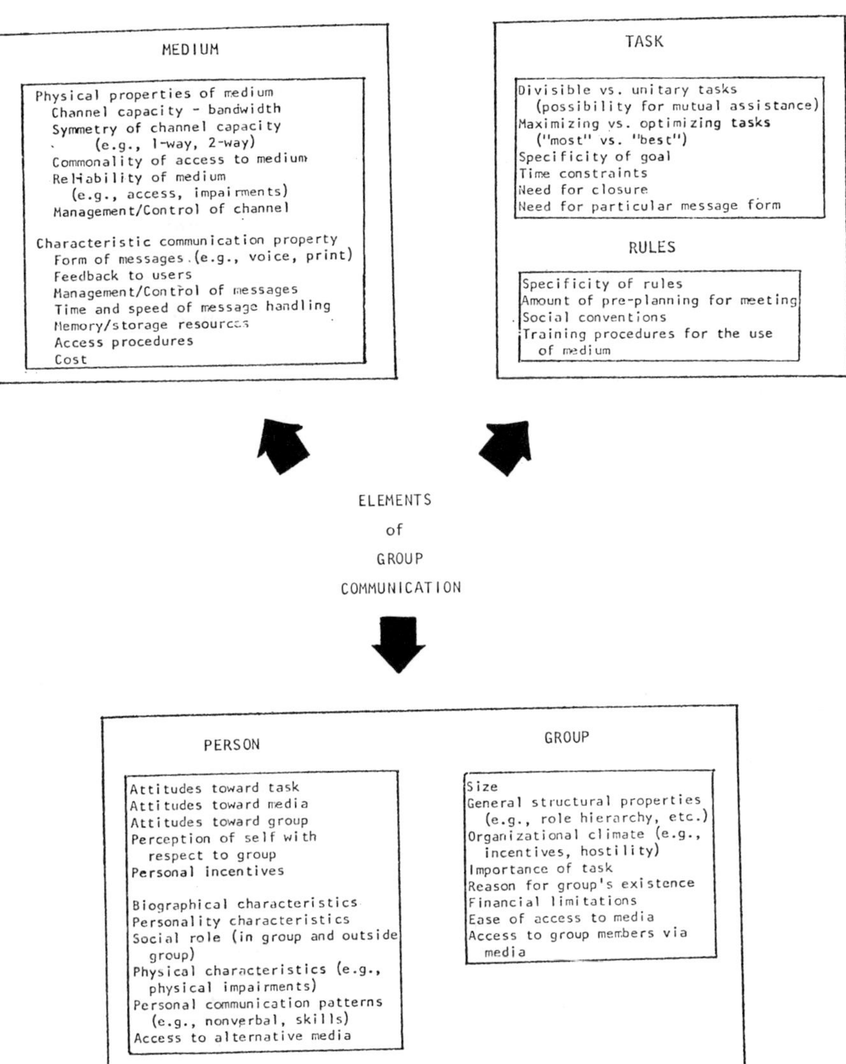

Fig. 4. Identification of elements in a group communication situation (before process actually begins).

for sending messages (e.g., audio/motion/visual, audio only, print only, etc.). Having established this taxonomy, he then traces the necessary decision points in making a choice of the simplest medium available to fill a specific instructional need [26]. When viewed within the context of the general taxonomy of group communication suggested in this article, the decision is based within the sections labeled Medium of Communication and Task. The dimensions of Individual, Group/Aggregate, and Rules for Conduct are thus not included in the Bretz taxonomy.

The question, of course, is determining the most fundamental data needed to make the best possible decision. The Bretz taxonomy moves away from generality in order to develop specific sequence of questions and typologies. This process seems inevitable, though the actual choice of fundamental factors remains debatable.

Another approach to selection of media for instruction was developed by C. Edward Cavert [27]. This is a very usable technique which operates from a matrix showing available media on one axis, with a modification of Bloom's taxonomy of learning on the other axis. (The latter is a generally accepted technique for classifying types of learning, from note memorization to ability to apply knowledge in new contexts.) Cavert then suggests that each available medium be matched to the level of learning with which it seems most appropriate, and that the overall media strategy of a school involces a distribution of available media across types of learning. Thus he has developed a simple technique for assessing the general choices of media for various needs within some overall design.

Another basic taxonomy, which differs somewhat in its approach to media, is that developed by Fred Lakin [28]. This taxonomy functions like a kind of coding sheet, which includes a general format for classification and possible exceptions. Though this approach is not yet developed for choice among media, it represents a good general format which is quite practical in doing initial classifications of media.

As was mentioned previously, the existing taxonomies of group process and most experimental data concerning communication process are based on two-person communication. Even though a number of experiments and field trials have been categorized as "group" communication, most of these tele-conference systems have dealt with the interconnection of two face-to-face groups (i. e., where an individual is in face-to-face contact with his own group and in contact via electronic media with a single distant group) [29].

With the exception of experiments involving conference telephone [30], little consideration has been given to the theoretical (i.e., in terms of taxonomy) or behavioral aspects of group interaction solely by means of a teleconferencing medium. This situation is analogous to the three-body or n-body problem in physics, in which description of two-body interactions does not reveal anything about three-body or n-body interactions.

Similarly, little research besides the communication network studies by persons such as Bavelas or Smith and Leavitt has dealt with any of the behavioral or theoretical aspects of the *n*-body problem of communications.

Future research with computer-assisted media such as FORUM demands investigation of this *n*-body problem, since the typical scenario includes several users, each connected to others by means of a terminal. The peculiarities of computer conferencing do not allow a facile transfer of theoretical or behavioral findings to other media due to differences in (*a*) time and synchronicity of interaction, and (*b*) the code (i.e., typewritten text) used for communication. Since it seems evident that the interaction of three or more individuals by means of different electronic media is affected by the limits and constraints of technology and human engineering, the problems of *n*-body communications should be a high priority subject for future investigation.

Another technological consideration in studying group interaction via electronic media is "medium memory"; that is, the capability of a medium to produce a record of an interaction which is available to the user. Information theory and communication theory have often dealt with "channels with memory," but this has been *only* on a technological/engineering basis. The various *behavioral* aspects of "medium memory" (e.g., its presence or absence, its availability to users, the extent of the record, etc.) may produce significant effects in the use and perception of teleconferencing media and thus must be examined in future research.

The question of taxonomies is frustrating, while still remaining basic to any hope for exchange of evaluations of various media. In pursuit of this goal, a small workshop was held at the Institute for the Future in February of 1974. One of its goals was to agree upon a general taxonomy for the assessment of group communication, with an emphasis on electronic media [31]. The workshop group was intentionally kept very small in order to promote maximum interaction, and thus was not inclusive of all persons doing key research in this field. However, it was hoped that this initial face-to-face workshop would build the basis for broader exchange of research approaches and results in a more continuous manner (probably using some telecommunications media).

An earlier draft of the elements shown in Fig. 2 was used as a basis for the workshop discussion of taxonomies. Our interpretation of the reactions expressed at that time is that the chart of elements identifies the basic issues—but doesn't deal with the vital factor of communication *process*. Though this was not the purpose of the chart, it is intended that the elements be used within a larger analysis of communication over time. It seemed that all the workshop participants were agreed that a sense of movement—including some analysis of before and after the actual meeting—was perhaps the most crucial need in current media research. Research methods have tended to be too static, and must somehow develop an ability to consider dynamic factors in group process.

Such a realization prompted several specific suggestions for the near future of research. One key suggestion was that our methodology must draw more heavily from fields such as anthropology, since the kind of input which is needed moves beyond the realm of traditional social-psychological research. It was also decided that a core set of questions (not a "universal questionnaire") should be developed and made available to anyone concerned with evaluating group communication situations. These questions could have an important effect on the ability to compare research results across media. Other methods of continued exchange among researchers were suggested, and specific conferences using the FORUM teleconferencing system have now begun.

Conclusion

The approaches outlined in this article all attempt various forms of systematic appraisal of media, using a range of formality. In the field of telecommunications, systematic evaluation of social effects has not been a broadly accepted practice. Rather, there has been a strong tendency toward initial expenditures on technology, with social impact analysts used only in later stages of implementation—and then only sparingly. The almost nonexistent literature on the sociology of the telephone is a prime example of what has now become a norm in the field of telecommunications. Thus, the approaches presented in this article represent a variety of techniques which—even when considered as a group—may amount to an (as yet) insignificant input to the future of telecommunications research. The range of activities described here, though, suggests that interest in social implications of telecommunications media is both growing and increasing in sophistication. As these techniques become more effective—and assuming that communications channels *within* the various research communities continue to develop—there is a hope for offering more intelligent answers to questions of media usage. The mysteries of human communication will remain dominant, but the ability to choose a complementary medium of communication seems likely to improve as the inherent "messages" of various media become known and more effectively directed.

References

1. Anna E. Casey-Stahmer and M. Dean Havron, *Planning Research in Teleconference Systems*, Human Sciences Research Institute (November 1973).
2. Joseph E. McGrath and Irwin Altman, *Small Group Research* (New York: Holt, Rinehart & Winston, 1966).
3. Alex Bavelas, *Teleconferencing*: Background Information, Research Paper P-106, Institute for Defense Analyses (1963), p. 4.
4. *Ibid.*, p. 12.
5. Alex Reid, *New Directions in Telecommunications Research*, a report prepared for the Sloan Commission on Cable Communications (June 1971).
6. Gerald Bailey, Peter Nordlie, and Frank Sistrunk, *Teleconferencing: Literature Review, Field Studies, and Working Papers*, Research Paper P-113, Institute for Defense Analyses (October 1963), p. 2.

7. *Ibid.*, p. 5.
8. W. Richard Kite and Paul C. Vity, *Teleconferencing: Effects of Communication Medium, Network and Distribution of Resources*, Study S-233, Institute for Defense Analyses.
9. Bailey, Nordlie, and Sistrunk, *op. cit.*, pp. 1–19.
10. E.g., Claude E. Shannon and Warren Weaver, *The Mathematical Theory of Communications* (Urbana, Illinois: University of Illinois Press, 1949).
11. See J. Douglas Carroll and Myron Wish, "Multidimensional Scaling: Models, Methods, and Relations to Delphi," Chapter VI, C, of this volume.
12. Communications Studies Group is a part of the Joint Unit for Planning Research of University College, London, and the London School of Economics.
13. Communications Studies Group, *Final Report, September 1973*, Vol.1, a report submitted to Management Services Division of the Civil Service Department and the Long Range Studies Division of the Post Office, p. 2.
14. Alphonse Chapanis, "The Communication of Factual Information Through Various Channels," *Information Storage and Retrieval* **9**, p. 215.
15. The classification of tasks is an essential problem in the research of media for group communication. Though no standard typology has yet been adopted by the entire field, the Communications Studies Group has developed one framework called Description and Classification of Meetings (DACOM). Eleven types of meeting tasks are identified, including tasks such as information exchange, problem-solving, conflict of interest, etc. Bell Labs at Murray Hill, New Jersey, has also done some work in this area and there is a standard typology of tasks which was developed at Harvard. Some agreement on typologies of task will be an important factor in data exchange among researchers in the future.
16. Donald T. Campbell and Julian C. Stanley, *Experimental and Quasi-Experimental Designs for Research* (Chicago: Rand McNally, 1963).
17. J. R. Weston, *Teleconferencing and Social Negentropy*, presented to International Communication Association, Montreal, Quebec (April 1973), p. 8. See also, J. R. Weston and C. Kristen, *Teleconferencing: A Comparison of Attitudes, Uncertainty and Interpersonal Atmospheres in Mediated and Face-to-Face Group Interaction*, The Social Policy and Programs Branch, Department of Communications, Ottawa, Canada (December 1973).
18. Computer teleconferencing is a medium of communication where people "talk" through typewriter keyboards connected to a central computer—often through a large computer network. The participants communicate in standard English, with the computer acting as a connecting device and storing the information in the form specified for the conference. The group may be large or small, and may be present simultaneously or whenever each individual prefers. This medium was first conceived by Dr. Murray Turoff, now at New Jersey Institute of Technology.
19. See *Group Communication through Computers,* Vols. 1 and 2, Institute for the Future (Summer, 1974). Volume 2 describes the current techniques being used in the analysis of FORUM conferences.
20. Personal correspondence from Anders Skoe, Scciological Analyst, British Columbia Telephone Company, September 25, 1973.
21. M. Dean Havron and Mike Averill, "Questionnaire and Plan for Survey of Teleconference Needs among Government Managers," prepared for the Socio-Economic Branch, Department of Communications, Canada, Contract OGR2-0303 (November 30, 1972), p. 1.
22. See James H. Kollen, "Transportation-Communication Substitutability: A Research Proposal," Bell Canada (February 1973).
23. Casey-Stahmer and Havron, *op. cit.*
24. Bailey, Nordlie, and Sistrunk, *op. cit.*
25. *Ibid.*, p. 14.
26. Rudy Bretz, *The Selection of Appropriate Communication Media for Instruction: A Guide for Designers of Air Force Technical Training Programs,* Report R-601-PR, the Rand Corporation (February 1971), pp. 30ff.
27. See C. Edward Cavert, *Procedures for the Design of Mediated Instruction*, State University of Nebraska Project (1972).
28. See Fred Lakin, *Media for a Working Group Display* (Fred Lakin, 218 Waverley St., #C, Palo Alto, California 94301).

29. For instance, the Canadian Department of Communication Audio-Graphic System, Bell Canada Conference TV, and the First National City Bank Audio-Video Teleconferencing System, all described in Casey-Stahmer and Havron, *op. cit* See also, Communications Studies Group, *The RMT Teleconference System*, P/72024/RD (1972).
30. See Communications Studies Group, *Progress in Current Experiment*, W/71132/CH (1971); Communications Studies Group, *Bargaining at Bell Laboratories*, E/71270/CH (1971); and Casey-Stahmer and Havron, *op. cit.*
31. The participants in this workshop were: Garth Jowett, Department of Communications, Canada; Lee McMahon, Bell Laboratories; Martin Elton, Communications Studies Group; Jay Weston, Carleton University; Christian Kristen, University of Montreal; Robert Johansen, Richard Miller, and Jacques Vallee, Institute for the Future; George Jull and James Craig, Communications Research Center, Canada; Mike Averill, National Film Board of Canada; Tony Niskanen, Arthur D. Little, Inc.; and Dean Havron, Human Sciences Research, Inc.

VII. D. Technology for Group Dialogue and Social Choice*

THOMAS B. SHERIDAN

Introduction

Usually the best way to discuss and resolve the choices that arise within groups of people is face-to-face and personally. For this reason, city planners and educators alike are calling for new kinds of communities for working, living, and learning, based more on familial relationships between people than on contractual relationships. When people get to know one another, conflicts have a way of being accommodated.

Beyond the circle of intimacy the problem of communication is obviously much greater; and while social issues can still be resolved more or less arbitrarily, it is more difficult to resolve them satisfactorily.

The "circle of intimacy" is constrained in its radius. One analyst has estimated that the average person in his lifetime can get to know, on a personal, face-to-face basis, only about seven hundred people—and surely one can know well only a much smaller number. The precise number is not important: the point is that it is dictated by the limitations of human behavior and is not greatly affected by urban population growth, by speed of transportation and communication, by affluence, or by any other technologically induced change in the human condition.

Indeed, these changes underlie the problem as we know it. Although the number of people with whom we have intimate face-to-face communication during a lifetime remains constant, we are in close proximity to more and more people.

We are, moreover, a great deal more dependent on one another than we used to be when American society was largely agrarian. We are all committed together in planning and paying for highways and welfare. We pollute each other's water and air. We share the risks and the costs of our military-industrial complex and the foreign policy which it serves. Technology, while aggravating the selfishly independent consumption of common resources, has made com-

*The research at M.I.T. described herein is supported on National Science Foundation Grant GT-16, "Citizen Feedback and Opinion Formulation" and a project "Citizen Involvement in Setting Goals for Education in Massachusetts" with the Massachusetts Department of Education. Reprinted, with permission, from vol. 39, Conference Proceedings, AFIPS Press, Montvale, N.J. 07645.

munications beyond the circle of intimacy both more awkward and more urgent.

Beyond the circle of intimacy, what kind of communications make sense? Surely most of us do not demand personal interactions with "all those other people." Yet in order to participate realistically in the decisions of industry and commerce, and in government programs to aid and regulate the processes which affect us intimately, we as citizens need to communicate with and understand the whole cross-section of other citizens.

Does technology help us in this? Can it help us do it better? We may now dial on the telephone practically anywhere in the world, to hear and be heard with relatively high fidelity and convenience. We may watch on our television sets news as it breaks around the world and observe our President as though he were in our living room. We can communicate individually with great flexibility; and at our own convenience we can be spectators en masse to important events.

But effective governance in a democracy requires more than this. It requires that citizens, in various ways and with respect to various public issues, can make their preferences known quickly and conveniently to those in power. We now have available two obvious channels for such "citizen feedback." First, we go to the polls roughly once a year and vote for a slate of candidates; second, we write letters to our elected representatives.

There are other channels by which we make our feelings known, of course—by purchasing power, by protest, etc. But the average citizen wields relatively little influence on his government in these latter ways. In terms of effective information transmitted per unit time, none of the presently available channels of citizen feedback rivals the flow from the centers of power outward to the citizens via television and the press.

What is it that stands in the way of using technology for greater public participation in the important compromise decisions of government, such as whether we build a certain weapon, or an S.S.T., or what taxes we should pay to fund what federal program, or where the law should draw the line which may limit one person's freedom in order to maintain that of others?

Somehow in an earlier day decisions were simpler and could involve fewer people—especially when it came to the use of technology. If the problem was to span a river and if materials and the skills were available, you went ahead and built the bridge. It would be good for everyone. Thus with other blessings of technology. There seemed little question that higher-capacity machines of production or more sophisticated weapons were inherently better. There seemed to be an infinite supply of air, water, land, minerals, and energy. Today, by contrast, every modern government policy decision is in effect a compromise—and the advantages and disadvantages have to be weighed not only in terms of their benefits and costs for the present clientele, but also for future generations. We are interdependent not only in space but in time.

Such complex resource-allocation and benefit-cost problems have been attacked by the whole gamut of mathematical and simulation tools of operations research. But these "objective" techniques ultimately depend upon subjective value criteria—which are valid only so far as there are effective communication procedures by which people can specify their values in useful form.

The Formal Social Choice Problem

The long-run prospects are bright, I think, that new technology can play a major role in bringing the citizenry together; individually or in small groups, communicating and participating in decisions, not only to help the decisionmakers but also for the purpose of educating themselves and each other. Hardware in itself is not the principal hurdle. No new breakthroughs are required. What is needed, rather, is a concerted effort in applying present technology to a very classical problem of economics and politics called "social choice"—the problem of how two or more people can communicate, compare values or preferences on a common scale, and come to a common judgment or preference ordering.

Even when we are brought together in a meeting room it is often very awkward to carry on meaningful communication due to lack of shared assumptions, fear of losing anonymity or fear of seeming inarticulate, etc. Therefore, a few excitable or most articulate persons may have the floor to themselves while others, who have equally intense feelings or depth of knowledge on the subject, may go away from the meeting having had little or no influence.

It is when we consider the electronic digital computer that the major contributions of technology to social choice and citizen feedback are foreseen. Given the computer, with a relatively simple independent data channel to each participant, one can collect individual responses from all participants and show anyone the important features of the aggregate—and do this, for practical purposes, instantaneously.

Much of technology for such a system exists today. What is needed is thoughtful design—with emphasis on how the machine and the people interact: the way questions are posed to the group participants; the design of response languages which are flexible enough so that each participant can "say" (encode) his reaction to a given question in that language, yet simple enough for the computer to read and analyze; and the design of displays which show the "interesting features" or "pertinent statistics" of the response data aggregate.

This task will require an admixture of experimental psychology and systems engineering. It will be highly empirical, in the same way that the related field of computer-aided learning is highly empirical.

The central question is, how can we establish scales of value which are mutually commensurable among different people? Many of the ancient philosophers wrote about this problem. The Englishmen Jeremy Bentham and John Stuart Mill first developed the idea of "utility" as a yardstick which could compare different kinds of things and events for the *same person*. More recently the American mathematician Von Neumann added the idea that not only is the worth of an event proportional to its utility, but that of an unanticipated event is proportional also to the probability that it will happen [1]. This simple idea created a giant step in mathematically evaluating combinations of events with differing utilities and differing probabilities—but again for a single person.

The recent history of comparing values for *different people* has been a discouraging one—primarily because of a landmark contribution by economist Kenneth Arrow [2]. He showed that, if you know how each of a set of individuals orders his preferences among alternatives, there is no procedure which is fair and will always work by which, from these data, the group as a whole may order its preferences (i.e., determine a "social choice"). In essence he made four seemingly fair and reasonable assumptions: (1) the social ordering of preferences is to be based on the individual orderings; (2) there is no "dictator" whom everyone imitates; (3) if every individual prefers alternative A to alternative B, the society will also prefer A to B; and, (4) if A and B are on the list of alternatives to be ordered, it is irrelevant how people feel about some alternative C, which is not on the list, relative to A and B. Starting from these assumptions, he showed (mathematically) that there is no single consistent procedure for ordering alternatives for the group which will always satisfy the assumptions.

A number of other theoreticians in the area have challenged Arrow's theorem in various ways, particularly through challenging the "independence of irrelevant alternatives" assumption. The point here is that things are never evaluated in a vacuum but clearly are evaluated in the context of circumstance. A further charge is a pragmatic one: while Arrow proves inconsistencies *can* occur, in the great majority of cases likely to be encountered in the real world they would not occur, and if they did they probably would be of minor significance.

There are many other complicating factors in social choice, most of which have not been, and perhaps cannot be, dealt with in the systematic manner of Arrow's "impossibility theorem"[2]. For example, there is the very fundamental question of whether the individual parties involved in a group-choice exercise will communicate their true feelings and indicate their uncertainties, or whether they will falsify their feelings so as to gain the best advantage for themselves.

Further difficulties arise when we try to include in the treatment the effects of differences among the participants along the lines of intensity-of-feelings *vs.* apathy, or knowledge *vs.* ignorance, or "extended-sympathy" *vs.* selfishness, or partial *vs.* complete truthfulness; yet these are just the features of the social-choice problem as we find it in practice.

To take as an ultimate goal the precise statement of social welfare in mathematical terms is, of course, nonsense. The differing experiences of individuals (and consequently differing assumptions) ensure that commensurability of values will never be complete. But this difficulty by no means relieves us of the obligation to seek value-commensurability and to see how far we can go in the quantitative assessment of utility. By making our values more explicit to one another we also make them more explicit to ourselves.

Potential Contributions of Electronics

Electronic media notwithstanding, none of the newer means of communication yet does what a direct face-to-face group meeting (town meeting, class bull session) does—that is, permit each participant to observe the feelings and gestures, the verbal expressions of approval or disapproval, or the apathetic silence—which may accompany any proposal or statement. As a group meeting gets larger, observation of how others feel becomes more and more difficult; and no generally available technology helps much. Telephone conference calls, for example, while permitting a number of people to speak and be heard by all, are painfully awkward and slow and permit no observation of others' reaction to any given speaker. The new Picture-Phone will eventually permit the participants in a teleconference to see one another; but experiments with an automatic system which switches everyone's screen to the person who is talking reveals that this is precisely what is not wanted—teleconferees would like most to observe the facial expressions of the various conferees who are *not* talking!

One can imagine a computer-aided feedback-and-participation system taking a variety of forms:

(1) A radio talk show or a television "issue" program may wish to enhance its audience participation by listener or viewer votes, collected from each participant and fed to a computer. Voters may be in the studio with electronic voting boxes or at home where they render their vote by calling a special telephone number. The NET "Advocates" program has demonstrated both.

(2) Public hearings or town meetings may wish to find out how the citizenry feel about proposed new legislation—who have intense feelings, who are apathetic, who are educated to the facts and who are ignorant—and correlate these responses with each other and with demographic data which participants may be asked to volunteer. Such a meeting could be held in the town assembly hall, with a simple pushbutton console wired to each seat.

(3) Several P.T.A.s or alternatively several eighth grades in the town may wish to sponsor a feed-back meeting on sex education, drugs, or some other subject where truthfulness is highly in order but anonymity may be desired. Classrooms at several different schools could be tied together by rented "dedicated" telephone lines for the duration of the session.

(4) A committee chairman or manager or salesman wishes to present some propositions and poll his committee members, sales representatives, etc., who may be stationed at telephone consoles in widely separated locations, or may be seated before special intercom consoles in their own offices (which could operate entirely independently of the telephone system).

(5) A group of technical experts might be called upon to render probability estimates about some scientific diagnosis or future event which is amenable to before-the-fact analysis. This process may be repeated, where with each repetition the distribution of estimates is revealed to all participants and possibly the participants may challenge one another. This process has been called the "Delphi Technique" after the oracle, and has been the subject of experiments by the Rand Corporation and the Institute for the Future, [3] and by the University of Illinois [4]. Their experience suggests that on successive interactions even experts tend to change their estimates on the basis of what others believe (and possible new evidence presented during the challenge period).

(6) A duly elected representative in the local, state, or national government could ask his constituency questions and receive their responses. This could be done through radio or television or alternatively could utilize a special van, equipped with a loudspeaker system, a rear-lighted projection/display device, and a number of chairs or benches which could be set up rapidly at street corners prewired with voter-response boxes and a small computer.

These examples point up one very important aspect of such citizen feedback or response-aggregation systems: that is, that they can *educate and involve* the participants without the necessity that the responses formally determine a decision. Indeed the teaching-learning function may be the most important. It demands careful attention to how questions are posed and presented, what operations are performed by the computer on the aggregated votes and what operations are left out, how the results are displayed, and what opportunity there is for further voting and recycling on the same and related questions.

Some skeptics feel that further technocratic invasion of participatory democracy should be prevented rather than facilitated—that the whole idea of the "computerized referendum" is anathema, and that the forces of repression will eventually gain control of any such system. They could be correct, for the system clearly presupposes competence and fairness in phrasing the questions and designing the alternative responses.

But my own fear is different. It is that, propelled by the increasing availability of glamorous technology and spurred on by hardware hucksters and panacea pushers, the community will be caught with its pilot experiments incomplete or never done.

The Steps in a Group Feedback Session

Seven formal steps are involved in a technologically aided interchange of views on a social-choice question:

(1) The leader states the problem, specifies the question, and describes the response alternatives from which respondents are to choose.

(2) The leader (or automated components of the system) explains what respondents must do in order to communicate their responses (including, perhaps, their degree of understanding of the question, strength of feeling, and subjective assessment of probabilities).

(3) The respondents set into their voting boxes their coded responses to the questions.

(4) The computer interrogates the voting boxes and aggregates the response data.

(5) Preselected features of this response-aggregate are displayed to all parties.

(6) The leader or respondents may request display of additional features of the response-aggregate, or may volunteer corrections or additional information.

(7) Based upon an a priori program, on previous results and/or on requests from respondents, the leader poses a new problem or question, re-starting the cycle from Step 1.

The first step is easily the most important—and also the most difficult. Clearly the participant must understand at the outset something of the background to any specific question he is asked, he must understand the question itself in nonambiguous terms, and he must understand the meaning of the answers or response alternatives he is offered. This step is essentially the same as is faced by the designer of any multiple-choice test or poll, except that there is the possibility that a much richer language of response can be made available than is usually the case in machine-graded tests. Allowed responses may include not only the selection of an alternative answer, but also an indication of intensity of feeling, estimates of the relative probability or importance of some event in comparison with a standard, specification of numbers (e.g., allowable cost) over a large range, and simple expressions of approval ("yea!") or disapproval ("boo!").

The leader may have to explain certain subtleties of voting, such as whether participants will be assumed to be voting altruistically (what I think is best for everyone) or selfishly (what I think is best for me alone, me and my family, etc.). Further, he may wish respondents to play roles other than themselves (if you were a person under certain specified circumstances, how would you vote?).

He may also wish to correlate the answers with informedness. He may do this by requesting those who do not know the answer to some test question to refrain from voting, or he can pose the knowledge test question before or after the issue question and let the computer make the correlation for him.

Ensuring the participants "play fair," own up to their uncertainties, vote as they really feel, vote altruistically if asked, and so on, is extremely difficult. Some may always regard their participation in such social interaction as an advocacy game, where the purpose is to "win for their side."

The next two steps raise the question of what equipment the voter will have for communicating his responses. At the extreme of simplicity a single on-off switch generates a response code which is easily interpreted by the computer, but limiting to the user. At the other extreme, if responses were to consist of natural English sentences typed on a conventional teletypewriter—which would certainly allow great flexibility and variety in response—the computer would have no basis for aggregating and analyzing responses on a commensurate basis (other than such procedures as counting key words). Clearly something in between is called for; for example, a voting box might consist of ten on-off switches to use in various combinations, plus one to indicate "ready," plus one "intensity" knob.

An unresolved question concerns how complex a single question can be. If the question is too simple, the responses will not be worth collecting and will provide little useful feedback. If too complex, encoding the responses will be too difficult. The ten switches of the voting box suggested above would have the potential (considering all combinations of on and off) for $2^{10} = 1024$ alternatives but that is clearly too many for the useful answers to any one question.

It is probably a good idea, for most questions, to have some response categories to indicate "understand question but am undecided among alternatives" or "understand question and protest available alternatives" or simply "don't understand the question or procedures," three quite different responses. If a respondent is being pressured by a time constraint, which may be a practical necessity to keep the process functioning smoothly, he may want to be able to say, "I don't have time to reach a decision"; this could easily be indicated if he simply fails to set the "done" switch. Some arrangement for "I object to the questions and therefore won't answer" would also be useful as a guide to subsequent operations and may also subsume some of the above "don't understand" categories. Figure 1 indicates various categories of response for a six-switch console.

The fourth step, in which the computer samples the voting boxes and stores the data, is straightforward as regards tallying the number of votes in each category and computing simple statistics. But extracting meaning from the data requires that someone should have laid down criteria for what is interesting; this might be done either prior to or during the session by a trained analyst.

It is at this point that certain perils of citizen-feedback systems arise, for the analyst could (either unwittingly or deliberately) distort the interpretation of the voting data by the criteria he selects for computer analysis and display. Though there has been much research on voting behavior and on methods of analyzing voting statistics, instantaneous feedback and recycling pose many new research challenges.

That each man's vote is equally important on each question is a bit of lore that both political scientists and politicians have long since discounted—at least in the sense that voters naturally feel more intensely about some issues than about others. One would, therefore, like to permit voters to weight their votes

Identification of self (note: if one of 1, 2, 3 not switched assume unregistered or other party; if one of 4, 5, 6 assume other or none)		1) Republican 2) Democrat 3) Independent 4) Protestant 5) Catholic 6) Jew
Expressions of feeling and experience		1) Am intensely interested 2) Am mildly interested 3) Am uninterested 4) Daily experience 5) Occasional experience 6) No experience
Four numerical categories Two administrative categories		1) Less than 10% 2) 10 to 30% 3) 30 to 60% 4) Greater than 60% 5) Don't know 6) Don't understand
Three alternatives plus Three administrative categories		1) I want plan 1 2) I want plan 2 3) I want plan 3 4) Undecided as to plans 5) Object to available plans 6) Confused by procedure
Rank ordering of three alternatives, A, B, C	first choice	1) A 2) B 3) C
	second choice	4) A 5) B 6) C
Response ɔ interpersonal communication of actors	as to	1) Miss Adams 2) Colonel Baker 3) Doctor Crank
	I	4) Agree 5) Disagree 6) Am bored

To select one of 8 on each of two questions

(dots under your answer indicate switches to be thrown)

		A	B	C	D	E	F	G	H
Question 1	1)					·	·	·	·
	2)			·	·			·	·
	3)		·		·		·		·
Question 2	4)					·	·	·	·
	5)			·	·			·	·
	6)		·		·		·		

Fig. 1. Sample categories of response for a six-switch console.

according to the intensity of their feeling. Can fair means be provided?

There are at least two methods. One long-respected procedure in government is bargaining for votes—"I'll vote with you on this issue if you vote with me on that one." But in the citizen-feedback context, negotiating such bargains does not look easy. A second procedure would be to allocate to each voter, over a set of questions, a fixed number of influence points, say 100; he would indicate the number of points he wished the computer to assign to his vote on each question, until he had used up his quota of 100 points, after which the computer would not accept his vote. (Otherwise, were votes simply weighted by an unconstrained "intensity of feeling" knob, a voter would be rather likely to set the "intensity of feeling" to a maximum and leave it there.)

A variant on the latter is a procedure developed at the University of Arizona [5] wherein a voter may assign his 100 points *either* among the ultimate choices *or among the other voters*. Provided each voter assigns *some* weight to at least one ultimate alternative an eventual alternative is selected, in some cases by a rather complex influence of trust and proxy.

Step 5, the display of significant features of the voting data, poses interesting challenges concerning how to convey distributional or statistical ideas to an unsophisticated participant, quickly and unambiguously.

The sixth step provides an opportunity for nonplanned feedback—informal exposition, challenges to the question, challenges to each other's votes, and verbal rebuttal—in other words a chance to break free of the formal constraints for a short time. This is a time when participants can seek to influence the future behavior of the leader—the questions he will ask, the response alternatives he will include, and the way he manages the session.

Experiments in Progress

Experiments to date have been designed to learn as much as possible as quickly as possible from "real" situations. Because the mode of group dialogue discussed above introduces so many new variables, it was believed not expedient to start with controlled laboratory experiments, though gradually we plan to make controlled comparisons on selected experimental conditions. But the initial emphasis has been on plunging into the "real world" and finding out "what works."

Experiments in a Semilaboratory Setting Within the University: In one set of experiments in the Man Machine Systems Laboratory at M.I.T. the group feedback system consists of fourteen hand-held consoles, each with ten on-off switches, a continuous "adjust" knob and a "done" switch. The consoles are connected by wire to a PDP-8 computer with a scope display output. Closed circuit television permits simulation of a meeting where questions are being posed and results aggregated at some distant point (e.g., a television station in another city) and where respondents may sit together in a single meeting room or may be located all at different places. Various aggregation display programs are available to the discussion leader, the simplest of which is a histogram display indicating how many people have thrown each switch. Other data reduction programs are also available, such as the one described above permitting voters to give a percentage of their votes to another voter. A variety of small group meetings, seminars and discussions have been held utilizing this equipment.

Two kinds of leadership roles have been tried. The first is where a single leader makes statements and poses questions. Here, among other things, we were concerned with whether respondents, if constrained to express themselves only in terms of the switches, can "stay with it" without too much frustration and can feel that they are part of a conversation. Thus far, for this type of meeting, we have learned the following:

(1) Questions must be stated unambiguously. We learned to appreciate the subtle ways in which natural language feedback permits clarification of questions or propositions. Often the questioner doesn't understand an ambiguity in his statement—where a natural language response from one or two persons chosen at random only for the purpose of clarifying the question is often well worth the time of others, though this by no means obviates the need to have some "I don't understand" or "I object" categories.

(2) The leader should somehow respond to the responses of the voters. If he can predicate his next question or proposition on the audience response to the last one, so much the better. Otherwise he can simply show the audience that indeed he knows how the vote on the last question turned out and freely express his surprise or other reaction. In cases where the leader seemed as though he was not as interested in the response and simply ground through a programmed series of questions, the audience quickly lost interest.

(3) Anonymity can be very important, and, if safe-guarded, permits open "discussion" in areas which otherwise would be taboo. For example, we have conducted sessions on drug use, in which students, faculty, and some total strangers quite freely indicated how often they use certain drugs and where they get them. Such discussions, led unabashedly by students (who knew what and how to phrase the key questions!) resulted in a surprising freedom of response. (We made the rule that voters had to keep their eyes on the display, not on each other's boxes, though a small voting box can easily be held close to the chest to obstruct others' view of which switches are being thrown.) It was found especially important, for this kind of topic, not to display any results until all were in.

In the same semilaboratory setting described above we have experimented with a second kind of leadership role. Here two or more people "discuss" or "act" and the audience continuously votes with "yea," "boo," "slow down and explain," "speed up and go on to another topic" type response alternatives. Voters were happy to play this less direct role but perhaps for a shorter time than in the direct-response role described above. Again it proved of great importance that the central actors indicate that they saw and were interested in how the voters voted.

Experiments with Citizen Group Meetings Using Portable Equipment: As of this writing five group meetings have been conducted in the Massachusetts towns of Stoneham, Natick Manchester, Malden, and Lowell to assist the Massachusetts Department of Education in a program of setting educational goals. In each case cross sections of interested citizens were brought together by invitation of persons in each community to "discuss educational goals." Four similar meetings were conducted with students and teachers at a high school in Newton, Massachusetts. (A similar meeting was also held in a church parlor in Newton to help the members of that church resolve an internal political crisis.) All groups ranged in size from twenty to forty, though at any one time only thirty-two could vote, since but that number of voting boxes have been built.

The portable equipment used for these meetings, held variously in church assembly halls, school classrooms and television studios, features small handheld voting boxes, each with six toggle switches, connected by wire to substations (eight boxes to a substation each of the latter containing digital counting logic) which in turn are series connected in random order to central logic and display hardware. The display regularly used to count votes is a "nixie tube" type display of the six totals (number of persons activating each of the six switches). The meeting moderator, through a three-position switch, can hold the numbers displayed at zero, set it in a free counting mode, or lock the count so that it cannot be altered. A second display, little used as yet, is a motorized bar graph to be used either to display histogram statistics or to provide a running indication of affective judgments such as "agree with speaker," "disagree," "too fast," "too slow," etc.

The typical format for these meetings was as follows. After a very brief introduction to the purpose of the meeting and the voting procedure itself several questions were asked to introduce members of the group to each other (beyond what is obvious from physical appearance), such as education, political affiliation, marital status, etc. An overhead projector has been used in most cases to ask the questions and record the answers and comments (on the gelatin transparency), since, unlike a blackboard, it need not be erased before making a permanent record. Following the introduction, the meetings proceeded through the questions, such as those illustrated by Fig. 2 and those posed by the participants themselves. The categories of "object to question" or "other" were used frequently to solicit difficulties or concerns people had with the question itself—its ambiguity, whether it was fair, etc. Asking persons who voted in prepared categories to identify themselves and state, after the fact, why they voted as they did, was part of the standard procedure. Roughly twenty questions, with discussion, can be handled in 1 1/2 hours.

After the meeting, evaluations by the participants themselves have suggested that the procedure does indeed serve to open up issues, to draw out those who would otherwise not say much, and generally to provide an enjoyable experience—in some cases for three hours' duration.

How are preschool children best prepared for school?	(as school now exists)	(as school should be)
1) lots of parental love	9	11
2) early exposure to books	2	1
3) interaction with other kids	14	8
4) by having natural wonders and esthetic delights pointed out	1	5
5) unsure	4	3
6) object	0	1

Salient comments after vote ("as school now exists" and "as school should be" not part of question then): One man object to "pointed out" in 4), as it emphasized "instruction" rather than "learning." Discussion on this point. Someone else wanted to get at "encouraging curiosity." Another claimed, "That's what question says," and another "discover natural wonders." Consensus: "leave wording as is." Then a lady violently objected that the vote would be different depending on whether voter was thinking of school as it now existed or as it should be. Others agreed. Two categories added. Above is final vote.

Student attendance should be:	
1) compulsory with firm excuse policy	11
2) compulsory with lenient excuse policy	0
3) voluntary, with students responsible for material missed	12
4) voluntary, with teachers providing all reasonable assistance to pupils who miss class	2
5) unsure	3
6) object	0

Comments centered on the feeling that some subjects require attendance more than others do. (Note the 0 vote on category 2) which is inappropriately self-contradictory.)

Fig. 2. Typical questions and responses from the citizen meetings on educational goals.

Extending the Meeting in Space and Time

The employment of such feedback techniques in conjunction with television and radio media appears quite attractive, but there are some problems.

A major problem concerns the use of telephone networks for feedback. Unfortunately telephone switching systems, as they presently work, do not easily permit some of the functions one would like. For example, one would like a telephone central computer to be able to interrogate, in rapid sequence, a large number of memory buffers (shift registers) attached to individual telephones, using only enough time for a burst of ten or so tone combinations (like touch-tone dial signaling), say about 1/2 second. Alternatively one might like to be able to call a certain number, and, in spite of a temporary busy signal, in a few seconds have the memory buffer interrogated and read over the telephone line. However, with a little investigation one finds that telephones were designed for random caller to called-party connections, with a busy signal rejecting the calling party from any further consideration and providing no easily employed mechanism for retrieving that calling party once the line is freed.

For this reason, at least for the immediate future, it appears that for a large number (much more than 1,000) to be sampled on a single telephone line in less than fifteen minutes, even for a simple count of busy signals, is not practical.

One tractable approach for the immediate future is to have groups of persons, 100 to 1,000, assembled at various locations watching television screens. Within each meeting room participants vote using hand-held consoles connected by wire to a computer, which itself communicates by telephone to the originating television studio.

Ten or more groups scattered around a city or a nation can create something approaching a valid statistical sample, if statistical validity is important, and within themselves can represent characteristic citizen groups (e.g., Berkeley students, Detroit hardhats, Iowa farmers, etc.) Such an arrangement would easily permit recycling over the national network every few minutes and within any one local meeting room some further feedback and recycling could occur which is not shared with the national network.

Cable television, because of its much higher band width, has the capability for rapid feedback from smaller groups or individuals from their individual homes. For example, even part of the 0-54 MHZ band (considered as the best prospect for return signals [6]) is more than adequate theoretically for all the cable subscribers in a large community, especially in view of time-sharing possibilities.

The above considerations are for extensions in space. We may also consider extensions in time, where a single "program" extends over hours or days and where each problem or question, once presented on television, may wait until slow telephone feedback or even mail returns of an IBM card or newspaper "issue ballot" [7], variety come in.

Development of such systems, fraught with at least as many psychological, sociological, political, and ethical problems as technological ones, will surely have to evolve on the basis of varied experiments and hard experience.

References

1. J. Von Neumann, O. Morgenstern, *Theory of Games and Economic Behavior*, Princeton University Press, Princeton, N.J., 2nd ed., 1947.
2. K. Arrow, *Social Choice and Individual Values*, John Wiley, New York, 1951.
3. N. Dalkey, O. Helmer, "An Experimental Application of the Delphi Method to the Use of Experts," *Management Science* **9** (1963).
4. C. E. Osgood, S. Umpleby, "A Computer-Based System for Exploration of Possible Systems for Mankind 2000," *Mankind 2000*, Allen and Unwin, London, pp. 346–59.
5. W. J. Mackinnon, M. K. Mackinnon, "The Decisional Design and Cyclic Cooperation of SPAN," *Behavioral Science* **14** No. 3 (May 1969), pp. 244–47.
6. "The Third Wire: Cable Communication Enters the City." Report by Foundation 70, Newton, Massachusetts, March 1971.
7. C. H. Stevens, "Citizen Feedback, the Need and the Response," *M.I.T. Technology Review*, pp. 39–45.

VII. E. Computerized Conferencing in an Educational System: A Short-Range Scenario

ROBERT JOHANSEN and JAMES A. SCHUYLER

The NUCLEUS—Annual Report for 1983

As most of you know, it was ten years ago (in 1973) that we formally began the NUCLEUS at Northwestern University, with somewhat limited goals and even more limited funding. We are happy to report that as remote computer uses have spread, the project has become an integral part of life at Northwestern, serving as a medium for many types of learning and communication. Let us turn back to the 1973 statement of purpose for NUCLEUS:

> The Northwestern University Computer-based Learning and Educational Utility System (NUCLEUS) will be designed to introduce educators and students to the computer as a tool common to all of their specialties.
>
> We deal first and foremost with the computer as a medium for presentation of new information, and second, as a utility to help with ancillary educational tasks and research.
>
> *Computer-based learning* will be explored and expanded by the NUCLEUS. It uses the computer as a "nucleus" around which educators from all specialties can gather. The teaching-computer can be used for Drill & Practice, tutorials, testing, information storage and retrieval, games, interactive conferencing or data analysis, but its most effective uses are in the classroom itself. NUCLEUS will explore new modes of dealing with educational computer technology and new interactive modes for educational computers.
>
> The *educational utility* is patterned after the concept of the computer utility; it is a set of computer programs, running on a "publicly" available computer, designed to augment the computer-based learning system; these programs are available interactively to all members of the University community. The utility includes programs of use to teachers in all disciplines, as opposed to data-analysis programs for specific research projects.
>
> NUCLEUS will attempt to be large enough to encompass hundreds of programs and lessons, yet simple enough to be used within five minutes by a new student. Foremost, it will attempt to bring together specialists from many disciplines by providing a common communication link among them.

This current (1983) report focuses specifically on one portion of the NUCLEUS—the ORACLE. ORACLE was our first attempt at using the

The Delphi Method: Techniques and Applications, Harold A. Linstone and Murray Turoff (eds.)
ISBN 0-201-04294-0; 0-201-04293-2

computer as the "common communication link" among specialists. Originally written as a computerized conferencing program, the ORACLE is now in everyday use for many other purposes, particularly those involving hybrids of computer conferencing and on-line Delphi conferencing. We will report on the evolution of ORACLE at Northwestern from 1973 to 1983, including some comments on successes and failures. Since ORACLE is now taken for granted by so many of us, it is hoped that our appraisal here can serve as a catalyst for discovering new ways to use the system more effectively.

Before describing the latest ORACLE, we should comment on the present physical state of the computer facility at Northwestern. Since 1973 the growth of "remote" computing activity has been extreme. In 1972 there were perhaps two dozen computers scattered across the Evanston and Chicago campuses of the university: primarily small minicomputers used to monitor data-gathering experiments, but including a large-scale CDC 6400 computer (called a "Super Computer" by its manufacturer at that time because of its speed and size). In 1983, however, computing power has been drastically centralized into two computer utility installations:

(1) There is a large central computer used for research, which is wired via cable to experiments taking place across the campus. In its spare time it processes "batch" computing jobs equivalent to ten times the 1973 load. It has connections to the campuses of several smaller colleges on the north shore of Lake Michigan, and to a junior college.

(2) There is a computer-based learning system, developed largely from ideas tested in the 1970s, when the University of Illinois' PLATO IV project was controlling about 2500 student data-terminals across the state. The computer-based learning system is the home of ORACLE.

The centralization of computing power in two computers was the result of the economically depressed period of the late 1960s and early 1970s, when it was found that human support requirements for a utility were far less than those of a dozen scattered computers.

In early 1976 there was a drive to locate inexpensive data terminals for our system. The PLATO IV plasma-display had brought the price down under $2,000 per device, but this was still beyond the means of many educational institutions. However, with the advent of cheaper television-technology, the administration made the decision to install terminals (at a cost of $600 each) in dormitory areas (one terminal per twenty-five students), department offices (one per three faculty members), in personal offices of administrators (one per office), board-of-trustees members (at their own expense), in study carrels in the library, and in the student union study areas. Most of the terminals were bought by the university, but some were provided through outside funding or private purchase and were often connected to time-sharing systems across the country. Some of the trustees who have installed terminals at their personal expense in their business offices use them to perform commerical computational tasks by connecting to commercial time-sharing. Increased use of computers in primary and secondary schools has helped to alleviate the uncomfortableness felt by many students and faculty members in the 1970s—our faculty is almost considering a proposal to require some computer experience of entering freshmen.

The ORACLE is a part of the computer-based learning utility. It is a computer program which connects students' data terminals to each other through the computer. This can be done in two ways: (1) students using the computer at the same time may be directly connected so that what one student types appears on the data terminals of the other students to whom he is connected, or (2) the ORACLE can set up ORACLE-groups in which "items" for consideration of the group are recorded in the computer (in the form of text) and are *later* typed for other students to see. The second is the more common of the two means of interacting.

Because of the multiplicity of interconnections, established by the computer's users themselves, ORACLE is extremely flexible. It treats messages from participants (e.g., new items to be entered for the consideration of other group members) as data to be stored for later examination. Comments entered by participants in a group are appended to the new items themselves. The ORACLE then presents a "menu" of items from a conference, and the participant asks it to retrieve the appropriate data. Thus ORACLE does not commit itself to certain topics in advance; students and faculty decide the topics, enter the items or events, and the ORACLE performs the data-handling function. Some of the areas of university life where ORACLE is now in common usage are as follows: (1) citizen sampling on current events and long-range alternative futures; (2) course and curricular evaluation; (3) committee work and long-range university planning; (4) conflict management and diagnosis; and (5) interface with computer-based learning. Each of these general areas will now be discussed in some detail.

1. Citizen Sampling on Current Events and Futures

In recent years there has been an increasing use of future-studies techniques in the operation of the university, much as systems analysis came into its own in the 1960s. Actually, the ORACLE itself offers various modifications of the Delphi technique for sampling and sharing opinions. These include options for exploring the *desirability* and *probability* of proposed events, as well as a *voting* (yes/no) option. When used in the citizen-sampling mode, ORACLE presents items to the participants, solicits judgements on desirability, probability and/or a vote (depending on the preference of the conference initiator or person who entered the new item), and then proceeds to the next item. Some of the citizen-samplings which have proven most popular in the past few years are:

"Alternative futures for the family"

"The university over the next fifty years"

"Possibilities for space travel and colonization"
"Conference on world simulation systems"
"World federations in the future"
"Expanding the ORACLE"
"Improving computer operations in the university"
"Social possibilities for the computer"

These ongoing samplings have been designed by a broad cross section of persons, many of whom have no programming experience. They are known as *public conferences* and are available to anyone who is interacting with the computer at any time. Though more long-range in focus, these conferences have prompted some fascinating dialogues which might not have otherwise occurred.

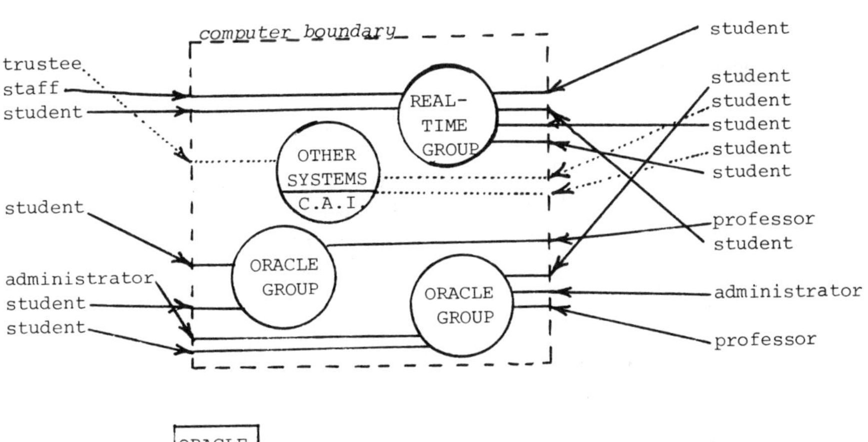

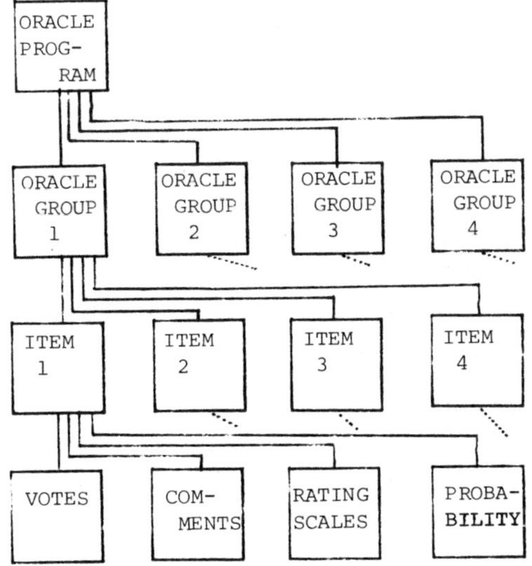

Fig. 1. ORACLE structure.

On a more immediate level, ORACLE is used by experimenters as a kind of automated suggestion box to test feelings about various ideas and current issues. This usage has proven especially helpful with regard to controversial issues where a quick (but broad) sampling of community opinion can add greatly to the potential for reaching creative solutions to pressing problems. The *Daily Northwestern* (our student newspaper) uses these immediate feedback methods to gather student opinion; often the editors will open a new public ORACLE group in the afternoon, and check the results the next morning. This is feasible because many students spend their evenings studying in the library or the student union, and will often take a study break by going to one of the data terminals and trying the latest ORACLE groups. For those who don't study, terminals are also available in the dormitories.

2. Course and Curricular Evaluation

Use of ORACLE in curricular evaluation was actually begun at the Garrett Theological Seminary (located on the Northwestern campus) at about the time the NUCLEUS began in 1973. Experimentation in curricular evaluation and feedback was done more easily at Garrett, since there were only a few hundred students involved, rather than the six thousand at Northwestern. The original motivation for the program centered around the construction of a computerized questionnaire which could give participants immediate feedback on their responses, rather than waiting months for data analysis and then getting only collective responses. Though this method raised immediate problems of reliability (especially when compared with paper-and-pencil methods), the program was eventually refined as an alternative questionnaire which students "took" from dormitory or apartment data terminals.

Another facet of this Garrett curriculum project involved a modification of the Delphi technique in which students were given a list of alternative futures for the school and asked to assess probability, desirability, and estimate of average *faculty* desirability ratings for each of the events. When these rating categories were then compared, participants were able to get some idea of how they perceived themselves in relation to the faculty—at least in regard to the cross section of futures under consideration. (For instance, an *index of perceived alienation from faculty* was computed by taking the student's own desirability ratings and comparing them with the ratings he thought the faculty would give —the difference would tell us how much he thought the faculty differed from his own desires.) Using the conferencing system, this feedback was both immediate and personal. Collective-data analysis was also performed to assess the differences between faculty and student views (the faculty also participated in the experiment). The *perceived* differences and accuracy of estimates from each group were also included. The result was an exploration of intragroup perception and stereotyped images, as well as a citizen-sampling of various alternative futures.

Emerging from these early efforts at curricular evaluation was the need for *more specific feedback* from students with regard to specific courses of study. Faculty evaluation such as this has, of course, now become generally accepted. But at that time (1972–73) such things as unlimited faculty tenure were commonly accepted traditions. Thus some caution was necessary in order to avoid approaches which might have proven overly threatening. This was partially alleviated by providing professors with the student feedback early in the course and only making the *final* data generally available. Thus professors got feedback early enough in a course to be able to revise their strategies if necessary.

The course evaluations have proven especially helpful in large section courses (more than thirty students). Feedback in such large groups had always been a problem and the availability of an immediate and constant feedback mechanism has promoted information exchange which was never before possible. It also provides (through directly recorded student comments) more open-ended *conversation*, rather than only the coded responses of a standardized testing instrument.

Now that so much of learning has moved out of the classroom (owing to television circuits, action research, independent study, etc.), this kind of feedback has become doubly important. The Chicago "TV College," run by one of the educational television stations, regularly uses the ORACLE for feedback through several remote centers, established in Chicago and suburbs. Students take the TV College courses in their homes, but report to the remote centers for testing and feedback. Up to 1975 this testing always took place by pencil and paper, at predetermined times and places; now it takes place whenever a student has finished a course, or even earlier if he desires, since the computer selects test questions at random from a rather large data base of questions presented by the teacher. Two students seldom get the same test. At each remote center there are several data terminals, connected to the computer via telephone lines. The student who finishes a test satisfactorily is often prompted to enter one of the ORACLE course-evaluation groups for the course in which he was just tested. Thus the professor gets constant feedback from his students which he could not get otherwise. Because Northwestern has adopted the computer utility standpoint, it also processes curricular feedback (at cost) for a growing number of local colleges and universities.

3. Committee Work and Long-Range Planning

Before data terminals reached their current low prices, it was uncommon to find ORACLE used for conferencing, because few participants would take the time to go to the library where the terminals were located. Those few users who owned their own terminals used ORACLE, but infrequently since there just weren't enough users to make a good ORACLE group. Now, with terminals in most offices, ORACLE is used extensively in planning the university's committee work. The University Computer Committee was the first to experiment in this direction; in 1973 they began prescreening their discussion subjects through ORACLE groups. A committee member first adds a suggested item to the ORACLE group anonymously. Later he may gather the comments made by other group members and decide to submit them with the item as an agendum. At this point the item is "voted" on by committee members (still through ORACLE). If it receives a substantial number of "yes" votes, it is placed on the agenda for the next committee meeting. Note that ORACLE does not take the place of face-to-face confrontations; it is primarily a prescreening device from which members may obtain soundings on the relative merits of various proposals before bringing them up for discussion.

The board of trustees also uses ORACLE to sound out proposals before they are actually discussed. A spin-off of this involves a special trustees' conference which has been created (it is not a public conference). Trustees enter items to be placed before the board. This was formerly done by writing letters or calling on the telephone but ORACLE has proven to be much faster and the trustee then knows that the proposal will be worded exactly as he intended.

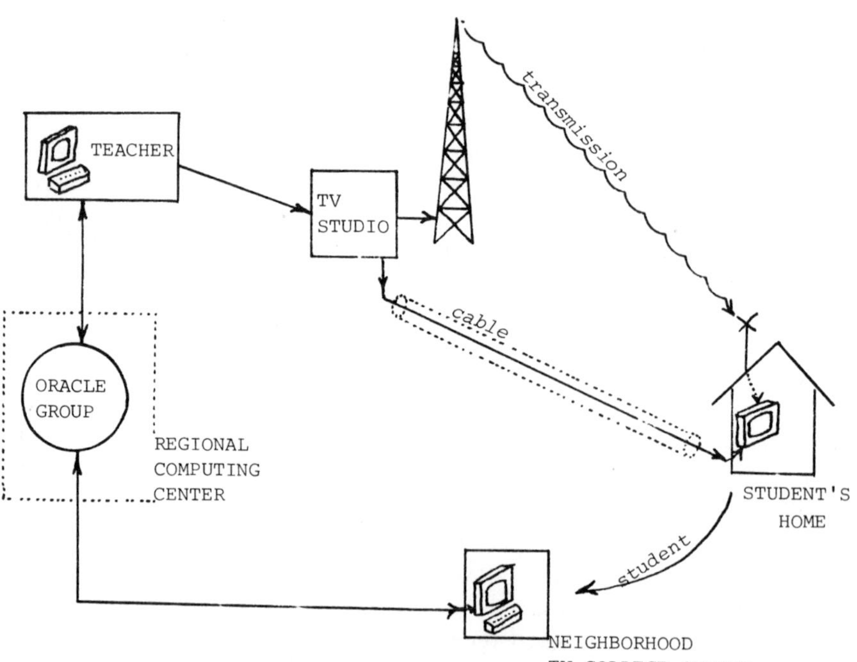

Fig. 2. A "feedback" TV College.

Network conferences with other universities around the country have also been implemented experimentally. One result of this usage has been to encourage administrators to move beyond a crisis-response orientation toward an examination of alternatives which might not have previously been part of the decisionmaking process.

4. Conflict Management and Diagnosis

The first uses of ORACLE to help professors and administrators deal with conflict began in the early 1974 school year. In this dialectic inquiry mode, ORACLE is used as a kind of blackboard on which opposing sides in a disagreement can sketch their positions. Usually the participants are "coached" in advance by a Conflict Manager as to what kinds of supportive evidence should be entered along with the "items" used to express the two groups' positions. Participants in each group enter items and evidence, as two teams, with each group working out its difficulties before entering its items. When the two ORACLE groups have been created, the tables are turned and each member of Group A is required to enter the ORACLE group B. Comments are recorded, along with desirability, probability, and perhaps votes (as in other ORACLE group conferences). We have also found especially helpful the use of "user-created" scales, in which the group itself decides exactly what wording to use in a specific question which is then asked of the participants—responses are given on a numeric scale, or as comments. Participants in conflict groups then return to their own original groups to view the comments made by the opposing group. This trading of groups continues until the two slates of items have been hammered into relatively congruent positions, at which point the human Conflict Mediators take over and attempt to resolve any remaining difficulties. ORACLE most often serves as a prelude to negotiation. Its most important function has been as an aid to understanding opposing positions and surfacing actual differences. Thus conflict may still exist, but it is more likely to be *real conflict* than simple misunderstanding. We have found that another advantage of ORACLE in conflict mediation is that the computer is viewed as an impersonal medium and, because it cannot take sides, both groups are often willing to work with it where they might be quite suspicious of a human mediator.

Building upon this potential, the conferencing system has also been used as an aid for small group communications. Limited experiments have been attempted in "T-group"-style labs, but the most intensive applications have focused on communication within various family arrangements. They first began with couples, but have recently been broadened to include extended families and other intimate living groups. In situations such as these, communication is both vital and elusive. ORACLE is used to encourage dialogue on future directions for the family, as well as to assess the accuracy of one's perception of other group members. For instance, one family member might be asked to register his opinion about particular alternatives being discussed and also to predict how each other family member will react to each idea. (These differences can then be analyzed statistically using an item-by-item index of dissimilarity, if this is desired.) Surprising misperceptions have often occurred which have served as a basis for reassessment of family relationships. Though an application such as this might have been perceived as dehumanizing in the 1970s, we have now come to see that human applications of technology can offer insights to even intimate relationships if care is taken continually to adapt the technology to human needs.

In summary, the great advantage of ORACLE in these conflict situations is that it provides an *open-ended* vehicle for expressing and examining the fundamental bases of some human relations.

5. Interface with Computer-Based Learning

The ORACLE is still a part of the NUCLEUS system, and therefore is embedded in a computer-based learning environment. The NUCLEUS is primarily used for giving students basic lessons and review work in connection with courses; there are nearly a thousand lessons (most from the early PLATO IV systems) available at any time of the day or night. Professors have now learned to coexist with computer-based learning; the best teachers rapidly shot into higher orbits, teaching their students more about the social implications of the technical subjects they were learning through the computerized lessons. Classrooms became hotbeds of discussion and criticism, and the more mundane technical problems were handled outside of class (e.g., students learned data-analysis techniques for sociology from the computer, but they came to class to discuss interesting experiments). The poor teachers, who were out of place in a classroom, moved into managerial and clerical positions in the computer-based networks, which are now employing more people than the educational systems of the late 1970s ever did.

ORACLE has been used from the start as a feedback tool for NUCLEUS. As a student finishes a lesson, he is advised that an ORACLE group exists for that lesson and is shifted into ORACLE if he elects to perform an evaluation. The results of student feedback are presented to the teacher who wrote the lesson periodically (normally once a week, since a thousand students may take a lesson during that period). However, the volume of such evaluations has grown so that there is now the necessity to prescan some of the students' comments and lump them together before presenting them to the teacher; the development of sophisticated English-language-understanding, computer-based learning systems in 1978 has made this almost feasible.

This feedback to computer-based lesson-writers is doubly important now that the State University systems of New York, Illinois, and California are bargaining for exchange rights on PLATO programs developed at their respective sites. The possibility now exists for interconnection of their total of fourteen computerized learning systems. By 1985 lessons developed in New York may be transported to computers in southern California and it will be important to be able to route feedback information to the author directly. This is just one of the more important problems these systems have yet to solve.

6. The Computer Learns to Talk

One of the truly remarkable applications of technology in the 1970s was the coupling of the computer with television. The early instructional uses of television were limited primarily to TV College (in the broadcast range) and to closed-circuit classrooms (in the closed-circuit range). Later instructional uses were developed as the cable-TV franchises began springing up around the country. Since every new home built after 1980 was required by law to have cable access as a utility, and since the signal carrying ability of cables could be divided up (or multiplexed) in several ways, it became possible to send different "programs" to each house in a city. The information-handling capabilities of the computer were harnessed to cable TV by assigning a device in each home a distinct *code number*, which was sent with the television picture for that home. Thus, the pictures were broadcast to all homes, but only the device with the proper code number would receive and activate the picture on its television screen. Ambitious experiments were undertaken to provide education in the home using computer-controlled cable TV. The computer would select the materials, generate display pictures, and send them to the appropriate homes. The students (both children and adults) would respond to the information displayed by calling the computer on their touch-tone telephone and using the numeric keys to communicate with the computer. Early systems of this sort used multiple-choice questions, or "menus" of possible responses identified by numbers. The trouble with these experiments was that they left the audio capabilities of the telephone/television communications link virtually unused.

We have recently begun experimenting with a "talking ORACLE" on a local computerized cable-TV system. Participants in the talking ORACLE see an item on the TV screen, generated by the computer, and sent to their home receiver. In some cases they may also request an oral reading of the item. They then press the "*" key on their touch-tone telephone to indicate that the computer may proceed. Our computer then plays recordings of questions, such as, "How desirable is this item by 1980? Please rate from -100 through $+100$." The participant punches the proper keys on his telephone to indicate his response and the computer continues to the next question. When the time comes for comments by the participant, the computer chooses a previously unused track on its recording disk and saves the participant's comments in digital form. Because they are recorded by a sampling technique, these comments retain the intonation and tonal quality of the participant, and can be played back for other participants. The computer can measure their length and compress them if necessary. This heightens the illusion that individuals are actually conversing with each other via the computer-controlled system. The prognosis for this system in our future looks quite good.

7. Possible Misuses of ORACLE

Though this report has been basically favorable to the current uses of ORACLE, computerized conferencing of this sort also has potential uses which the authors of this report would regard as improper. In general, this judgment is grounded in the original purposes of ORACLE, which focused on its use as a catalyst for human communication. It was never intended to be a replacement for human interaction, except in regard to the most mundane matters. Thus a primary goal is to facilitate communication, not replace it.

Still, ORACLE is only a framework for communication and can be developed in many different ways. Thus it is quite possible for it to be used as a *barrier* to separate persons from each other and encourage only automated interaction. For instance, computer conferencing could be used by decisionmaking bodies to filter out dissenting opinions and discourage the consideration of controversial issues. By requiring a high number of approving votes in conference screening sessions, many of the issues could simply be kept off the agenda for actual meetings. The consensus-encouraging potential of computer conferencing is appealing, but not if this is a false consensus, forced by the form of the interaction.

It would also .be possible for decisionmaking bodies to use ORACLE as a voting device for issues which demanded more consideration than would be possible in a hundred years of ORACLE-ing. The system contains a voting mechanism, but this is intended only for preliminary sampling, *not* for actual decisionmaking. It is also true that voting connotes majority rule and a number of decisionmaking bodies do not function along majority-rule lines. We do not want to enforce majority rule on those who have other ways of deliberation. For these processes, ORACLE serves to introduce some of the issues and opposing views; it is not a substitute for dialogue.

Perhaps predictably, the increasing availability of data terminals at low cost has promoted both positive and negative applications of computer conferencing. Thus we find a growing need to examine continually the more subtle implications of using ORACLE in communications systems. These considerations must involve more than cost and productivity analysis. Potential users should be aware (to the degree to which this is possible) of the possible effects which computer conferencing could have on their particular groups. Such information can be available only if current users are able to pool their experiences in a form which enlightens the newcomer, without limiting his perspective on potentials. A continuous ORACLE group with this "introspective" purpose has been established and is available to all users. In this arena, both criticisms and possibilities are surfaced and discussed.

Written summaries of ideas raised in these conferences are now available in the form of a newsletter distributed monthly. This newsletter reaches many nonusers of the system, and a major function of the publication is to encourage new applications. It serves as a permanent forum for debate on uses and misuses of ORACLE. The newsletter complements the public conference (where most of the ideas are initially raised) and makes the dialogue more generally available. It is our feeling that the dialogue on the effects of computer conferencing may be just beginning, since it is only now that the system can be given adequate tests.

Getting Oriented in Space/Time

This scenario has been extrapolated from work already in progress at Northwestern University. It is a projection of very possible futures, perhaps even very probable futures. The time referent is short range—approximately ten years. But most of the uncertainties are social and political, rather than technological (the computer-based learning systems LINGO and PLATO IV have already been developed, the NUCLEUS exists, the cable-TV system is being tested by MITRE corporation, and the ORACLE has been a reality for over two years).

The scenario revolves around a computerized conferencing system called ORACLE, which the authors built and operated during the 1971–72 period at Northwestern University. ORACLE operates on a Control Data 6400 computer and is written in the LINGO programming language, developed specifically for computer-based learning experiments. As a "utility" program, ORACLE serves many users at once, each in unique ways. Since it is not sensitive to the content of ORACLE groups, it provides a Delphi-conferencing framework on which experimenters may build questionnaires, citizen-sampling, feedback networks, and other types of conferences.

It is possible for persons with no programming experience to establish ORACLE groups (or conferences) on whatever subjects they may desire. ORACLE is centered around numerous "groups" of participants, each group considering a number of "items" (alternative futures). The items may change, and new items may be added by participants. The options available to initiators of ORACLE groups and to participants or "eavesdroppers" are outlined below:

Conference-related options (set by the initiator of the conference or the item):

- Anonymous conference (alternative is names recorded)
- Mandatory events for each participant to look at (How many?)

- Can new items be added by participants, or is the conference to remain as initiated?
- Additional items or event to be deleted by the initiator

Item-related options (set by the person who enters an item):

- Nonvoting items (comments only)
- Voting items
  (*a*) comments recorded
  (*b*) no comments
  (*c*) secret ballot or names recorded
- Delphi items
  (*a*) probability only (What scale?)
  (*b*) desirability only (What scale?)
  (*c*) both (What scales?)
  (*d*) by what date will each item take place?
- Other
  (*a*) design your own scales (for questionnaires, etc.)
  (*b*) Likert-type scales (agree, neutral, disagree)

Feedback-related options:

- No feedback
- Feedback after each event is completed by a participant
- Feedback only after all events have been completed by participant
- Eavesdropper (feedback only, no participation)
  (*a*) complete printout on high-speed printer of conference
  (*b*) comments only since a certain date
  (*c*) comments on a particular event only
  (*d*) comments from a particular person only
  (*e*) printout of data for an analysis program

References

Alpert, D. and Bitzer, D. L. "Advances in Computer-based Education," *Science* **167**., 1970

Johansen, Robert and Schuyler, James. "ORACLE: Computerized Conferencing in a Computer-Assisted-Instruction System," *Proceedings International Conference on Computer Communications*, October 1972.

Schuyler, James A. A Proposal for NUCLEUS; Computer-based Learning and Educational Computer Utility, Vogelback Computing Center, Northwestern University, Evanston, Illinois (1972).

Schuyler, James A. LINGO, Computer-Aided-Instruction, Vogelback Computing Center, Northwestern University, Evanston, Illinois (1972).

Sheridan, Thomas B. "Technology for Group Dialogue and Social Choice," Chapter VII, D, of this volume.

Stetten, Kenneth J. *Interactive Television Software for Cable Television Application*, MITRE Corporation, Washington, D.C. (June 1971).

VII. F. Meeting of the Council on Cybernetic Stability: A Scenario

MURRAY TUROFF

Editor's Comment: *The leader by Murray Turoff, our Advisory Board Member, uses a fictional format to focus on an issue of mounting significance for long range planners. Scenarios have been used extensively in technological forecasting as a vehicle for summarizing the results of fairly involved studies. In a sense one may look on those types of scenarios as positive ones. Much less used, but no less useful, are what one may term as the negative scenarios. This is the process of extrapolating certain trends and situations to their extreme, but often logical, conclusions. While few people would rate the probability of such scenarios actually occurring as high, their real objective is that of highlighting current problems and thereby maintaining low probabilities of occurrence.*

Subject:	Three Hundred and Seventy Third Meeting of the *Council on Social and Economic Cybernetic Stability* (CSECS-373, Extraordinary Session)
Topic:	Transcription of the Remarks of Dr. Murray Turoff (Senior Coordinator, Level One)
Key Index:	INSTABILITY, PRIORITY, ACTION
Date:	February 13, 2011
To:	Master Control Computer Records
Viewer Authorization:	*Level One Control*

Content:

My fellow members of level one and senior council members, I called this extraordinary session of the council immediately after scanning a report that came in view on my terminal. This report provides a high level of confirmation that this council has grown considerably lax in carrying out its mission. Because of this, I am submitting this report, and the supplemental material that will be gleaned from our joint efforts, to the MASTER CONTROL COMPUTERS for a full stability analysis.

However, in order to familiarize you with the seriousness of this matter, I will briefly highlight material from this report.

The subject under review in this report is a resource allocation manager (SS-553-71-7915) who, in fifteen years, has held five occupational positions

---

Dr. Turoff is associated with the Systems Evaluation Division of the Office of Emergency Preparedness, Washington, D.C. The views contained in this article do not necessarily reflect the policy or position of the Office of Emergency Preparedness.

Reprinted from *Technological Forecasting and Social Change* 4 (1972) with permission of American Elsevier.

with various organizations and has moved from an initial level ten to a level two position. The subject has never moved his place of residency and has always, of course, operated in his job function from the computer terminal setup in his home. For those of you not familiar with this occupational classification, it was the duty of this individual's occupational group to solve nonlinear analytical matrix problems involving the requirement to balance a given set of organizational resources against a required set of organizational objectives.

The individual's history file, until last year, showed absolutely no abnormalities other than a higher than average ratio of pride in his job function. At no time in the routine monitoring of his activity was the fact uncovered that he manually transcribed, from his transient view screen to permanent paper, the final matrix solution for each resource allocation problem he had ever solved. It seems he kept these as some sort of sentimental scrap book of his personal accomplishments. A review of this document uncovered some 991 solutions he had generated over his occupational period.

In January of last year, the subject was given a 200 by 300 matrix problem (60,000 elements, type number 735) which he successfully completed and for which he received proper commendation from his level one supervisor. Apparently, the instability developed when the subject was browsing in his scrap book and discovered that every entry of the solution matrix, for his recent problem, agreed exactly (to five places) with a problem he had solved ten years earlier.

When he appropriately called this to the attention of his level one supervisor he was given an explanation that it was probably a fluke of chance, but would be investigated. A special investigation team was immediately set up, but before it could begin operation the subject, without previously notifying anyone, took the unprecedented step of traveling across country to visit the supposed location of his company's major operational site. Since the subject visit was unauthorized and unscheduled, this particular site has signs and labels, at the time, intended to simulate an organization different from the one employing the subject.

Since the subject was becoming quite distraught and probably reaching a partially accurate evaluation of the truth, the special investigation team decided, on the basis of an incomplete background evaluation, to provide him with an OPTION C EXPLANATION.

The OPTION C EXPLANATION is specifically tailored for those subjects who, upon discovering their job function is in reality the playing of a game with the MASTER ORGANIZATIONAL COMPUTER, would probably not be content to join the 1.031 percent of the population (level one supervisors) responsible for creating games for the majority of the work force. In the C option the subject is given the choice either of joining the level one supervisor group, or of joining a secret research institute dedicated toward investigating the causes of the current society structure and proposing steps to correct the situation.

It was justifiably felt by the investigation team that, because of the subject's high job pride ratio, he would not be able to immediately accept the validity of the current societal makeup, and would therefore need a transition period at the research institute to adapt to the inherent logic of PSM—the PRIME STABILITY MODEL.

Except for the unusual behavior of the subject in making an unscheduled organizational site visit, there was no abnormality in the events that occurred to this point as compared with the case studies of the 2.193 percent of the population which ultimately discovers the gaming nature of current occupational functions.

As was expected, the subject, being too young for retirement and completely distraught over his recently acquired knowledge, immediately chose to join the research institute.

As you all know, these institutions are carefully programmed to insure that subjects of this type, after spending three to four years reviewing the literature provided, will arrive at the logical conclusion of which we are all aware, that is:

> "The current social-economic structure is the only possible cost-effective one exhibiting a high stability index."

But, fellow level oners, the subject, in less than a year, has come to a position and set of conclusions completely contrary to the view intended. The research unit, in this case, proved to be a complete failure. In fact, the other 83 subjects at this particular institution may have been contaminated beyond help, with views I can only refer to as emotional, illogical, and completely unscientific.

It is our most urgent concern to determine why this occurred and to reprogram all such institutions so this does not happen again. I can best alert you to the seriousness of this problem by quoting to you the subjects' evaluation of what brought about the current situation. I should remind you, once again, that the subject reached these observations in about nine months time—considerably below the two year norm for which the research institute was set up. Now, quoting from the subject's personal note file:

"In the sixties, governmental, corporate, and institutional organizations had become large and highly structured. These structures evolved to meet what appeared to be independent missions of these respective organizations. However, in the late sixties and early seventies these organizations were confronted with a new set of problems which cut across structural and even organizational divisions. Most of these problems were a product of the increasing complexity of the urban industrial environment of the time.

"Organizations found that the typical response of reorganizing their structure was insufficient to handle meaningfully the diversity of problems encountered. As a result of this, more and more emphasis was placed upon the construction of computer models of the society and various processes within it. A new generation of specialists evolved who were adept at constructing such models. A major development occurred in the late seventies when large scale efforts were started to tie many of these models together into what was the beginning of what we now know as PSM, the PRIME STABILITY MODEL.

"Through the eighties, this model (PSM) began to exercise direct control over the process of matching the supply and demand for materials and resources. It appears that by the late eighties no group of individuals could lay claim to an understanding of the operational details of PSM or to having any direct control of the model as a whole. The only constraint on the model to assuming greater control was the built-in requirement to assure jobs for the work force. As near as can be determined, the model itself introduced in the early nineties the first virtual jobs. This was done for the purpose of maintaining stability in the occupational supply and demand category. (This capability was adopted from the model's programmed educational and exercise components.) With PSM's creation of the virtual job concept, the number of actual jobs dropped to less than 10% of the total in a ten-year period. Since PSM no longer had to worry about putting people out of work, there was no constraint upon its assuming responsibility for the real jobs."

Of course, my fellow controllers, none of this is news to us. However, the subject, instead of realizing that this was a beneficial outcome in terms of the overall efficiency and stability of the society, reaches an entirely different conclusion; and I further quote:

"The result today is an artificial society. The original intent of the modeling effort was to represent in quantitative terms the functioning of the society, so that *humans* could better understand how to modify it for the general good. However, when the models were allowed to exercise the judgment or decision process the structure of the *model* now became the template for molding the society. What we have today is a society forced to conform to a computer model, and not a computer model which reflects a human society."

Fellow professionals, have you ever heard such a degree of deviant thought? The dangers of this are evident. I wish you to further note that this subject was even able to establish, from the literature available at the research center, a quasiscientific basis for his view. The types of material which must be weeded out of the research files so that this unfortunate incident does not reoccur are indicated from the following remarks by the subject:

"The primary emphasis of applying objective model structures as the vehicles to solve the social-economic-technical problems of the 70's resulted from the attempt on the part of the social sciences to emulate the physical sciences by adopting a Leibnitzian view of the world. Such a view held that there existed "valid" or true models of the world which were independent of data or the representation of the data associated with any particular problem. Counterattempts by individuals such as Churchman and Mitroff to establish other philosophical foundations for the scientific investigation of these problems, were ignored."

As you see, my colleagues, it is evident we must eliminate the literature by Churchman and Mitroff which deals with non-Leibnitzian Inquiring Systems. The additional assertions, put forth by the subject, provide a complete misrepresentation of the reasons for the introduction of large scale computer model efforts in the 70's. For, the subject goes on to say,

"Since the standard organizational structures could no longer deal with the problems facing society, many humans came to view their role in the organization as an ineffectual game they were forced to play. The primarily hierarchical nature of these organizations prevented the type of effective lateral communication, within and without the organization, which humans would have needed to deal with these problems.

"Since the Leibnitzian models and their associated Management Information Systems were data independent, they were highly responsive to the humans who began to utilize them. This computer responsiveness became a subconscious surrogate for effective human communication so that many humans began to give up attempts to communicate with other humans and turned their efforts instead to communicating with the computer."

My friends, we cannot help but admire the ingenuity of the subject's rationalizations. However, this in no way changes the totally erroneous nature of his assertions. The subject does not even hesitate to attack that segment of the society which in the seventies provided the expertise for the efforts that led to PSM. I quote:

"The most disheartening aspect of the early seventies was the inflexibility of the university-type organizations. They completely failed to break down their strong department structures, which in turn forced individuals to constrain their approaches to problems along somewhat antiquated disciplinary lines, which had no real relationship to the scope of any of the major problems of the period. The typical university was no better at fostering lateral communications among humans than were governments or corporations."

Furthermore, the subject had the audacity to suggest that there existed in the 70's an alternative approach to the utilization of computers to deal with the situation. He states that:

"The last concerted effort to correct the situation was the attempt by a small segment of the professional community to introduce the Delphi technique, as a mechanism to allow large groups to communicate meaningfully about the complex problems of the period. In essence this was an attempt to put human judgment on a par with a page of computer output. There is even one instance in the literature where this effort got to the point of automating a Delphi communication structure on a computer so that individuals spread over wide geographical distances could meaningfully communicate with each other by conducting continuous conferences over weeks or months."

As you see, it is quite clear that our original decision to include literature on the Delphi process, as an example of a technique that failed to deal with the problems of the late 20th century was a serious mistake. I strongly urge that this council recommend the purging of such items from the research literature. The subject was furthermore able to attribute the failure of the Delphi Conferencing concept to a completely different rationale than the one programmed. He claims that:

"The concerned organizations of the period went to great lengths to distinguish between communication systems as things to be regulated, and computer systems as things not to be regulated. Since the Delphi Conference capability did not fit neatly in either pigeon hole both government and industry did their best to ignore its existence, lest the established divisions between computer and communication systems be disturbed.

As a result, the ability to do conferencing via a computer was never implemented on a sufficient number of computers to have any major impact. Since man-to-computer communication was not viewed as a communication process, there was no corresponding inhibition in creating the Leibnitzian models or the associated Management Information Systems."

The subject in his notes continues to ramble on about such things as recommending a resurrection of the Delphi process as a mechanism for forming a collective human intelligence capability and using such a mechanism to take back control of society from PSM, and re-establishing the control along human values. However, all of you have been provided with a complete transcript of the subject's notes, so there is no need for me to continue with these idiocies. I think the excerpts I've quoted above are sufficient to impress upon you the seriousness of the deficiencies in our current operation.

The subject is expected to realize within the next six months that the research institute is intended as a nonsurgical reconditioning unit. At that time he will have to be committed for surgical reconditioning. However, because of the seriousness of this case, a special three month delay in reconditioning will be granted, during which time tests may be conducted on the subject's emotional characteristics to determine if we can turn up subsequent deviates of this type before they have gone so far.

It is indeed fortunate that it was decided to destroy all documentation on PSM's programs and structure a decade ago. Thus we may safely rest assured that no deviates, such as this subject, would have an opportunity to tamper with the current stability and efficiency of our society.

Fellow controllers, since we in effect hold the only real jobs available in our present society, it is our sacred trust to see that we execute our jobs with the utmost of our abilities. I am sure you will give this matter the consideration it deserves. I thank you for your attention. I've certainly enjoyed this opportunity to exchange our professional views on this matter.

Subject: Prime Control Analyses

Topic: Evaluation of CSECS Activity

Date: February 13, 2011

Viewer Authorization: Prime Control Only (PSM)

Content:

Objective One

0.92 probability that members of CSECS will be job-occupied for at least three months with current exercise

Objective Two

Virtual data for council analyses effort established to provide 0.89 probability of CSECS adopting recommendations desired by PRIME CONTROL.

Objective Three

Analysis of emotional syntax of remarks by senior coordinator indicates possible subconscious agreement with subject's position. Increase monitoring level on senior coordinator to 0.8.

Objective Four

Purging of following items and references to them from research files:

1. The Delphi conference, *Futurist* **V**, No. 2 (April 1971).
2. Immediate access and the user revisited, *Datamation Magazine* (May 1969).
3. The design of a policy Delphi, *Technological Forecasting* **2**, No. 2, 149 (1970). †
4. Delphi conferencing (i.e. computer-based conferencing with anonymity) *Technological Forecasting* **3**, No. 2, 159 (1971).
5. Implementation of an interactive conference system, *Spring Joint Computer Conference Proceedings* (1971).
6. Delphi and its potential impact on information systems, *Fall Joint Computer Conference Proceedings* (1971).
7. "Party-line" and "discussion"—computerized conference systems, International Conference on Computers and Communications, Washington, D. C. (October 1972).
8. EMISARI: An on-line management system in a dynamic environment, International Conference on Computers and Communications, Washington, D. C. (October 1972).

END OF TRANSMISSION

†*Technological Forecasting and Social Change*

VIII. Eight Basic Pitfalls: A Checklist

VIII. Eight Basic Pitfalls: A Checklist

HAROLD A. LINSTONE

"New technological knowledge creates new ignorance."
Joseph F. Coates

A significant new approach inevitably spawns criticism. If it serves a constructive purpose the criticism is healthy indeed; if it inspires literary talent (e.g., Hoos and Adams on Systems Analysis[(1)]) it can even be entertaining. The most extensive critique of Delphi available at present, the Sackman Report,[(2a)] fails both tests. Ironically this study from the RAND Corporation raises its voice in righteous indignation at the offspring of its own seed:

> "The future is far too important for the human species to be left to fortune tellers using new versions of old crystal balls. It is time for the oracle to move out and science to move in." (2a)

Science to Sackman means psychometrically trained social scientists. His tradition-bound attitude is not uncommon; it is in the same vein as the illusion that science is "objective", that only Lockean or Leibnizian inquiring systems are legitimate, and that subjective or Bayesian probability is heretical. Orthodoxy faced with new paradigms often responds with sweeping condemnations and unwitting distortions. Poorly executed applications are brought forth to censure the entire method, quotations are taken out of context, the basis for criticism is left vague, significant supportive research and new directions are ignored, and irrelevant "standards" are applied. A case in point is Sackman's comparison of Delphi with standards for psychological testing developed by the American Psychological Association: procedures designed to evaluate the testing of individuals are assumed to be meant for evaluation of opinion questionnaires.[(2b)]

Coates offers the view that Delphi is a method of last resort in dealing with extremely complex problems for which there are no adequate models. As such,

> "...one should expect very little of it compared to applicable analytical tech-

[1]I. Hoos, "Systems Analysis in Public Policy: A Critique," University of California Press, Berkeley, California, 1972. J. G. U. Adams, "You're Never Alone with Schizophrenia," *Industrial Marketing Management*, 4 (1972), p. 441, Elsevier Publishing Co., Amsterdam.

[2a]H. Sackman, "Delphi Assessment: Expert Opinion, Forecasting, and Group Process," The RAND Corporation, R-1283-PR, April 1974.

[2b]P. G. Goldschmidt, Review of Sackman Report, *Technological Forecasting and Social Change*, Vol. 7, No. 2 (1975), American Elsevier Publishing Co., New York.

The Delphi Method: Techniques and Applications, Harold A. Linstone and Murray Turoff (eds.)
ISBN 0-201-04294-0; 0-201-04293-2

> niques. One should expect a great deal of it as a technique of last resort in laying bare some crucial issues on a subject for which a last resort technique is required....
> If one believes that the Delphi technique is of value not in the search for public knowledge but in the search for public wisdom; not in the search for individual data but in the search for deliberative judgment, one can only conclude that Sackman missed the point."[2c]

In this chapter we shall attempt to set criticism of Delphi in a constructive key. In particular, we draw the problems together in a checklist of pitfalls to serve as a reminder which the Delphi designer should keep in clear view.

1. Discounting the Future[3]

If Delphi is used to elicit value judgments or other subjective opinions involving the future, a unique difficulty arises: the universal practice of intuitively applying a discount rate to the future.

A bitter lesson which every forecaster and planner learns is that the vast majority of his clientele has a very short planning horizon as well as a short memory. Most people are really only concerned with their immediate neighborhood in space and time. Occurrences which appear to be far removed from the present position are heavily discounted. Uncertainty increases as we move progressively further from the present and it is uncomfortable. Fear of the unknown generates resistance to change; in Hamlet's words, we "rather bear those ills we have than fly to others that we know not of". We shy away from considerations which might endanger our individual or group status (i.e., our economic security, social prestige, peace of mind).

Decision making becomes more difficult as uncertainty grows. First, the range of alternatives becomes large and cumbersome. Second, the possibility of accidents (low probability events) and "irrational" actions increases. Consider the large impact of the assassinations of the Archduke Franz Ferdinand and John Kennedy, the discovery of the Watergate burglars, and the decision to bomb North Vietnam in the face of massive evidence of the ineffectiveness of such a strategy in preceding wars.

By ignoring the longer time horizon we may hope that additional options or solutions to currently unsolved problems will materialize, that the need to make a decision will vanish, or that the responsibility for a decision will be in other hands. Furthermore, the Western incentive and reward system strongly favors

---

[2c] J. F. Coates, Review of Sackman Report, *Technological Forecasting and Social Change*, Vol. 7, No. 2 (1975), American Elsevier Publishing Co., New York.

[3] H. Linstone, "On Discounting the Future," *Technological Forecasting and Social Change*, Vol. 4 (1973), pp. 335–338, American Elsevier Publishing Co., New York.

discounting of the future. The politician's chances of re-election depend on his near term achievements ("But what have you done for me lately?") and long term federal debts are blissfully ignored. Americans, in particular, are nurtured on immediate material gratification through installment buying and "fly-now-pay-later" exhortations which discount future costs. Corporate management is judged by near term sales growth or profits to its stockholders and its long range planning activity is a ritual with little substance. Donald Michael quotes Ewing: "Reward systems generally favor the man who turns in a good *current* showing...salary, bonus, and promotion rewards tend to be based on this month's, this season's, this year's performance—not contributions to goals three, four, or more years off". And Michael adds that "rewarding *present* payoff makes it impossible by any known means to simultaneously reward concern with a future that would interfere with immediate payoff".[4]

The degree of discounting may well vary with the individual's cultural and social status. A person at the bottom of Maslow's human values pyramid will discount environmental pollution much more heavily than someone near the top. The poor, for whom survival is a daily challenge, are hardly going to lose much sleep over a pollution or population crisis twenty years in the future. A similar difference applies to the spatial dimension: a slum dweller worries about rats he can see, the jet set worries about depletion of wild game in distant Africa. To further complicate matters we should note that discounting operates in both directions—past as well as future. The former also impacts on forecasting in a number of ways. Disregard of the past is evident in the rare use of historical analogy (see Pitfall 4 below). In the context of Delphi we see that evaluation of subjective probability of likelihood by a Delphi participant (or any forecaster) is influenced more strongly by recent events than those in his more distant past. The phenomenon is the same as that observed by Tversky in his experiments: drivers who have just passed the scene of an accident forecast a higher likelihood of being themselves involved in an accident than those who have not had this experience (and hence they reduce their speed temporarily).[5a] Thus the individual's time perception distorts his own data base as he integrates it to develop an "intuitive" judgment or forecast.

The massive impact of such a personal discounting process is vividly illustrated by reference to the Forrester-Meadows World Dynamics model.[5b] Application of an annual discount rate equal to, or greater than, 5 percent

[4]D. Michael, "On Learning to Plan—and Planning to Learn", Jossey-Bass Publishers, San Francisco, 1973, p. 99. D. Ewing, "The Human Side of Planning: Tool or Tyrant?", Macmillan, N.Y., 1969, p. 47.

[5a]A. Tversky and D. Kahneman, "Judgment under Uncertainty: Heuristics and Biases," *Science*, Sept. 27, 1974, pp. 1124–1131.

[5b]J. W. Forrester, "World Dynamics", Wright-Allen Press, Cambridge, Mass., 1971. D. H. Meadows et al., "The Limits of Growth", Universe Books, New York, 1972.

reduces the future population and pollution crises in Meadows' "standard" case to minor significance, i.e., no dramatic worsening of the current situation is perceived by today's observer (Fig. 1). It is not surprising, therefore, that cries of crises fall on deaf ears and questioning involving future goals or values can prove exceedingly frustrating. Alternatively, use of a small elite group may lead to gross misconceptions since the differences in planning horizon between such a group and a truly representative population spectrum may cause major distortions of Delphi results.

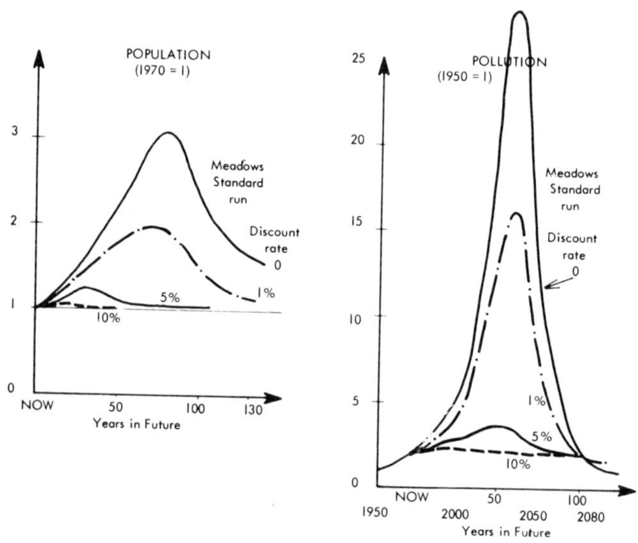

Fig. 1. The discounting effect in crisis perception.

Unfortunately this space-time discounting phenomenon is usually poorly understood by both the futures researcher and the Delphi designer. Rarely do they try to come to grips with the basic perception difficulty. Ultimately the pitfall may be avoided in two ways (schematically shown in Figure 2): (*a*) moving the distant crisis or opportunity well within the participant's current field of perception or planning horizon, or (*b*) extending the participant's field of perception or planning horizon.

Communications have been successful in drastically foreshortening the space dimension (e.g., bringing the distant Apollo landing and the Kennedy assassination events vividly into the living room). Technology has been far less

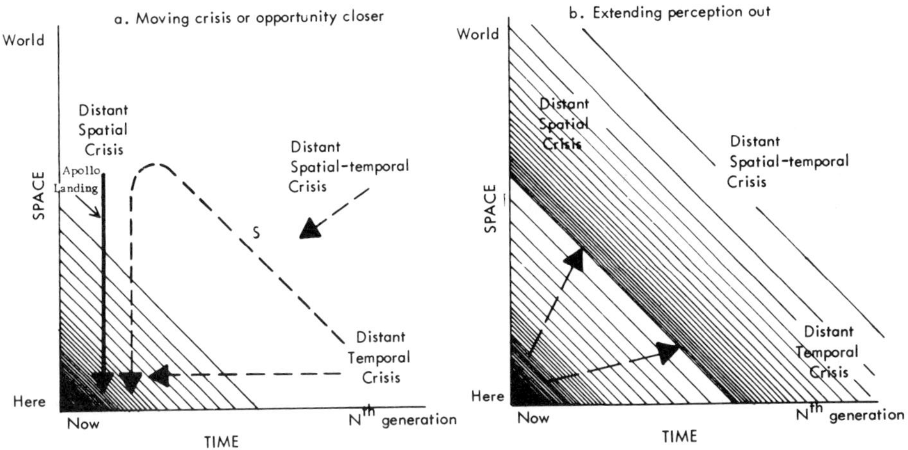

Fig. 2. Space-time perception.

effective in foreshortening the time dimension (Orson Welles' broadcast of H. G. Wells' *War of the Worlds* is a rare example). In some instances it may be possible to substitute space for time and then compress the time dimension (see arrow *S* in Fig. 2a). A future crisis or lifestyle for us may already exist somewhere on the earth today (e.g., overcrowded India, communal living in the Kibbutz). Communications can then bring such "scenarios" within our planning horizon. The arts also have great potential, as Orwell, Kafka, and Burgess/Kubrick have demonstrated.

The other approach, extension of the planning horizon, suggests the exploration of altered states of consciousness to facilitate the imaging capability of the individual. Masters and Houston have been experimenting with this concept (without the use of drugs). They point out that:

> [time distortion] is a very common phenomenon that we all experience, at least to some extent. We've all known five minutes to pass as if they were an hour or *vice versa*. ... There's no known limit to the amount of subjective experience that can occur within just an instant of clock-measured time.[6]

Following some of Scheele's concepts in Chapter II, C, above, one way of overcoming this problem is for the Delphi designer to attempt, in his design, to bring the respondents into the future by taking on or playing roles in the future —a Delphi with overtones of psychodrama and other role-playing techniques.

---

[6]Interview in *Intellectual Digest*, March 1973, p. 18. See also R. Masters and J. Houston, *Mind Games*, Viking Press, New York, 1972.

These approaches do not suggest a "quick fix." But they provide clues and may suggest some modest and practical steps to the Delphi designer.

2. The Prediction Urge

Most human beings have a strong predilection for certainty and a dislike of uncertainty. The oracle at Delphi, Nostradamus, Jeanne Dixon, and Edgar Cayce have all been popular because, in effect, they dispelled uncertainty about the future. The intellectual establishment may be startled to learn that "in 1967 the most widely read view of the year 2000 was that of [fundamentalist prophet] Edgar Cayce as popularized in Jess Stearn's best-seller, *The Sleeping Prophet*," and not Kahn and Wiener's *The Year 2000*, which was published in the same year.[7] Most people would prefer a precise prediction ("California will sink into the ocean at 6:34 A.M. on August 14, 1988") to descriptions of alternative scenarios of the future.[8] The striving for certainty has been strongly reflected in the dogmatism of the Judeo-Christian religious heritage (e.g., exclusivity, infallibility) and to some extent in the orthodoxy of much traditional science (e.g., search for the "truth," the single "best" model).

The same tendency is often seen in the interpretation and use of Delphi studies. A probability distribution of the date of occurrence of an event, together with reasons for panel disagreement, is transformed into a statement that "there is a 50-50 chance that *X* will occur in Year *Y*," or, even more blatantly, into the statement that "*X* will occur in year *Y*." Results which exhibit a high degree of convergence are often accepted, while those which involve wide differences after the final iteration are considered unusable.

Such suppression of uncertainty can mask the real significance of the Delphi results. As shown in Goldstein's paper (Chapter III) the ability to expose uncertainty and divergent views is an inherent strength of the Delphi process. Her Delphi, laying bare the uncertainties, gave a much different picture than the conventional homogenized panel report produced by another group for the same user.

Maruyama projects a societal change from a homogenistic to a heterogenistic logic.[9a] Jantsch and Ozbekhan are stressing the growing significance of normative planning. In both cases prediction is far less important than alternatives and differences in views of the future. Several of the articles in this volume have stressed that exploration of differences through Delphi is feasible. If the technique is viewed as a two-way communication system rather than a device to produce consensus it fits this evolving culture admirably.

---

[7]W. I. Thompson, *At the Edge of History*, Harper & Row, New York, p. 123.

[8]There are some clues to suggest that tolerance of uncertainty, like the degree of discounting, is correlated to an individual's position on Maslow's scale of human values.

[9a]M. Maruyama, "Symbiotization of Cultural Heterogeneity: Scientific, Epistemological, and Esthetic Bases". Paper presented to American Anthropological Association Conference, 1972.

3. The Simplification Urge

As certainty is preferred to uncertainty, so simplicity is preferred to complexity. Flushed by the triumphs of science and technology we expect to use the same reductionist approach on social/behavioral systems as we have applied to technological ones. We are convinced that the complexity of social systems can be reduced for purposes of analysis without sacrificing realism.[9b] It is often a quixotic quest. The mathematics may look elegant, the models meaningful, yet, in actuality we may be developing "superficial caricatures".

Complex systems frequently exhibit strongly counter-intuitive behavior. Forrester has noted that "intuition, judgment, and argument are not reliable guides to the consequences of an intervention into [complex social] systems behavior".[10] The secondary effects which are ignored often prove determinative (and highly undesirable) in the long run.

Holling finds that in urban systems all policies considered "have 'unexpected consequences'...The basic trap is that people conceive of a small fragment of a whole and the very best policy for that fragment can produce the reverse through interaction with other parts of the system".[11a]

In other words, everything interacts with everything and the tools of the classical hard sciences are usually inadequate. And certainly most of us cannot deal mentally with such a multitude of interactions. *Typically we forecast by taking one or a few innovations and fitting them in a mental image into an environment set in the familiar structure context of the past and present.* We do not visualize a future situation in its own holistic pattern. Cross-impact analysis (Chapter V) should be of some help, although it does not by any means eliminate the problem. Unless the components of the system are autonomous we should never expect to forecast the behavior of the whole by forecasting the behavior of its parts.

The weakness in visualizing future situations also applies to the past. Oversimplification of the future matches oversimplification of the past. Only now are attempts being made to develop complex interactive dynamic simulations—retrospective futurology—in historic human systems such as Athens and other city-states. What Forrester and Meadows have done through systems dynamics to economics, Bellman, Wilkinson, and Zadeh may do to the study of history.[10]

[9b]Reductionism inevitably leads to success in the eyes of the traditional researcher, as reflected in Heinz Von Foerster's Theorem Number One: "The more profound the problem that is ignored, the greater are the chances for fame and success". (Cf. "Responsibilities of Competence", Journal of Cybernetics, 1972, **2**, 2, p. 1)

[10]J. W. Forrester, op. cit., p. 97. Also see J. Wilkinson, R. Bellman, and R. Garaudy, "The Dynamic Programming of Human Systems", The Center for the Study of Democratic Institutions, MSS Information Corp., New York, 1973, pp. 22, 29.

[11a]C. S. Holling and M. A. Goldberg, "Resource Science Workshop", University of British Columbia, Vancouver, B.C., p. 20.

There are other psychological difficulties. An individual asked to list his preferences on a sheet of paper may well develop responses significantly different from those he would actually give in a real-life/real-time setting. His preferences in an artificial setting may indicate the characteristics of a bold risk-taker; however, in an actual situation ("when the chips are down") the same person may be quite conservative.

Intuitive procedures such as Delphi usually lean heavily on subjective probability assessments. And most human beings exhibit a tenacious tendency to simplistic misjudgments and biases in dealing with such probabilities. We are all familiar with the common preference to bet on a coin coming up Tails after a long string of Heads. Tversky[11b] notes that prior probabilities are blissfully ignored when worthless or irrelevant information is added, sample size is casually disregarded in favor of probability data, and representativeness or desirability is confused with predictability. Example: a reasonable sounding (e.g., surprise-free) scenario is judged to be "more likely" to occur than an unfamiliar one even if there is no evidence to support such a differential forecast evaluation.

The simplification urge is also evident in the frequent effort to suppress conflict (see Section 2, above). Dialectic inquiry with confrontation between conflicting theories is still relatively rare, although it may prove exceedingly fruitful for an understanding of ill-structured systems.

The means of communication present another facet of oversimplification. Language itself can be a major pitfall! Just as a linear progression of words fails to communicate a Rembrandt painting, so a panelist may be unable to communicate his views or insights by means of a concise sentence or even by diagrams. We also observe that different cultural groups communicate in diverse ways; forcing them into a conventional Delphi format may destroy their message. New means to communicate the gestalt of complex systems and to deal with patterns are needed for all aspects of Delphi.[12]

Furthermore, as W. T. Weaver has noted, "persons with different kinds of 'self-structures' (needs, attitudes, beliefs, etc.) would hold different perceptions about the present as well as the future, and thus produce different kinds of forecasts about the future."[13]

Finally we are confronted with a problem here which arises also in the use of subjective weightings and utility theory. Luce and Raiffa note that "since neither the zero nor the unit of a utility scale is determined, it is not meaningful in this theory to compare utilities between two people."[14] In an analogous

---

[11b]A. Tversky and D. Kahneman, op. cit.

[12]Techniques such as multidimensional scaling, time-lapse-oriented displays, and multimedia concepts appear to have considerable potential. See also the work of Adelson *et al.* (Chapter VI, D).

[13]W. T. Weaver, "The Delphi Forecasting Method," *Phi Delta Kappan*, January 1971, p. 270.

[14]R. D. Luce and H. Raiffa, *Games and Decisions*, J. Wiley & Sons, New York, 1957, p. 38.

manner it is dangerous to compare two persons' estimates of a future event when each views the past, present, and future in his own subjective way.

4. Illusory Expertise

In the application of Delphi to forecasting, reliance is almost invariably placed on panels of experts or specialists. As those familiar with forecasting have learned, the specialist is not necessarily the best forecaster. He focuses on a subsystem and frequently takes no account of the larger system. Reciprocating-engine experts in the 1930s forecast that propeller aircraft would be standard up to 1980. Military aircraft experts forecast a succession of manned bombers beyond the B-52 as primary weapon systems and for many years did not consider the replacement of manned bombers by missiles. These experts concentrate on a single logistic curve rather than on the envelope of a series of such curves.

A panel consisting of experts on the various body subsystems (e.g., circulation, respiration, reproduction) does not constitute expertise on human behavior and group dynamics. Similarly, a group of experts, each knowledgeable about one aspect of a complex system, does not necessarily comprise expertise about the total system.

We have seen many examples in recent years of the failure of group expertise. Economists have misjudged the impact of fiscal policies on inflation. Moynihan has noted that the warriors in the war on poverty of the 1960s mistook hypotheses for scientific answers.[15]

Technologists have consistently underestimated the complexity and cost of new aircraft and electronics equipment. S. H. Browne has shown that, based on about fifty recent military programs, cost overrun has averaged nearly 50 percent.[16]

United States military experts have for years grossly miscalculated the enemy's capability to fight in Vietnam.[17] Despite the sobering experience of World War II, the effectiveness of strategic conventional bombing was again vastly overrated. The use of Delphi to elicit judgments from such experts would only have systematically reproduced error.

There is a remarkable degree of ahistoricity in the outlook of most scientists and technologists. This is reflected in the rarity with which historical analogy is used in forecasting as well as the lack of interaction with historians. The tremendous value of such interactions is clearly seen in the superb address by

[15]D. Moynihan, *Maximum Feasible Misunderstanding: Community Action in the War on Poverty*, Free Press, Glencoe, Ill., 1968.

[16]S. H. Browne, "Cost Growth in Military Development and Production Programs," December 1971, unpublished.

[17]Cf. D. Halberstam, *The Best and the Brightest*, Random House, New York, 1972; and C. Fair, *From the Jaws of Victory*, Weidenfeld and Nicolson, London, 1971.

Lynn White on "Technology Assessment from the Stance of a Medieval Historian."[18]

Conversely, historians and archaeologists traditionally take a nonsystematic view of their subject. In the words of John Wilkinson, "most historical accounts of the year 2000 B.C. seem as implausible as the pseudo-scientific auguries and 'scenarios' that all of us read nearly everyday concerning the year 2000."[19] He cites the German historian Mommsen as an example: he was a great historian of Rome, but his account of Rome resembled his contemporary Berlin more than ancient Rome.

A dogmatic drive for conformity, the "tyranny of the majority," sometimes threatens to swamp the single maverick who may actually have better insight than the rest of the "experts" who all agree with each other. This is not unknown in science; it is, in fact, a normal situation in the arduous process of creating new paradigms, i.e., scientific revolutions. In short, a consensus of experts does not assure good judgment or superior estimates.

There are, of course, uses of Delphi for which it is obvious that no experts exist. Consider quality-of-life criteria as a subject for Delphi. The panel selections must be made to ensure representation of all relevant social and cultural groups. But the analysts who carry out the study themselves constitute a highly select group (middle class, college educated, urban, young or middle aged, mobile, etc.). They may thus find it difficult, if not impossible, to enlist the multitudes who are suspicious of intellectuals, hostile to the establishment, or fearful of disclosing their views to "investigators." Sometimes the analysts attempt to solve the problem by having other analysts play the role of the poor or the old. This inbreeding is a dangerous practice and can yield highly misleading results. In terms of the Singerian mode, we cannot hope to divorce the exercise from the psychology and experience of either respondent or designer.

Complete objectivity is an illusion in the eye of the beholder. Neither layman nor expert should be expected to be free of bias. Robert Machol recalls that more than a generation ago Morse and Kimball, the godfathers of operations research, stressed the limitations of "expert opinion" and asserted that such opinion is "nearly always unconsciously biased." Or, as Rufus Miles has put it, "where you stand depends on where you sit."

5. Sloppy Execution

The blame for this occurrence may lie with either analyst or participant. First, the analyst. Poor selection of participants (e.g., friends recommending each

[18] L. White, Jr., "Technology Assessment from the Stance of a Medieval Historian," *American Historical Review*, January 1974.

[19] J. Wilkinson, R. Bellman, and R. Garaudy, *op. cit.*, p. 20.

other for panel membership) can produce a cozy group of like-thinking individuals which excludes mavericks and becomes a vehicle for inbreeding. Poor interaction between participant and analyst can give the former the impression that he is in a poll or will receive nothing of value to him from the process. He also resents being "used" to educate the analyst. It is incumbent upon the analyst that he provide the atmosphere of a fruitful communication process among peers, that time is not wasted on obvious aspects, that subtleties in responses are appreciated and understood.

In Chapter IV we have pointed to the importance of the proper formulation of Delphi statements. Excessive specification or vagueness in the statement reduces the information produced by the respondents.

Superficial analysis of responses is a most common weakness. Agreement about a recommendation, future event, or potential decision does not disclose whether the individuals agreeing did so for the same underlying reasons. Failure to pursue these reasons can lead to dangerously false results. Group agreement can be based on differing, or even opposing, assumptions; they might also be subject to sudden changes with the passage of time. In this case, an individual attempting to utilize the results later will not be aware that the results are now invalid. It is clearly essential that the potential user be able to examine the underlying assumptions for their current validity. In particular, forecasting as a professional endeavor is defined by many practitioners not as the formulation of predictions but of conditional statements about the future, i.e., if A, then B.

Perhaps the most serious problem associated with execution for which we can offer no remedies is a basic lack of imagination by the designer. A good designer must be able to conceptualize different structures for examining the problem. He must be able to perceive how different individuals may view the same problem differently and he must develop corresponding designs which allow these individuals the opportunity to make their inputs. Whatever it is, imagination and/or creativity, it is the rare quantity we cannot formulate for you in concrete terms and which represents the artistic component of Delphi design.

Sloppy execution on the part of the respondents often takes the form of impatience to "get the job over with." Answers are hastily given without adequate thought. Obvious contradictions are permitted to creep into the responses and possible cross-impacts are ignored. But here, too, the fault may lie with the designer. He may have used little discretion and created a seemingly endless questionnaire weighted down with trivial, superficially unrelated, or repetitious statements.

We are really past the stage in the evolution of Delphi where an excuse exists for this pitfall. Most of the common errors have been amply demonstrated in a significant number of poorly conducted Delphis.

6. Optimism - Pessimism Bias

A common occurrence is a bias toward overpessimism in long-range forecasts and overoptimism in short-range forecasts.[20] Our brains are filled with experiential data. As already noted (see section 3 above) we tend to project selectively, superimposing the new elements onto the vast experiential data base. Thus we fail to recognize entirely new approaches to achieving a solution and consequently tend to be overly pessimistic. Tied too much to our experience, we suffer a failure of imagination.

For the near term, the bias is often in the opposite direction, particularly in the area of technological achievements. A new system may be at hand "in principle" when the applied research on all components is completed. The Delphi respondent assumes that system development, production, and marketing present no major stumbling blocks. The fact that complexity of a system is not a linear function of its subsystems is ignored. It is assumed that if each subsystem is made more complex by a factor of 2, then the total system increases in complexity by the same factor. But, in fact, the interactions greatly compound the complexity.

These tendencies are complicated by individual characteristics—some participants are inherently optimistic, others pessimistic.[21] However, insight into this type of bias can minimize its intrusion into the Delphi process through selective adjustments.

7. Overselling

In their enthusiasm some analysts have urged Delphi for practically every use except cure of the common cold. The first major Delphi study was published only ten years ago. Much progress has been made but improper applications have also mushroomed. We seem unable to resist faddism and it gets in the way of solid progress.

Inbreeding is one consequence of overuse. Repeated Delphi studies on the same subject quickly achieve a point of diminishing returns.[22] Either the same experts are used or the respondents are familiar with the earlier studies and regurgitate the same ideas.

A person who wants to introduce a new communication system, such as Delphi, into a group setting must also ascertain that he is not acting under false premises concerning the psychology of the potential user community.

---

[20] R. Buschmann, "Balanced Grand-Scale Forecasting," *Technological Forecasting*, **1** (1969), p. 221.

[21] See J. Martino, "The Optimism/Pessimism Consistency of Delphi Panelists," *Technological Forecasting and Social Change*, **2**, No. 2 (1970).

[22] J. M. Goodman, "Delphi and the Law of Diminishing Returns," *Technological Forecasting and Social Change*, **2**, No. 2 (1970).

Possible fallacy a: The user really wants a more effective and different system than he now employs.

Do the user's statements reflect mere lip service to progress? Does he know what benefits to expect from Delphi other than the "prestige" of having done it?

The typical successful executive experienced in running conferences which culminate in decisions may expect the same result from a Delphi. The conscientious Delphi manager thus senses great pressure to assure a consensus to satisfy such a client.

Anonymity may be a disadvantage in certain organizational settings. In diplomatic communications the source of a statement may be far more significant than its substance. Consensus of several participants may be of less value than knowledge of their identity. Credibility of the response may hinge on the identification of the respondent (see IIIB1).

Possible fallacy b: The more individuals are involved with a Delphi as users, the more effective it will be.

An unfamiliar and anonymous communication system can develop into a threat to established individuals and intraorganizational relationships. Like other analytical tools, it can serve in an advocacy role as well as in an inquiry role.

Possible fallacy c: The goals of the organization are the same as those of the individuals in the organization.

The Delphi designer is always faced with the problem of understanding the user and his organization. It is the same problem that has confronted efforts in all aspects of management science—operations research, systems analysis, cost benefit studies, etc. The presuppositions on the part of the analyst about the utility and correctness of his methods and their 'good' goals may be entirely unwarranted. The possible advantage of Delphi in a circumstance of this sort is that it can be oriented, if allowed, to expose the existence of the fallacies. If an organization is to function over the long run, misconceptions must at least be held within reasonable bounds and their exposure can serve to sharpen those boundaries.

8. Deception

Today the least acknowledged hazard in connection with Delphi is its potential use for deceptive, manipulative purposes. Welty reaches back to the Greek myth of Ino, the wife of King Athamus of Orchomenus.[23] When the King dispatched a messenger to the Oracle of Delphi, Ino bribed him to return with a falsified story. In a second round of consultation at Delphi the Oracle based

[23]G. Welty, "Plato and Delphi", *FUTURES*, Vol. 5, No. 3, June 1973.

its pronouncements on the false version of the first utterances. In other words, the Oracle did not recognize the deception.

Cyphert and Gant have conducted a Delphi experiment[(24)] where false information was introduced during the analysis of the first round responses and sent out in the second round. The participants did not ignore the false information but distorted their own subsequent responses, to reflect acceptance of this new input.

There is a vital lesson here. The Delphi process is not immune to manipulation or propaganda use. The anonymity in such a situation may even facilitate the deception process: how can the participants in a Policy Delphi possibly detect distortion of the feedback they receive? One answer can be inferred from our earlier discussion of holistic experimentation (Chapter VII, A) based on the work of Mitroff and Blankenship. The "subjects" themselves can insist on taking a major role in the staff activities, including monitoring and analysis of the responses in each round.

All of these pitfalls exist to greater or lesser degree no matter what communication process we choose to utilize in approaching a problem. However, since an honestly executed Delphi makes the communication process and its structure explicit, most pitfalls assume greater clarity to the observer than if the process proceeds in a less structured manner. While the Delphi designer in the context of his application may not be able to deal with, or eliminate, all these problems, it is his responsibility to recognize the degree of impact which each has on his application and to minimize any that might invalidate his exercise. The strength of Delphi is, therefore, the ability to make explicit the limitations on the particular design and its application. The Delphi designer who understands the philosophy of his approach and the resulting boundaries of validity is engaged in the practice of a potent communication process. The designer who applies the technique without this insight or without clarifying these boundaries for the clients or observers is engaged in the practice of mythology.

---

[24]F. Cyphert and W. Gant, "The Delphi Technique", *Journal of Teacher Education*, Vol. 21, No. 3, 1970, p. 422.

Appendix: Delphi Bibliography

Subject Index

Appendix: Delphi Bibliography

HAROLD A. LINSTONE and MURRAY TUROFF

The following bibliography is complementary to the references associated with each of the articles in the book. We have broken the reference material into a number of separate sections. It is felt this particular breakdown will make the reference list of greater utility to the user. The sections in their order of appearance and the page on which each begins, are:

There are four journals which, during the past five years, have served as the principal source of articles on Delphi and its applications:

(1) *Technological Forecasting and Social Change* (cited as *TFSC*)
(2) *FUTURES*
(3) *Long Range Planning*
(4) *Socio-Economic Planning Sciences*

The first reference list (Journal Articles) is an alphabetical listing by author of Delphi articles, or articles involving Delphi, that have appeared in these four journals.

For approximately a decade almost all work on Delphi was carried on at the Rand Corporation where the concept was born. It was not until the mid-sixties that it began to catch on elsewhere. We have, therefore, listed in chronological order the Rand papers on Delphi. This list provides some insight into the evolution of the technique. Several of these papers do appear in other sections of the bibliography when they were published outside Rand in the same or slightly altered form.

In the mid-sixties a new nonprofit organization, the Institute for the Future (IFF), was formed by a group comprised, in part, of Rand people. One of the goals of this organization was the application of a range of futures methodologies, including Delphi, to social and technological problems. A list of its

The Delphi Method: Techniques and Applications, Harold A. Linstone and Murray Turoff (eds.)
ISBN 0-201-04294-0; 0-201-04293-2

publications in the Delphi area is also included. Another group, spun off from IFF, is the Futures Group, a profit concern willing to do Delphi studies on a proprietary basis. While some of its work is not available to the public, we have included a list of its recent efforts to indicate the scope of application for Delphi in this sort of environment.

In the next section we find essentially all other articles on Delphi not encompassed by these journals or organizational headings. One can see from this list the beginnings of university, corporate, and governmental use of the technique. Unfortunately, many users have not published their results and are not, therefore, represented in this list.

For a great many years Delphi has been associated with the subject of Technological Forecasting. We have provided a separate list of articles on technological forecasting which do discuss Delphi. Our caution to the reader about this set of references is that some of these writings tend to define and consider Delphi solely within the scope of that one application area. It has been our intent in this book to present Delphi in the wider context of an alternative communication form.

There is evident in the literature a set of new directions having to do with the application of the Delphi concept in such areas as computerized conferencing, management information systems, citizen feedback, participatory democracy, etc. For convenience we have placed these references in a separate list.

For those seeking references in other languages we have compiled a separate list of some foreign language articles on Delphi.

The final section of this bibliography is perhaps the most important to the potential Delphi designer. It was gathered by perusing the papers on Delphi and asking what writings in other disciplines have been referenced there. The papers in this list represent work in the fields of economics, operations research, philosophy, planning, psychology, sociology, and statistics. In essence, this provides a guide to the techniques and knowledge in other areas of utility to the study and use of Delphi. As one would expect, psychology and sociology make up a large part of the referenced material.

The lists which follow represent a search of materials readily available to the editors through 1974. We apologize in advance to those authors we have missed. If anyone has additions to make, and so informs us, we will compile them for an update in a future edition of this book.

Distribution of Bibliography

The Distribution of articles in the bibliography is expressed in the following table. While Delphi has been around since the early fifties, much more material has been published on the subject in the last four years than in all the years before 1970.

	Before 1970	1970–1974	Total
Journal Articles	14	54	68
Rand Papers	34	4	48
IFF and Futures Group	13	67	80
Other Delphi Articles	45	95	140
Technological Forecasting	19	25	44
New Directions	3	61	64
Foreign Language Articles	6	39	45
Total of Delphi References	134	355	489
Related Work	136	45	181
Total	270	400	670

Journal Articles

Alderson, R. C. *et al.*, "Requirements Analysis, Need Forecasting, and Technology Planning Using the Honeywell PATTERN Technique," *TFSC* **3**, No. 2 (1972).

Amara, Roy C., "A Note on Cross-Impact Analysis," *FUTURES* **4**, No. 3 (1972).

——, and Andrew J. Lipinski, "Some Views on the Use of Expert Judgment," *TFSC* **3**, No. 3 (1972).

——, and Gerald R. Salancik, "Forecasting: From Conjectural Art Toward Science," *TFSC* **3**, No. 4 (1972).

Ament, R., "Comparison of Delphi Forecasting Studies in 1964 and 1969," *FUTURES* **2**, No. 1 (March 1970).

Beaton, A. E., "Scaling Criterion of Questionnaire Items," *Socio-Economic Planning Sciences* **2**, Nos. 2, 3, 4 (1969).

Bernstein, G. B., and Marvin Cetron, "SEER: A Delphic Approach Applied to Information Processing," *TFSC* **1**, No. 1 (June 1969).

Bender, Strack, Ebright, and von Haunalter, "Delphic Study Examines Developments in Medicine," *FUTURES* **1**, No. 4 (June 1969).

Bjerrum, C. A., "Forecast of Computer Development and Applications, 1968-2000," *FUTURES* **1**, No. 4 (June 1969).

Blackman, A. Wade, "A Cross-Impact Model Applicable to Forecasting for Long Range Planning," *TFSC* **5**, No. 3 (1973).

——, "The Use of Bayesian Techniques in Delphi Forecasts," *TFSC* **2**, No. 3/4 (1971).

Campbell, George S., "Relevance of Signal Monitoring to Delphi Cross-Impact Studies," *FUTURES* **3**, No. 4 (December 1971).

Cetron, M. J., and G. B. Bernstein, "SEER: A Delphic Approach Applied to Information Processing," *TFSC* **1**, No. 1 (Spring 1969).

Cole, H. S. D., and J. Metcalf, "Model Dependent Scale Values for Attitude Questionnaire Items," *Socio-Economic Planning Sciences* **5**, No. 9 (1971).

Currill, D. L., "Technological Forecasting in Six Major U. K. Companies," *Long Range Planning* **5**, No. 1 (March 1972).

Dalkey, Norman C., "An Elementary Cross-Impact Model," *TFSC* **3**, No. 3 (1972).

——, "An Experimental Study of Group Opinion: The Delphic Method," *FUTURES* **2**, No. 3 (September 1969).

——, "Analyses from a Group Opinion Study," *FUTURES* **1**, No. 6 (December 1969).

——, B. Brown, and S. Cochran, "Use of Self-Ratings to Improve Group Estimates," *TFSC* **1**, No. 3 (March 1970).

Day, Lawrence H., "Long Term Planning in Bell Canada," *Long Range Planning* London, U. K. (submitted for publication, 1973).

Decker, Robert L., "A Delphi Survey of Economic Development," *FUTURES* **6**, No. 2 (April 1974).

Derian, J. C., and F. Morize, "Delphi in the Assessment of R&D Projects," *FUTURES* **5**, No. 5 (October 1973).

Dobrov, G. M., and L. P. Smirnov, "Forecasting as a Means for Scientific and Technological Policy Control," *TFSC* **4**, No. 1 (1972).

Doyon, L. R., T. V. Sheehan, and H. I. Zayor, "Classroom Exercises in Applying the Delphi Method for Decision-Making," *Socio-Economic Planning Sciences* **5**, No. 4 (1971).

Dressler, Fritz R. S., "Subject Methodology in Forecasting," *TFSC* **3**, No. 4 (1972).

Enzer, Selwyn, "A Case Study Using Forecasting as a Decision-Making Aid," *FUTURES* **2**, No. 4 (December 1970).

——, "Cross-Impact Techniques in Technology Assessment," *FUTURES* **4**, No. 1 (March 1972).

——, "Delphi and Cross-Impact Techniques—An Effective Combination for Systematic Futures Analysis," *FUTURES* **3**, No. 1 (March 1971).

Goodman, Joel, "Delphi and the Law of Diminishing Returns," *TFSC* **2**, No. 2 (1970).

Goodwill, Daniel F., "A Look at the Future Impact of Computer-Communications on Everyday Life," *TFSC* **4**, No. 2 (1972).

Gordon, T. J., "Cross-Impact Matrices: An Illustration of Their Use for Policy Analysis," *FUTURES* **1**, No. 6 (1969).

——, "Potential Changes in Employee Benefits," *FUTURES* **2**, No. 2 (1970).

——, S. Enzer, and R. Rochberg, "An Experiment in Simulation Gaming for Social Policy Studies," *TFSC* **1**, No. 3 (March 1970).

——, and H. Hayward, "Initial Experiments with the Cross-Impact Matrix Method of Forecasting," *FUTURES* **1**, No. 2 (December 1968).

——, R. J. Wolfson, and R. C. Sahr, "A Method for Rapid Reproduction of Group Data and Individual Estimates in the Use of the Delphic Method," *FUTURES* **1**, No. 6 (December 1969).

Grabbe, Eugene M., and Donald L. Pyke, "An Evaluation of the Forecasting of Information Processing Technology and Applications," *TFSC* **4**, No. 2 (1972).

Helmer, O., "Cross-Impact Gaming," *FUTURES* **4**, No. 2 (June 1972).

Huckfeldt, Vaughn E., and Robert C. Judd, "Issues in Large Scale Delphi Studies," *TFSC* **6**, No. 1 (1974).

Johnson, Howard E., "Some Computational Aspects of Cross-Impact Matrix Forecasting," *FUTURES* **2**, No. 2 (June 1970).

Judd, Robert C., "Use of Delphi Methods in Higher Education," *TFSC* **4**, No. 2 (1972).

Kane, Julius, "A Primer for a New Cross-Impact Language—KSIM," *TFSC* **4**, No. 2 (1972).

Lachmann, Ole, "Personnel Administration in 1980: A Delphi Study," *Long Range Planning* **5**, No. 2 (June 1972).

Martino, Joseph P., "The Lognormality of Delphi Estimates," *TFSC* **1**, No. 4 (Spring 1970).

——, "The Optimism/Pessimism Consistency of Delphi Panelists," *TFSC* **2**, No. 2 (1970).

——, "The Precision of Delphi Estimates," *TFSC* **1**, No. 3 (March 1970).

Overbury, R. E., "Technological Forecasting: A Criticism of the Delphi Technique," *Long Range Planning* **1**, No. 4 (June 1969).

Pavitt, Keith, "Analytical Techniques in Government Science Policy," *FUTURES* **4**, No. 1 (March 1972).

Pill, Juri, "The Delphi Method: Substance, Context, A Critique and An Annotated Bibliography," *Socio-Economic Planning Science* **5**, No. 1 (1971).
Pyke, Donald L., "A Practical Approach to Delphi," *FUTURES* **3**, No. 2 (June 1970).
Rochberg, Richard, "Convergence and Viability because of Random Numbers in Cross-Impact Analyses," *FUTURES* **2**, No. 3 (September 1970).
——, "Information Theory, Cross-Impact Matrices, and Pivotal Events," *TFSC* **2**, No. 1 (1970).
Rowland, D. G., "Technological Forecasting and the Delphi Technique—A Reply," *Long Range Planning* **2**, (December 1969).
Salancik, J. R., "Assimilation of Aggregated Inputs into Delphi Forecasts: A Regression Analysis," *TFSC* **5**, No. 3 (1973).
——, William Wenger, and Ellen Helfer, "The Construction of Delphi Event Statements," *TFSC* **3**, No. 1 (1971).
Sandow, Stuart, "The Pedagogy of Planning: Defining Sufficient Futures," *FUTURES* **3**, No. 4 (December 1971).
Schneider, Jerry B., "The Policy Delphi: A Regional Planning Application," *TFSC* **3**, No. 4 (1972).
Smil, Vaclav, "Energy and the Environment—A Delphi Forecast," *Long Range Planning* **5**, No. 4 (December 1972).
Stover, John, "Improvements to Delphi/Cross-Impact," *FUTURES* **5**, No. 3 (June 1973).
Sulc, Oto, "A Methodological Approach to the Integration of Technological and Social Forecasts," *TFSC* **2**, No. 1 (June 1969).
——, "Interaction between Technological and Social Changes: A Forecasting Model," *FUTURES* **1**, No. 5 (September 1969).
Teeling-Smith, George, "Medicine in the 1990's: Experience with a Delphi Forecast," *Long Range Planning* **3**, No. 4 (June 1971).
Turoff, Murray, "An Alternative Approach to Cross-Impact Analysis," *TFSC* **3**, No. 3 (1972).
——, "Delphi Conferencing: Computer-Based Conferencing with Anonymity," *TFSC* **3**, No. 2 (1970).
——, "The Design of a Policy Delphi," *TFSC* **2**, No. 2 (1970).
——, "Meeting of the Council on Cybernetic Stability: A Scenario," *TFSC* **4**, No. 2 (1972).
Umpleby, Stuart, "Is Greater Citizen Participation in Planning Possible and Desirable?" *TFSC* **4**, No. 1 (1971).
Welty, Gordon, "Plato and Delphi," *FUTURES* **5**, No 3 (June 1973).
Wills, Gordon, "The Preparation and Development of Technological Forecasts," *Long Range Planning* **2**, No. 3 (March 1970).

Rand Papers in Chronological Order

Sackman, H., *Delphi Assessment: Expert Opinion, Forecasting and Group Process*, R-1283-PR, April 1974.
Brown, Thomas, *An Experiment in Probabilistic Forecasting*, R-944-ARPA, 1972.
Dalkey, Norman C., *An Elementary Cross-Impact Model*, R-677-ARPA, May 1971.
——, and Bernice Brown, *Comparison of Group Judgment Techniques With Short-Range Predictions and Almanac Questions*, R-678-ARPA, May 1971.
Schmidt, D. L., *Creativity in Industrial Engineering*, P-4601, March 1971.
Dalkey, Norman C., and Daniel L. Rourke, *Experimental Assessment of Delphi Procedures With Group Value Judgments*, R-612-ARPA, February 1971.
Quade, E. S., *On the Limitations of Quantitative Analysis*, P-4530, December 1970.
Farquhar, J. A., *A Preliminary Inquiry into the Software Estimation Process*, RM-6271-PR, August 1970.
——, R. J. Lewis, and D. Snyder, *Measurement and Analysis of the Quality of Life: With Exploratory Illustrations of Applications to Career and Transportation Choices*, RM-6228-DOT, August 1970.
Quade, E. S., *An Extended Concept of "Model,"* P-4427, July 1970.

Brown, T. A., *Probabilistic Forecasts and Reproducing Scoring Systems*, RM-6299-ARPA, June 1970.

Quade, E. S., *Cost Effectiveness: Some Trends in Analysis*, P-3529-1, March 1970.

Dalkey, Norman C., H. Brown, and S. W. Cochran, *The Delphi Method, IV: Effect of Percentile Feedback and Feed-In of Relevant Facts*, RM-6118-PR, March 1970.

Paxson, Edwin, *A Delphi Examination of Civil Defense*, RM-6247-ARPA, February 1970.

Dalkey, Norman C., H. Brown, and S. W. Cochran, *The Delphi Method, III: Use of Self Rating to Improve Group Estimates*, RM-6115-PR, November 1969.

Rescher, N., *Delphi and Values*, P-4182, September 1969.

Dole, S. H., G. H. Fischer, E. D. Harris, and J. String, Jr., *Establishment of a Long-Range Planning Capability*, P-3540, September 1969.

Brown, Bernice, S. Cochran, and N. Dalkey, *The Delphi Method, II: Structure of Experiments*, RM-5957-PR, June 1969.

Pinkel, B., *On the Decision Matrix and the Judgment Process: A Developmental Decision Example*, P-3620, June 1969.

Dalkey, Norman C., *The Delphi Method: An Experimental Study of Group Opinion*, RM-5888-PR, June 1969.

Quade, E. S., *The Systems Approach and Public Policy*, P-4053, March 1969.

Dalkey, Norman C., *Predicting the Future*, P-3948, October 1968.

Brown, Bernice B., *DELPHI Process: A Methodology Used for the Elicitation of Opinions of Experts*, California, P-3925, September 1968.

Quade, E. S., and W. I. Boucher, *Systems Analysis and Policy Planning: Applications in Defense*, California, June 1968.

Dalkey, Norman C., *Experiments in Group Prediction*, P-3820, March 1968.

——, *Quality of Life*, P-3805, March 1968.

Rochberg, R., *Some Comments on the Problem of Self Affecting Predictions*, P-3735, December 1967.

Helmer, Olaf, *Systematic Use of Expert Opinions*, P-3721, November 1967.

Dalkey, Norman C., *DELPHI*, P-3704, October 1967.

Helmer, Olaf, *Prospects of Technological Progress*, California, P-3643, August 1967.

——, *Methodology of Societal Studies*, California, P-3611, June 1967.

——, *The Future of Science*, P-3607, May 1967.

Rescher, N., *The Future as an Object of Research*, P-3593, April 1967.

Helmer, O., *New Developments in Early Forecasting of Public Problems: A New Intellectual Climate*, P-3576, April 1967.

Haydon, Brownlee, *The Year 2000*, P-3571, March 1967.

Helmer, Olaf, *Analysis of the Future: The Delphi Method*, P-3558, March 1967.

Quade, E. S., *Cost-Effectiveness: Some Trends in Analyses*, P-3529, March 1967.

Neiswender, R., *"The Exploration of the Future," Réalités*, No. 245, June 1966. (Translated in P-3540, February 1967.)

Helmer, Olaf, *The Use of the Delphi Technique in Problems of Educational Innovations*, P-3499, December 1966.

Downs, Anthony, *Bureaucratic Structure and Decision Making*, RM-4646-1-PR, October 1966.

Jouvenel, B. de, *Futuribles*, P-3045, January 1965.

Helmer, Olaf, and Bernice Brown, *Improving the Reliability of Estimates Obtained from a Consensus of Experts*, P-2986, September 1964.

Gordon, Theodore J., and Olaf Helmer, *Report on a Long-Range Forecasting Study*, P-2982, September 1964.

Helmer, Olaf, *Convergence of Expert Consensus Through Feedback*, P-2973, September 1964.

Quade, E. S., *Analysis for Military Decisions*, R-387-PR, November 1964.

Helmer, Olaf, *The Systematic Use of Expert Judgment in Operations Research*, P-2795, September 1963.

——, and E. S. Quade, *An Approach to the Study of a Developing Economy by Operational Gaming*, P-2718, March 1963.

Dalkey, Norman C., and Olaf Helmer, *The Use of Experts for the Estimation of Bombing Requirements—A Project Delphi Experiment*, RM-727-PR, November 1951.

Kaplan, A., A. L. Skogstad, and M. A. Sirshick, *The Prediction of Social and Technological Events*, P-93, April 1949.

Publications of the Institute for the Future*

Amara, Roy C., "A Framework for National Science Policy Analysis," P-18 (published in *IEEE Transactions on Systems, Man and Cybernetics*, Vol. SMC-2, No. 1, January 1972).
——, "The Social Responsibilities," P-16, April 1972.
——, "Toward a Framework for National Goals and Policy Research," P-20 (published in *Policy Sciences*, Vol. 3, 1972).
——, and Andrew J. Lipinski, "Some Views on the Use of Expert Judgment," P-19 (published in *Technological Forecasting and Social Change*, Vol. 3, No. 3, 1972).
——, and J. R. Salancik, "Forecasting from Conjectural Art Toward Science," P-14 (published in *Technological Forecasting and Social Change*, Vol. 3, No. 4, 1972).
Ament, Robert H., "Comparison of Delphi Forecasting Studies in 1964 and 1969," P-9 (published in *FUTURES*, Vol. 2, No. 1, March 1970) (reprint).
Baran, Paul, *The Future of Newsprint, 1970–2000*, R-16, December 1971.
——, *Japanese Competition in the Information Industry*, P-17, May 1972.
——, *Notes on a Seminar on Future Broad-Band Communications*, WP-1, February 1970.
——, "On the Impact of the New Communications Media upon Social Values," P-3 (published in *Law and Contemporary Problems*, Vol. 34, No. 2, Spring 1969) (reprint).
——, *Potential Market Demand for Two–Way Information Services to the Home, 1970–1990*, R-26, December 1971.
——, and Andrew J. Lipinski, *The Future of the Telephone Industry, 1970–1985*, R-20, September 1971, a special industry report.
Becker, Harold S., *A Framework for Community Development Action Planning, Volume I: An Approach to the Planning Process*, R-18, February 1971.
——, *A Framework for Community Development Action Planning, Volume II: Study Procedure, Conclusions, and Recommendations for Future Research*, R-19, February 1971.
——, *A Method of Obtaining Forecasts for Long-Range Aerospace Program Planning*, WP-7, April 1970.
——, and Raul de Brigard, *Considerations on a Framework for Community Action Planning*, WP-9, June 1970.
Brigard, Raul de, and Olaf Helmer, *Some Potential Societal Developments: 1970-2000*, R-7, April 1970.
Enzer, Selwyn, *A Case Study Using Forecasting as a Decision-Making Aid*, WP-2, December 1969.
——, *Cross-Impact Techniques in Technology Assessment*, P-15.
——, *Delphi and Cross-Impact Techniques: An Effective Combination for Systematic Futures Analysis*, WP-8, June 1970.
——, *Federal/State Science Policy and Connecticut: A Futures Research Workshop*, R-24, October 1971.
——, *Some Commercial and Industrial Aspects of the Space Program*, P-5, November 1969.
——, *Some Developments in Plastics and Competing Materials by 1985*, R-17, January 1971.
——, *Some Prospects for Residential Housing by 1985*, R-13, January 1971.
——, Wayne I. Boucher, and Frederick D. Lazar, *Futures Research as an Aid to Government Planning in Canada: Four Workshop Demonstrations*, R-22, August 1971.
——, Wayne I. Boucher, and Frederick D. Lazar, *Futures Research as an Aid to Government Planning in Canada: Four Workshop Demonstrations—Supporting Appendices*, R-22A, August 1971.
——, and Raul de Brigard, *Issues and Opportunities in the State of Connecticut: 1970-2000*, R-8, March 1970.
——, Raul de Brigard, and Frederick D. Lazar, *Some Considerations Concerning Bankruptcy Reform*, IFF Report R-28, March 1973.
——, Dennis L. Little, and Frederick D. Lazar, *Some Prospects for Social Change by 1985 and Their Impact on Time/Money Budgets*, R-25, March 1972.
——, Gordon J. Slotsky, Dennis L. Little, James E. Doggart, and David A. Long, *The Automobile Insurance System: Current Status and Some Proposed Revisions*, WP-18, December 1971.

---

* *Reports*—Formal documentation of Institute Studies
Working Papers—Preliminary contributions to the Institute's work for its sponsors
Papers—Individual contributions by Institute staff members to the professional literature

Gordon, Theodore J., *A Study of Potential Changes in Employee Benefits, Volume I: Summary and Conclusions*, R-1, April 1969.

——, *A Study of Potential Changes in Employee Benefits, Volume II: National and International Patterns*, R-2, April 1969.

——, *A Study of Potential Changes in Employee Benefits, Volume III: Delphi Study*, R-3, April 1969.

——, *A Study of Potential Changes in Employee Benefits, Volume IV: Appendices to the Delphi Study*, R-4, April 1969.

——, *Potential Institutional Arrangements of Organizations Involved in the Exploitation of Remotely Sensed Earth Resources Data*, P-6 (published as AIAA Paper No. 70-334, March 1970).

——, *Some Possible Futures of American Religion*, P-4, May 1970.

——, *The Current Methods of Futures Research*, P-11, August 1971.

——, and Robert H. Ament, *Forecasts of Some Technological and Scientific Developments and their Societal Consequences*, R-6, September 1969.

——, Selwyn Enzer, Richard Rochberg, and Robert Buchele, *A Simulation Game for the Study of State Policies*, R-9, September 1969.

——, Olaf Helmer, Selwyn Enzer, Raul de Brigard, and Richard Rochberg, *Development of Long-Range Forecasting Methods for Connecticut: A Summary*, R-5, September 1969.

——, Dennis L. Little, Harold L. Strudler, and Donna D. Lustgarten, *A Forecast of the Interaction between Business and Society in the Next Five Years*, R-21, April 1971.

——, Richard Rochberg, and Selwyn Enzer, *Research on Cross-Impact Techniques with Applications to Selected Problems in Economics, Political Science, and Technology Assessment*, R-12, August 1970.

Helmer, Olaf, *Long-Range Forecasting—Roles and Methods*, P-7, May 1970.

——, *Multipurpose Planning Games*, WP-17, December 1971.

——, *On the Future State of the Union*, R-27, May 1972.

——, *Political Analysis of the Future*, P-1, August 1969.

——, *Report on the Future of the Future-State-of-the-Union Reports*, R-14, October 1970.

——, *Toward the Automation of Delphi*, International Technical Memorandum, Institute for the Future, March 1970.

——, *et al.*, *Development of Long-Range Forecasting Methods for Connecticut: A Summary*, IFF Report, R-5, September 1969.

——, and Helen Helmer, *Future Opportunities for Foundation Support*, R-11, June 1970.

Institute for the Future, "Development of a Computer-Based System to Improve Interaction Among Experts," First Annual Report, National Science Foundation, Grant GJ-35 326X, August 1973.

Institute for the Future, "Social Assessment of Mediated Group Communication: A Workshop Summary," March 1974.

Johansen, Robert, Richard H. Miller, Jacques Vallee, "Group Communication through Electronic Media: Fundamental Choices and Social Effects," Working Paper, March 1974.

Kramish, Arnold, *The Non-Proliferation Treaty at the Crossroads*, P-2, October 1969.

LeCompte, Gare, *Factors in the Introduction of a New Communications Technology into Syria and Turkey: Background Data*, WP-10, August 1970.

Lipinski, Andrew J., *Toward a Framework for Communications Policy Planning*, WP-19, December 1971.

Little, Dennis L., *Models and Simulation: Some Definitions*, WP-6, April 1970.

——, *STAPOL: Appendix to the Simulation Game Manual*, WP-14, October 1970.

——, and Raul de Brigard, *Simulation, Futurism, and the Liberal Arts College: A Case Study*, WP-15, April 1971.

——, and Richard Feller, *STAPOL: A Simulation of the Impact of Policy, Values, and Technological and Societal Developments upon the Quality of Life*, WP-12, October 1970.

——, and Theodore J. Gordon, *Some Trends Likely to Affect American Society in the Next Several Decades*, WP-16, April 1971.

——, Richard Rochberg, and Richard Feller, *STAPOL: Simulation-Game Manual*, WP-13, March 1971.

Nagelberg, Mark, *Simulation of Urban Systems: A Selected Bibliography*, WP-3, January 1970.

——, and Dennis L. Little, *Selected Urban Simulations and Games*, WP-4, April 1970.

Rochberg, Richard, *Information Theory, Cross-Impact Matrices, and Pivotal Events*, P-8 (published in *Technological Forecasting and Social Change* **2**, No. 2, 1970)(reprint).

——, Theodore J. Gordon, and Olaf Helmer, *The Use of Cross-Impact Matrices for Forecasting and Planning*, R-10, April 1970.

Sahr, Robert C., *A Collation of Similar Delphi Forecasts*, WP-5, April 1970.
Salancik, J. R., Theodore J. Gordon, and Neale Adams, *On the Nature of Economic Losses Arising from Computer-Based Systems in the Next Fifteen Years*, R-23, March 1972.
Vallee, J. *et al.*, *Group Communication Through Computers*, R-32 and R-33, July 1974.
Wilson, Albert, and Donna Wilson, *Toward the Institutionalization of Change*, WP-11, August 1970.

Delphi Studies by the Futures Group

Since some of these studies are not available for general public distribution, a short description is included of some of the items.

Boucher, W. I., *Quantifiable Goal Statements for the U.S. Criminal Justice System: A Preliminary Assessment*, Report 53-30-01, May 1972.
Done for the National Advisory Commission on Criminal Justice Standards and Goals; members of the Commission served as respondents. May eventually be published as part of the report on the Commission's work. Time horizon: to 1980–1985.
——, *Report on a Hypothetical Focused Planning Effort (FPE)*, Report 43-01-06, February 1972.
——, and Theodore J. Gordon, *The Literature of Cross-Impact Analysis: A Survey*, Report 41-24-01/2, February 1972.
——, and J. Talbot, *Report on a Delphi Study of Future Telephone Interconnect Systems*, Report 49-26-01, April 1972.
The first Delphi conducted entirely through interviews; also the first (to our knowledge) to have as its goal the conceptual design of a physical system.
Becker, H. S., and J. Incerti, *Future Technological and Social Changes: A Delphi Study of Opportunities and Threats for an Industrial Firm*, Report 76-35-02, January 1973.
Done for a multidivisional, hard-technology corporation; the aim was to identify prospective social and technological developments relevant to the diversions, determine their probability, indicate who might be most instrumental in making them happen, assess their impact if they did happen, and define policies appropriate for the company to take. Time horizon: to 1982 +.
Gordon, T. J., and Harold A. Becker, *Analysis of Ailing Products—It's Decisions That Count (A Case Study of R & D Methodology)*, Report 24-15-01, February 1972.
——, and J. Cohen, *An Investigation of Future Technological and Social Developments and Their Implications for Entry into the Produce Market*, Report 66-36-01, October 1972.
The third "interview Delphi"; here the attempt was to determine whether, in view of possible changes in technology and society generally, it made sense for a particular company to consider becoming a produce supplier. Time horizon: to 1982.
——, William L. Renfro, and W. I. Boucher, *Challenges and Opportunities in the Photographic Industry: Report on a Focused Planning Effort*, Report 85-42-03, June 1973.
Weaver, W. T., *A Delphi Study of the Potential of a New Communications System*, Report 58-26-02, August 1972.
The second "interview Delphi": again, the focus fell on the conceptual design of a specific physical product. Time horizon: to 1975–76.

Other Delphi Articles

Adelman, I., and Morris, C. T., "Society, Politics and Economic Development: A Quantitative Approach," Johns Hopkins Press, Baltimore, 1967.
Adelson, M., M. Alkin, C. Carey, and O. Helmer, "The Education Innovation Study," *American Behavioral Scientist* **10**, No. 7 (1967).

Allnut, Bruce, *A Study of Consensus on Psychological Factors Related to Recovery from Nuclear Attack*, Human Sciences Research, Inc., HSR-RR-71/3 Office of Civil Defense, May 1971.

——, *A Review of the Development and Application of the Delphi Method*, Human Sciences Research, Inc., September 1970.

Bahary, E. *et al.*, *Communication Needs of the Seventies and Eighties*. AT&T Marketing and Western Electric Engineering, December 1971.

Barach, J. L., and R. J. Twery, "The Application of Forecasting Techniques in Research and Development Planning," *Presented at AICHE meeting*, November 1969.

Bayne, C. Dudley, Jr., and Walter Price, "Technological Forecasting," in Richard F. Vancil (ed.), *Formal Planning Systems—1971*, Harvard Business School, Cambridge, Mass.

Bedford, Michael, T., "The Future of Communications Services in the Home," *Bell Canada*, September 1972.

Bell, Daniel, "The Oracles at Delphi," *The Public Interest*, Fall 1965.

Bender, A. D., A. E. Strack, G. W. Ebright, and G. von Haunalter, *A Delphic Study of the Future of Medicine*, Smith, Kline, and French Laboratories, January 1969.

Berghofer, D. E., "General Education in Post-Secondary Non-University Institutions in Alberta," Research Studies in Post-Secondary Education, No. 9 (Edmonton, Alberta: Alberta College Commission, April 1970).

Bernstein, George, *et al.*, *A Fifteen Year Forecast of Information Processing Technology*, Naval Supply Systems, AD681752, January 1969.

Borko, Harold, "Predicting Needs in Librarianship and Information Science Education," Proceedings of the ASIA Annual Meeting, Vol. 7, 1970.

Brech, Ronald, *Britain 1984: Unilever's Forecast: An Experiment in Economic History of the Future*, Dartan, Longman & Todd Ltd., London, 1963.

Brown, Bernice, "Delphi Process: A Methodology Used for the Elicitation of Opinions of Experts," *ASTME Vectors* **3**, No. 1 (1968).

——, "Technological Forecasting by Iterative Guesstimation," *Product Design and Value Engineering* **13**, (October 1968).

Campbell, Robert M., "A Methodological Study of the Utilization of Experts in Business Forecasting," Ph.D. dissertation, University of California, Los Angeles, California, 1966.

——, and David Hitchin, "The Delphi Technique: Implementation in the Corporate Environment," *Management Services*, Report Number 11, Graduate School of Business Administration, U. S. C., March 1969.

Cetron, Marvin J., and Don Overly, "Toward a Consensus on the Future," *Innovation*, No. 31, May 1972.

Chambers, J. C. *et al.*, "Catalytic Agency for Effective Planning," *Harvard Business Review*, January-February 1971.

Chapman, R. L., and F. N. Cleaveland, *Meeting the Needs of Tomorrow's Public Service*, Report of the National Academy of Public Administration, January 1973.

Clarke, S. C. T., and H. T. Coutts, "The Future of Teacher Education," unpublished manuscript, Edmonton, Alberta: Faculty of Education, University of Alberta, 1971.

Craver, J. Kenneth, "A Practical Method for Technology Assessment," paper presented at the First International Conference on Technology Assessment, held in The Hague, The Netherlands, May 27-June 2, 1973.

——, "Technology Assessment by Cross-Impact," paper presented at the Engineering Foundation Conference on Technology Assessment held at New England College, Henniker, N. H., July 1972.

Cyphert, Frederick, and Walter L. Gant, "The Delphi Technique," *Journal of Teacher Education* **21**, No. 3 (1970).

——, and Walter L. Gant, "The Delphi Technique: A Case Study," *PHI DELTA KAPPAN*, January 1971.

——, and Walter L. Gant, "The Delphi Technique: A Tool for collecting Opinions in Teacher Education," *Journal of Teacher Education* **21**, No. 3, (1970).

Dalby, James F., "Practical Refinements to the Cross-Impact Matrix Technique," in *Industrial Applications of Technological Forecasting*, Cetron and Bartocha (eds.), Wiley-Interscience, New York, 1971.

Dalkey, Norman C., *Studies in the Quality of Life* (Delphi and Decision-making), Lexington Books, 1972.

——, and Olaf Helmer, "An Experimental Application of the Delphi Method to the Use of Experts," *Management Science* **9**, (1968).

Darling, Charles M., III, *An Experimental 1968 Public Affairs Forecast*, Available from: National Industrial Conference Board, New York, N. Y.

Davidson, Frank P., "Futures Research: A New Scientific Discipline," *Proceedings of the Social Statistics Section*, American Statistical Association, 1969.

Day, Lawrence H., "The Future of Computer and Communications Services," AFIPS Conference *Proceedings* **43**, (1973): 723–34.

Dean, B. V., S. Mathis, *et al.*, *Analysis of the Exploratory Development Project Evaluation Experiment*, Case Western Reserve University School of Management, TM 165, November 1969.

Delbecq, A. C., and A. Van DeVen, *A Group Process Model for Problem Identification and Program Planning*, University of Wisconsin, Madison, October 1970.

Doak, Nick, "Future Rubber Processing," *Rubber Journal*, November 1972.

Doyle, Frank J., and Daniel Z. Goodwill, "An Exploration of the Future in Educational Technology," *Bell Canada*, January 1971.

——, and Daniel Z. Goodwill, "An Exploration of the Future in Educational Technology," in H. A. Stevenson, R. M. Stamp, J. D. Wilson (eds.), *The Best of Times, The Worst of Times—Contemporary Issues in Canadian Education*, Holt, Rinehart and Winston, Montreal, 1972.

——, and Daniel Z. Goodwill, "An Exploration of the Future of Medical Technology," *Bell Canada*, 1971.

Dyck, H. J., and G. J. Emery, *Social Futures for Alberta*, Westrede Institute, Edmonton, Alberta, 1970.

Emlet, H. E., and John Williamson, "Alternative Methods for Eliciting Estimates for Health-Care Benefits and Required Resources," presented at Joint ORSA-TIMS meeting, Atlantic City, November 1972.

Enzer, Selwyn, "Applications of Futures Research to Society's Problems," in Nigel Hawkes (ed.), *International Seminar on Trends in Mathematical Modeling*, Venice, December 13–18, 1971 (New York: Springer-Verlag, 1973), pp. 243–87.

Feldman, Philip A., "Technology Assessment of Computer-Assisted-Instruction," *Bell Canada*, Business Planning, Montreal, August 1972 (internal document).

Firschler, Martin A., and L. S. Coles, "Forecasting and Assessing the Impact of Artificial Intelligence on Society," Proceedings from International Joint Conference on Artificial Intelligence, Stanford Research Institute, August 1973.

Folk, Michael, "A Critical Look at the Cross-Impact Method," Research Report RR-5, Educational Policy Research Center, Syracuse University Research Corporation, August 1971.

——, "Computer and Educational Futures Research," *The Potential of Educational Futures*, Warren Ziegler and Michael Marien (eds.). Part of NSSE Series on Contemporary Education. Charles A. Jones, Worthington, Ohio, 1972.

Fulmer, Robert M., "Forecasting the Brave New World of 1984," *Manage*, March 1971.

——, "Forecasting the Future," *Managerial Planning*, July/August 1972.

——, *Managing Associations for the 1980's*, report for the American Society of Association Executives, January 1972.

——, "On Forecasting the Future," Inspection News, May 1971.

Glazier, Frederick, P., "A Multi-Industry Experience in Forecasting Optimum Technological Approach," in James R. Bright and M. E. F. Shoeman (eds.), *A Guide to Practical Technological Forecasting*, Prentice-Hall, Englewood Cliffs, N. J., 1973.

Golding, Edwin, "A Technological Forecast in Transportation 1975–2000," 39th ORSA meeting, May 1971.

Goodwill, D. Z., "An Exploration of the Future in Business Information," *Bell Canada*, October 1971.

Gordon, P., and MacReynolds, W., "Optimal Urban Forms," *Journal of Regional Science* **14**, No. 2, August 1974.

Gordon, Theodore J., "Forecasters Turn to Delphi," *Futurist*, February 1967.

——, "Forecasting Policy Impacts," in Walter A. Hahn and Kenneth F. Gordon (eds.), *Assessing the Future and Policy Planning*, Gordon and Breach Science Publishers, New York, 1973.

——, "New Approaches to Delphi," in *Technological Forecasting for Industry and Government*, James Bright (ed.), Prentice-Hall, Englewood Cliffs, N. J., 1968.

——, and Harold S. Becker, "The Cross-Impact Matrix Approach to Technology Assessment," *Research Management*, July 1972.

——, and R. E. LeBleu, "Employee Benefits 1970–1985," *Harvard Business Review*, January–February 1970.

Gustafson, D., and R. Ludke, *Information Manual on Computer-Aided Medical Diagnosis Study for Physician Likelihood Ratio Estimators*, University of Wisconsin, Madison, February 1970.

——, R. Shulka, A. Delbecq, and G. Walster, *A Comparative Study of Differences in Subjective Likelihood Estimates Made by Individuals*, University of Wisconsin, Madison, January 1971.

Hallan, Jerome B., and Benjamin Harriss III, "Estimation of a Potential Hemodialysis Population," *Medical Care* **8**, No. 3 (May–June 1970).

Hammonds, Timothy M., and David L. Call, *Utilization of Protein Ingredients in the U. S. Food Industry*, Part II, Department of Agricultural Economics, Cornell University, July 1970.

Helmer, Olaf, *Accomplishments and Prospects of Futures Research*, Center for Futures Research, Graduate School of Business Administration, University of Southern California, Los Angeles, California, August 1973.

——, "Analysis of the Future: The Delphi Method," in *Technological Forecasting for Industry and Government*, James Bright (ed.), Prentice-Hall, Englewood Cliffs, New Jersey, 1968.

——, "The Delphi Method—An Illustration," in *Technological Forecasting for Industry and Government*, James Bright (ed.), London, 1968.

——, *The Delphi Method for Systematizing Judgments about the Future*, Institute of Government and Public Affairs, University of California, Los Angeles, MR-61, April 1966.

——, "The Delphi Technique and Educational Innovation," *Inventing Education for the Future*, W. Hirsch (ed.), Chandler, 1967.

——, "New Development in Early Forecasting on Public Problems, a New Intellectual Climate," *Vital Speeches* **33**, (June 1967).

——, *Social Technology*, Basic Books, New York, 1966.

——, "The Use of Expert Opinion in International Relations Forecasting," Center for Futures Research, Graduate School of Business Administration, University of Southern California, Los Angeles, July 1973.

——, and T. J. Gordon, "Probing the Future," *News Front Magazine* **9**, (April 1965).

Huckfeldt, Vaughn E., "A Forecast of Changes in Postsecondary Education," Western Interstate Commission for Higher Education, Boulder, Colorado, 1972.

——, and Robert C. Judd, "Methods for Large Scale Delphi Studies," Western Interstate Commission for Higher Education, Boulder, Colorado, 1972.

Hudson, Lawrence, "Uses of the Cross-Impact Matrix in Exploring Alternatives for the Future," Educational Policy Research Center, Syracuse University Research Corporation, December 1969.

Hudspeth, Delayne, and Stuart Sandow, *A Long-Range Planning Tool for Education: The Focus Delphi*, Educational Policy Research Center, Syracuse University, April 1970.

Johnson, Howard E., "An Experience in Teaching the Delphi Technique," Graduate School of Business, the University of Texas at Austin, 1971.

Jolson, Marvin, and Gerald L. Rossow, "The Delphi Process in Marketing Decision Making," *Journal of Marketing Research* **8**, (November 1971).

Judd, Robert C., "Delphi Applications in Decision Making," *Planning and Changing* **2**, No. 3, 1971.

——, "Delphi Method: Computerized 'Oracle' Accelerates Consensus Formation," *College & University Business* **53**, No. 1 (July 1972).

——, "Forecasting to Consensus Gathering, Delphi Grows Up to College Needs," *College and University Business* **53**, No. 1 (July 1972).

Kiefer, David, "Rubber Industry: A Glimpse of the Future," *Chemical and Engineering News*, April 1972.

Kimble, Robert, *Delphic Forecasting of Critical Personnel Requirements*, U. S. Army Electronics Command, Fort Monmouth, N. J., November 1969.

Knock, Richard T., *Electro-Optical Technology*, Long Range Planning Service Report 468, Stanford Research Institute, Menlo Park, Calif., August 1972 (Client Proprietary).

Lamont, Valarie, W. Pearson, and S. Umpleby, *The Future of the University*, Educational Research Laboratory, University of Illinois, May 1971.

Locke, H. B., "NRDC and Industrial Chemistry," *Inventions for Industry*, No. 32, July 1968.

Maddox, M. M., "Delphi Survey of Critical Coal Issues," Hanover: Dartmouth College, 1973.

Marien, Michael, *The Hot List Delphi*, The Educational Policy Research Center at Syracuse, ER-6, 1972.

Martino, Joseph P., *An Experiment with the Delphi Procedure for Long-Range Forecasting*, IEEE, February 1967.

——, "What Computers May Do Tomorrow," *Futurist*, October 1969.

McGlauchlin, Laurence D., "Technological Audits: An Aid to Research Planning," in James R. Bright and M. E. F. Schoeman (eds.), *A Guide to Practical Technological Forecasting*, Prentice-Hall, Englewood Cliffs, N. J., 1973.

Milholland, Arthur V., *et al.*, "Medical Assessment by a Delphi Group Opinion Technique," *The New England Journal of Medicine*, June 14, 1973.

Milkovich, George T., Anthony Annoni, and Thomas A. Mahoney, "The Use of Delphi Procedures in Manpower Forecasting," *Management Science* **19**, No. 4 (December 1972).

Moore, C. G., and H. P. Pomrehn, *Technical Forecasting of Marine Transportation Systems 1970-2000*, U. S. C., June 1969, [reprinted in *Technological Forecasting and Social Change* **3**, (1971), p. 99].

Morrow, James Earl, *A Delphi Approach to the Future of Management*, Ph.D. Dissertation, Georgia State University, Atlanta, 1971.

Naill, R. F., "Long Term Natural Resource Availability: Environmental and Political Implications in the United States," Washington NSF-RANN, 1973.

Nanus, Burt, Michael Wooton, Harold Borko, "Social Impacts of the Multinational Computer," National Computer Conference Exposition, New York City, June 1973.

North, Harper Q., *A Probe of TRW's Future*, TRW Corporation, July 1966.

——, "Delphi Forecasting in Electronics," Thompson Ramo Woodridge, Corporate Report, 1968 (unpublished).

——, and Donald L. Pyke, "Probes of the Technological Future," *Harvard Business Review*, May–June 1969.

——, and Donald L. Pyke, "Technological Forecasting in Planning for Company Growth," *IEEE Spectrum*, January 1969.

Parker, E. F., "British Chemical Industry in the 1980's—A Delphi Method Profile," *Chemistry and Industry*, January 1970.

——, "Some Experience with the Application of the Delphi Method," *Chemistry and Industry*, September 1969.

Pill, J., *The Delphi Method: Substance, Context, a Critique, and an Annotated Bibliography*, Technical Memorandum No. 183, Case-Western Reserve University School of Management, May 1970.

Pyke, D. L., *A Practical Approach to Delphi, Technological Forecasting and Long-Range Planning*, American Institute of Chemical Engineers, November 1969.

——, *Long-Range Technological Forecasting: An Application of the Delphi Process*, Fifth Summer Symposium of Energy Economic Division of ASEE, U. C. L. A., June 21, 1968.

——, *The Role of Probe in TRW's Planning, Technological Forecasting—An Academic Inquiry*, Graduate School of Business, University of Texas, April 1969.

——, *TRW's Probe II*, presented at Advanced Technological Forecasting: Experiences and Workshops, Industrial Management Center, Inc., September 2–5, 1969.

Reilly, Kevin D., "Prospects for Use of the Delphi Method in Information Science Research," *Proceedings of the American Society for Information Services*, July 25–26, 1970.

Reisman, A., and Martin Taft, *On a Computer-Aided System Approach to Personnel Administration*, Dept. of OR, Case Western Reserve University, December 1968.

——, *et al.*, *Evaluation and Budgeting Model for a System of Social Agencies*, Dept. of OR, Case Western Reserve University, TM-167, November 1969.

Rescher, Nicholas, "A Questionnaire Study of American Values by 2000 A.D.," *Values and the Future*, Baier and Rescher (eds.), Free Press, Glencoe, Ill., 1969.

Riesenfeld, W., *et al.*, "Perceptions of Public Service Needs," *The Gerontologist*, Summer 1972.

Rosove, Perry E., "A Trend Impact Matrix for Societal Impact Assessment," Center for Futures Research, Graduate School of Business Administration, University of Southern California, Los Angeles, April 1973.

Sandow, Stuart, *A Survey of Continuing Education Goals for Degree Granting Post-Secondary Institutions in New York State Research with the Focus Delphi Process*, EPRC, Syracuse, 1972.

——, and Michael Folk, *Report of a Planning Study of the United Ministers in Higher Education. Research with the Focus Delphi Process*, Educational Policy Research Center, Syracuse, 1972.

Scheele, S., "The Future of Leisure and Recreation (Proprietary Study), SET, Inc., 1972.

——, "The Role of the Mentally Retarded in Society" (Proprietary Study), SET, Inc., 1970.

Scott, Randall F., and Dick B. Simmons, "Programmer Productivity and the Delphi Technique," *DATAMATION*, May 1974.

Sheldon, A., and Curtis McLaughlin, "The Future of Medical Care: A Delphi Study," Meeting of the Society for General Systems Research, December 1970.

Shimada, Syozo, *A Note on the Cross-Impact Matrix Method*, Hitachi Ltd., Research Report No. HC-70-029, March 1971.

Shimmen, Toru, "A Short Paper on Cross-Impact Analysis: A Basic Cross-Impact Model," Institute for Future Technology, Tokyo, 1973.

Teeling-Smith, J., *Medicine in the 1990's*, Office of Health Economics, London, England, October 1969.

Turoff, Murray, *A Summary of the Rand Civil Defense Delphi*, Office of Emergency Preparedness, Systems Evaluation Division, Technical Memorandum-122, February 1970.

——, "A Synopsis of Innovation by the Delphi Method," ORSA Meeting in Detroit, 1970.

Voyer, Robert, *Future Developments in Canada*, Science Forum, Science Council of Canada, October 1969.

Waldron, J. S., *An Investigation into the Relationship among Conceptual Level, Time Delay of Information Feedback, and Performance in the Delphi Process*, Doctoral Dissertation, Syracuse University, 1970.

Weaver, Timothy, "Delphi: A Forecasting Tool," *Notes on the Future of Education*, 1 Issue 3 (1970), Educational Policy Research Center, Syracuse University.

——, "The Delphi Forecasting Method," *Phi Delta Kappan*, January 1971.

——, "The Delphi Forecasting Method: Some Theoretical Considerations," in *The Potential of Educational Futures*, Ziegler and Marien (eds.), Charles A. Jones, Worthington, Ohio, 1972.

——, *The Delphi Method: Background and Critique*, Educational Policy Research Center, Syracuse University, June 1970.

——, *The Delphi Method: Research Design and Findings*, Educational Policy Research Center, Syracuse University, June 1970.

——, *An Exploration into the Relationship between Conceptual Level and Forecasting Future Events*, Doctoral Dissertation, Syracuse University, 1969

Welty, Gordon, "A Critique of the Delphi Technique," *Proceedings of the American Statistical Association*, Washington, D. C., 1972.

——, "A Critique of Some Long-Range Forecasting Developments," 38th Session of the International Statistical Institute, Washington, D. C., August 1971.

——, "Problems of Selecting Experts for Delphi Exercises," *Academy of Management Journal* **15**, (1972):1.

——, "The Selection and Deflection of Expertise in Delphi Exercises," joint ORSA-TIMS meeting, Atlantic City, November 1972.

Wennerberg, Ulf, "Using the Delphi Technique for Planning the Future Libraries," Unesco Bulletin for Libraries, V. 26 (5), 242–246, September–October 1972.

Williamson, John, *Prognostic Epidemiology of Absenteeism*, Department of Medical Care, Johns Hopkins University, March 1971.

——, *Prognostic Epidemiology of Breast Cancer*, Department of Medical Care, Johns Hopkins University, 1971.

Technological Forecasting

Arnfield, R. V. (ed.), *Technological Forecasting*, Edinburgh University Press, Edinburgh, 1969.

Barach, J. L., and R. J. Twery, "The Application of Forecasting Techniques in Research and Development Planning," presented at the AIChe meeting, November 1969.

Bayne, C. Dudley, Jr., and Walter Price, "Technological Forecasting," in *Formal Planning Systems—1971*, Richard F. Vancil (ed.), Harvard Business School, Cambridge, Mass., 1971.

Bestuzhev-Lada, I. V., "Bourgeois 'Futurology' and the Future of Mankind," *Political Affairs*, September 1970.

——, "How to Forecast the World's Future: Asimov Disputed," *The Current Digest of the Soviet Press* **19**, No. 20 (June 1967).

——, "Social Forecasting," *Soviet Cybernetics Review* **3**, No. 5 (May 1969).

——, *Window into the Future*, Mysl, Moscow, 1970.

Bright, James R., *A Brief Introduction to Technological Forecasting—Concepts and Exercises*, Pemaquid Press, Austin, Texas, 1972.

——, "Evaluating Signals of Technological Change," *Harvard Business Review* **48** (1970):1.

——(ed.), *Technological Forecasting for Industry and Government, Methods and Applications*, Prentice-Hall, Englewood Cliffs, New Jersey, 1968.

——and M. E. F. Schoeman (ed.), "A Guide to Practical Technological Forecasting" (a second collection of papers), Prentice-Hall, Englewood Cliffs, N. J., 1973.

Chambers, J. C., *et al.*, "How to Choose the Right Forecasting Technique," *Harvard Business Review*, July–August 1971.

Davidson, Frank, "Futures Research: A New Scientific Discipline?" *Proceedings of the Social Statistics Section*, American Statistical Association meeting, August 1969.

Davis, Richard C., "Organizing and Conducting Technological Forecasting in a Consumer Goods Firm," in James R. Bright and M. E. F. Schoeman (eds.), *A Guide to Practical Technological Forecasting*, Prentice-Hall, Englewood Cliffs, N. J., 1973.

De Houghton, Charles, William Page, and Guy Streatfeild,...*and Now the Future: A PEP Survey of Future Studies*, PEP Broadsheet 529, **37** (London: PEP, August 1971).

Ehin, I., "Some Methodological Problems of Forecasting," *Transactions*, AS Estonian SSR, **20**, No. 4 (1971).

Gerstenfeld, Arthur, "Technological Forecasting," *The Journal of Business* **44**, No. 1 (January 1971).

Girshick, M., A. Kaplan, and A. Skogstad, "The Prediction of Social and Technological Events," *Public Opinion Quarterly* **13**, (Spring 1953); Rand P-93, April 1949.

Hall, P. D., "Technological Forecasting for Computer Systems," in *Technological Forecasting and Corporate Strategy*, G. S. C. Wills, *et al.* (eds.), American Elsevier, New York, 1969.

Hayden, Spencer, "How Industry Is Using Technological Forecasting," *Management Review*, May 1970.

Heiss, K., K. Knorr, and O. Morgenstern, *Long Term Projections of Political and Military Power*, Mathematica, Princeton, N. J., January 1973.

Kiefer, David M. (ed.), "The Futures Business" (survey article), *Chemical and Engineering News*, August 11, 1969.

Lanford, H. W., *A Synthesis of Technological Forecasting Methodologies* (Wright-Patterson Air Force Base, Ohio: Headquarters, Foreign Technology Division, Air Force Systems Command, U. S. Air Force, May 1970).

Lätti, V. I., *A Survey of Methods of Futures Research*, Report NORD-3 (Pretoria: Institute for Research Development, South African Human Sciences Research Council, 1973).

Martino, Joseph, "The Paradox of Forecasting," *The Futurist*, February 1969.

——, *Technological Forecasting for Decision Making*, American Elsevier, New York, 1972.

——, "Technological Forecasting Is Alive and Well in Industry," *The Futurist*, August 1972.

——, "What Computers May Do Tomorrow," *The Futurist*, October 1969.

——, "Tools for Looking Ahead," *IEEE Spectrum*, October 1972.

——, *et al.*, *Long Range Forecasting Methodology*, A Symposium Held at Holman AFB, Alamogordo, New Mexico, October 11, 1967.

Massenet, M., "Methods of Forecasting in the Social Sciences," in *Three Papers Translated from the Original French for the Commission on the Year 2000*, Brookline, Mass., American Academy of Arts and Sciences.

McLoughlin, William G., "Technological Forecasting LTV," *Science and Technology*, February 1970.

Mitroff, Ian, and Murray Turoff, "Technological Forecasting and Assessment: Science and/or Mythology?" *IEEE Spectrum*, March 1973.

North, Harper Q., "Technological Forecasting in Planning for Company Growth," *IEEE Spectrum*, January 1969.

——, "Technological Forecasting to Aid Research and Development Planning," *Research Management* **12**, No. 4 (1969).

——, "Technology, the Chicken—Corporate Goals, the Egg," *Technology Forecasting for Industry and Government*, James R. Bright (ed.), Prentice-Hall, Englewood Cliffs, N. J., 1968.

——, and Donald L. Pyke, *Technological Forecasting in Industry*, NATO Defense Research Group Conference, National Physical Laboratory, England, November 1968.
Pry, Robert, *et al.*, *Technological Forecasting and Its Application to Engineering Materials*, National Materials Advisory Board, NMAB-279, March 1971.
Quinn, J. B., "Technological Forecasting," *Harvard Business Review*, March–April 1967.
Rosen, Stephen, "Inside the Future," *Innovation Magazine*, No. 18, 1971.
Thiesmeyer, Lincoln R., "The Art of Forecasting the Future Is Losing Its Amateur Status," *Financial Post*, June 26, 1971.
——, "How to Avoid Bandwagon Effect in Forecasting: The Delphi Conference Keeps Crystal Ball Clear," *Financial Post*, October 23, 1971.
Turoff, Murray, "Communication Procedures in Technological Forecasting," *IEEE 73 INTERCON Proceedings*, Session 28: Technological Forecasting Methodologies, March 1973.
Wills, Gordon, *et al.*, *Technological Forecasting*, Penguin Books, Baltimore, 1972.

New Directions

Amara, Roy, and Jacques Vallee, "FORUM: A Computer-Based System to Support Interaction among People," IFIP Congress, *1974 Proceedings*, August 1974.
Bahm, Archie J., "Demo-Speci-ocracy," *Policy Sciences* **3**, No. 1 (1972).
Baran, Paul, "Voice-Conferencing Arrangement for an On-Line Interrogation," Institute for the Future, March 1973.
——, Hubert M. Lipinski, Richard H. Miller, and Robert H. Randolph, "ARPA Policy-Formulation Interrogation Network," Semiannual Technical Report, April 1973.
Baver, Raymond, "Societal Feedback," *The Annals* **2**, September 1967.
Billingsley, Ray V., "System Simulation as an Interdisciplinary Interface in Rural Development Research," unpublished, Texas Agricultural Experiment Station Technical Article 9915, College Station, Texas (April 1972).
Carter, George E., "Computer-Based Community Communications," mimeographed, Computer-Based Education Research Laboratory, University of Illinois, Urbana, Illinois (1973).
——, "Second Generation Conferencing Systems," Computer-Based Educational Research Laboratory, University of Illinois (November 1973).
Conrath, David W., "Teleconferencing: The Computer, Communication, and Organization," *Proceedings* of the First International Conference on Computer Communication (1972).
——, and J. H. Bair, "The Computer as an Interpersonal Communication Device: A Study of Augmentation Technology and its Apparent Impact on Organizational Communication," Second International Conference on Computers and Communications, Stockholm, August 1974.
Englebart, D. C., "The Augmented Knowledge Workshop," *AFIPS Conference Proceedings* **42**, 1973 Joint Computer Conference.
——, "Coordinated Information Services for a Discipline-or-Mission-Oriented Community," Second Annual Computer Communications Conference, San Jose, Calif. (January 1973).
——, "Intellectual Implications of Multi-Access Computer Networks," Interdisciplinary Conference on Multi-Access Computer Networks, Austin, Texas (April 1970).
Etzioni, Amitai, "Minerva: An Electronic Town Hall," *Policy Sciences* **3**, No. 4 (1972).
Ewald, William R., Jr., *ACCESS To Regional Policymaking*. A report to the National Science Foundation, July 27, 1973.
——, *GRAPHICS for Regional Policymaking, A Preliminary Study*. A report to the National Science Foundation, August 17, 1973.
Gusdorf, *et al.*, *Puerto Rico's Citizen Feedback System*, Technical Report 59, OR Center, M. I. T. (April 1971).
Hall, T. W., "Implementation of an Interactive Conference System," Spring Joint Computer Conference, Vol. 38, AFIPS Press, 1971.
Havron, M. Dean, and Anna E. Casey Stahmer, "Planning Research in Teleconference Systems," Department of Communications, Canada, November 1973.

Johansen, Robert, and Richard H. Miller, "The Design of Social Research to Evaluate a New Medium of Communication," a working paper prepared for the American Sociological Association Annual Meeting, February 1974.

Johnson, Norman, and Edward Ward, "Citizen Information Systems," *Management Science* **19**, No. 4, Part II (December 1972).

Kimbel, Dieter, "Computers and Telecommunication Technology," International Institute for Management of Technology, 1973.

Krendel, Ezra S., "A Case Study of Citizen Complaints as Social Indicators," *IEEE Transactions of Systems, Science, and Cybernetics*, October 1970.

Kupperman, R. H., and R. H. Wilcox, "Interactive Computer Communications Systems—New Tools for International Conflict Deterrence and Resolution," Second International Conference on Computers and Communications, Stockholm, August 1974.

Leonard, Eugene, *et al.*, "Minerva: A Participatory Technology System," *Bulletin of Atomic Scientists*, November 1971.

Licklider, J. C. R., "The Computer as a Communication Device," *International Science and Technology*, April 1968.

Linstone, H. A., "Communications in Futures Research," paper presented at the Conference of Research Needs in Futures Research, January 10–11, 1974, sponsored by the Futures Group, to be published in W. I. Boucher (ed.), *The Study of the Future: An Agenda for Research.*

Lipinski, A. J., H. M. Lipinski, and R. H. Randolph, "Computer-Assisted Interrogation: A Report on Current Methods Development," International Conference on Computer Communications, IEEE, October 1972.

Lipinski, Hubert M., and Robert H. Randolph, "Conferencing in an On-line Environment: Some Problems and Possible Solutions," *Proceedings* of the Second Annual Computer Communications Conference, 1973.

Little, J. K., C. H. Stevens, and P. Tropp, "Citizen Feedback System: The Puerto Rico Model," *National Civic Review*, April 1971.

Macon, N., and J. D. McKendree, "EMISARI Revisited: The Resource Interruption Monitoring System," Second International Conference on Computers and Communications, Stockholm, August 1974.

Miller, Richard H., "Trends in Teleconferencing by Computer," *Proceedings* of the Second Annual Computer Communications Conference, 1973.

Noel, Robert, "Polis: A Computer-Based System for Simulation and Gaming in the Social Sciences" (September 1972), Department of Political Sciences, University of California at Santa Barbara.

Renner, Rod, *Conference Systems Users Guide*, TM-225, Office of Emergency Preparedness, AD 756–813, November 1972.

——, *Delphi Users Guide*, TM-219, Office of Emergency Preparedness, September 1971.

——, *et al.*, "EMISARI: A Management Information System Designed to Aid and Involve People," *COINS* (International Symposium on Computer and Information Science), December 1972.

Santi, V. De., *et al.*, *Genie*, Office of Economic Opportunity, June 1971.

Schuyler, James, and Robert Johansen, "ORACLE: Computer Conferencing in a Computer-Assisted Instruction System," International Conference on Computer Communications, IEEE, October 1972.

Sheridan, Thomas B., "Citizen Feedback: New Technology for Social Choice," *Technology Review*, M. I. T., January 1971.

——, "Technology for Group Dialogue and Social Choice," *AFIPS Conference Proceedings* **39** (1971), 1971 Fall Joint Computer Conferences.

Stevens, Chandler H., "Citizen Feedback and Societal Systems," *Technology Review*, M. I. T., January 1971.

Thompson, Gordon, "Moloch or Aquarius," *THE* **4** (February 1970), Bell Northern Research.

——, "A Socially Responsible Approach to Communication System Design," Second International Conference on Computers and Communications, Stockholm, August 1974.

Turoff, Murray, "Computerized Conferencing and Real Time Delphis," Second International Conference on Computers and Communication, August 1974, Stockholm.

——, "Conferencing via Computer," Proceeding of *NERM*. IEEE, November 1972.

——, "Conferencing via Networks," *Computer Decisions*, January 1973.

——, "DELPHI and Its Potential Impact on Information Systems," Conference Proceedings *FJCC* **39**, AFIPS Press, 1971.

——, "DELPHI + Computers + Communications = ?" in *Industrial Applications of Technological Forecasting*, Cetron and Ralph (eds.), Wiley-Interscience, New York, 1971.

——, "The DELPHI Conference," *Futurist*, April 1971.

——, "Human Communication," *EKISTICS* **35**, No. 211 (June 1973).

——, "*Party-Line* and *Discussion*, Computerized Conference Systems," International Conference on Computer/Communications, IEEE, October 1972.

——, "Potential Applications of Computerized Conferencing in Developing Countries," Rome Special Conference on Futures Research, 1973.

——, "Potential Applications of Computerized Conferencing in Developing Countries," *Herald of Library Sciences* **12**, No. 2–3 (April–July 1973).

Umpleby, Stuart, *Citizen Sampling Simulations: A Method for Involving the Public Social Planning* ERL, University of Illinois, June 1970.

——, "Citizen Sampling Simulations: A Method for Involving the Public in Social Planning," *Policy Sciences* **1**, No. 3 (1970).

——, *The DELPHI Exploration, A Computer-Based System for Obtaining Subjective Judgements on Alternative Futures*, University of Illinois, CERL Report F-1, August 1969.

——, "Structuring Information for a Computer-Based Information System," *FJCC* Conference Proceedings **39**, AFIPS Press, 1971.

——, and Valarie C. Lamont, "Final Report—Social Cybernetics and Computer-Based Communications Media," mimeographed, Computer-Based Education Research Laboratory, University of Illinois, Urbana, Illinois (1973).

Vallee, Jacques, "Network Conferencing," *Datamation*, May 1974.

Weston, J. R., and C. Kristen, "Teleconferencing: A Comparison of Attitudes, Uncertainty and Interpersonal Atmospheres in Mediated and Face-to-Face Group Interaction," prepared for the Social Policy and Programs Board, the Department of Communications, Canada, December 1973.

Wilcox, Richard, "Computerized Communications, Directives, and Reporting Systems," Information and Records Administration Conference (IRAC), May 19, 1972.

——, and R. Kupperman, "EMISARI: An On-Line Management System in a Dynamic Environment," International Conference on Computer Communications, IEEE, October 1972.

Wilson, Stan, "A Test of the Technical Feasibility of On-Line Consulting Using APL," APL Laboratory, Texas A&M University, 1973.

Some Foreign-Language Articles

Albach, H., "Informationsgewinnung durch strukturierte Gruppenbefragung—Die Delphi-Methode,"ZfB, Suppl., 40th Year, (1970), pp. 11–26.

Atlas Copco MCT AB, "The Driving of Tunnels—Needs and Technologies," Nacka, Sweden, 1972.

Dvořák, K. *et al.*, "The Application of the Delphi Method to the Forecasting of Machinery Technology," *Trend* **3**, (1972).

Ehrlemark, G., and L-G Remstrand, "A Modified Delphi-Experiment—Performance and Feasibility," from *Some Papers on Futures Studies*, Konsultkollegiets kompendier 1, Beckmans, Stockholm, Sweden, 1971.

Ejerhed, A., G. Hulten, M. Johnsson, Y. Karlsson, M. Olander, "Futures Studies in Public Health," report from the Swedish Planning and Rationalization Institute of Health and Social Services (SPRI), Stockholm, Sweden, 1971.

The Federation of Swedish Industries, "Political Values and Structural Developments in the Environment of the Federation of Swedish Industries," Stockholm, Sweden, 1971.

Gardborn, I., "An Experiment with the Delphi-Technique," Stockholm School of Economics and Business Administration, Stockholm, Sweden, 1971.

Gidai, E., "On the Practical Use of Intuitive Methods," paper presented at the Summer School of Prognostics, Warsaw, September 1972.

Grojer, J-E, "A Futures-Study with Delphi-Technique Applied to the Urbanization Process," University of Stockholm, Sweden, 1971.

Halberštátová, R., and I. Žádný, "The Application of the Delphi Method in the Research of Iron Metallurgy," *Trend* **4**, (1972).

Horn, J., "The World in the Future—A Delphi-Experiment," Naturvetenskapliga Forskningsradet/The Swedish National Science Research Council, Stockholm, Sweden, 1972.

Husen, T., "Education in the Year 2000—A Futures Study," Stockholm, Sweden, 1971.

Institute of Management, "The Use of some Delphi Method Principles for Research in the Development of a Management System of Socialist Industrial and Business Organizations," Research Report, Prague, 1972.

Jacobsson, S., "Build for the Future—The Future for the Builders," Byggnadsindustrin (Building Industry) journal, No. 18 (1969), Stockholm, Sweden.

Krauch, Helmut, and Georg Rüdinger, "Ein Rückkoppelungs—System zwischen Rundfunkanstalt and Zuhörerschaft," Studiengruppe für Systemforschung, Heidelberg, Germany, November 1970.

Lachman, O., "Personnel Administration in 1980: A Delphi-Study," Danish Institute of Graduate Engineers, Naerum (Copenhagen), Denmark, 1971.

Landergren, U., S. Fredriksson, and A. Littke-Persson, "An Application of the Delphi Technique to a Forecast of the Field of Ceramics—A Methodological Experiment," Forsvarets Forskningsanstalt/The Swedish National Defense Research Institute (FOA), Stockholm, Sweden, 1968.

Liden T., and L. O. Teinvall, "Convenience 1985—The Prediction of a Selling-Argument," University of Lund, Faculty of Business Administration, Lund, Sweden, 1972.

Machálek, J., and O. Šulc, "A Combination of Delphi and Cross-Impact Matrix Methods," *Trend* **5** (1970).

The Nordic Museum, "The Future of the Museums on the History of Civilization," Nordic Museum, Stockholm, Sweden, 1973.

The Nordic Summer-University, "New Sciences?—A Delphi Study," *Information Fra Nordisk Sommeruniversitet*, No. 2 (1969), Copenhagen, Denmark.

Oxelösund Iron Works AB, "Delphi 1985," Oxelösund, Sweden.

Perstorp AB, "The Future and Its Importance for Perstorp," Perstorp, Sweden, 1969.

Platmanufaktur AB (producers of cans, bottles, and other packaging), "Technological Forecasting in the Beer Field," Malmö, Sweden.

——, "Values and Their Changes in Sweden," Malmö, Sweden.

Research Institute for Construction Materials, "A Forecast of Key Tendencies in the Development of Construction Materials," Research Report, Brno, Czechoslavia, 1973–74.

The Royal College of Forestry, "The Future Role of the Forest—A Delphi Study," Stockholm, Sweden, 1973.

Salancik, G. R., T. J. Gordon, and N. Adams, for Skandia AB, "On the Nature of Economic Losses Arising from Computer-Based Systems in the Next Fifteen Years," Institute for the Future, Menlo Park, Calif., Report R-23, March 1972.

Schollhammer H., "Die Delphi-Methode als betriebliches Prognose- und Planungsverfahren," ZfB, M.S., 22nd Year (1970), pp. 128–137.

Sjölind, S., "A Future Study of the Medicine Market (Drug Market) in Sweden," Stockholm School of Economics and Business Administration, Stockholm, Sweden, 1971.

Skandia AB (Insurance Company), "The Internal Delphi-Study of Skandia," Stockholm, Sweden, 1971.

Svenska Kullagerfabriken AB (SKF), "Technology Evolution in Bearing Manufacturing for the Coming 15 Years," Göteborg, Sweden, 1970.

Swedish Agency for Administrative Development (SAFAD), "Information, Documentation and Media—Report on a Delphi Study," Allmanna Förlaget, Stockholm, Sweden, 1971.

The Swedish Board for Technical Development, "Priorities for Research in Medical Techniques in Sweden," Stockholm, Sweden, 1969.

The Swedish Central Bureau of Statistics, "The Future Development of Computer Usage and Its Effects on the Labour Market," Stockholm, Sweden, 1973.

Swedish Lloyd, "The Future Transportation Environment," Göteborg, Sweden, 1969.

The Swedish Newspaper Publishers Training Board, "Swedish Newspaper Industry—A Futuristic Vision by the Press," Stockholm, Sweden.

The Swedish Savings Banks Association, "The Future Environment of the Swedish Savings Banks," Stockholm, Sweden.

The Swedish Shipbuilding Experimental Tank, "The Gasturbine Technique in Merchant Ships," Göteborg, Sweden, 1972.

Swedish Unilever AB, "Scandinavian Environmental Study," Stockholm, Sweden, 1971.

Ullman, A., "Research and Development in the Field of Materials," Naval Material Department, Stockholm, Sweden, 1971.

Volvo AB, "Delphi—An Experiment with a Futuristic Methodology," Göteborg, Sweden, 1970.

Westerlund, S., "The Delphi-Technique—Swedish Practice," *Ekonomen*, No. 20 (1968), Stockholm, Sweden.

Wikstrom, S., "The Distribution of Our Every Day Commodities—A Futures Study," University of Stockholm, Faculty of Business Administration, Bonniers Publishers, Stockholm, Sweden, 1973.

Ytongbolagen, Hallabrottet, "Product Innovations at Ytong," Hallabrottet, Sweden, 1968.

Acknowledgment

The editors wish to thank Göŗan Axelsson and Jan Wisén of the Swedish Agency for Administrative Development and Ota Šulc of the Czechoslavak Academy of Sciences, Prague, for their assistance in compiling portions of this bibliography.

Related Work

Ackoff, Russell L., "Towards a Behavioral Theory of Communication," *Management Sciences* **4**, No. 3 (1958).

——, "Towards a System of Systems Concepts," *Management Sciences* **17**, No. 11 (July 1971).

Adelson, M., *et al.*, "Planning Education for the Future," *American Behavioral Scientist*, March 1967.

Adorno, T. W., E. Frenkel-Brunswik, D. J. Levinson, and R. N. Sanford, "The Authoritation Personality, Part One," Wiley, New York, 1950.

Allen, Allen D., "Scientific Versus Judicial Fact Finding in the United States," *IEEE Transactions on Systems, Man & Cybernetics*, SMC-2, No. 4, 1972.

Allen, Vernon L., "Situational Factors in Conformity," *Advances in Social Psychology*, in Vol. II by Berkowitz, Academic Press, New York, 1965.

Arrow, K. J., "Alternative Approaches to the Theory of Choice in Risk-Taking Situations," *Econometrica* **19**, No. 4 (1951).

Asch, S. E., "Effects of Group Pressure upon the Modification and Distortion of Judgments," *Readings in Social Psychology*, Holt, Rinehart and Winston, New York, 1958.

——, "Studies of Independence and Conformity: A Minority of one against a Unanimous Majority," *Psychological Monographs* **70** (1956).

Back, K. W., "Time Perspective and Social Attitudes," paper presented at APA annual meeting, symposium on Human Time Structure, Chicago, September 1965.

Baier, K., and N. Rescher, *Values and the Future*, Free Press, New York, 1969.

Bailey, Gerald, Peter Nordlic, and Frank Sistrunk, "Literature Review, Field Studies, and Working Papers," (from IDA Teleconferencing Studies), Human Sciences Research, October 1963, Institute for Defense Analyses, Research Paper P-113. NTS #AD-480 695.

Bass, B. M., "Authoritatianism or Acquiescence?" *Journal of Abnormal and Social Psychology* **51**, (1955), pp. 616–23.

Bavelas, Alex, "Teleconferencing: Background Information," Institute for Defense Analyses, Research Paper P-106, 1963.

Bell, D., "Twelve Modes of Prediction," in Julius Gould (ed.), *Penguin Survey of the Social Sciences*, Penguin Books, Baltimore, 1965.

Belnap, Nuel, *An Analysis of Questions*, System Development Corporation Technical Memorandum 1287/000/00, Santa Monica, Calif., 1963.

Berelson, Bernard, and Gary Steiner, *Human Behavior*, Harcourt and Brace, New York, 1964.

Berger, R. M., J. P. Guilford, and P. R. Christensen, "A Factor-Analytic Study of Planning Abilities," *Psychological Monographs* **71** (Whole No. 435), 1957.

Beum, Corlin D., and Everett G. Brundage, "A Method for Analyzing the Sociomatrix," *Sociometry* **13**, No. 2 (May 1950).

Bonier, R. J., *A Study of the Relationship between Time Perspective and Open-Closed Belief Systems*, M.S. thesis, Michigan State University Library, 1957.

Boocock, Saran S., and E. O. Shild, *Simulation Games in Learning*, Sage Publishing Co., Beverly Hills, Calif., 1968.

Bottenberg, R. A., and R. E. Christal, "Grouping Criteria—A Method which Retains Maximum Predictive Efficiency," *Journal of Experimental Education* **36**, No. 4 (Summer 1968), pp. 28–34.

——, and J. H. Ward, Jr., *Applied Multiple Linear Regression*, PRL-TDR-63-6, AD-413 128, Lackland AFB, Texas: Personnel Research Laboratory, Aerospace Medical Division, March 1963.

Bretz, Rudy, "The Selection of Appropriate Communication Media for Instruction: A Guide for Designers of Air Force Technical Training Programs," Rand Report R-601-PR, February 1971, pp. 30 ff.

Brockhoff, K., "On Determining Relative Values," *Zeitschrift für Operations Research* **16** (1972), pp. 221–32.

Brown, Judson S., "Gradients of Approach and Avoidance Responses and Their Relation to Levels of Motivations," *Journal of Comparative and Physiological Psychology* **41**, No. 6 (1948).

Caldwell, C. H. Coombs, Schoeffler, and R. M. Thrall, "A Model for Evaluating the Output of Intelligence Systems," *Naval Research Logistics Quarterly* **8** (1961).

——, and R. M. Thrall, "Linear Model for Evaluating Complex Systems," *Naval Research Logistics Quarterly* **5** (1958).

Campbell, Donald T., and Julian C. Stanley, *Experimental and Quasi-Experimental Designs for Research*, Rand McNally, Chicago, 1963.

Cantril, H., "The Prediction of Social Events," *Journal of Abnormal Psychology* **33** (1938), pp. 364–89.

——, "The World in 1952: Some Predictions," *Journal of Abnormal and Social Psychology* **38** (1943), pp. 6-47.

Carbonell, Jaime R., "On Man-Computer Interaction: A Model and Some Related Issues," *IEEE Transactions on Systems Science and Cybernetics*, January 1969.

Cavert, C. Edward, "Procedures for the Design of Mediated Instruction," State University of Nebraska Project, 1972.

Chapanis, Alphonse, "The Communication of Factual Information through Various Channels," *Information Storage and Retrieval* **9**.

Christal, R. E., "JAN: A Technique for Analyzing Group Judgment," *Journal of Experimental Education* **36**, No. 4 (Summer 1968), pp. 25–7.

——, *Officer Grade Requirements Project: I. Overview*, PRL-TDR-65-15, AD-622 806, Lackland AFB, Texas: Personnel Research Laboratory, Aerospace Medical Division, September 1965.

——, "Selecting a Harem—And Other Applications of the Policy-Capturing Model," *Journal of Experimental Education* **36**, No. 4 (Summer 1968), pp. 35–41.

Churchman, C. W., *Challenge to Reason*, McGraw-Hill, New York, 1968.

——, *The Design of Inquiring Systems*, Basic Books, New York, 1971.

——, *Theory of Experimental Inference*, Macmillan, New York, 1948.

Clark, Charles H., *Brainstorming*, Doubleday, Garden City, N. Y., 1958.

Cohen, Arthur R., "Some Implications of Self-Esteem for Social Situations," in Persuasion and Persuasibility, Hovland and Janis (eds.), Yale University Press, New Haven, 1959.

Coleman, J., E. Katz, and H. Menzel, "The Diffusion of an Innovation," *Sociometry* **20** (1957), pp. 253–70.

Collaros, P. A., and L. R. Anderson, "Effect of Perceived Expertness upon Creativity of Members in Brainstorming Groups," *Journal of Applied Psycnology* **53**, No. 2 (1969).

Collins, B. E., and H. Guetzkow, *A Social Psychology of Group Processes for Decision-Making*, Wiley, New York, 1964.

Delbecq, André, and Andrew Van de Ven, "A Group Process Model for Problem Identification and Program Planning," *Journal of Applied Behavioral Science* **7**, No. 4 (1971).

Dickson, Paul, *Think Tanks*, Atheneum, New York, 1971.

Drucker, Peter, *Technology, Management, and Society*, McGraw-Hill, New York, 1970.

Dunnette, M. D., "Are Meetings Any Good for Solving Problems?", *Personnel Administration*, 1964.

Eckenrode, R. T., "Weighting Multiple Criteria," *Management Science* **12**, No. 3 (1965).

Edwards, A. L., *Techniques of Attitude Scale Construction*, Appleton-Century-Crofts, New York, 1957.

——, and L. L. Thurstone, "An Internal Consistency Check for Scale Values Determined by the Method of Successive Intervals," *Psychometrika* **17**, No. 2 (1952).

Emlet, Harry, *et al.*, "Selection of Experts," 39th ORSA meeting, May 1971.

Eysenck, H. J., "The Validity of Judgments as a Function of the Number of Judges," *Journal of Experimental Psychology* **25**, No. 6 (1939).

Farmer, Richard, and Barry M. Richman, *Comparative Management and Economic Progress*, Richard Irwin Inc., 1965.

Fisher, Lloyd, "The Behavior of Bayesians in Bunches," *The American Statistician*, December 1972.

Garner, W. R., *Structure and Uncertainty as Psychological Concepts*, Wiley, New York, 1962.

Gerardin, Lucien A., "Topological Structural Systems Analysis: An Aid for Selecting Policy and Actions for Complex Sociotechnological Systems," paper presented at the IFAC/IFORS Conference on Systems Approaches to Developing Countries, Algiers, May 1973.

Gordon, William, "Operational Approach to Creativity," *Harvard Business Review* **34**, No. 6 (November–December 1956).

Guttman, L. A., "A Basis for Scaling Qualitative Data," *American Sociological Review* **9**, No. 2 (1944).

Hainer, Raymond M., Sherman Kingsbury, and David Gleicher, *Uncertainty in Research, Management and New Product Development*, Reinhold, New York, 1967.

Hall, E. J., "Decisions, Decisions, Decisions," *Psychology Today*, November 1971.

——, Jane S. Mouton, and Robert R. Blake, "Group Problem Solving Effectiveness under Conditions of Pooling vs. Interaction," *Journal of Social Psychology* **59** (1963), pp. 147–57.

Hare, A. P., "A Study of Interaction and Consensus in Different Sized Groups," *American Sociological Review* **17**, No. 3 (1952).

——, "Handbook of Small Group Research," Free Press, Glencoe, Illinois, 1962.

Harvey, O. J., "Conceptual Systems and Attitude Change," in M. Sherif and C. Sherif (eds.), *Attitude, Ego Involvement and Change*, Wiley, New York, 1967.

—— (ed.), *Experience, Structure and Adaptability*, Springer Publishing Co., 1966.

——, "System Structure, Flexibility, and Creativity," in O. J. Harvey (ed.), *Experience, Structure, and Adaptability*, Springer Publishing Co., 1966.

——, D. E. Hunt, and H. M. Schroder, *Conceptual Systems and Personality Organization*, Wiley, New York, 1961.

——, and J. A. Kline, *Some Situational and Cognitive Determinants of Role Playing: A Replication and Extension*, Tech. Rep. No. 15, Contract Nonr. (07), University of Colorado, 1965 (cited by Harvey [1967]).

Havron, M. Dean, and Mike Averill, "Questionnaire and Plan for Survey of Teleconference Needs among Government Managers" for the Socio-Economic Branch, Department of Communications, Canada, Contract OGR2-0303, November 30, 1972.

Haythorn, William, "The Influence of Individual Members on the Characteristics of Small Groups," *Journal of Abnormal and Social Psychology* **48**, No. 2 (1953).

Helmer, Olaf, and N. Rescher, "On the Epistemology of the Inexact Sciences," *Management Science* **6**, No. 1 (1959).

Hoffman, L. R., and G. G. Smith, "Some Factors Affecting the Behaviors of Members of Problem Solving Groups," *Sociometry* **23**, No. 3 (1960).

Hoffman, P. J., "The Paramorphic Representation of Clinical Judgments," *Psychological Bulletin* **47** (1960), pp. 116–31.

House, Robert J., "Merging Management and Behavioral Theory: The Interaction between Span of Control and Group Size," *Administrative Science Quarterly* **14**, No. 3 (September 1969).

Huber, George, and André Delbecq, "Guidelines for Combining the Judgements of Individual Members in Decision Conferences," *Academy of Management Journal* (June 1972).

——, V. Sahney and D. L. Ford, "A Study of Subjective Evaluation Models," *Behavioral Science* **14**, No. 6 (1969).
——, J. Wardrop, and T. Herr, *An Analysis of Errors Obtained When Aggregating Judgments*, Paper #6704, Social Systems Research Institute, University of Wisconsin, Madison, 1967.
Israeli, N., *Abnormal Personality and Time*, Science Press, New York, (1936).
——, "Attitudes to the Decline of the West," *Journal of Social Psychology* **4** (1933).
——, "Group Estimates of the Divorce Rate for the Years 1935–1975," *Journal of Social Psychology* **4**, No. 1 (1933).
——, "The Psychopathology of Time," *Psychological Review* **39**, No. 5 (1932).
——, "Some Aspects of the Social Psychology of Futurism," *Journal of Abnormal and Social Psychology* **25**, No. 2 (1930).
——, "Wishes Concerning Improbable Future Events: A Study of Reactions to the Future," *Journal of Applied Psychology* **16** (1932), pp. 584–88.
Janis, Irving L., "Groupthink," *Psychology Today*, November 1971.
Johnson, S. C., "Hierarchial Clustering Schemes," *Psychometrika* **32**, No. 3 (1967).
Jouvenel, B. de, "L'Art de la Conjecture," in the *Futuribles Series*, edition du Roucher, Monaco, 1964. (English translation: *The Art of Conjecture*, Basic Books, New York, 1967.)
Kahneman, O., and A. Tversky, "Subjective Probability: A Judgment of Representativeness," Oregon Research Institute, *Research Bulletin II*, No. 2 (1971).
Kaplan, A., A. Skogstad, and M. Girschick, "The Prediction of Social and Technological Events," *Public Opinion Quarterly* **14**, No. 1 (1950).
Kastenbaum, R. J., "A Preliminary Study of the Dimensions of Future Time Perspective," Doctoral Dissertation, University of Southern California. Ann Arbor, Mich., University Microfilms, No. 60–394, 1960.
Katz E., "Communication Research and the Image of Society: Convergence of Two Traditions," *American Journal of Sociology* **65** (1960).
Kelly, H., and J. Thibaut, "Experimental Studies of Group Problem Solving and Process," *Handbook of Social Psychology* **2**, Addison-Wesley, Reading, Mass., 1954.
Kite, Richard W., Paul C. Vity, "Teleconferencing: Effects of Communication Medium, Network and Distribution of Resources," IDA Study S-233.
Kleinmuntz, B., *Formal Representation of Human Judgment*, Wiley, New York, 1968.
Kotler, Philip, "A Guide to Gathering Expert Estimates: The Treatment of Unscientific Data," *Business Horizons*, October 1970.
Lassey, William R., *Leadership and Social Change*, University Associates Publishers, Iowa City, 1971.
Leavitt, H. J., "Some Effects of Certain Communication Patterns on Group Performance," *Journal of Abnormal and Social Psychology* **46**, No. 1 (1951).
Levinson, D. J., "An Approach to the Theory and Measurement of Ethnocentric Ideology," *Journal of Psychology* **28** (1949), (cited by Rokeach [1960]).
Lichtenstein, Sarah, and J. Robert Newman, "Empirical Scaling of Common Verbal Phrases Associated with Numerical Probabilities," *Psychometric Science* **9**, No. 10 (1967).
Likert, R. A., "A Technique for the Measurement of Attitudes," *Archives of Psychology* **22** (1932), pp. 55.
Linstone, H. A., "On Discounting the Future," *Technological Forecasting and Social Change* **4** (1973), pp. 335–38.
——, "Planning: Toy or Tool," *IEEE Spectrum* **11**, No. 4 (April 1974).
Lund, F. H., "The Psychology of Belief," *Journal of Abnormal Psychology* **20**, No. 1 (1925).
Mackay, D. M., "Towards an Information Flow Model of Human Behavior," *British Journal of Psychology* **47** (1956), pp. 30.
Madden, J. M., "An Application to Job Evaluation of a Policy-Capturing Model for Analyzing Individual and Group Judgment," *Journal of Industrial Psychology* **2**, No. 2 (1964), pp. 36–42.
Madden, J. M., and M. J. Giorgia, "Identification of Job Requirement Factors by Use of Simulated Jobs," *Personnel Psychology* **18**, No. 3 (Autumn 1965), pp. 321–31.
Maier, Norman R. F., "Assets and Liabilities in Group Problem Solving," *Psychological Review* **74**, No. 4 (July 1967).
——, *Problem Solving and Creativity in Individuals and Groups*, Brooks/Cole Press, 1970.
Mansfield, Edwin, "The Speed of Responses of Firms to New Techniques," *Quarterly Journal of Economics* **77** (May 1963).
Martin, James, *Design of Man-Computer Dialogues*, Prentice-Hall, Englewood Cliffs, N. J., 1973.

Maslow, A., "The Further Reaches of Human Nature," Viking Press, New York, 1972.

Mason, Richard O., "A Dialectical Approach to Strategic Planning," *Management Science*, April 1969.

McGrath, Joseph E., and Irwin Altman, *Small Group Research*, Holt, Rinehart and Winston, New York, 1966.

McGregor, D., "The Major Determinants of the Prediction of Social Events," *Journal of Abnormal Psychology* **33**, No. 2 (1938).

Meadow, Arnold, "Evaluation of Training in Creative Problem Solving," *Journal of Applied Psychology* **43**, No. 3 (June 1959).

——, "Influence of Brainstorming and Problem Sequence in a Creative Problem Solving Test," *Journal of Applied Psychology* **43**, No. 6 (December 1959).

Meister, David, and Gerald F. Rabideau, *Human Factors Evaluation in System Development*, Wiley, New York, 1965.

Menzel, H., "Innovation, Integration and Marginality," *American Sociological Review* **25**, No. 5 (1960).

Miller, D. W., and M. K. Starr, *The Structure of Human Decision*, Prentice-Hall, Englewood Cliffs, N. J., 1967.

Mitroff, Ian I., "A Communication Model of Dialectrical Inquiring Systems—A Strategy for Strategic Planning," *Management Science* **14**, No. 10 (June 1971).

——, and Frederick Betz, "Dialectical Decision Theory: A Meta-Theory of Decision Making," *Management Science* **19**, No. 1 (September 1972).

Mulder, Mauk, and Henk Wilke, "Participation and Power Equalization," *Organizational Behavior and Human Performance* **5**, No. 5 (1970).

Naylor, J. C., and R. J. Wherry, Sr., *Feasibility of Distinguishing Supervisors' Policies in Evaluation of Subordinates by Using Ratings of Simulated Job Incumbents*, PRL-TR-64-25, AD-610 812, Lackland AFB, Texas: Personnel Research Laboratory, Aerospace Medical Division, October 1964.

Newell, Allen, and Herbert Simon, "Computer Simulation of Human Thinking," *Science*, December 22, 1961.

Nowakowska, Maria, "Perception of Questions and Variability of Answers," *Behavioral Science* **18**, No. 2 (March 1973).

Nylen, Donald, Robert Mitchell, and Anthony Stout, *Handbook of Staff Development and Human Relations Training*, Institute for Applied Behavioral Sciences, National Education Association, Washington, D. C., 1967.

Osborn, A. F., *Applied Imagination*, Scribners, New York, 1957.

Peters, William S., and George Summers, *Statistical Analysis for Business Decisions*, Prentice-Hall, Englewood Cliffs, N. J. 1968.

Peterson, C., and L. R. Beach, "Man as an Intuitive Statistician," *Psychological Bulletin* **68**, No. 1 (1967).

Pettigrew, T. F., "The Measurement and Correlates of Category Width as a Cognitive Variable," *Journal of Personality* **26**, No. 4 (1958).

Pfeiffer, J. L., "Preliminary Draft Essays and Discussion Papers on a Conceptual Approach to Designing Simulation Gaming Exercises," Technical Memorandum #1 (preliminary draft), Syracuse, N. Y.: Educational Policy Research Center, October 1968.

Pfeiffer, J. William, and John E. Jones, *A Handbook of Structured Experiences for Human Relations Training*, Vols. I, II, and III, University Associates Press, Iowa City, 1972.

Pierce, J. R., and J. E. Karlin, "Reading Rates and the Information Rate of a Human Channel," *Bell System Technical Journal* **36**, No. 2 (1957).

Reid, Alex, "New Directions in Telecommunications Research," a report prepared for the Sloan Commission on Cable Communications, June 1971.

Rescher, A., "The Inclusion of the Probability of Unforeseen Occurrences in Decision Analysis," *IEEE Transactions on Engineering Management*, Vol EM-14, December 1967.

Roberts, A., "Dogmatism and Time Perspective: Attitudes Concerning Certainty and Prediction of the Future," unpublished doctoral dissertation, University of Denver, 1959.

——, and R. S. Herrmann, "Dogmatism, Time Perspective, and Anomie," *Journal of Individual Psychology* **16** (1960), pp. 67–72.

Rogers, E. M., Diffusion of Innovations, Free Press, New York, 1962.

Rokeach, M., *Beliefs, Attitudes, and Values*, Jossey Bass Inc., San Francisco, 1968.

——, "A Method for Studying Individual Differences in 'Narrowmindedness,'" *Journal of Personality* **20** (1951), pp. 219–33.

——, *The Open and Closed Mind, Investigations into the Nature of Belief Systems and Personality Systems*, Basic Books, New York, 1960.

——, and B. Fruchter, "A Factorial Study of Dogmatism and Related Concepts," *Journal of Abnormal and Social Psychology* **53**, No. 3 (1956).

Romney, A. Kimball, *et al.*, *Multidimensional Scaling*, Vol. I (Theory), Vol. II (Applications), Seminar Press, New York, 1972.

Rubenstein, Albert H., "Research on Research: The State of the Art in 1968," *Research Management* **11**, No. 5 (September 1968).

Samuelson, P. A., "Consumption Theory in Terms of Revealed Preference," *Economics* **15** (1948).

——, "Probability and the Attempts to Measure Utility," *The Economic Review*, Hitotsubshi University, Tokyo, 1950.

Schroder, H. M., M. J. Driver, and S. Streufert, *Human Information Processing Individuals and Group Functioning in Complex Social Situations*, Holt, Rinehard and Winston, 1967.

Shepard, R. N., and Tegtsoonian, "Retention of Information Under Conditions Approaching a Steady State," *Journal of Experimental Psychology* **62** (1961).

Sherif, M., *The Psychology of Social Norms*, Harper Brothers, New York, 1936.

Shull, Fremong, André Delbecq, and L. Cummings, *Organizational Decision Making*, McGraw-Hill, New York, 1970.

Siegel, S., *Nonparametric Statistics for the Behavioral Sciences*, McGraw-Hill, New York, 1956.

Sigford, J. V., and R. H. Parvin, "Project PATTERN: A Methodology for Determining Relevance in Complex Decision Making," *IEEE Transactions on Engineering Management*, Vol. EM-12, No. 1, March 1965.

Souder, W. E., "The Validity of Subjective Probability of Success Forecasts by R&D Managers," *IEEE Transactions on Engineering Management*, Vol. EM-16, No. 1, February 1969.

Steiner, Ivan D., *Group Process and Productivity*, Academic Press, New York, 1972.

Stevens, S. S., "Ratio Scales of Opinion," in D. K. Whitla (ed.), *Handbook of Measurement and Assessment in Behavioral Sciences*.

Stolber, Walter B., "The Objective Function in Program Budgeting, Some Basic Outlines," *Zeitschrift für die Gesamte Staatswissenschaft*, 127 Band/Z Heft, May 1971.

Taylor, C. W., Widening Horizons in Creativity, Wiley, New York, 1964.

——, and R. Ellision, "Biographical Predictors of Scientific Performance," *Science* **155**, No. 3766 (March 3, 1967).

Thomas, Hugh, "On the Feedback Regulation of Choice Behavior," *International Journal of Man-Machine Studies* **2** (1970).

Thurstone, L. L., and E. J. Chave, *The Measurement of Attitude*, University of Chicago Press, Chicago, 1959.

Torgeson, W. G., *Theory and Methods of Scaling*, Wiley, New York, 1958.

Treisman, A. M., "Monitoring and Storage of Irrelevant Message in Selective Attention," *Journal of Verbal Learning Behavior* **3** (1964).

——, "Selective Attention in Man," *Brain Medical Bulletin* **20**, No. 12 (1964).

Van de Ven, Andrew, and André L. Delbecq, "The Comparative Effectiveness of Applied Group Decision-Making Processes," *Academy of Management Journal* (1974).

——, and André Delbecq, "Nominal and Interacting Group Processes for Committee Decision Making Eftectiveness," *Academy of Management Journal* **14**, No. 2 (1971).

Vickers, Sir Geoffrey, *The Art of Judgment: A Study in Policy Making*, Basic Books, New York, 1965.

Walker, Evan H., "The Nature of Consciousness," *Mathematical Bioscience* **7**, No. 1/2 (February 1970).

Wallach, M. A., and N. Kogan, "The Roles of Information, Discussion, Consensus in Group-Risk Taking," *Experimental Social Psychology* **1** (1965).

Ward, J. H., Jr., *Hierarchical Grouping to Maximize Payoff*, Lackland AFB, Texas: Personnel Laboratory, Wright Air Development Division, March 1961.

——, "Hierarchical Grouping to Optimize an Objective Function," *Journal of the American Statistical Association* **58** (March 1963), pp. 236–44.

Warfield, John N., "An Assault on Complexity," a Battelle Monograph, No. 3, April 1973.

Webb, E. J., and J. R. Salancik, *The Interview, or the Only Wheel in Town*, University of Texas Press, Austin, 1966.

Webber, Melvin, *Societal Contexts of Transportation and Communication*, Working Paper No. 220, Institute of Urban and Regional Development, University of California-Berkeley, November 1973.

——, "The Urban Place and the Nonplace Urban Realm," in *Explorations into Urban Structure*, University of Pennsylvania Press, Philadelphia, 1963, pp. 79ff.

Whisler, Thomas L., *The Impact of Computers on Organizations*, Praeger, New York, 1970.

White, Ralph K., "Selective Inattention," *Psychology Today*, November 1971.

Wilson, A. G., "Research for Regional Planning," *Regional Studies* **3** (1969).

Winkler, Robert, "The consensus of Subjective Probability Distributions," *Management Science* **15**, No. 2 (October 1968).

——, "The Quantification of Judgment: Some Methodological Suggestions," *American Statistical Association* **62**, No. 320 (December 1967).

Zipf, George Kingsley, *Human Behavior and the Principle of Least Effort*, Addison-Wesley, Reading, Mass., 1949.

Subject Index